Handbook of Indigenous Foods Involving Alkaline Fermentation

FERMENTED FOODS AND BEVERAGES SERIES

Series Editors

M.J.R. Nout and Prabir K. Sarkar

Cocoa and Coffee Fermentations (2014)
Editors: Rosane F. Schwan and Graham H. Fleet

Handbook of Indigenous Foods Involving Alkaline Fermentation (2014)
Editors: Prabir K. Sarkar and M.J.R. Nout

Solid State Fermentation for Foods and Beverages (2013)
Editors: Jian Chen and Yang Zhu

Valorization of Food Processing By-Products (2013)
Editor: M. Chandrasekaran

Fermented Foods and Beverages Series

Handbook of Indigenous Foods Involving Alkaline Fermentation

Edited by

Prabir K. Sarkar • M.J. Robert Nout

CRC Press
Taylor & Francis Group
Boca Raton London New York

CRC Press is an imprint of the
Taylor & Francis Group, an **informa** business

CRC Press
Taylor & Francis Group
6000 Broken Sound Parkway NW, Suite 300
Boca Raton, FL 33487-2742

Printed on acid-free paper
Version Date: 20140320

International Standard Book Number-13: 978-1-4665-6529-6 (Hardback)

Library of Congress Cataloging-in-Publication Data

Handbook of indigenous foods involving alkaline fermentation / editors, Prabir K. Sarkar and M. J. Robert Nout.
p. ; cm. -- (Fermented foods and beverages series)
Includes bibliographical references and index.
ISBN 978-1-4665-6529-6 (hardcover : alk. paper)
I. Sarkar, Prabir K. (Prabir Kumar), editor of compilation. II. Nout, M. J. Robert, editor of compilation. III. Series: Fermented foods and beverages series.
[DNLM: 1. Fermentation. 2. Food Microbiology--methods. 3. Food Preservation--methods. QW 85]

RA601.5
664.001'579--dc23 2014004207

Visit the Taylor & Francis Web site at
http://www.taylorandfrancis.com

and the CRC Press Web site at
http://www.crcpress.com

Contents

Series Preface

Natural fermentation precedes human history, and since ancient times, humans have been controlling the fermentation process. Fermentation, the anaerobic way of life, has attained a wider meaning in the biotransformations resulting in a wide variety of fermented foods and beverages.

Fermented products made with uncontrolled natural fermentations or with defined starter cultures achieve their characteristic flavor, taste, consistency, and nutritional properties through the combined effects of microbial assimilation and metabolite production, as well as from enzyme activities derived from food ingredients.

Fermented foods and beverages span a wide diversity range of starchy root crops, cereals, pulses, vegetables, nuts, and fruits, as well as animal products such as meats, fish, seafood, and dairy.

The science of chemical, microbiological, and technological factors and changes associated with manufacture, quality, and safety is progressing and is aimed at achieving higher levels of control of quality, safety, and profitability of food manufacture.

Both producers and consumers benefit from scientific, technological, and consumer-oriented research. Small-scale production needs to be better controlled and safeguarded. Traditional products need to be characterized and described to establish, maintain, and protect their authenticity. Medium- and large-scale food fermentation required selected, tailor-made, or improved processes that provide sustainable solutions for the future conservation of energy and water, and responsible utilization of resources and disposal of by-products in the environment.

The scope of the CRC book series "Fermented Foods and Beverages" shall include (1) globally known foods and beverages of plant and animal origin (such as dairy, meat, fish, vegetables, cereals, root crops, soybeans, legumes, pickles, cocoa and coffee, wines, beers, spirits, starter cultures, and probiotic cultures), their manufacture, chemical and microbiological composition, processing, compositional and functional modifications taking place as a result of microbial and enzymic effects, their safety, legislation, development of novel products, and opportunities for industrialization; (2) indigenous commodities from Africa, Asia (South, East, and Southeast), Europe, Latin America, and the Middle East, their traditional and industrialized processes and their contribution to livelihood; and (3) several aspects of general interest such as valorization of food-processing by-products, biotechnology, engineering of solid-state processes, modern chemical and biological analytical approaches (genomics, transcriptomics, metabolomics and other –omics), safety, and health and consumer perception.

The third book born in the series is *Handbook of Indigenous Foods Involving Alkaline Fermentation*. This treatise, edited by Professor Prabir K. Sarkar and Professor M.J. Robert Nout, does not deal with conventional alcoholic or acidic fermentation but deals with a third major kind of fermentation where the microbial and/or endogenous proteolytic activity leads to the generation of products having alkaline pH in most of the cases. Since alkaline fermented foods are widely known and used in Asia as well as in Africa, they are of importance to a large segment of the world population.

This happens to be the first book dedicated explicitly to alkaline fermentations. This treatise with contributions from experts from all over the world will serve students, teachers, researchers, and members of the general public as a source of information and inspiration.

Preface

Fermentation of food has been practiced since the early ages and still remains an aspect of food processing that cannot be ignored. Often advocated for their preservative effect, fermented foods remain popular with people of all generations, because of the flavor they get from microbial and enzymatic modifications of food components such as sugars, fats, and proteins.

Many of us are well aware of acidic products such as yogurt and vinegar or alcoholic beverages obtained through fermentation. These have been widely described and documented. With this book on alkaline-fermented foods, we introduce for the first time insight into the diversity of this relatively unknown group of fermented foods that are especially popular in Asia and Africa. The term "alkaline" refers to the pH of the fermented products, which tends to increase during the fermentation due to the formation of ammonia as a result of protein degradation.

The concept of the present book stems from the series editors' experience teaching food fermentations. In this project, the book editors offer state-of-the-art information on the various plant- and animal-based ingredients used, the culinary usage of the resulting products, as well as insight into the microbiology and chemistry of the fermentation processes. Hence, the book offers a perfect blend of art with science.

Although a number of alkaline-fermented foods are now produced on an industrial scale and some even for export markets, a large majority are still manufactured at household- or small-scale. Considering their cultural importance, these products require an upgrade of their quality, presentation, and hygiene process to warrant their continued popularity with future generations.

The present book opens a window to the future and looks at the technological challenges ahead, thereby attempting to broaden the food horizon. This book is directed at graduate students as well as researchers of food science and technology, nutrition and dietetics, food microbiology, and related disciplines. The text has a logical structure and a narrative flow. Although the majority of the references cited are very recent and relevant, some older, classic texts have also been recalled.

The content of this book is the result of a worldwide effort of contributors who have shared their expert knowledge of processing, cultural, historical, chemical, microbiological, and many other aspects of well-known as well as rather unknown alkaline-fermented products. Although the full list of contributors is given on the following pages, we would like to express our gratitude to our seventeen colleagues from Asia, thirteen colleagues from Africa, and the twelve colleagues from other parts of the world, who made this all happen.

The editors are equally grateful to the editorial staff of Taylor & Francis Group for their technical assistance.

Prabir K. Sarkar
M. J. Robert Nout

Editors

Prabir K. Sarkar has been a professor at the University of North Bengal, India since 2000. He earned his bachelor's (1972), master's (1974), and doctoral (1981) degrees in botany, with specialization in microbiology, from the University of Burdwan, India. He carried out postdoctoral research on food fermentation at the University of Reading in the UK, Queensland Health Scientific Services Laboratory in Australia, and Wageningen University in the Netherlands. He has 34 years of experience in teaching microbiology, published about 70 research articles in impact journals, and supervised 12 doctoral candidates. His research focus is fermentation and microbiological safety of indigenous foods. Prabir was honored with Burdwan University Gold Medal and International Foundation for Science (IFS) Silver Jubilee Award. He headed the Department of Botany and Department of Microbiology in the university for about 8 years. He has been acting as an IFS scientific adviser.

M. J. Robert Nout graduated as a food technologist at Wageningen University, the Netherlands, in 1970. His interest in food fermentation evolved during his career of university teaching and research in several African and Asian countries. He defended his doctural thesis at Wageningen University in 1981 on the topic of indigenous fermented beverages of Kenya. Since 1983, he has been an associate professor at Wageningen University, a visiting professor at China Agricultural University since 2005, and has published over 250 scientific articles and supervised 26 doctoral candidates. Robert's long-time involvement as a senior adviser of IFS made the link with Prabir Sarkar, who was an IFS grantee at the time. Retired since 2011, he is now active as a consultant in the area of food fermentation research.

Contributors

Paulin Azokpota
Department of Nutrition and Food Science
University of Abomey-Calavi
Gégamey, Benin

Genevieve Bardwell
Rising Creek Bakery
Mount Morris, Pennsylvania

Soottawat Benjakul
Department of Food Technology
Prince of Songkla University
Hat Yai, Thailand

Susan Ray Brown
Rising Creek Bakery
Mount Morris, Pennsylvania

Siriporn Chaikaew
Food Biotechnology Research Unit
National Science and Technology Development Agency
Pathumthani, Thailand

Chanya Chuenarrom
Department of Prosthetic Dentistry
Prince of Songkla University
Hat Yai, Thailand

Ekachai Chukeatirote
School of Science
Mae Fah Luang University
Chiang Rai, Thailand

Matthijs Dekker
Food Quality and Design Group
Wageningen University
Wageningen, the Netherlands

Bréhima Diawara
Food Technology Department
University of Ouagadougou
Ouagadougou, Burkina Faso

Palanivel Ganesan
Department of Food Science and Technology
Universiti Putra Malaysia
Selangor, Malaysia

Egon Bech Hansen
Department of Systems Biology
Technical University of Denmark
Lyngby, Denmark

Jenneke K. Heising
Food Quality and Design Group
Wageningen University
Wageningen, the Netherlands

Philippe Jacques
ProBioGEM Lab
Lille 1 University, Science and Technology
Villeneuve d'Ascq, France

Yemisi Adefunke Jeff-Agboola
Department of Biological Sciences
Ondo State University of Science and Technology
Okitipupa, Nigeria

Ji-Kang Jeong
Department of Food Science and Nutrition
Pusan National University
Busan, South Korea

Lene Jespersen
Department of Food Science
University of Copenhagen
Frederiksberg, Denmark

Donatien Kaboré
Food Technology Department
University of Ouagadougou
Ouagadougou, Burkina Faso

Elmer Nayra Kpikpi
Department of Applied Biology
University for Development Studies
Navrongo, Ghana

C. Ganesh Kumar
Biotechnology Laboratory
CSIR-Indian Institute of Chemical Technology
Hyderabad, India

Cherl-Ho Lee
Korea Food Security Research Foundation
Korea University
Seoul, South Korea

Ggot-Im Lee
Korea Food Security Research Foundation
Korea University
Seoul, South Korea

Jang-Eun Lee
Korea Food Research Institute
Seongnam, South Korea

Huub L. M. Lelieveld
Global Harmonization Initiative (GHI)
Bilthoven, the Netherlands

Li-Te Li
College of Food Science and Nutritional Engineering
China Agricultural University
Beijing, China

Anita R. Linnemann
Food Quality and Design Group
Wageningen University
Wageningen, the Netherlands

Yan-Li Ma
College of Food Science and Nutritional Engineering
China Agricultural University
Beijing, China

Toshirou Nagai
Genebank
National Institute of Agrobiological Sciences
Tsukuba, Japan

M. J. Robert Nout
Laboratory of Food Microbiology
Wageningen University
Wageningen, the Netherlands

Adewale Olusegun Obadina
Department of Food Science and Technology
Federal University of Agriculture
Abeokuta, Ogun State, Nigeria

Folarin Anthony Oguntoyinbo
Department of Microbiology
University of Lagos
Akoka, Lagos State, Nigeria

Maureen-Theodore Chinwe Ojinnaka
Department of Food Science and Technology
Imo State University
Owerri, Imo State, Nigeria

Labia Irène Ivette Ouoba
Thornton Heath
London, United Kingdom

Olusola Bandele Oyewole
Department of Food Science and Technology
Federal University of Agriculture
Abeokuta, Ogun State, Nigeria

Kun-Young Park
Department of Food Science and Nutrition
Pusan National University
Busan, South Korea

Charles Parkouda
Food Technology Department
University of Ouagadougou
Ouagadougou, Burkina Faso

Abiodun Isiaka Sanni
Department of Microbiology
University of Ibadan
Ibadan, Oyo State, Nigeria

Prabir K. Sarkar
Department of Botany
University of North Bengal
Siliguri, India

Aly Savadogo
CRSBAN/UFR-SVT
University of Ouagadougou
Ouagadougou, Burkina Faso

Shawn Mark Somerset
School of Allied Health
Australian Catholic University
Brisbane, Australia

Kwaku Tano-Debrah
Department of Nutrition and Food Science
University of Ghana
Accra, Ghana

Line Thorsen
Department of Food Science
University of Copenhagen
Frederiksberg, Denmark

Kris Herawan Timotius
Faculty of Medicine
Krida Wacana Christian University
Jakarta, Indonesia

Wonnop Visessanguan
Food Biotechnology Research Unit
National Science and Technology Development Agency
Pathumthani, Thailand

Hongjiang Yang
Key Laboratory of Industrial Fermentation Microbiology
Ministry of Education, College of Biotechnology
Tianjin University of Science and Technology
Tianjin, China

1

Introduction

Prabir K. Sarkar and M. J. Robert Nout

"Tell me what you eat, and I'll tell you who you are" – the statement made by the renowned French gastronome Jean Anthelme Brillat-Savarin in 1825 indicates that food is the signature character of a culture. Food is one of the key parameters of knowing a culture and learning about the people who nurture it. The values and traditions transcend through generations in the context of cultural dishes. Food shapes our culture; what we eat, how we get it, who prepares it, who shares the table, how it is served, and who eats first give a reflection of that culture. Food is more than a necessity of physical survival; the philosophy centering what we eat and with whom we eat breaks the barriers between individuals, communities, and even countries. Food can influence the social relationships and act as a method of cultural exchange. As the world gets smaller, regional delicacies get nearer, less foreign.

Food by its nature begins to spoil the moment it is harvested or prepared. Food preservation enabled ancient man to settle and live in one place and form a community. Foods were preserved by a range of "ancient" methods such as sun-drying, salting, cooling, canning, smoking, and fermentation. Each culture preserved their local food sources using the same basic methods of food preservation. These basic approaches are still being practiced in principle although significant technical advancements have been made that enable us to preserve food and obtain high quality, in accordance with the requirements of the present-day consumer. We no longer consume all of the kill or harvest immediately, but preserve some for later use.

Fermentation is one of the oldest and most valued technologies used for food preservation. Probably no other process has had such an impact on the nutrition habits and food culture of mankind.

A food is considered fermented when one or more of its constituents have been acted upon by microorganisms to produce a considerably altered final product acceptable for human use (Van Veen, 1957). Campbell-Platt (1994) re-defined fermented foods as "those foods that have been subjected to the action of microorganisms or enzymes so that desirable biochemical changes cause significant modification to the food."

The word "fermentation" is derived from the Latin meaning "to boil," since the bubbling and foaming of early fermenting beverages seemed closely akin to boiling. Fermentation was not invented, rather discovered. Indigenous or traditional fermentation processes are those that have been used for centuries, even before those were recorded in history (Hesseltine and Wang, 1980). Most of these processes were developed long before the existence of microorganisms was recognized. The history of fermented foods goes back as far in time as inscriptions were available. The Sumerians, Egyptians, Babylonians, and Assyrians had a culture of using barley to produce beer. A cuneiform inscription on a Babylonian rock from 2800 BC showed a recipe for the production of beer (Borgstrom, 1968). Records of shoyu (soy sauce) and miso (soy paste) production in China go back to around 1000 BC with the transfer of knowledge of these production processes to Japan occurring around 600 AD (Yokotsuka and Sasaki, 1998). Early Europeans were known to be making flat sour dough bread from rye in 800 BC. Around 100 BC, there were 250 bread bakeries operating in Ancient Rome (Pederson, 1979). The practice of preparing and consuming dahi (curd), dadhavat (milk product), and kali (fermented rice) in India are as old as 2000 BC (Prakash, 1961).

The objectives of developing indigenous fermentation technology were to carry over supplies from the time of plenty to those of want. It transpires an essence of knowledge and wisdom, gained by experience and based on trial and error. People might not be able to explain what was going on during storage and processing in terms of scientific language, but they certainly knew what they had to do to get the desired product (Dietz, 1984). The indigenous fermentation techniques pass as a trade from older to younger generations in the families. These methods were based on interdependent factors such as available raw materials, climatic zones, proximity to the sea, available energy source, topography, culture, and religion.

Besides preservation (extension of shelf life) indigenous fermented foods are prized for the following advantages: (a) enhancement of nutritional quality (such as vitamins, sterols, and essential amino acids and fatty acids during the production of Japanese natto), (b) increase in digestibility (for example, during Indonesian tempeh and natto production proteins are broken down to amino acids), (c) destruction of undesirable components (such as oligosaccharides, trypsin inhibitors, and hemagglutinin during the production of Indian kinema), (d) changing the physical state of a substrate (as in making tempeh, loose soybean cotyledons are bound together to make a solid cake, or in making shoyu, solid substrates are converted to a liquid product), (e) production of color (as in making Chinese angkak, rice is fermented with mold resulting in a brilliant purple-red color), (f) providing dietary variety where choice of food is limited, (g) masking of undesirable flavor (as in natto making the objectionable beany flavor gets removed) or production of improved flavor and aroma (as in cheese making), (h) fuel efficiency (where little or no heat is required, as in Korean kimchi production), (i) enhancement of medicinal values (as in Russian koumiss and kvass, which are used to treat pulmonary tuberculosis and cancer, respectively), and (j) ease in transport.

Over the centuries, fermentation has evolved and been refined and diversified. Today, a wide variety of food products is derived from this technology in households, small-scale food industries as well as in large enterprises.

In the case of alcoholic or acidic fermentations, the alcohol or acid content along with pH determines which microorganisms can survive, multiply, and carry out the fermentation. The presence or absence of oxygen also exerts an essential steering influence on which microorganisms develop sequentially. However, in case of alkaline fermentations, an alkaline pH in combination with ammonia controls the fermentation (Steinkraus, 1991). The increase in pH is due to the extensive hydrolysis of proteins to peptides to amino acids and finally liberating ammonia (Figure 1.1). While alcoholic fermentations are dominated by yeasts and acidic fermentations are dominated by lactic or acetic acid bacteria, respectively, alkaline fermentations are dominated by *Bacillus* spp., principally *Bacillus subtilis*. The alkaline pH makes the substrate adverse to the undesirable microorganisms (Steinkraus, 1996). Although in most cases the final pH would reach

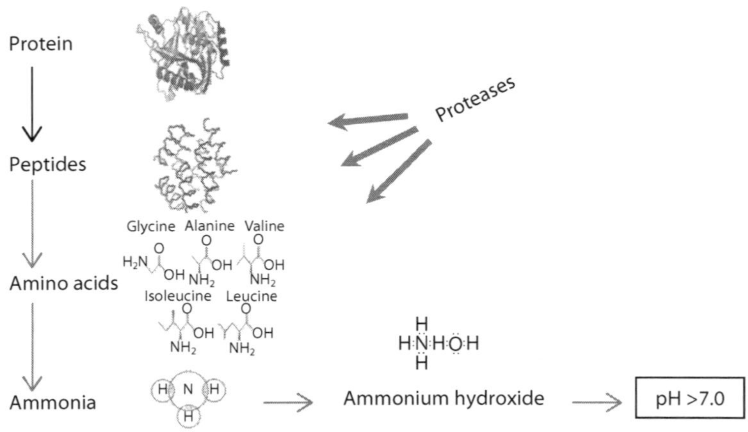

Figure 1.1 The principle of alkaline food fermentations.

values higher than 7, some foods, for example, fermented fish pastes and salt-rising bread, the pH remains restricted within 7.0 despite extensive hydrolysis of proteins because of a strong buffering capacity of the substrate and production of profuse amount of organic acids.

Alkaline-fermented foods (AFFs) constitute a group of less-known food products that are widely consumed almost exclusively in Asian and African countries (Figure 1.2). These are made using a range of protein-rich substrates. Interestingly, soybean-derived AFFs are exclusively confined to Asia; hence, Asia is considered as the

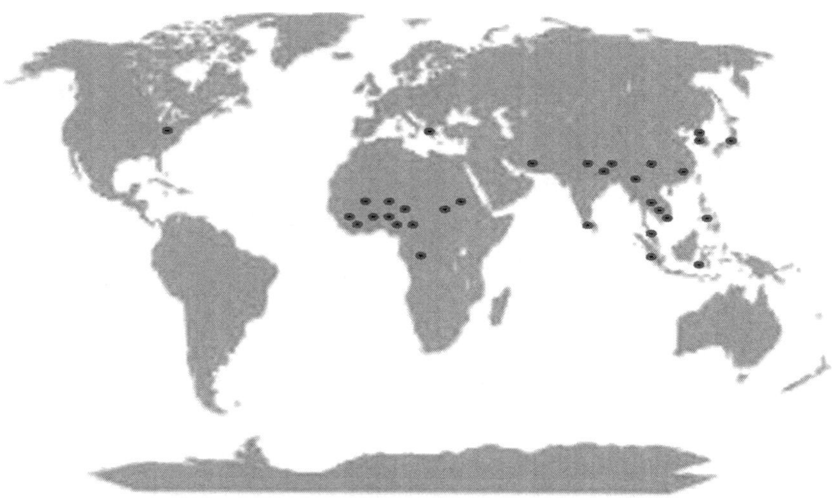

Figure 1.2 Locations indicating use of indigenous foods involving alkaline fermentation.

soybean-derived AFF belt. This is because the first domestication of soybean has been traced to the eastern half of North China in 1000 BC or perhaps a bit earlier. Soybeans were grown for centuries in Asia mainly for their seeds. These were used in preparing a large variety of fresh, fermented, and dried food products that were considered indispensable to oriental diets. In Africa, in the absence of soybeans, seeds of several wild as well as cultivated legumes and non-leguminous plants and leaves began to be used as the source of AFFs. Fish serves as an ingredient for AFF as well; fermented fish pastes and sauces are used in the coastal areas of mainly Asian countries.

The present treatise is a reference-cum-textbook documenting the AFFs that we have spotted, their indigenous preparation processes, properties, culinary use, nutritional quality, safety, and (potential) benefits. Here, many original and reference works from the authorities of their respective areas have been coordinated. This handbook bears the character of a textbook because it clarifies the principles of the processes, properties and role of the microbiota involved, the biochemical modifications, and their impact on perceived quality and benefits. Also, we have designed the book with a view on the future, challenges ahead on the road to up-scaling of production and industrial control of safety and quality. This happens to be the first book dedicated solely on AFFs, expanding the idea of Steinkraus (1996), who, in the preface of the second edition of his book, remarked "Since the first edition, it has become clear that there is a new, very important category of fermentations – those involving an alkaline reaction. I have prepared a chapter on the subject because these fermentations are likely to become even more important as scientists realize that, in addition to acidity and alcohol content, alkalinity is another way of controlling fermentations."

The content of the book is presented in eight more chapters. The diversity of indigenous fermented foods involving an alkaline reaction has been discussed in Chapters 2 and 3. The following Chapter 4 deals with the taxonomy, ecology, physiology, and genetics of predominant microorganisms occurring in AFFs. In Chapter 5, an attempt has been made to elucidate how the microorganisms or enzymes transform the raw ingredients into flavorsome, nutritious, and healthy products. Chapter 6 discusses the safety aspects of AFFs. What are the challenges associated with the technological aspects in modernizing AFFs and how those hurdles could be crossed are focused in Chapter 7. The following chapter discusses the valorization of AFFs or the associated dominant microorganisms. Chapters 9 and 10 are concerned with the prospects and outlook (research needs), respectively, on AFFs.

References

Borgstrom, G. 1968. *Principles of Food Science, vol. 2. Food Microbiology and Biochemistry*. New York: Macmillan.

Campbell-Platt, G. 1994. Fermented foods – a world perspective. *Food Research International* 27: 253–257.

Dietz, M. 1984. Fermented dried vegetables and their role in nutrition in Nepal. *Proceedings of the Institute of Food Science and Technology (UK)* 17: 208–213.

Hesseltine, C.W., and Wang, H.L. 1980. The importance of traditional fermented foods. *BioScience* 30: 402–404.

Pederson, C.S. 1979. *Microbiology of Food Fermentations*, 2nd ed. Westport, CT: AVI Publishers.

Prakash, O. 1961. *Foods and Drinks in Ancient India*. Delhi, India: Munshiram Manoharlal Publishers.

Steinkraus, K.H. 1991. African alkaline fermented foods and their relation to similar foods in other parts of the world. In *Traditional African Foods: Quality and Nutrition*, eds. A. Westby, and P.J.A. Reilly, pp. 87–92. Stockholm, Sweden: International Foundation for Science.

Steinkraus, K.H., ed. 1996. *Handbook of Indigenous Fermented Foods*, 2nd ed. New York: Marcel Dekker.

Van Veen, A.G. 1957. *Fermented Protein-Rich Foods*: FAO Report No. FAO/57/3/1966.

Yokotsuka, T., and Sasaki, M. 1998. Fermented proteins in the Orient: Shoyu and miso in Japan. In *Microbiology of Fermented Foods*, 2nd ed., ed. B.J.B. Wood, pp. 351–415. London: Blackie Academic and Professional.

2

Diversity of Plant-Based Food Products Involving Alkaline Fermentation

One of the hallmarks of alkaline fermentation is the rich diversity of products. The broad range of alkaline-fermented foods (AFFs) attributes to the variety of substrates used and the metabolic activities of the microorganisms involved in various fermentation processes. The knowledge of microbial diversity in AFFs will help in establishing their genetic resources and preparing knowledge databases.

AFFs are essential components of the diet in a number of countries in Asia and Africa and are consumed as condiments. They are prepared from both plant and animal materials, using processes in which proteolytic microorganisms or endogenous proteases play an active role in the physical, nutritional, and organoleptic modifications of the starting materials.

Plant parts, particularly seeds of cultivated as well as local wild plants, are important resources of a wide variety of indigenous AFFs. The seeds of the selected plants are rich in protein along with energy, essential fatty acids, and some minerals such as calcium and potassium.

2.1 Legume Products

Legumes occupy an important place in human nutrition, as in many countries they are one of the staple foods. Besides being a cheap source of valuable proteins, carbohydrates, and several micronutrients including minerals and vitamins, they are known as rich in dietary fiber and low in fat. The contribution of legumes in the daily diet has many beneficial physiological effects because it prevents common metabolic diseases, such as diabetes mellitus, coronary heart disease, and cancer.

2.1.1 Soybean Products

Among the cultivated crops of the world, soybean (*Glycine max* (L.) Merr.) has the highest protein content (approximately 40%) and the highest gross output of vegetable oil (approximately 20%). Soybean cultivation originated in China. About 2000 years ago, soybean was introduced to Japan, and eventually soybean cultivation spread in Korea, Indonesia, Thailand, Vietnam, and other countries. Fermented soybean products have been widely used as condiments for hundreds of years in Asian countries. AFFs are produced using soybean as the sole or main ingredient, and they are widely consumed in different countries of Asia. Not surprisingly, almost all indigenous AFFs are organoleptically different and have different culinary functions because processing methods differ. Only a few AFFs are produced using pure starter cultures, whereas the majority undergoes natural uncontrolled fermentation.

2.1.1.1 Natto

Toshirou Nagai

Natto is fermented soybeans mainly produced and consumed in Japan and has a characteristic odor (Kanno and Takamatsu, 1987; Tanaka et al., 2003) and a viscous polymer, poly-γ-glutamate (PGA, see Section 7.2). Natto tastes umami and slightly bitter and sweet, coming from amino acids and peptides after digestion of soy proteins (Nagai et al., 1994). Although breakdown of proteins occurs during natto fermentation, digestion and absorption of natto are about equal to those of boiled soybeans (Science and Technology Agency of Japan, 1979). In Japan, two types of natto are produced: a non-salted type, which is described in detail in this section, and a salted type, like miso and soy sauce, produced by *Aspergillus* fungus. The salty one is called shiokara-natto (salty natto), tera-natto (temple natto), or hama-natto (natto produced around the Hamana Lake), and it is the same type of fermented soybeans as salty douchi in China (see Section 2.1.1.1.2). Non-salted natto is often called "itohiki-natto (stringy natto)," or simply "natto," and hereafter, referred as natto. Natto is classified into three types, typical natto fermented from whole soybeans (often referred as tsubu-natto; Figure 2.1a), hikiwari-natto (Figure 2.1b), and processed natto. Hikiwari-natto is produced by fermentation of split soybeans instead of

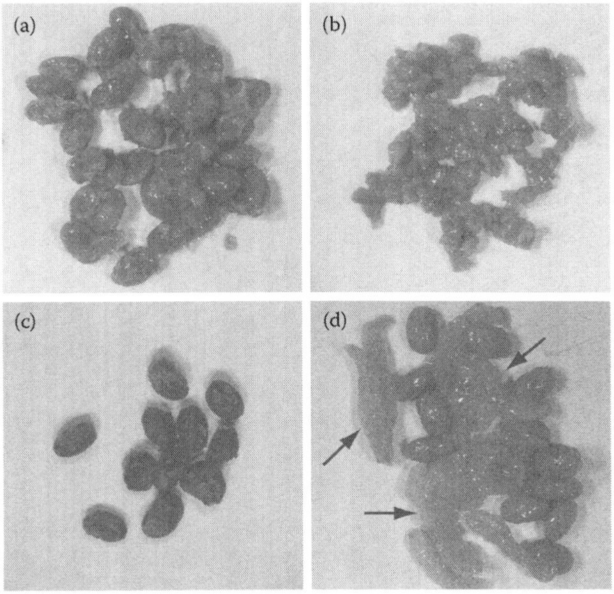

Figure 2.1 Some kinds of natto. (a) Tsubu-natto (typical whole natto); (b) hikiwari-natto (fermented split soybeans); (c) hoshi-natto (dried natto); (d) soboro-natto (natto with pieces of dried Japanese radish indicated by arrows).

whole ones. Using natto, a variety of processed natto is also produced: hoshi-natto (dried natto; Figure 2.1c), soboro-natto (mixed with sliced dried radish; Figure 2.1d), and goto-natto (further fermented with *Aspergillus* or koji mold in the presence of salt). Incidentally, another type of natto, ama-natto, is also produced in Japan. However, ama-natto is not fermented soybeans but candied beans.

Typically, natto is filled in white plastic containers (Figure 2.2a) or cups, and sachets of sauce and mustard are put on a sheet on natto (Figure 2.2b). After fermentation, a highly viscous material is produced in natto. Natto gives threads between the grains when picking it up (Figure 2.2c). As described below, natto was produced originally by wrapping boiled soybeans in rice straw (Figure 2.2d). However, the method is not employed by natto manufacturers any longer because of hygienic reasons. Before modernization of natto factories, inadequate sterilization of rice straw caused food poisoning. Some manufacturers make natto using starter strains and rice straw that has been sterilized, as nostalgic souvenirs.

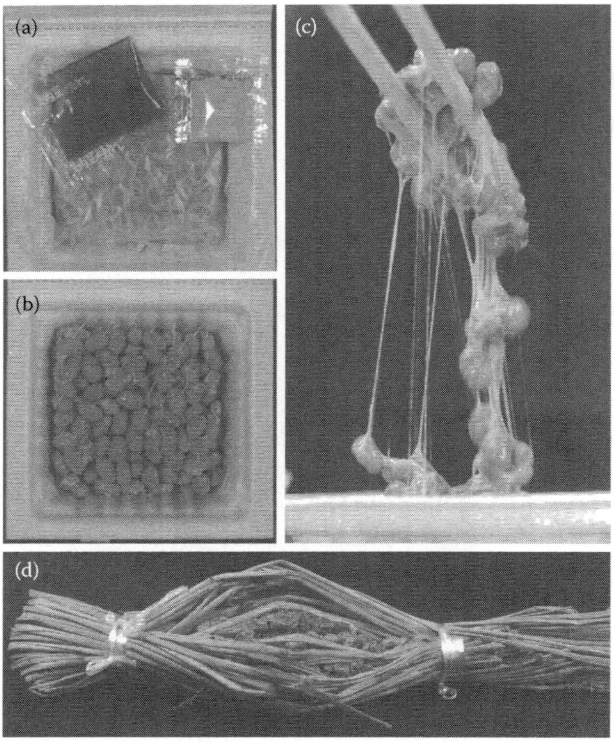

Figure 2.2 Natto. (a) Inside of a package of natto. Sachets of sauce based on soysauce and mustard put on a sheet covering natto. (b) Natto under the sheet of panel A. (c) Picking up natto with chopsticks. Strings between natto and the package are a viscous polymer, poly-γ-glutamate. (d) An ancient-type natto in appearance. This natto is made as a souvenir, and the rice straw has been sterilized for hygienic reasons.

2.1.1.1.1 History The origin of natto remains legendary, however, obscure. The first record of natto appeared in the book *Shin-sarugou-ki*, written by Akihira Fujiwara around 1052 AD. However, it is possible that the natto mentioned in the book was salted natto. The second record of natto is in an anonymous tale, *Shojin gyorui monogatari*, written before 1495. In the tale, natto was described as a samurai warrior, Taro-Itoshige Natto, with an anthropomorphic image of natto. "Ito" in his name means a string, which conjures an image of strings of viscous polymer on natto (Figure 2.2c). A sentence in the tale, "Taro-Itoshige Natto had an afternoon nap in rice straws, and woke up driveling," expresses the process of natto production at the time and strings of natto are compared with drivel. There are many legends of the origin of natto production in Japan. The most famous one is that boiled soybeans packed in rice straws for provisions were naturally fermented during a civil war in northern Japan in 1083. Although the exact year when people began to produce natto cannot be determined, it is thought that natto was produced and consumed at least by around the 11th century.

Nakao (1972) proposed his idea of the natto triangle, three apexes of which are set on Japan (natto), Nepal (kinema), and Indonesia (tempeh) (non-salted fermented soybean foods produced in each country in parentheses; Figure 2.3). These areas are the outermost areas where non-salted fermented soybeans are produced by *Bacillus* (natto and kinema) or mold (tempeh). The center of the triangle is in Yunnan province of China, and he speculated that Yunnan province was the place of the origin of natto or non-salted fermented soybeans. The technique of natto production might have spread toward the three

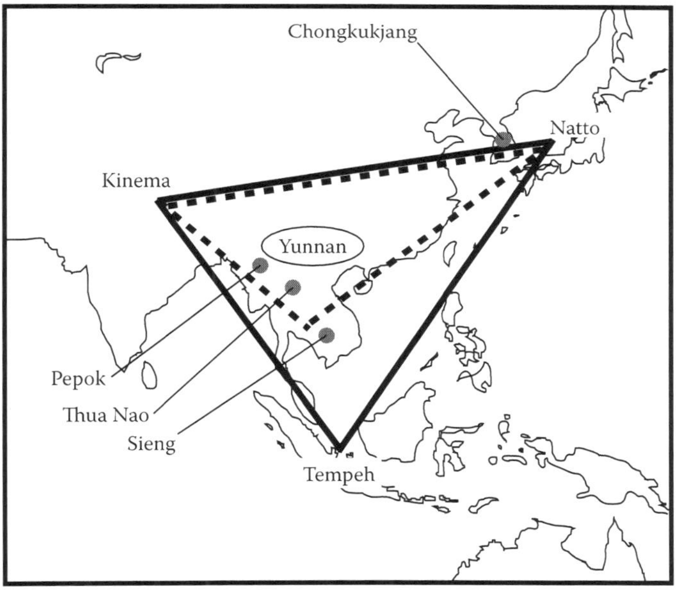

Figure 2.3 Natto triangle (proposed by S. Nakao). Yunnan province in China is depicted as an open circle at the center of the triangle with solid lines. Natto (produced in Japan), kinema (Nepal), chongkukjang (Korea), pepok (Myanmar), thua nao (Thailand), sieng (Cambodia) are non-salted fermented soybeans produced by *Bacillus subtilis*, and tempeh (Indonesia) produced by *Rhizopus oligosporus* (a mold) also is included. The KNT (kinema-natto-thua nao) triangle proposed by Tamang (2010) is also shown as a dotted triangle. (Modified from Nagai, T., and Tamang, J.P., Fermented legumes: Soybean and non-soybean products, in *Fermented Foods and Beverages of the World*, eds. J.P. Tamang, and K. Kailasapathy, pp. 191–224, CRC Press, Boca Raton, FL, 2010.)

apexes of the natto triangle from the center. Tamang (2010) proposed a KNT (kinema-natto-thua nao) triangle based only on non-fermented soybeans produced by *Bacillus subtilis* (Figure 2.3). On the other hand, Sayuki Nikkuni, a Japanese food microbiologist, pointed out that natto and other non-salted fermented soybeans could arise independently at some areas in Asia because of their simple process of fermentation, i.e., "only packaging boiled soybeans in rice straws or something." Packaging is a basic and simple technique for preservation of foods. It is no wonder that many non-salted fermented soybeans arose independently at many areas in Southeast or East Asia as a result of daily meal preparation.

2.1.1.1.2 Starter Culture A *Bacillus* strain was isolated from a traditionally prepared natto and was named *Bacillus natto* by Sawamura (1906). However, *B. natto* was included in *Bacillus subtilis* in the *Bergey's Manual of Determinative Bacteriology*, eighth edition (Gibson and Gordon, 1974). This re-classification is supported by phylogenetic analyses using 16S rRNA genes (Tamang et al., 2002; Kubo et al., 2011). Instead of *B. subtilis*, an informal name *B. subtilis* (*natto*) is often used for *B. subtilis* strains producing natto on soybeans because typical *B. subtilis* strains including Marburg cannot produce natto from soybean at all.

Three corporate companies (Miura, Naruse, and Takahashi) in Japan provide natto manufacturers with *B. subtilis* (*natto*) starters as spore suspensions. From the starters, the key strains for natto production were isolated and characterized (Kiuchi et al., 1987; Sulistyo et al., 1988). Although they found some non-*B. subtilis* (*natto*)-type strains in the starters at the same time, the strains seemed to be natural mutants of the key strains caused by insertional mutation by ISs (insertion sequences) (Nagai et al., 2000).

The three natto starter-producing companies keep the quality of their starters protected from being mutated naturally by the ISs or by others, which might be a very tough job for smaller natto manufacturers. Nowadays, however, some manufacturers have come to prepare their own starters, which have excellent characteristics that differ from the commercial starters, like high productivity of PGA and proteases (Nagai et al., 1994), elastase (Muramatsu et al., 1995), and nattokinase (Kiuchi and Suzuki, 1991), low productivity of vitamin K (Sumi, 1990), resistance to low temperatures (low productivity of ammonia)

(Tamura, 1988), non-productivity of branched short-chain fatty acids or offensive smells (Takemura et al., 2000), and ability to soften natto (Takahashi et al., 2005).

2.1.1.1.3 Fermentation Originally, natto was produced by wrapping boiled soybeans in rice straw treated with boiling water for killing microorganisms other than spores of *B. subtilis* (*natto*) and then incubating under soil or above an oven for 2 or 3 days (Figure 2.2d). Before modernization of natto production, natto was produced after harvest of rice during the winter. After discovery of *B. subtilis* (*natto*), the process changed drastically at two points: one is usage of seed culture instead of resident *B. subtilis* (*natto*) in rice straw and the other is usage of clean packages for fermentation instead of rice straw. The process of natto production in modern factories is very simple and basically the same as the old process using rice straw. The process described in Figure 2.4 is a standard and normalized example, and the conditions are varied in each natto factory and depend on temperature or season (Ueda, 1989; Kiuchi and Watanabe, 2004). Undamaged and non-discolored soybeans are selected, cleaned from

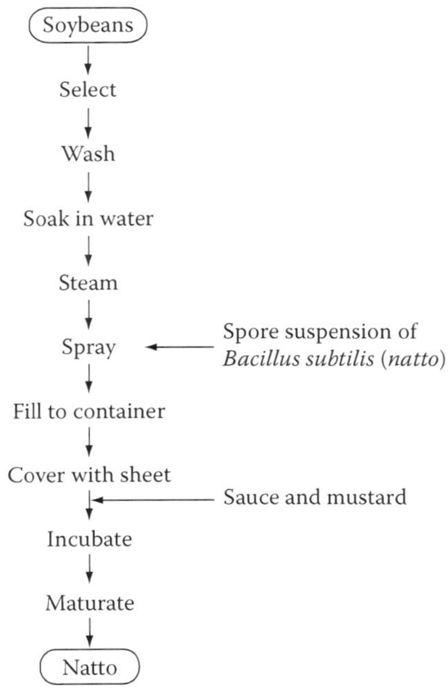

Figure 2.4 Production process of natto.

dust or metal debris, washed, and soaked in water of 10°C for 18 h. After soaking, the weight of soybeans is about twice the weight of raw soybeans. Longer soaking time causes a loss of soluble nutritional components from the soybeans. The soybeans are cooked in a large rotary pressure cooker at 2 kg/cm^2 and for 1–1.5 h. As soon as cooking is finished, to prevent the natto from being contaminated by heat-labile microorganisms and bacteriophages (Nagai and Yamasaki, 2009), heat-tolerant spore suspension of *B. subtilis* (*natto*) is sprayed on the cooked beans at a density of 10^8 cfu/60 kg soybeans. Batches (30–50 g each) of inoculated soybeans are put into containers made of polystyrene and covered with a vinyl sheet on which small pouches of seasonings, mustard, and sauce in many cases, are placed (Figure 2.2a). The packages are transferred into an incubation room with computer-controlled temperature, humidity, and aeration. At first, incubation is carried out at 48°C–50°C and a high humidity for 16–18 h, and then maturation of the natto is carried out at 3°C–10°C and a low humidity for 8 h. The packages are labeled and delivered to retail shops or supermarkets by cold chain systems.

Natto produced from small-sized soybeans (diameter, 4.9–7.3 mm) is preferable to that from larger sized ones, and the soybeans are imported mainly from the United States, Canada, and China. Soybeans with higher sugar and lower fat content seem to be suitable for natto production (Kiuchi, 2000).

2.1.1.1.4 Nutritional and Chemical Components Nutritional composition of natto, boiled soybeans, and raw soybeans for reference are shown in Table 2.1. The most prominent change in the compositions after natto production is observed in the content of vitamin K. This remarkable increase is due to vitamin K_2 (menaquinone-7 or MK-7), produced by *B. subtilis* (*natto*). Vitamin K is essential for blood coagulation factors II, VII, IX, and X produced in liver. Although hydrophobic by nature, MK-7 produced by *B. subtilis* (*natto*) exists as a water-soluble mixed micelle with 2-kDa glycopeptides (Ikeda and Doi, 1990; Ikeda, 1992). The glycopeptide, vitamin K_2-binding factor, is composed of four kinds of amino acids, glutamic acid, aspartic acid, leucine, and valine. Infants, fed with breast milk, contract vitamin K deficiency bleeding (VKDB) at a relatively high rate (1/1500), and the patients often die due to intracranial hemorrhage. One cause

Table 2.1 Nutritional Composition of Raw Soybean, Boiled Soybean, and Natto

PARAMETER	SOYBEAN	BOILED SOYBEAN	NATTO
Proximate			
Energy (kcal)	477	493	494
Protein (g)	40.3	43.8	40.7
Lipid (g)	21.7	24.7	24.7
Carbohydrate (g)	32.2	26.6	29.9
Ash (g)	5.7	4.9	4.7
Mineral (mg)			
Sodium	1	3	5
Potassium	2200	1600	1600
Calcium	270	190	220
Magnesium	250	300	250
Phosphorus	660	520	470
Iron	10.7	5.5	8.1
Zinc	3.7	5.5	4.7
Copper	1.12	0.66	1.51
Vitamin			
A, retinol (μg)	(0)	(0)	(0)
A, α-carotene (μg)	0	0	–
A, β-carotene (μg)	7	8	–
A, cryptoxanthin (μg)	0	0	–
D (μg)	(0)	(0)	(0)
E, α-tocopherol (mg)	2.1	2.2	1.2
E, β-tocopherol (mg)	0.8	0.8	0.5
E, γ-tocopherol (mg)	16.5	16.4	14.6
E, δ-tocopherol (mg)	9.4	9.3	8.1
K (μg)	21	19	1500
B_1 (mg)	0.95	0.60	0.17
B_2 (mg)	0.34	0.25	1.38
Niacin (mg)	2.5	1.4	2.7
B_6 (mg)	0.61	0.30	0.59
B_{12} (μg)	0	(0)	Tr
Folate (μg)	260	110	300
Pantothenic acid (mg)	1.74	0.79	8.89
C (mg)	Tr	Tr	Tr
Fatty acid (g)			
Saturated	2.96	3.34	3.63
Monounsaturated	4.18	4.74	4.69
Polyunsaturated	11.90	13.51	13.31
Cholesterol (mg)	Tr	(Tr)	Tr

(*continued*)

Table 2.1 (Continued) Nutritional Composition of Raw Soybean, Boiled Soybean, and Natto

PARAMETER	SOYBEAN	BOILED SOYBEAN	NATTO
Dietary Fiber (g)			
Soluble	2.1	2.5	5.7
Insoluble	17.5	16.7	10.9

Source: These data (per 100 g dry matter) were converted from the data in *Food Composition Database.* Data from Ministry of Education–Culture–Sports–Science and Technology, *Standard Tables of Food Composition in Japan: Food Composition Database*, http://fooddb.mext.go.jp, 2010.

Note: –, Not determined; (), estimated; Tr, trace. Water contents of raw soybean, boiled soybean, and natto were 12.5%, 63.5%, and 59.5%, respectively.

of VKDB is a low level of vitamin K in breast milk. Administration of natto to lactating women increases the level of vitamin K in breast milk and prevents the infants from VKDB (Ryo and Baba, 1981). On the other hand, natto rich in vitamin K is a contraindication to administration with warfarin, especially after artificial cardiac valve replacement operation (Kudo, 1990). Warfarin competitively antagonizes vitamin K, and therefore, inhibits biosynthesis of vitamin K-dependent clotting factors in livers. A high concentration of vitamin K in blood negates the inhibitory effect of warfarin. Recently, new anticoagulants not antagonistic to vitamin K, rivaroxaban and dabigatran, have become available, so these drugs could replace warfarin.

From GC-MS analysis on commercial natto products and other investigations, 10 volatile compounds were identified as key components of the smell of natto: ethanol, 2,3-butanedione (diacetyl), pyrazine, 2-methylpyrazine, 3-hydroxy-2-butanone (acetoin), 2,5-dimethylpyrazine, 2,3,5-trimethylpyrazine, 2-methylpropanoic acid (isobutyric acid), 2-methylbutanoic acid, and 3-methylbutanoic acid (isovaleric acid) (Tanaka and Tomiyasu, 1973; Mitsuboshi and Kiuchi, 2007). Ammonia is also detected especially from aged natto, but it undermines the quality of natto.

Isoflavones, which are known to have estrogenic activity, are contained in soybeans as free isoflavones, isoflavone glycosides, and acetylated isoflavone glycosides. During the fermentation, free isoflavones and acetylated isoflavone glycosides decrease, and succinylated isoflavone glycosides concomitantly increased (Toda et al., 1999; Shimakage et al., 2006). Polyamines, which are concerned with the growth and differentiation of cells, are contained in soybeans. Although polyamines are degraded to some degree during fermentation of natto, intake of natto increases the polyamine concentration in human blood (Soda et al., 2009). Other bioactive components in natto are also described in this book.

On the surface of aged natto, white crystals of tyrosine and struvite ($MgNH_4PO_4.6H_2O$) are often deposited because of their low solubility in water (Tanaka and Tomiyasu, 1973; Muramatsu et al., 1997, 2000). The crystals are not harmful to human health, but they impair the quality of natto greatly due to their sand-like texture.

2.1.1.1.5 Consumption Japanese people eat natto after mixing it with seasonings (mustard, soy sauce, finely chopped raw leek, or raw egg) and placing it on steamed rice (Figure 2.5). One pack of natto (30–50 g) is used for one bowl of steamed rice (150 g). Natto is also used as an ingredient with spaghetti, Chinese fried rice, sushi, soup, and so on. One pack (40 g) of natto is currently (April, 2014) selling for about 30 yens (= US$0.29) in Japan.

Consumed soybeans for natto production amounted to 130,000 tons in 1999. Natto is more consumed in the northeast areas of Japan than the southwest areas. The percentages of families consuming natto once or more a week are 90.6, 93.7, 81.5, 82.2, 76.3, 69.9, 69.7, and 79.1, ordered from the northeast to the southeast Japan (Japan Natto Cooperative Society Federation, 2011). The regional characteristic in consumption of natto might reflect the ideal place of the origin of natto production, that is, northern Japan, as described above. However, this regional characteristic in consumption is likely to fade in the future because of the movement of people all over Japan for

Figure 2.5 Natto on cooked rice after mixing with soy sauce and chopped raw leek, the most popular way to eat natto in Japan.

business or university entrance, and homemaker's recognition of natto as a healthy and cheap foodstuff for everyday meals.

2.1.1.1.6 Conclusion Natto, fermented by *B. subtilis* (*natto*), is mainly produced and consumed in Japan. Natto has characteristics of the production of a highly viscous polymer (PGA) and fermentation odor (branched short-chain fatty acids and ammonia). Formerly, natto was produced by wrapping boiled soybeans with straws where *B. subtilis* (*natto*) inhabits naturally. After modernization, pure cultures of *B. subtilis* (*natto*) and clean containers have been used. It takes about 2 days to prepare natto: cleaning and washing soybeans, soaking the soybeans in water, boiling the soybeans, inoculating *B. subtilis* (*natto*) on the boiled soybeans, filling the containers with soybeans, and incubating them. Natto takes over the functional and nutritional components from soybeans themselves, and *B. subtilis* (*natto*) produces bioactive substances, some of which are described in the other chapters of this book. Now some non-Japanese people pay attention to natto because of its benefit on human health, and worldwide consumption of natto will increase if its unique viscosity and odor can be overcome through further research.

2.1.1.2 Douchi

Li-Te Li and Yan-Li Ma

2.1.1.2.1 Origin of Douchi Douchi (Figure 2.6), also known as salted black bean, owing to its black appearance after fermentation, is a food fermented by molds or bacteria on whole soybean. It originated in China before the Han Dynasty (206 BC) and has been consumed there for more than 2000 years.

Figure 2.6 Douchi as sold in a supermarket in China.

Douchi ginger was found as a residue buried in Mawangdui West Han Grave in Changsha of Hunan province, central China. This discovery proved that douchi had been a favorite food in China since 200 BC. Douchi was the first soybean food to be described in written records and was first recorded in the *Ben-Cao-Gang-Mu* (a well-known Chinese medicine material book) compiled by Li Shi-zhen in 1597. In ancient China, douchi was called "youshu," which means boiled soybean fermented in an airtight environment. Since the Qin Dynasty, it has been referred to as douchi.

Douchi is made from soybeans by pretreatments (soaking followed by steaming) and a two-step fermentation process (pre-fermentation and post-fermentation), some of which include a subsequent drying process. More recently, soy paste and soy sauce have been developed from douchi. Since ancient times, douchi has been considered a nutrient-rich and delicious fermented soybean food in China. It is widely used as a seasoning to enhance the appetite and is commonly consumed in Zhejiang, Fujian, Sichuan, Hunan, Hubei, Jiangsu, and Jiangxi provinces, and northern areas of China. Due to the diffusion of Chinese Buddhism during the era of Tang Dynasty, the processing technology of douchi was introduced to many countries such as Japan, Indonesia, Korea, and Philippines. The differences in climate, material, and eating habits have resulted in the evolution of douchi into different products, like natto in Japan, tempeh in Indonesia, and chongkukjang in Korea. Natto and tempeh are now internationally accepted foods and have increasing presence in the worldwide market.

Besides being a popular seasoning ingredient with a deep, rich, and savory flavor, douchi has been used for medicinal purposes in China since ancient times. For example, the *Pen Ts'ao Kang Mu/ Ben-Cao-Gang-Mu*, China's most famous *Materia Medica* compiled by Li Shi-Zhen (1518–1593), described how douchi could be used to prevent and treat problems such as stroke, diarrhea, irritability, restlessness, and insomnia (Fan et al., 2009). *Zhang Zhong-Jing* of the Han Dynasty recorded that soup cooked with cape jasmine (*Gardenia jasminoides* J.Ellis) and douchi were useful to relieve tiredness, weakness, insomnia, and poor appetite. Even today, it is still added to some Chinese traditional medicines.

2.1.1.2.2 Classification There are many kinds of douchi in China, and they are classified in a number of ways (Figure 2.7).

1. According to the types of raw materials used in douchi production, it can be divided into yellow soybean douchi and black bean douchi.
2. According to the dominant microorganisms used in fermentation, there are four types of douchi, namely *Aspergillus*-type douchi, *Mucor*-type douchi, *Rhizopus*-type douchi, and bacterial-type douchi, which are fermented by *Aspergillus*, *Mucor*, *Rhizopus*, and bacteria, respectively. Among them, *Aspergillus*-type douchi originated first, with origins of production in China identifiable at least 2000 years ago. It has been the most popular type and is still produced widely in China. *Aspergillus*-type douchi is popular especially in Hunan and Guangdong provinces in China, for example, Liuyang douchi and Yangjiang douchi. *Mucor*-type douchi is the main type consumed in Sichuan and Chongqing provinces, such as Yongchuan douchi. On the other hand, bacterial-type douchi is mainly produced in Shandong, Yunnan, and Guizhou provinces. The bacterial-type douchi is made by natural fermentation. This kind of spontaneous and uncontrolled fermentation may lead to problems related to hygiene safety, and

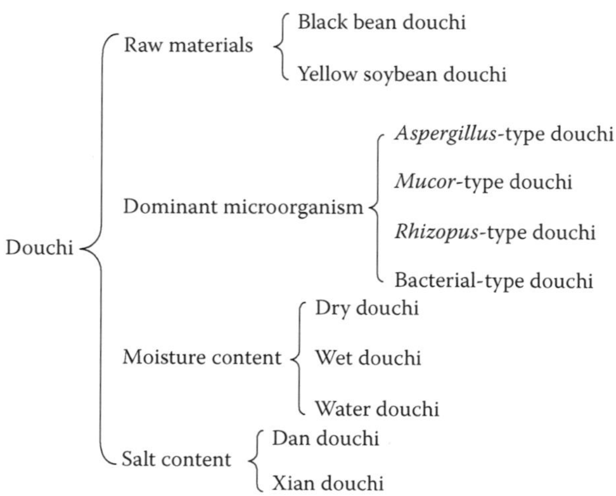

Figure 2.7 Classification of douchi according to different standards.

inconsistent sensory and nutritional properties. Therefore, application of pure microbial cultures for bacterial-type douchi production is essential to accelerate the industrial production of douchi.

3. Douchi is also divided into two types based on the amount of NaCl used, namely non-salt douchi (Dan douchi) and salt-fermented douchi (Xian douchi). Dan douchi, a naturally fermented product used as a Chinese medicine, is primarily produced by bacteria and *Aspergillus*. It is produced without the post-fermentation step and could be thought as a semi-product of douchi. Other than Dan douchi, natto and tempeh are also non-salt-fermented soybean foods. Similar products are consumed in southern China, the continental northern areas of Southeast Asia, and West Africa. In many of these areas, non-salt-fermented soybean products are used in a manner similar to the way shrimp paste is used in Asian countries.

   In China, most douchi samples are salt-fermented and are used as flavoring. Salt plays multiple roles in douchi production. It imparts a salty taste and controls microbial growth and enzyme activity of douchi. However, epidemiologic studies suggested that there was a positive correlation between dietary NaCl intake and blood pressure. Therefore, in recent years, some researchers committed to the development of low-salt douchi (Hu et al., 2011). However, whether the NaCl in fermented soybean foods caused health problems or not is still controversial. Watanabe et al. (2006) investigated the effects of miso (Japanese salt-fermented soybean food) and NaCl on the blood pressure of NaCl-sensitive Dahl rats. The results showed that blood pressure of Dahl rats increased significantly on the NaCl diet whereas for those on the miso diet with the same NaCl content, their blood pressure remained constant. Some Chinese dishes use douchi instead of NaCl to increase saltiness, which may be a way to meet the dual need of being healthy and tasty.

4. Douchi is classified into dry douchi, wet douchi, and water douchi, according to moisture content present. Douchi is dried after fermentation to reduce moisture content and improve

Figure 2.8 Types of douchi: (a) dan douchi, which is non-salted fermented; (b) dry douchi, which experiences a drying process in the later stages of the production process; and (c) water douchi, which is generally soaked in water.

preservation, which is thus named as dry douchi and is mainly produced in southern China. Wet douchi, on the other hand, is not dried after fermentation, and is mainly produced in North China. It is currently the most common type of douchi in the market. As for water douchi, it refers to homemade douchi soaked in water. These different types of douchi are distributed in different regions of China, which are created by the Chinese people during long-term labor practice. Figure 2.8 shows some types of douchi, which include dan douchi, dry douchi, and water douchi.

2.1.1.2.3 Production Methods Douchi has been produced in small workshops in China for thousands of years. It is usually prepared in cottages or on a home-scale production level and has since expanded into cottage industry with increasing consumer demands. The preparation methods of douchi for fermentation and brine composition may vary from one region to another, but the essential features are similar. The manufacturing process includes three main steps: pre-treatment, pre-fermentation, and post-fermentation. The schematic diagram for production of douchi is as shown in Figure 2.9. Fermentation, dressing, aging, and drying process of douchi production in Shaanxi province in China are shown in Figure 2.10.

2.1.1.2.3.1 Pre-Treatment Douchi is made from soybeans, preferably using small-seeded ones with high protein content. Black soybeans are often used since they have thick epidermis and loose particle shape, which make it easy for the beans to maintain its integrity during

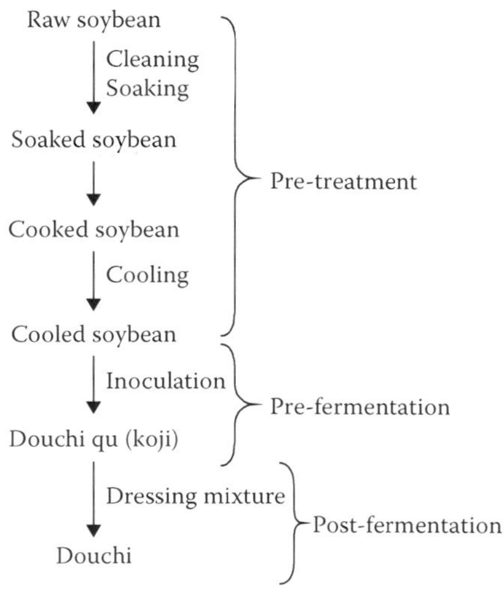

Figure 2.9 Production process of douchi.

Figure 2.10 Homemade douchi production process followed in Shaanxi province of China.

the production process. In addition, black beans are rich in melanin, giving douchi a bright black color and high nutritional value.

Choice of soybeans: Soybeans for douchi production should be carefully selected. Generally, fresh soybeans with high protein content and full uniform particles are good choices. At the same time, soybeans should be matured, non-worm-eaten, non-rotten, and have minimal impurities.

Soaking: Soybeans are washed, and soaked in tap water for about 8 h at ambient temperature (24°C ± 2°C) to take up sufficient moisture. Soaking time is dependant on room temperature; in general, the higher the temperature, the shorter the soaking time needed. Halfway through the soaking process, it is advised for the water to be changed. When 90% of the beans have wrinkle-free skin and the moisture content is about 45%, the soaking process is completed.

Boiling or steaming: In the past, when producing douchi on home scale, soaked soybeans were usually boiled in water, whereas steaming is now widely used. The steaming of soybean requires around 4 h under atmospheric pressure, while only 1 h is needed under 0.1 MPa pressure when using rotary pressure cooking pot in the industrial production scale. After steaming or cooking, a moisture content of 45% in soybean is appropriate. The cooked or steamed soybeans, which are tender enough to be mashed by fingers, are then drained and cooled to about 40°C.

2.1.1.2.3.2 Pre-Fermentation Pre-fermentation is the process of making douchi qu (also known as "koji"). It is a microbial solid-state fermentation process on soybeans, generating complex enzymes and abundant metabolites. To make douchi qu by the traditional method, cooked soybeans are fermented by microorganisms that are naturally present in the environment, which is closely related to environmental conditions of different regions. In industrial production, cooked soybeans are inoculated with a pure microbial culture to make douchi qu. There would be differences between the methods used to make douchi qu due to different microorganisms used.

Aspergillus-type douchi qu. Natural koji: Steamed or boiled soybeans are encased into bamboo dustpans when they are cooled to about 35°C, which are then moved to the fermentation room. The temperature of fermentation room and the soybeans are kept at 26°C–30°C and 25°C–35°C, respectively. After 24 h of fermentation, the temperature of soybeans would begin to increase and a few small agglomerations would appear. After about 48 h of fermentation, the temperature of soybeans may increase to about 37°C. The soybeans are now covered with mycelia and many big agglomerations are observed. At this time, loosening the agglomeration-shaped soybeans and mixing the douchi qu are in favor of propagating spores and decreasing the temperature of soybeans to about 32°C. The secondary turnover should be carried out when agglomerations appear again and yellow-green spores emerge. The soybeans are kept at about 35°C–37°C for 2 days and then cooled to 28°C–30°C through ventilation. After 6–7 days of fermentation, matured koji is obtained. The moisture content of matured koji is about 21%. Yellow-green spores and mycelia can be seen on the surface and interior of soybean, respectively.

Pure inoculum douchi qu: Pure culture inocula are used to produce douchi qu on industrial scale, which are beneficial in reducing the production period and controlling the quality of douchi. Many kinds of *Aspergillus* strains including Huliang 3.042 and *Aspergillus* 3.798 have been developed recently and are widely used. When a strain of *Aspergillus* is used to make douchi qu, boiled or steamed soybeans are cooled to about 35°C, inoculated with 0.3% *Aspergillus* mass, mixed, loaded into bamboo dustpans with a thickness of 2 cm, and then moved to the fermentation room. The temperature of fermentation room and soybeans are kept at 25°C and 25°C–35°C, respectively. After about 22 h of fermentation, soybeans are agglomerated and covered with white mycelia. The temperature would probably increase to about 35°C. At this stage, the soybeans should be flipped back and forth. When fermented for about 72 h, mycelium and yellowish green spores would cover the soybeans. The final product, douchi qu, is obtained.

Mucor-type *douchi qu: Natural koji:* The production of *Mucor*-type douchi qu is usually carried out in winter. Steamed or boiled soybeans are cooled to 30°C–35°C, put on dustpans or mats with a thickness of 3–5 cm, and then moved to the fermentation room. The temperature of fermentation room and the soybeans are kept at 2°C–6°C and 5°C–12°C, respectively. After 3–4 days of fermentation, white mildews can be seen on the soybean surface. After fermentation for 8–12 days, orderly mycelia and a few brown spores can be found in soybeans. When fermented for 16–21 days, the mycelia would have a tight texture and change from white to gray. Mycelia erect to a height of 0.3–0.5 cm and there are green thalli attached tightly to the soybeans. At this moment, matured koji can be obtained. The period of making *Mucor*-type douchi qu is generally 15–21 days, depending on climate changes.

Pure inoculum douchi qu: The production of *Mucor*-type douchi qu requires low temperature and long period, which are not conducive for large-scale industrial production. Pure culture inocula are generally chosen to make *Mucor*-type douchi qu in modern industrial production. When pure *Mucor* is used to make douchi qu, boiled or steamed soybeans are cooled to 30°C, inoculated with 0.5% pure *Mucor* mass, mixed, and loaded in dustpans having a thickness of 3–5 cm. The soybeans are then kept at 23°C–27°C and fermented in the cultivation room. White *Mucor* colonies can be seen on the soybeans surface after fermented for 24 h. After fermentation for 36 h, soybeans are covered with mycelium. After fermented for 48 h, mycelium would erect and it changes from white to gray. At the same time, spores gradually increase. It usually needs 3 days to obtain douchi qu with pure *Mucor*.

Bacteria-type douchi qu: Water douchi and most homemade douchi are produced using bacteria. Currently, bacteria-type douchi qu is rarely made by pure bacterial artificial inoculation. The process is to make use of bacteria in the environment to inoculate naturally. Soybeans, boiled or steamed, are wrapped with bags or straw and left to incubate naturally for 5–6 days in a warm place (20°C–40°C) until the soybeans are covered with a stringy, mucilaginous coating and obtain a typical douchi-like flavor. Microorganisms in bacterial-type douchi qu are complex, where *Bacillus subtilis* and lactic acid bacteria are dominant.

2.1.1.2.3.3 Post-Fermentation Post-fermentation is a crucial step for the creation of taste, flavor, and special texture as well as nutritional properties of douchi. It is a complex biochemical reaction process utilizing a variety of enzymes secreted by microorganisms during production of koji. The douchi qu is first mixed with the desired amounts of salt and some seasoning and then aged for several weeks or months, according to local ways. The finished products are blackish and have a salty taste, and their flavor resembles that of soybean sauce. Some of them include a subsequent drying process. There are significant differences in post-fermentation methods according to types and producers. For instance, Liuyang douchi was fermented at a temperature of 28°C–30°C for 4 days, and the post-fermentation period is less than 30 days. While Sichuan douchi was fermented at a temperature of 6°C–12°C for 5–12 days, the post-fermentation period is up to about 8 months. The following is the introduction of post-fermentation process in accordance with douchi types.

Aspergillus-type douchi: The first step of producing *Aspergillus*-type douchi is washing koji, which is a special process different from that of other types of douchi. After washing koji, some spores, mycelia, and enzymes attached to the surface of soybeans can be washed away to limit the excessive hydrolysis of raw materials and to avoid the formation of bitter flavor. Washing of koji is performed as follows: douchi qu is poured into a pool with warm water to wash conidiophores and mycelia, then moved to bamboo baskets to rinse repeatedly until there is no mycelia and spores present. The process of washing koji should be controlled into about 10 min to avoid the fester of fermented soybeans. The moisture of washed douchi qu should be within the range of 33%–35%. Subsequently, washed douchi qu is piled and intermittently sprinkled with water to adjust the moisture content to about 45%. It is important to control moisture content because if the moisture content is too high, soybeans will rot easily and lose luster; in contrast, low moisture content will lead to the increase in hardness of final products. Douchi qu with 45% moisture is covered with plastic film to increase the temperature to about 55°C. When there are mycelia on surface of douchi qu again and douchi-like flavor can be smelled, some spices such as capsicum, ginger, and so on, may be added and mixed. After mixing with seasoning, douchi qu is pressed tightly layer by layer in the jar and covered with salt on surface to ferment for 4–6 months. *Aspergillus*-type douchi is thus obtained. Sometimes, to extend the storage period, mature douchi is dried until its moisture content is below 30%, whereas others are sun-dried for long-term preservation. The quality of douchi depends on the effectiveness of sun drying. Rainy seasons often result in spoilage. Controlled fermentation and mechanical drying seem to be potential ways to commercialize the douchi products.

Mucor-type douchi: Mucor-type douchi qu is poured into mixing pool, where a certain amount of salt and water are added and mixed. After being stewed for one day, liquor, fermented rice, or some spices are added and mixed. The douchi qu are pressed layer by layer in a jar and sealed to ferment for 10–12 months for aging.

Bacterial-type douchi: Bacterial-type douchi qu is poured into mixing pool, added with a certain amount of salt, spices, and so on, before mixing with water left over from boiling soybean. The douchi qu is encased in a jar and sealed to ferment for 1 month before bacterial-type douchi can be obtained.

As there are no defined microorganisms and standard procedures for douchi processing, a wide range of douchi production processes exist in the various regions of China. For example, while Tongchuan douchi usually ripens within 8–12 months, the ripening period of Liuyang douchi is no more than 30 days. Therefore, a great variation in quality and functional activity exists among douchi from different regions (Fan et al., 2009).

2.1.1.2.4 Chemical and Nutritional Components Douchi is a popular product among the Chinese community worldwide and has been consumed for centuries as a protein source and flavoring ingredient.

Chemical and nutritional components in soybeans are changed during douchi production by a series of complex biochemical reactions. During fermentation, protein, lipid, and starch are converted to peptides, amino acids, fatty acids, reducing sugars, ethanol, and various aroma components by salt-tolerant microorganisms. Amino-type nitrogen content of douchi products are much higher than that of corresponding douchi qu, which suggests protein hydrolysis occurred mainly during the post-fermentation process (Zhang et al., 2007).

The concentration of free amino acids increases gradually during the fermentation process (Table 2.2). The ratio of total free amino acids in douchi qu is much lower than that in douchi, which suggests that most of the free amino acids are produced during the post-fermentation process. Contrary to most of the free amino acids, the concentrations of threonine and phenylalanine in douchi qu are more than half of that in douchi, and the concentration of arginine in douchi qu is more than that in douchi. With the exception of aspartic acid and proline, free amino acid concentration of douchi that is fermented

Table 2.2 Free Amino Acid Content (mg) per Gram Douchi Prepared Using *Aspergillus oryzae*

AMINO ACID	PRE-FERMENTED TIME (h)				POST-FERMENTED FOR 15 DAYS	RATIO[a] (%)
	0	24	48	72		
Asp	0.42	0.57	0.55	1.07	6.42	16.7
Thr	0.39	0.52	1.62	3.14	4.90	64.1
Ser	ND	0.24	0.95	1.71	5.96	28.7
Glu	0.58	1.16	2.19	3.75	25.92	14.5
Pro	ND	ND	0.35	0.95	7.08	13.4
Gly	0.06	0.07	0.26	0.45	2.68	16.8
Ala	0.18	0.31	1.05	1.74	7.20	24.2
Val	0.11	0.26	0.99	2.00	7.00	28.6
Met	0.09	0.02	0.49	0.77	2.09	36.8
Ile	0.03	0.10	0.64	1.27	6.32	20.1
Leu	0.06	0.23	2.06	3.82	11.20	34.1
Tyr	0.36	0.49	1.73	2.13	6.33	33.6
Phe	0.39	0.85	3.07	4.90	8.68	45.4
Lys	0.12	0.22	0.90	1.97	5.81	33.9
His	0.06	0.23	0.27	0.48	1.88	25.5
Arg	1.58	1.58	1.83	2.92	0.08	D
Total	4.43	6.84	18.95	33.07	109.54	30.2

Note: ND, not detected; D, decreased.

[a] Ratio of 72-h concentration to 15-day concentration, expressed as percentage.

for 15 days is greater than, or very similar to, that of some douchi product with a post-fermentation period of 2 months.

As a condiment, douchi has a unique flavor that is loved by many people. The flavor ingredients of douchi are generated by a series of complex biochemical reactions catalyzed by the enzymes during fermentation. Carbohydrases and proteinases are the two main groups of enzymes produced by fermentation microorganisms during douchi production. Lipase activity has also been reported. These major enzymes hydrolyze carbohydrates and proteins to sugars, amino acids, and low-molecular-weight peptides, respectively. Many flavor compounds have been identified from douchi, and there are many factors affecting the flavor and volatile compounds of douchi such as raw materials, soaking time, cooking time, fermentation microorganisms, inoculation rate, fermentation temperature, and the addition of certain ingredients (such as salt, alcohol, and ginger) during post-fermentation. Wang et al. (2010) identified the volatile components in three samples. There were a total of 80, 55, and 79 components, respectively. Combining the total number of volatiles in all the three douchi samples, 131 compounds were identified, of which only 25 were common in all brands. The volatiles belonged to 13 classes of compounds including 29 kinds of esters, 18 kinds of acids, 16 kinds of alcohols, 14 kinds of pyrazines, 13 kinds of ketones, 12 kinds of aldehydes, 6 kinds of phenols, 5 kinds of hydrocarbons, 5 kinds of furans, 5 kinds of sulfur-containing compounds, 4 kinds of pyridines, 2 kinds of pyrimidines, and 2 kinds of miscellaneous compounds. According to the results obtained and the volatile compounds of other fermented soybean foods, pyrazines, esters, acids, and alcohols are the important contributors to the characteristic flavor of douchi.

The configuration of isoflavone is changed during douchi production. The aglycones (daidzein, genistein, and glycitein) increased, while the β-glucosides (daidzin, genistin, and glycitin) decreased (Wang et al., 2007b). Isoflavones in the form of aglycones exceeding 90% shows higher bioactivity than the isoflavones that is not transformed. Further, isoflavones in the form of aglycones exceed 90% of total content during the post-fermentation. Fibrinolytic enzyme has been found in douchi (Peng et al., 2003), which is a serine proteinase produced by microorganism screened from douchi. The enzyme is safer, easier to be absorbed by human, and can decompose fibrin directly. It also has a strong capability to resolute the stubborn thrombus. Douchi fibrinolytic enzyme could be used to design fibrinolytic medicine with lower cost by fermentation. It can be seen that the nutrition and functionality of soybean are improved by the production of douchi.

2.1.1.2.5 Functional Properties Douchi has been appreciated by consumers as a health food due to its valuable nutritional and medicinal attributes. Many historical medical books, e.g., *Zhang Zhong-Jing*, have described douchi being able to prevent and cure diseases (Chen et al., 2007).

In recent years, douchi has received much attention due to revelations of its various functionalities. Modern studies have shown that douchi exhibited α-glucosidase inhibitory activity, and the anti-diabetic properties containing in douchi were able to prevent and cure patients with diabetes (McCue and Shetty, 2003). A water-soluble extract of douchi was found to exert a strong inhibitory activity against rat intestinal α-glucosidase, and in vivo test demonstrated the antihyperglycemic effect and may have potential use in the management of patients with non-insulin-dependent diabetic mellitus (Fujita et al., 2001). Some studies have suggested the antioxidant properties of douchi. Wang et al. (2007a) found that the antioxidative compounds were released during the douchi production, and that they are inhibited by high salt addition. Angiotensin-converting enzyme (ACE) inhibitory activities of douchi have also been extensively studied (Zhang et al., 2006; Li et al., 2009; Wang et al., 2011), and it was proved that fermentation was a useful step for improving the ACE inhibitory activity in douchi. The acetylcholinesterase (AChE) inhibitory activity were investigated in 19 representative commercial douchi products (Liu et al., 2009). It was found that douchi in China exhibited various extents of AChE inhibitory activity, and *Aspergillus*-type douchi extracts exhibited significantly higher AChE inhibitory activity than that of *Mucor*- and *Bacillus*-type douchi.

2.1.1.2.6 Consumption and Perspectives In some areas of China, douchi is a regular part of the local diet and plays an important role in economic, social, and cultural development. Most douchis in China are salty and savory, while some are spicy. They are often used as condiments to flavor foods such as stir-fries, stews, and soups. Douchi can be used as an appetizer for consumption with bland foods such as rice gruel, or can be cooked as a flavoring agent with vegetables, meats, and seafoods. It has been used as a savory enhancer by most vegetarians as a substitute for meat. Several dishes have also been prepared from douchi (Figure 2.11).

Recently, workshop productions of douchi are always at a small scale. Although small-scale manufacturing has advantages such as short distribution lines and income generation for families, the growing demand for ready-to-consume and high-quality foods requires larger-scale controlled industrial production. There has been an increase in douchi consumption in recent years. Currently, there is an increasing interest of douchi industrialization in China, thus scientific studies of microbial communities are being carried out to improve production methods, extend shelf life, and expand production scale. Furthermore, quality control of douchi will safeguard the product hygiene and provide predictable organoleptic properties. Upgrading of traditional home-scale processes is necessary so that they can

Figure 2.11 Dishes prepared from douchi in China.

continue to maintain and strengthen the cultural heritage and can compete successfully with imported products.

In conclusion, douchi is one of the traditional alkaline-fermented foods, which has been widely consumed as a relish by the Chinese people for more than 2000 years. Although there has been some research on douchi, further study needs to be taken to improve various aspects in the production of douchi and supply people with more scientific information. If innovations in processing technology, functionality, and storage are made, douchi may become more popular not only in China but also in the world.

2.1.1.3 Kinema and Similar Products

Prabir K. Sarkar and M. J. Robert Nout

Kinema is a unique flavorsome delicacy to many of the people of Nepal and the neighboring countries. This indigenous soybean-fermented condiment is a low-cost meat substitute and a source of income for many rural households.

2.1.1.3.1 Origin and Culture The word "kinema" is derived from "kinambaa" of the Yakthung or Limbu (a tribe indigenous to Limbuwan comprising nine districts within Mechi and Kosi zones of east Nepal) dialect where "ki" means fermented and "nambaa" means flavor. Although the history of origin of kinema is lost in antiquity, it is certain that among the Nepalese, the Limbu started its production and consumption. After unification of Limbuwan and Nepal in the 18th century, there was creation of a mixed society of multiethnic communities. This might have resulted in the spread of kinema culture to the allied Nepali communities, such as the Rai, Tamang, Gurung, and Mangar (Tamang, 2001). Because of trans-boundary movement of the people, kinema is also popular in the adjacent Darjeeling district of the State of West Bengal and the State of Sikkim in India, and also in Bhutan (Tamang et al., 1988). In northeast India, kinema is traditionally consumed by the non-Nepalis and called by various names (Table 2.3; Figure 2.12). On visiting a local periodic market (called "haat"), one might catch a glimpse of small, neatly packed, green-colored (wrapped in large-sized locally available leaves) bundles, tied loosely using a piece of straw. This is how these foods are commonly sold by women of the ethnic communities (Figure 2.13).

Table 2.3 Kinema Variants, Used by Non-Nepali Communities in Northeast India

PRODUCT NAME	LOCATION	USER TRIBE	MAIN PROCESSING STEPS	MODE OF CONSUMPTION
Aagya	Basar town in West Siang district of the State of Arunachal Pradesh	Galo	Whole beans are boiled and fermented for 12–14 days.	Mixed with boiled vegetables and chutney
Axone/akhone/aakhuni/kheuha/sabrocha/sadocha	State of Nagaland	Sema	Whole beans are boiled and fermented for 5–8 days.	Ground with tomato, chillies, and salt to make chutney
Bekang/bekanthu	State of Mizoram	Mizo	Crushed whole beans are boiled and fermented for 3 days.	Mixed with herbs and boiled leafy vegetables
Bhari/bari	State of Sikkim	Bhutia	Beans are soaked, boiled, crushed to grits, and fermented for 4–5 days.	Ground with vegetables and herbs to make chutney or cooked to curry
Chukchoro	Shergaon area of West Kameng district of the State of Arunachal Pradesh	Sherdukpen	Beans are soaked, dehulled, boiled, and fermented for 10–12 days (15–30 days in winter).	Mixed with vegetables, or ground with vegetables and yak cheese to make chutney
Hawaijar/hawaizaar	State of Manipur	Meitei	Whole beans are soaked, boiled, and fermented for 2–5 days.	Ground to make chutney, or cooked with rice, vegetables, spices, and prawn/fish to make curry
Peron namsing	East Siang district of the State of Arunachal Pradesh	Adi	Whole beans are boiled and fermented for 10–12 days (25–30 days in winter).	Mixed with boiled vegetables and herbs or ground to make chutney
Satlyangser	State of Sikkim	Lepcha	Beans are soaked, boiled, crushed to grits, and fermented for 2–3 days.	Cooked with vegetables to make curry
Tungrymbai	State of Meghalaya	Khasi	Whole beans are soaked, dehulled, boiled, and fermented for 3–5 days.	Mixed with herbs and spices to make curry or pickle
Yanni perung/peruyyan	Lower Subansiri district of the State of Arunachal Pradesh	Apatani	Whole beans are boiled and fermented for 3–4 days (10–12 days in winter).	Mixed with herbs to make chutney

Source: Tamang, J.P. et al., *Journal of the Science of Food and Agriculture*, 44, 375–385, 1988; Tamang, J.P. et al., *Indian Journal of Traditional Knowledge*, 8, 122–126, 2009; Singh, A. et al., *Indian Journal of Traditional Knowledge*, 6, 79–94, 2007; Premarani, T. and Chhetry, G.K.N., *Assam University Journal of Science & Technology: Biological and Environmental Sciences*, 6, 82–88, 2010.

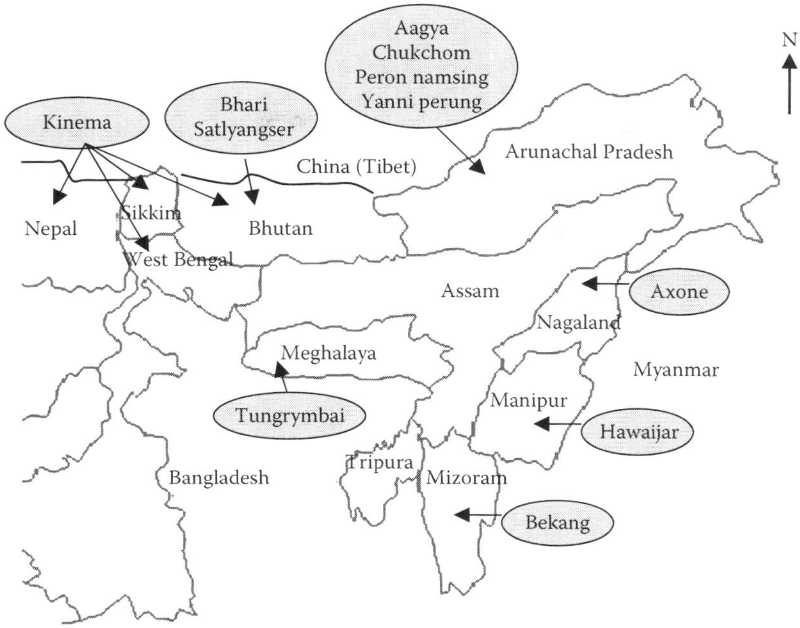

Figure 2.12 Localities of use of kinema and its similar products in eastern part of Nepal, Bhutan, and northeast Indian States.

Figure 2.13 Packages of fresh kinema, as sold in a market of Kalimpong, India.

2.1.1.3.2 Indigenous Preparation of Kinema and Its Socio-Economic Impact The preparation of kinema (Figure 2.14) is exclusively practiced by women. The skill of production of this delicacy has been protected as a hereditary right and passed through generations. Small sized (~6 mm dia) soybeans (locally called "bhatamas") of "yellow" cultivar ("dark brown" variety in Nepal) are washed, dipped in spring water where the level of water is twice than that of the beans, and left overnight (8–12 h) at ambient temperature (10°C–25°C). The beans are drained and cooked by boiling (~95°C) until (2–3 h) they can be crushed easily between fingertips. The beans are drained and crushed lightly using a wooden pestle (locally called "muslo") and mortar (locally, "okhli") to make grits of mainly half-cotyledons with a view to increase the surface area of the fermenting substrate resulting in a speedy fermentation by the aerobic sporeformers. The indigenous practice is not to clean the mortar and pestle before undertaking another batch of crushing. The grits are then transferred to a bamboo basket, lined with sack-cloth and a layer of locally available fresh fern (*Athyrium* sp.) fronds or leaves of banana (*Musa x paradisiaca* L.), *Leucosceptrum canum* Sm., *Macaranga denticulata* (Blume) Müll.Arg., *Ficus hookeriana* Corner, or

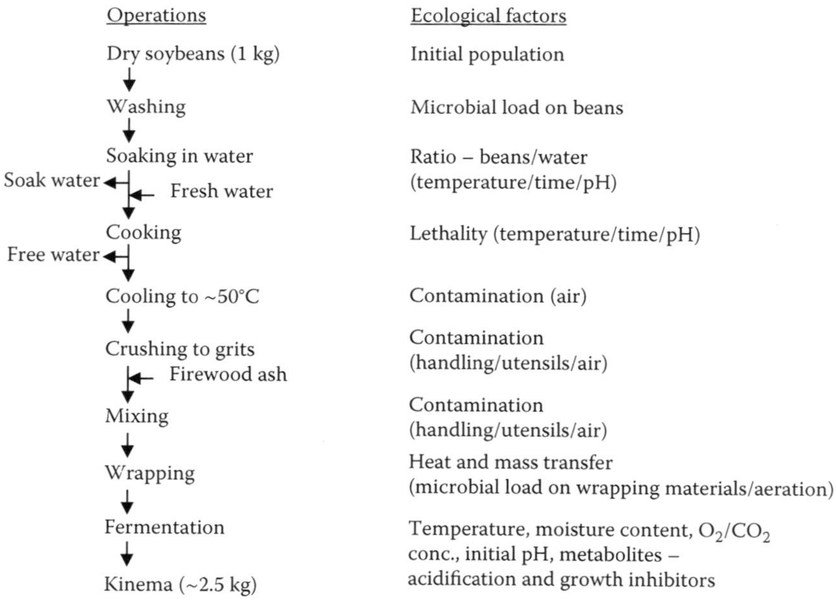

Figure 2.14 Indigenous method of preparing kinema and factors influencing microbial composition in each of the processing stages.

Bauhiania vahlii Wight & Arn. At this stage the beans are still quite hot (50°C–60°C). To maintain the condition alkaline, a small amount (~1%) of firewood ash is often sprinkled over the grits which are then wrapped with those leaves and sackcloth to insulate the fermenting beans and left in a warm (35°C–25°C) room, usually kitchen, for 1–3 days. A combination of porous fern fronds with dimensioning of the bean mass are of importance in ensuring appropriate aerobic condition during the fermentation. The desired state of fermentation is determined primarily by a typical kinema flavor dominated by ammonia as well as the development of a rough, white viscous fluid on the beans. This fluid has the property of forming long stringy threads when touched and stretched with thumb and middle finger; the longer the thread, the better is the quality of kinema. Generally speaking, fresh kinema (Figure 2.15a) of good quality gives a nutty flavor accompanied by a mild smell of ammonia, a grayish brown color, and is semi-hard, like raisins. When fresh kinema is fried in oil, within a few seconds, the pungent ammoniacal smell disappears, leaving a persistent nutty odor. After a brief frying, kinema is added with vegetables (grated onion, tomatoes, and green chillies), spices, salt, and a little amount of water and cooked for 5 mins to make a thick curry, which is eaten as a sidedish of steamed rice. Fresh kinema keeps for 2–3 days during summer and 1 week during winter. The shelf life is often lengthened to several months by drying kinema (Figure 2.15b) in the sun or keeping it above an earthen oven in kitchen (Tamang et al., 1988).

Kinema production is a means of marginal income for some rural households. Usually, it is sold by volume measured using a small silver mug containing 150–200 g kinema, which is wrapped with *Ficus* or banana leaves and then tied loosely using a piece of straw. In Darjeeling hills, 1 kg of fresh and sun-dried kinema costs INR 100

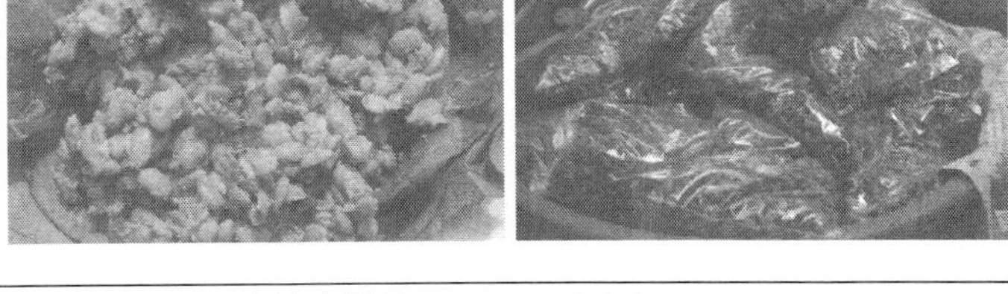

Figure 2.15 Fresh (a) and dried (b) kinema, as marketed.

(= US$1.6) and INR 240 (= US$3.8), respectively (April, 2014). Per capita daily consumption of kinema was recorded during 1997–1998 as 3.3 g with annual production of 326 tons in Darjeeling hills and 2.2 g with annual production of 326 tons in Sikkim (Yonzone and Tamang, 1998). An average of 5 kg is sold per day by each seller and brings a profit of approximately 40%. Although there is a good market for kinema, processing is still restricted to the individual households as an unorganized sector.

2.1.1.3.3 Changes in the Substrate during Kinema Production

2.1.1.3.3.1 Stage 1: Soaking Soybeans soaked in excess water absorb approximately their own weight of water and undergo a natural microbial acidification. During soaking, there is no significant change in the frequency of coliforms and *Bacillus subtilis* cells present on raw soybeans, but a population of lactic acid bacteria appears, resulting in a decline in pH of soaked beans from 6.6 to 5.3 (Sarkar, 2000). The microbiota residing in soaking vessels acts as a starter for the mixed lactic/acetic acid fermentations, which lower the pH (Nout and Rombouts, 1990). The microbial load and composition of soak water seem to depend on the bean/water ratio, time, and temperature of soaking, and the initial microbial composition of the dry beans (Nout et al., 1987). Soaking of beans leads to a significant decrease in thiamine content; however, the content of riboflavin remains unchanged (Sarkar et al., 1998). The remarkable decrease in thiamine content may be due to increased thiaminase activity or complex formation.

2.1.1.3.3.2 Stage 2: Cooking Most of the bacteria in soaked beans are killed during the subsequent process of steaming, which leaves a near pure culture of *Bacillus*, which survives as spores and can grow without competition (Nout and Rombouts, 1990; Leejeerajumnean et al., 2000). Because of draining of soak water and cooking the soaked beans in fresh water, the pH of the cooked beans rises from 5.3 up to 6.2–6.6. However, soaking is found to be essential because bypassing this hydration step yields kinema of poor organoleptic quality due to significantly lower sensory scores in the body and texture attributes (Sarkar, 2000). The acidification of beans must also be considered important in restricting the growth of spoilage bacteria (Ashenafi and Busse, 1991). Cooking has no significant influence on thiamine content, although riboflavin and niacin levels increase (Sarkar et al., 1998). The increase in riboflavin contents after cooking (Table 2.4) may result from a more complete extraction of coenzyme forms, e.g., FMN, FAD.

2.1.1.3.3.3 Stage 3: Fermentation

Microbiota: Fermentation of cooked soybeans to produce kinema is natural, i.e., no deliberate inoculation of the substrates with starter cultures is practiced. It is likely that the environment

Table 2.4 Nutrient Contents of Unfermented Soybeans and Kinema

PARAMETER	RAW BEANS	COOKED BEANS	KINEMA
Moisture (g/kg)	121b ± 0	749a ± 4.0	752a ± 0
Essential amino acids (free) (mg/kg dry wt)			
Valine	–	0.08b ± 0	7.05a ± 0.12
Leucine	–	1.0b ± 0	9.50a ± 0.13
Isoleucine	–	0.06b ± 0	6.22a ± 0.33
Threonine	–	0.13b ± 0	2.79a ± 0.19
Lysine	–	<dl (0.01)	8.23 ± 0.28
Histidine	–	0.09b ± 0	4.52a ± 0.07
Arginine	–	0.27a ± 0	0.22b ± 0.02
Phenylalanine	–	0.15b ± 0	11.82a ± 0.09
Tryptophan	–	<dl (0.01)	<dl (0.01)
Methionine	–	0.04b ± 0	3.15a ± 0.06
Essential fatty acids (free) (mg/kg dry wt)			
C18:2 (linoleic acid)	3838b ± 42	3758b ± 28	4774a ± 5
C18:3 (linolenic acid)	511b ± 11	524b ± 13	580a ± 13
Phytosterols (mg/kg dry wt)			
Campesterol	169b ± 11	170b ± 9	255a ± 14
Stigmasterol	146b ± 10	144b ± 7	225a ± 14
β-Sitosterol	492b ± 30	488b ± 30	788a ± 29
B-group vitamins (mg/kg dry wt)			
B_1 (thiamine)	16.8a ± 0.4	5.8b ± 0.3	8.4b ± 0.6
B_2 (riboflavin)	3.4c ± 0.2	6.8b ± 0.5	11.6a ± 0.3
B_3 (niacin)	<dl (14.7)	36.4 ± 4.8	44.8 ± 5.6

Source: Sarkar, P.K., and Tamang, J.P., *Food Microbiology*, 11, 317–325, 1994; Sarkar, P.K., and Tamang, J.P., *Food Microbiology*, 12, 317–325, 1995; Sarkar, P.K. et al., *Letters in Applied Microbiology*, 24, 337–339, 1997a; Sarkar, P.K. et al., *Food Chemistry*, 59, 69–75, 1997b; Sarkar, P.K. et al., *Journal of the Science of Food and Agriculture*, 71, 321–328, 1996; Sarkar, P.K. et al., *Journal of the Science of Food and Agriculture*, 78, 498–502, 1998.

Note: dl, detection limit. Values with standard error were obtained from three replicates. Means within a row sharing a common letter are not significantly different ($p < 0.05$).

and utensils become highly contaminated with a "house flora" that functions as an inoculum. Market samples of kinema contain six species of *Bacillus*, namely *B. subtilis*, *B. licheniformis*, *B. cereus*, *B. circulans*, *B. thuringiensis*, and *B. sphaericus* (Sarkar et al., 2002). In addition, *Enterococcus faecium*, *Candida parapsilosis*, and *Geotrichum candidum* occur in 100%, 50%–80%, and 40%–50%, respectively, of kinema samples (Sarkar and Tamang, 1994). A study with 83 isolates of *Bacillus* from kinema revealed *B. subtilis* as the most abundant among the six species (Figure 2.16) (Sarkar et al., 2002). Further, in pure culture experiments *B. subtilis* has been shown to be capable of the fermentation as the sole organism; the members of the accompanying flora are probably opportunists (Sarkar and Tamang, 1994; Sarkar et al., 1994).

A number of interesting changes occurs during processing of soybeans to produce kinema. Spores of *B. subtilis*, which are normal inhabitants of soybeans, pass through soaking and cooking treatments to initiate and carry out fermentation. The mean viable counts of *B. subtilis*, *E. faecium*, and *C. parapsilosis* increase in fermenting beans. In monoculture fermentation (Figure 2.17), the count of *B. subtilis* cells increases from an initial 10^5/g up to 10^9/g wet beans after 48 h at 37°C (Sarkar et al., 1993, 1994).

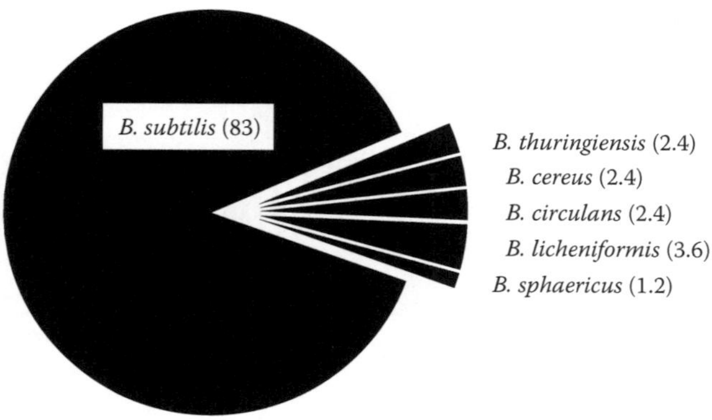

Figure 2.16 Relative abundance of *Bacillus* spp. in marketed kinema. Values within parentheses indicate %occurrence out of 83 isolates. (Modified from Sarkar, P.K. et al., *International Journal of Food Microbiology*, 77, 175–186, 2002.)

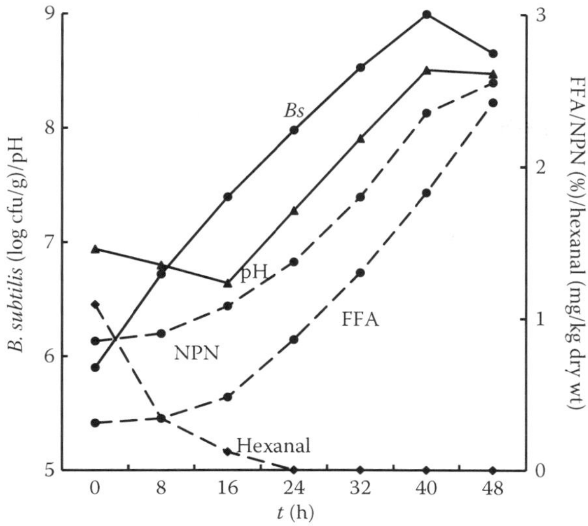

Figure 2.17 Changes in *Bacillus subtilis* cell count (*Bs*), pH, and contents of free fatty acid (FFA), non-protein nitrogen (NPN), and hexanal accompanying fermentation of soybeans to produce kinema. (Modified from Sarkar, P.K., *Journal of Food Science and Technology*, 37, 477–483, 2000.)

Proximate Composition: It is interesting that before 16 h of fermentation there is no significant increase in the contents of free fatty acids (FFA) and non-protein nitrogen, suggesting that sugars, not fats or proteins, are initially used as substrates for metabolism and growth of *B. subtilis*. This conjecture gets support from the fact that while the sucrase activity sharply rises from the onset of fermentation and reaches its peak after 12 h, amylase activity starts after 6 h and then continues to increase until the end of fermentation (Figure 2.18). The lipase activity is maximum after 24 h only (Sarkar, 2000).

The average moisture content of market kinema is 615 g/kg. Kinema contains 477 g protein, 169 g fat, 281 g carbohydrate, 72 g ash, and 20 MJ energy per kg dry weight (Sarkar and Tamang, 1994). Although the titratable acidity (0.1%) in kinema is about 10 times higher than in raw soybeans, the product has a high pH value (>7.8; even up to 8.5). This is due to the high buffering capacity of the legume beans and the proteolytic activities of *B. subtilis* leading to ammonia release, characteristic of most vegetable protein fermentations (Hesseltine, 1965). The pH drops from an initial 6.9 to 6.4

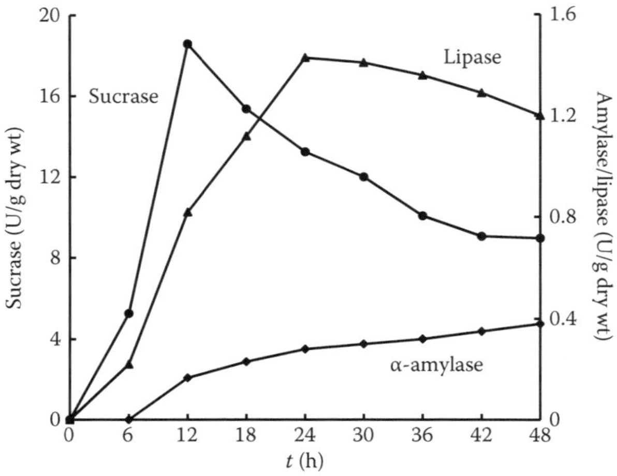

Figure 2.18 Changes in the content of some hydrolases accompanying fermentation to produce kinema. (Modified from Sarkar, P.K., *Journal of Food Science and Technology*, 37, 477–483, 2000.)

after 8 h, and then rises to 8.6 at the end of fermentation (Figure 2.17), with a coincident increase in proteolytic activity (Sarkar et al., 1993).

Amino Acids: The very high acidic amino acid content of kinema is of particular interest. Glutamic acid is the most abundant amino acid, followed by aspartic acid, together representing 31% of total residues (Sarkar et al., 1997b). The strong proteolytic activity results in a significant decline in protein nitrogen content, with a concomitant increase in non-protein and soluble nitrogen contents (Sarkar et al., 1993; Sarkar and Tamang, 1995). The free amino acid content of unfermented soybeans, which is only 0.18% of the total dry mass, increases 60-fold in kinema. The percent liberation of amino acids (based on the ratio of sum of the free amino acid contents to that of the total amino acid contents) at the time of fermentation was 37 (Sarkar et al., 1997b). Although *E. faecium* has no influence, the presence of yeasts (*C. parapsilosis* and *G. candidum*) significantly reduces the levels of several free amino acids, suggesting their preferential uptake by the yeasts. So, in case of traditionally prepared kinema the percent liberation of amino acids was 11.4 only (Nikkuni et al., 1995). The net decrease in some individual amino acid levels after fermentation suggests that they are metabolized to a greater extent than they are replaced by proteolytic activity. The decrease is particularly pronounced for arginine and cyst(e)ine. In addition, levels of some amino acids (alanine, isoleucine, serine, aspartic acid, asparagine, arginine, tyrosine, methionine, and hydroxyproline) are significantly depleted when yeasts contribute to the *Bacillus* fermentation of soybeans (traditional fermentation). Interestingly, kinema proteins are a good source of almost all the essential amino acids (EAA; Table 2.4) (Sarkar et al., 1997b). Kinema has EAA_7 and EAA_{10} values of 41.4% and 52.2%, respectively, of total protein. The EAA profile of kinema is comparable with two high-quality protein foods (eggs and cow's milk) and also with that of the reference pattern established by FAO/WHO/UNU (1985) and FAO/WHO (1990).

Lipids: Although there is no significant change in the crude lipid (CL) content of soybeans after soaking and cooking, the increase (20%–30%) is significant after fermentation. This is probably due to active assimilation of carbohydrates and limited consumption of lipids, resulting in an enrichment of CL at the end of fermentation. In terms of CL content, neither the *Enterococcus* nor the yeasts have any influence on the *Bacillus* fermentation of soybeans (Sarkar et al., 1996).

In both raw soybeans and kinema, the content of unsaturated fatty acids (oleic, linoleic, and linolenic) is six times higher than that of saturated fatty acids, and linoleic acid (the essential fatty acid in animal diets) is the predominant fatty acid, comprising 57% of the total fatty acids (Table 2.4). The major fatty acids are liberated during fermentation by microbial lipases. The overall level of FFAs does not change significantly upon soaking and cooking but increases significantly after fermentation. The increase in FFA is 16%–27% during fermentation. In kinema, the FFA content is 3.1%–3.7% of CL. The increased fatty acid levels in fermentation products may be due to the synthesis and accumulation of extra lipids or hydrolysis of glycerides during fermentation. Another possibility is dissociation of lipoprotein complexes in soybeans during fermentation, resulting in the release of ether-extractable FFAs (Wang et al., 1975). Since soybeans are heated (inactivating the intrinsic enzymes), liberation of FFAs during fermentation is most likely due to the action of lipases of the kinema microorganisms. A possible negative aspect of kinema is that FFAs can inhibit glycolytic and gluconeogenic enzymes, some hydrolases including trypsin, and fatty acid synthesis.

The unsaponifiable lipids of unfermented soybeans, identified as campesterol, stigmasterol and β-sitosterol (1.2:1:3.4) increase by 58% in kinema (Table 2.4), which have a cholesterol-lowering effect (Sarkar et al., 1996). Mattson et al. (1977) found a 30% reduction in cholesterol absorption by fortifying dietary fat (containing 10 g cholesterol/kg) with twice this amount of phytosterols.

Vitamins: Bacillus fermentation (37°C, 48 h) enhances thiamine, riboflavin, and niacin levels by 45, 71, and 23%, respectively (Table 2.4). This shows that *B. subtilis* has a great synthetic capacity for all these vitamins. These levels decline by 31, 18, and 74%, respectively, in the presence of *E. faecium*, indicating that this bacterium uses readily available vitamins for growth and metabolism. *Enterococcus* requires vitamin-rich media for growth and would be expected to decrease vitamin levels in foods when it is present during fermentations. Since thiamine is susceptible to heat, kinema prepared at 45°C leads to 33% decrease compared with that prepared at 37°C (Sarkar et al., 1998). In terms of overall B-group vitamin content, *Bacillus*-fermented kinema produced at 37°C for 48 h is superior to that prepared by other fermentation conditions. Increase in these vitamins as a result of fermentation has important nutritional implications in communities where this food contributes to the formation of the only side dish of rice.

Minerals: Kinema contains significantly lower levels of minerals than found in raw beans. In the kinema production process, soaking as well as cooking water is discarded, and this may be responsible for the 6-fold depletion in mineral levels. Of course, the mineral content is influenced by that of the water used for soaking beans. Despite such a large loss in minerals during processing, kinema (prepared in laboratory using distilled water for soaking and cooking and without adding firewood ash) still contains appreciable quantities (4, 10, 13, 18, 683, 494, 1257, and 2077 mg/kg dry weight) of copper, manganese, zinc, iron, calcium, magnesium, phosphorus, and potassium, respectively. It is noteworthy that potassium is the most abundant mineral in kinema, and sodium content in kinema is below its detection limit (0.5 mg/kg dry weight) (Sarkar et al., 1998). Since in practice spring water is used for soaking and firewood ash is often added, mineral content of traditional kinema may not be less than that observed and may even be higher in some instances than in raw beans. In fact, Nikkuni et al. (1995) estimated the amount of copper, manganese, zinc, iron, calcium, magnesium, phosphorus, and potassium as 17, 541, 45, 177, 4320, 2520, 7290, and 17,680 mg/kg dried kinema collected from Nepal. In industrialized countries, people generally consume too much sodium and not enough potassium. The high ratio of sodium to potassium (Na/K) has been linked to high blood pressure, heart disease, and stroke (Cook et al., 1998). Higher ratios of Na/K were independently linked to a higher risk of cardiovascular diseases and to higher all-cause mortality (Yang et al., 2011). Public health recommendations have stressed that a reduction in sodium intake should be accompanied by a simultaneous increase in potassium intake. The net effect would be to reduce the ratio of Na/K in the diet. Hence, kinema is not only a valuable source of minerals particularly where intake from other sources is marginal but also an important potassium-rich diet.

Digestibility: An in vitro model was used by Kiers et al. (2000) to demonstrate a change in the digestibility of soybeans during fermentation by using *B. subtilis* strains from kinema. During fermentation for 48 h of autoclaved soybeans the quantity of soluble and dialyzable matter increased from 22% and 6% up to 65% and 40%, respectively. The *Bacillus* strains were able to degrade soybean macromolecules to a large extent resulting in water-soluble low-molecular-weight compounds, being small enough to pass the dialysis membrane. Soybeans contain two main storage proteins, β-conglycinin (consisting of three components – α, α′, and β subunits) and glycinin (consisting of two components – an acidic [A] and a basic [B] polypeptide). A complete breakdown of all the subunits from these proteins was observed after fermentation of cooked soybeans by a kinema strain (*B. subtilis* DK-W1). After 24 h of fermentation, substantial quantities of polymeric carbohydrate matter were degraded into compounds <5 kDa. In vitro digestibility was increased from 29% to 33%–43% after 48 h of fermentation. The ratio between absorbability and digestibility was more than 80% after fermentation for 24 h increasing up to 95% after fermentation for 48 h. In vitro digestion of *Bacillus*-fermented soybean using gastrointestinal enzymes only slightly increased the amount of dialyzable matter. The results clearly demonstrated that after 18–24 h of fermentation considerable substrate modification had occurred leading to kinema with high nutrient availability in which the need for degradation of nutrients by gastrointestinal enzymes is minimal.

Flavor: In the course of fermentation, the objectionable beany flavor in cooked soybeans gradually diminishes and finally disappears. Gradually, a nutty flavor begins to develop and eventually becomes masked by the smell of ammonia (Sarkar and Tamang, 1995). The volatile compounds detected in cooked soybeans include 1-hexanol, 1-octane-3-ol, hexanal, nonanal, decanal, 2-heptanone, 3-octanone, decylbutanoate, low concentrations of several pyrazines, 2-pentylfuran, vinylbenzene, benzaldehyde, 2-methoxyphenol, 4-vinyl-2-methoxyphenol, and indoline (Owens et al., 1997). The content of hexanal (the major cause of beany flavor), which accounts for 7.5% (quantitatively the second highest) of the total volatiles of unfermented soybeans, decreases steadily and goes below the limit of detection within 24 h of fermentation (Figure 2.17). On the other hand, the ammonia content of unfermented soybeans shows a 34-fold increase in kinema (Owens et al., 1997; Sarkar et al., 1997b; Sarkar, 2000).

Antioxidative Activity: Fermentation of soybeans with *B. subtilis* to kinema causes 144% increase of total phenol content (TPC) and enhances free radical-scavenging activity, metal-chelating ability, reducing power, and lipid peroxidation inhibitory activity. All these antioxidant activities of kinema are significantly higher than those of unfermented soybean, suggesting the role of fermentation in enhancing these attributes, which are positively correlated with the TPC. While unfermented soybean shows 36% inhibition of linoleic acid peroxidation, kinema exhibits antioxidant activity with 44% inhibition. However, the peroxidation inhibition of kinema extracts declines with time and reaches merely 26% after 72 h. Thus, the kinema strain (*B. subtilis* DK-W1) is an effective producer of antioxidant activities. The sharp decline of the peroxidation inhibition in kinema beyond 24 h can be attributed to the shelf life of the product. After about 12 h of traditional fermentation, the surfaces of the beans are covered with a rough, white viscous mass when kinema is ready for cooking. Kinema can be shelved for only two days, after which a rapid softening along with a strong off-flavor develops in the product, rendering it unacceptable for consumption. In unfermented soybean, the highest correlation is between the radical-scavenging activity and the metal-chelating activity, while in kinema, it is a radical-scavenging activity with lipid peroxidation inhibitory activity, being closely followed by metal-chelating activity (Moktan et al., 2008). The higher level of antioxidant activity in kinema compared with unfermented soybean could be attributed to the extensive hydrolysis of proteins and an increase of 58% in the overall content of phytosterols that happens during the fermentation (Sarkar et al., 1996; 1997b). Because the sequence and composition of amino acids in the peptides are critical, several peptides in soy protein hydrolysate have good antioxidant activity while the others have marginal activity (Quinn and Tang, 1996).

Anti-Nutritional Factors: Although soaking and subsequent cooking of soybeans cause a 73%–88% reduction in the content of raffinose family oligosaccharides (RFOs) in raw soybeans, monoculture fermentation to produce kinema lowers the levels below their detection limit (Table 2.5), thus eliminating a potential cause of flatulence for consumers (Sarkar et al., 1997a).

Table 2.5 Anti-Nutrient Contents of Unfermented Soybeans and Kinema

PARAMETER	RAW BEANS	COOKED BEANS	KINEMA
Moisture (g/kg)	121b ± 0	749a ± 4.0	752a ± 0
RFOs (g/kg dry wt)			
Raffinose	19.2a ± 1.0	2.3b ± 0.1	<dl (1.0)
Stachyose	43.2a ± 0.6	10.6b ± 0.4	<dl (1.0)
Phytic acid (g/kg dry wt)	8.4a ± 0.6	4.7b ± 0	1.9c ± 0.3
Trypsin inhibitor activity (U/g dry wt)[a]	3373a ± 139	1907b ± 5	1139c ± 83
Hemagglutination activity (U/g dry wt)[b]	427a ± 106	20b ± 4	0c

Source: Sarkar, P.K. et al., *Letters in Applied Microbiology*, 24, 337–339, 1997; A. Sharma and P. K. Sarkar (unpublished data).

Note: dl, detection limit. Values with standard error were obtained from three replicates. Means within a row sharing a common letter are not significantly different ($p < 0.05$).

[a] One unit is defined as the amount of trypsin inhibitor required to cause 50% inhibition of trypsin (50 μg) under the assay condition.

[b] One unit is defined as the least amount of hemagglutinin, which produces positive evidence of agglutination under the assay condition.

Phenolic compounds and phytate in plant foods have been shown to interfere with the availability of minerals such as iron (Svanberg et al., 1993). Low bioavailability of iron from vegetable diets is recognized to contribute to anemia. Among the phenolic compounds, galloyl, resorcinol, and catechol groups are known to have an inhibitory effect on iron availability. Several simple processing methods, such as cooking and fermentation, were found to reduce the amounts of phenolic compounds in soybeans during kinema production (Table 2.5).

Phytic acid, widely distributed in mature legume grains, stores most of the grain phosphorus. Phytic acid chelates several minerals and thereby reduces their bioavailability (Nolan and Duffin, 1987). During kinema production, after soaking and cooking there was 44.9% reduction in the phytic acid content of raw soybeans (8.44 g/kg dry weight); monoculture fermentation caused a further reduction of the content in cooked beans by 58.4% (Table 2.5). This indicates that the reduction in phytate content during soaking and cooking was due to its hydrolysis by endogenous phytases in seeds, leaching out in the medium, and thermal degradation, and during fermentation was due to the phytases produced by *B. subtilis* (A. Sharma and P.K. Sarkar, unpublished data).

Trypsin inhibitor (TI) has been implied to be one of the factors responsible for reducing protein digestibility, pancreatic hypertrophy, and poor growth performance in rats, mice, and chicks (Gumbmann et al., 1989). Therefore, reduction of TI activity is useful in improving nutritional quality of soybean with respect to protein digestibility. During kinema production, the TI activity reduced by 43.5% after soaking and cooking, and 66.2% after monoculture fermentation (Table 2.5). A reduction in TI activity during *Bacillus* fermentation of soybeans has been reported by other investigators (Teng et al., 2012).

Soybean agglutinin (SBA, a tetramer with a 30-kDa subunit) belongs to the lectin family and has the unique property of binding to carbohydrates. SBA, by binding itself with the carbohydrate moiety of cell surfaces, causes cells to agglutinate. This activity is referred to as hemagglutination activity (Liener, 1994). Ma and Wang (2010) demonstrated after heat denaturation at 100°C for 20 min, SBA had 60% of hemagglutination activity remaining; after single enzyme (such as trypsin, chymotrypsin, and thermolysin) hydrolysis, SBA still had 50%–60% activity, and the activity of SBA was fully eliminated by multienzyme hydrolysis. They explained that on heating the tetramer SBA is denatured, the activity is reduced due to a three-dimensional structure change; however, hydrolyzing by single enzymes, bioactive peptides are released, displaying some activity; after hydrolyzing by a combination of enzymes, the active peptides are fully destroyed, losing active sites and their binding capacity. During the production of kinema only 5% of the hemagglutination activity remained after soaking and cooking soybeans, and the activity was fully destroyed after fermentation (Table 2.5).

Biogenic amines are toxic nitrogenous compounds and produced mainly by microbial decarboxylation of amino acids in many foods, especially in protein-rich foods. The major biogenic amines found in food products are putrescine, cadavarine, histamine, tyramine, tryptamine, 2-phenyl-ethylamine, spermine, agmatine, and spermidine. Common intoxication symptoms of biogenic amines in human are nausea, respiratory distress, hot flushes, sweating, heart palpitation, headache, a bright red rash, oral burning, and hypotension or hypertension. Oral intake of biogenic amines generally provokes no adverse reactions, because amine oxidases in the intestine rapidly metabolize and thereby detoxify them. However, the capacity of amine oxidases can be saturated by ingesting high doses of biogenic amines, which may lead to food intoxication. Therefore, upper limit of total biogenic amines for human consumption has been suggested by Ten Brink et al. (1990) as 1000 mg/kg food. During the production of kinema although the level of total biogenic amines in raw soybeans increased significantly in the product, the final level is below the safe upper limit prescribed (Table 2.5).

2.1.1.3.4 Safety of Kinema Like any other food, kinema harbors some food-borne bacterial pathogens, including enterotoxigenic strains of *B. cereus* and fecal coliforms. In the presence of functional bacterium (*B. subtilis*), the growth of *B. cereus* is adversely affected and the production of enterotoxin is ceased. Kinema is free from *Staphylococcus aureus*. However, the presence of high numbers of Enterobacteriaceae in fresh kinema is an indication of preparing and/or selling the product in an unhygienic condition. It can serve as a vehicle for contamination of other foods by cross-contamination routes. Since *B. cereus* enterotoxin is heat-sensitive, the traditional way of making kinema and its culinary use in curries is safe. However, if we consider the utilization of kinema in novel applications, e.g., as a dried powdered ingredient for protein or flavor enrichment of starchy foods, adequate measures must be taken to avoid the carry-over of enterotoxin or *B. cereus* endospores, which might germinate, proliferate, and produce enterotoxin in fabricated foods (Nout et al., 1998).

2.1.1.3.5 Plasmid for γ-Polyglutamate Production A γ-polyglutamate (γ-PGA)-producing *Bacillus* strain, isolated from kinema, was found to harbor a plasmid species (Hara et al., 1995). The hybridized region on the kinema plasmid, pNKH, with the γ-PGA production-stimulating factor (PSF) gene (*psf*) on natto plasmid pUH1, which is responsible for γ-PGA production in *B. subtilis* (*natto*), was identified, and the nucleotide sequence was determined. There was found only one possible 1281-bp-long open reading frame (ORF) that could potentially encode a polypeptide consisting of 427 amino acids with a molecular weight of 50,251 Da. The polypeptide was found to be 7 residues longer than that of pUH1 PSF. The percent match of the amino acid sequence between pNKH ORF and pUH1 PSF was 92. Their results suggested that the plasmid pNKH of kinema *Bacillus* might be responsible for γ-PGA production in kinema. Hara (1990) revealed that pUH1 resembled pNKH in partial sequences more than pTNH4, isolated from γ-PGA-producing *Bacillus* strain responsible for thua nao production. Based on pattern similarity of free amino acid compositions of kinema, thua nao, and natto using the complete linkage cluster analysis, Nikkuni et al. (1995) also indicated that kinema is similar to natto rather than thua nao. However, they suggested that the hydrolytic products of the soybean proteins of kinema, thua nao, and natto are affected by their fermentation process rather than the characteristics of the *Bacillus* strains.

2.1.1.3.6 Process/Product Improvisation Organoleptically, the monoculture fermentation of soybeans by *B. subtilis* DK-W1, isolated from kinema from the market of Darjeeling, produced the best quality kinema (Sarkar and Tamang, 1994; Sarkar et al., 1994). In this study, *B. subtilis* DK-W1 was selected for further improvement of kinema quality. Since the optimum temperature for the growth of this strain is 45°C (Sarkar and Tamang, 1994), an attempt was made to determine the optimum time for fermentation at that temperature following *Bacillus* inoculation. The 18-h period was the optimum time for fermentation at 45°C. That kinema had a very pleasant nutty flavor associated with a mild ammonia odor and a highly sticky texture, scoring "excellent." During the production of natto, which is also produced by *B. subtilis*, the fermentation lasts for 18–20 h at 40°C–43°C (Fukushima, 1979).

The monoculture fermentation of soybeans has many advantages over the conventional natural fermentation. First, the load of sporeformers in monoculture fermentation can be manipulated to make a good start. Second, the desirable fermentation is completed within a much shorter time with the organoleptically excellent product development. Third, at high temperature and short period of incubation, the chance of the growth of contaminants is remote. Fourth, because of controlled fermentation, the biochemical changes up to 18 h are desirable. The pH initially goes down significantly within 6 h of incubation and then increases significantly up to 7.5 at the end of fermentation. This indicates that the production of ammonia during monoculture fermentation is not so high as in natural fermentation. The FFA content in kinema produced by monoculture fermentation is 140% less than that in kinema produced by the conventional method. Monoculture fermentation results in the formation of kinema without undesirable odor. The desirable increase in non-protein and soluble nitrogen content within an 18-h fermentation time, compared with that of the conventional method, makes the product acceptable and probably contributes to the improved digestibility.

2.1.1.3.7 Development of Kinema Starter Tamang (1999) attempted to develop pulverized starter using a selected strain of *B. subtilis*, previously isolated from traditionally prepared kinema. Kinema prepared using this strain, harvested after growing in soybean extract broth, was dried in an oven at 70°C for 10 h and ground aseptically. The pulverized starter was packed aseptically in a polyethylene bag, where it can be kept for at least 6 months.

Attempts have been made to search for high-quality kinema starter strains (Tamang and Nikkuni, 1996; Sarkar et al., 2002); however, these were only preliminary studies. Strain selection should depend on individual performance of strains under practical conditions, also considering additional factors, such as the production of desirable aroma and competitiveness in the presence of contaminating microbiota in case of natural fermentations (Sarkar et al., 2002).

2.1.1.3.8 Development of Kinema-Fortified Cookies Although highly nutritious, fresh kinema has some limitations, especially its ammoniacal odor and slimy appearance, which deter its acceptability to a broader dimension of public. Hence, product diversification has been attempted for better acceptance and usefulness of kinema. Biscuits, supplemented with kinema flour, were prepared by Shrestha and Noomhorm (2002). Kinema, prepared using dehulled soybeans and a *B. subtilis* isolate from kinema, was dried at 65°C–70°C for 10–12 h and ground to flour, which was incorporated at the 15% level into biscuit formulation. Kinema supplementation resulted in decreased hardness but increased weight and spread ratio (an indicator of cookie density) in fortified red colored biscuits, which had greater organoleptic acceptability compared with full-fat soy-flour-supplemented biscuits. Protein content of the product was more than 17%.

2.1.1.3.9 Conclusion Traditional knowledge of tribal women in preparing kinema has relevance to the sustainable food and nutritional security and subsistence economy of the kinema-practicing societies. To promote commercialization of kinema, the product must maintain superior quality. It is necessary to increase efforts to introduce consumer preferences in terms of taste and produce high-quality products, packaging, and design as well as safety management. Given the concerted efforts of academia, industry, and government and private organizations, kinema has a potential to become an enjoyable part of the menu.

2.1.1.4 Doenjang

Kun-Young Park and Ji-Kang Jeong

Korea has a long history of eating fermented soybean foods, and doenjang is one of the representative Korean traditional fermented soybean foods. Historically, soybean and processed soybean foods have been one of the most important protein sources in the Korean diet because Koreans mainly eat vegetables. Doenjang has also been used as a seasoning for enhancing the taste of foods (Park and Jung, 2005). As a traditional Korean fermented soybean food, doenjang has unique qualities and characteristics in terms of its flavor and taste compared with those of other similar products from Asian countries. This is due to the fermentation process and Korea's unique natural environment (Park et al., 2000; Kim, 2002).

Doenjang is recognized as a nutritious food that provides essential amino acids that are lacking in cereal and vegetable diets as well as fatty acids, organic acids, minerals, and vitamins, which are highly related to its quality and diverse nutritional benefits (Lim et al., 1999). Proximal composition of doenjang is 54.7% water, 13.8% crude protein, 8.0% crude lipid, 14.4% titratable acidity, and 11.8% salt. Free amino acids are comprised 3.8% of the nitrogen content, of which glutamic acid content (25%) is the highest, but leucine, alanine, histidine, lysine, proline, and valine are also present in relatively high quantity. Lactic acid was the most abundant organic acid, but acetic, malic, citric, and oxalic acids were also detected. Linoleic acid (52.2%), oleic acid (20.7%), and linolenic acid (8.7%) are the major unsaturated fatty acids, comprising 81.6% of total fatty acids. Doenjang also contains oligosaccharides and 3.1% dietary fiber. The digestibility of raw soybean is approximately 55% and that for the cooked soybean is 65%. Digestibility of doenjang increased up to 85% (Park et al., 2000). Doenjang had been used as a folk medicine. It has various health functionalities of antioxidative, anticancer, antimutagenic, antiatherosclerosis, and antiobesity (Park and Jung, 2005).

2.1.1.4.1 Manufacturing Procedures Doenjang (Figure 2.19a) has been traditionally made from meju, which is a fermented rectangular block of cooked and crushed soybeans (Figure 2.19b). Other Korean traditional fermented soybean foods such as ganjang and kochujang are also made from meju. The standardized process to prepare traditional doenjang is shown in Figure 2.20. To prepare traditional meju, which is the main ingredient and starter of doenjang, raw soybeans are soaked, cooked, crushed, and then molded into a rectangular block. Next, it is dried for 3 days in the air, tied up with rice straw and hung at the edge of an eave for its natural fermentation (Figure 2.19b). In general, traditional meju is prepared only with cooked soybeans and fermented with various beneficial microorganisms, typically *Bacillus subtilis* and *Aspergillus oryzae*, which are naturally harbored in rice straw. The duration for traditional meju making generally requires 1 or 2 months (Park and Kim, 1970; Kim et al., 1993, 2011).

After 1 or 2 months, the mejus are washed with water to remove any grown molds, straw, and dirt from the surface. Then, those are placed in an earthen jar, and brine is added at the ratio of meju, salt, and water of approx. 1:1:4 w/w/w (Kim et al., 2010). Some dried red

Figure 2.19 Appearance of traditional doenjang (a) and meju (b).

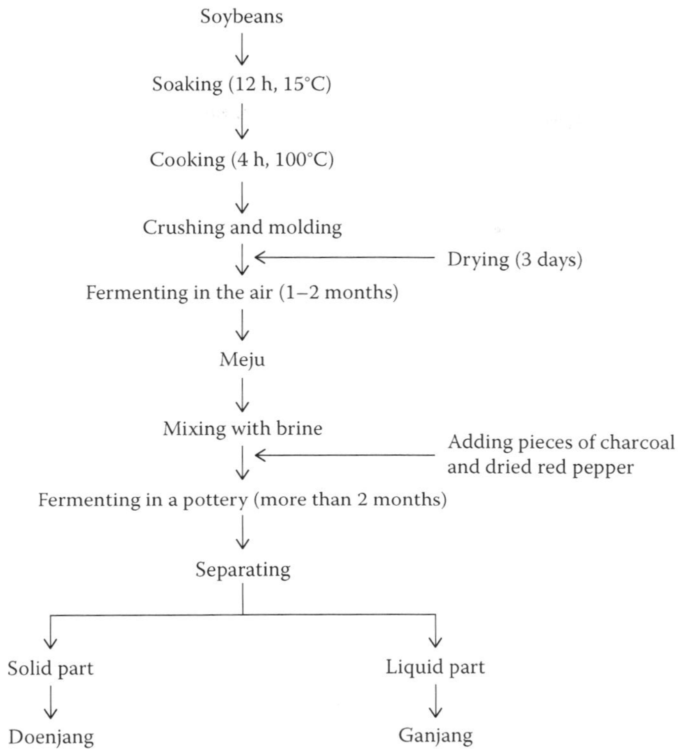

Figure 2.20 Traditional manufacturing method for meju and doenjang.

peppers and pieces of charcoal are placed on top of the mixture and fermented for 2 to 3 months in a jar. After fermentation, the meju–brine mixture is filtered. The solid part of the mixture becomes doenjang, and the crushed solid part is put into another earthen jar. Then, the doenjang undergoes an additional 2 to 6 months of fermentation. The long fermentation time of more than 1 year gives high quality and health functionality to doenjang. The liquid part of the mixture is ganjang (soy sauce), which can be boiled to destroy harmful microorganisms and remove off-flavors.

Recently, traditional doenjang has been commercially produced by several Korean manufacturing companies, and the traditional method of making doenjang has to be adapted for mass production (Kim et al., 2010). Compared with traditional meju fermented with many natural microorganisms, only *A. oryzae* is generally used as fermentation starter when meju is manufactured industrially. Koji, a modified meju, is made of wheat flour or rice by inoculating with only *A. oryzae*. Koji is widely used to produce doenjang, especially commercial products. Because using the koji is more convenient than the traditional method, it can be easily applied to mass production and it has been widely used (Kim et al., 2006; Choi et al., 2009). However, it is known that the taste and quality of traditional doenjang are generally higher than those of industrially manufactured doenjang, and one of its reasons is the microorganisms related to fermentation. These microorganisms are more diverse in traditional doenjang than in commercial one. Therefore, to develop the fermented soybean products that have reduced fermentation time and are convenient to prepare, as well as, improve the quality and functionalities, various methods have been tried.

Attempts have been made to develop doenjang manufacturing methods that take advantage of both traditional and industrial methods (Kim et al., 2000; Kim, 2002; Ko et al., 2010), and one of such methods is the use of effective and beneficial microorganisms as starter cultures. In the most recently conducted study, Jeong (2012) used mixed starter cultures to manufacture doenjang. Numerous microorganisms were isolated from traditional meju; then three microorganisms, *A. oryzae*, *B. subtilis*, and *Lactococcus lactis*, which showed the highest enzyme activity and functionality in the isolates were selected as starters. Grain-type meju (industrial method) was prepared with the inoculation of this mixed starter cultures, and doenjang was manufactured using this meju. This doenjang showed excellent fermentation qualities, organoleptic qualities, and health functionalities compared with industrial doenjang and was comparable to traditional one. Therefore, the doenjang manufacturing method of using beneficial mixed starter cultures can improve the qualities and functionalities of doenjang. In addition, it is expected to be applicable to the fermented soybean product industry due to its great food safety, health functionality, and sensorial characteristics.

2.1.1.4.2 Microorganisms in Doenjang Microorganism is the most important factor in the fermentation of doenjang, and various kinds of microorganisms participate in the fermentation. The traditionally manufactured meju has various fermenting microbiota depending on the geographical location. Commercially manufactured mejus, where

Aspergillus spp. are generally used as a starter, also have different microbiota by fermentation condition (Lee et al., 2010). Many studies have reported that *Aspergillus*, *Mucor*, and *Rhizopus* species were detected in meju and doenjang, and *B. subtilis* and *Bacillus licheniformis* are the dominant microorganisms in them (Yoo et al., 1999; Kang et al., 2000; Cho and Seo, 2007).

These earlier studies were based on culture media and methods and were limited to identifying the strains that grow on specific media. Recently, several molecular biological methods such as denaturing gradient gel electrophoresis (DGGE) have been used to investigate, identify, and quantify microorganisms, and it has also been applied to analyze the microbial community in doenjang (Kim et al., 2009). As a result, some microorganisms such as *Staphylococcus equorum* and some lactic acid bacteria (LAB), which have been thought to be unrelated to doenjang production, are newly considered as important microorganisms in doenjang production (Yoo et al., 1999; Kang et al., 2000; Cho and Seo, 2007).

Recent studies have shown that the strains of bacteria and fungi existed in meju and doenjang are quite different by the region and manufacturing method. Lee et al. (2010) reported that a variety of fungi including *Absidia corymbifera*, *Aspergillus* spp., and *Candida rugosa* were the main fungi in meju; besides, *Mucor* spp., *Pichia farinosa*, and *Penicillium* spp. were also detected in meju. *Bacillus subtilis* is the major *Bacillus* spp. present in meju, and *B. licheniformis* is detected in lower amounts. LAB are the major fermenting microorganisms in meju. *Enterococcus durans* is a very common LAB species, and *Pediococcus* spp., *Lactobacillus* spp., *Leuconostoc* spp., and *Lactococcus* spp. are also detected in meju. They also reported that LAB including *Enterococcus faecium*, *Leuconostoc mesenteroides*, and *Tetragenococcus halophilus* and fungi and yeasts including *A. oryzae*, *Mucor plumbeus*, and *Debaryomyces hansenii* were generally present in doenjang. *Aspergillus oryzae* was the most representative fungus in meju and is used generally as a starter in commercially manufactured doenjang (Kim et al., 2009).

Although traditionally manufactured fermented soybean products including doenjang have high qualities and health functionalities, they can also produce toxic substances that are harmful to human health. Doenjang once had been suspected of being contaminated with mycotoxin, especially aflatoxins from the growth of *Aspergillus flavus* and

Aspergillus paprasiticus, and the contamination by these strains occurred during the procedures of meju fermentation. Aflatoxin is known as a potent carcinogen linked to liver cancer. However, it was demonstrated that the aflatoxins can be destroyed by ammonia, sunlight, mixed culture conditions, melanoidin formation, addition of charcoal, etc. during the fermentation period. In addition, attempts to reduce contamination possibilities have been made by controlling the microorganisms related to meju fermentation and using bacteriocin-producing microorganisms. Thus, doenjang may not have any significant harmful effect, despite the possible contamination by aflatoxins (Park and Lee, 1985, 1989; Park and Kweon, 1987; Park et al., 1988, 1989).

2.1.1.4.3 Functional Properties Many epidemiological studies suggest that the consumption of doenjang provides various health benefits in human. Doenjang is rich in flavonoids, vitamins, and unsaturated fatty acids and especially, one of the most important active compounds is isoflavone. Genistein and daidzein are the two major soy isoflavones, and they principally occur in nature as their glycosylated or methoxylated derivatives, which are cleaved in the intestine to yield free aglycones and further metabolites. It has been demonstrated that isoflavone glucosides in soybean foods are decomposed to aglycone by microorganisms during fermentation. Genistein and daidzein are absorbed better than their glucosides in human, indicating that isoflavone aglycone-rich foods may be more effective than glucoside-rich foods in preventing chronic diseases such as cancer and cardiovascular disease (Grun et al., 2001; Park et al., 2008). Other active compounds in doenjang are linoleic acid, trypsin inhibitor, saponins, etc., and it seems that the fermented soybean paste contained more bioactive compounds than the soybean itself (Choi et al., 2001).

It has been reported that eating traditional doenjang demonstrated strong antimutagenic activities against various carcinogens and mutagens including aflatoxin B_1. Park et al. (1996) reported that strong antimutagenic activity toward aflatoxin B_1 (AFB_1) was observed by treatment with a methanol extract of doenjang. The AFB_1-mediated mutagenesis was completely inhibited at the level of 25 mg/plate of the traditional doenjang extract, and it showed the highest antimutagenic activity compared with other fermented soybean food such as miso and chongkukjang (Figure 2.21).

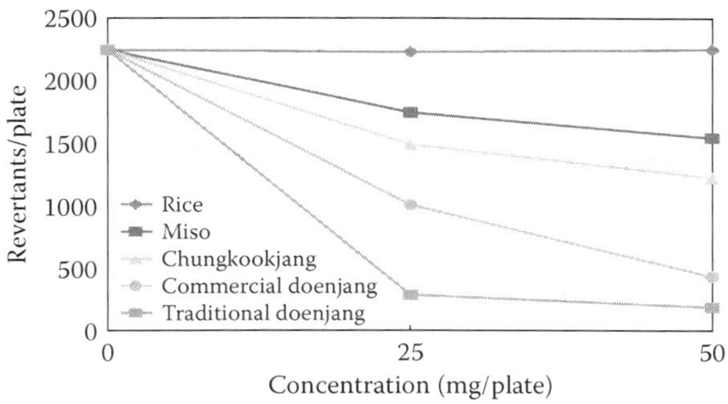

Figure 2.21 Effect of methanol extract of various fermented soybean foods on the mutagenesis of aflatoxin B_1 in *Salmonella typhimurium* TA100.

The antimutagenic effects of doenjang were not limited to AFB_1, but were also exhibited toward other mutagens and carcinogens such as the indirect mutagens, benzo(*a*)pyrene (BaP) and *N*,*N*-dinethylnitrosamine (DMN), and the direct mutagens, *N*-methyl-*N*′-nitro-*N*-nitrosoguanidine (MNNG) and 4-nitroquinoline 1-oxide (4-NQO) (Park et al., 1996). Trypsin inhibitor, genistein, soya saponin, α-tocopherol, β-sitosterol, and linolenic acid are believed to be active compounds in doenjang that possess antimutagenic effects. All active compounds exerted a strong antimutagenic effect against AFB_1. Genistein and linolenic acid were identified as the most active antimutagenic compounds in doenjang (Jung et al., 2006) (Table 2.6).

Table 2.6 Effect of Various Active Compounds (1.25 mg/plate) on the Mutagenicity Induced by Aflatoxin B_1 in *Salmonella typhimurium* TA100

	AFB_1	
SAMPLE	REVERTANT/PLATE	INHIBITION RATE (%)
Spontaneous	104 ± 2	
Control (AFB_1)	1218a ± 9	
Genistein	307c ± 31	82
Genistin	579b ± 37	57
Linoleic acid	324c ± 28	80
α-Tocopherol	786b ± 104	39
β-Sitosterol glucoside	549b ± 99	60
Soya saponin	621b ± 142	54
Trypsin inhibitor	646b ± 87	51

Note: Means (± SD), with the different letters beside data, are significantly ($p < 0.01$) different by Duncan's multiple range test.

Doenjang is also known to exhibit anticancer activities. The growth of several human cancer cells such as AGS gastric adenocarcinoma cells, HT-29 colon cancer cells, and MCF-7 human breast cancer cells was significantly reduced by treatment with the methanol extract of doenjang. Choi et al. (1999) also indicated that the hexane fraction from methanol extract of doenjang suppressed various tumor cells. The doenjang hexane fraction (DHF) was evaluated for its effects on the cell cycle progression in MCF-7 human breast cancer cells. DHF induced a G1-phase arrest of the cell cycle in MCF-7 cells, which correlated with the accumulation of hypophosphorylated form of retinoblastoma protein (pRB) and enhanced association of pRB with the transcription factor E2F1. As shown in Figure 2.22, the expression of D-type cyclins was decreased by DHF in a time-dependent manner, but DHF did not affect the level of cyclins E and A. However, the activity of Cdk2 and cyclin E–associated kinase was decreased in a time-dependent manner. The tumor suppressor, p53, and Cdk inhibitor, p21, which is a known downstream effecter of p53, and association of p21 with Cdk2 were markedly induced in DHF-treated cells (Figure 2.23). Thus, DHF inhibited cancer cell growth by inducing an

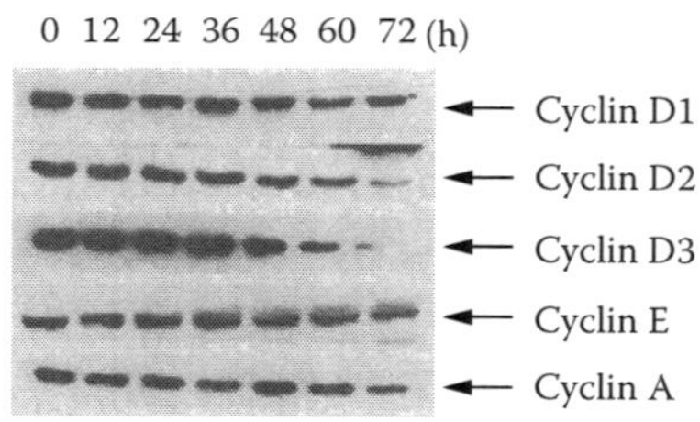

Figure 2.22 Western blot analysis of cyclins from MCF-7 cells after increasing period of exposure to doenjang hexane fraction.

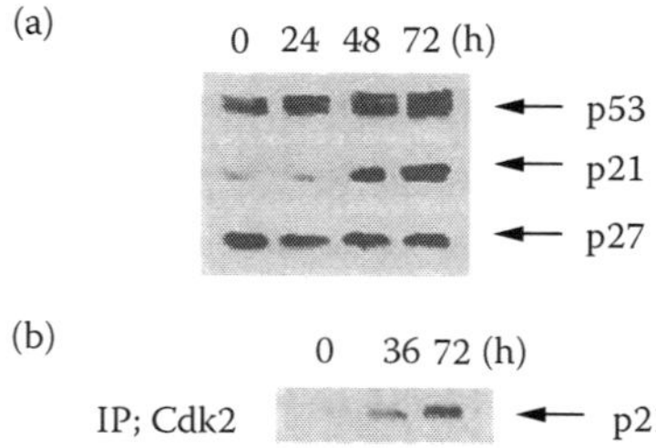

Figure 2.23 Induction of Cdk inhibitor p21 (a) and association of p21 with Cdk2 (b) by doenjang hexane fraction in MCF-7 cells.

inhibition of pRB phosphorylation, decreasing expression of D-type cyclins, and increasing of Cdk inhibitor p21 that appears to be responsible for the observed G1 arrest (Choi et al., 2001).

Anticancer activity of doenjang was also demonstrated in an animal study by inhibiting the solid tumor growth from transplanted sarcoma 180 tumor cells (Park et al., 1999). In a recent study, doenjang exerted the inhibitory effect on chemically induced colitis associated colon cancer in mice (Jeong, 2012). In this study, two kinds of doenjang were used to investigate and compare their functionalities. One was existing commercial doenjang and the other was newly developed doenjang, which was prepared using mixed beneficial starter cultures. Mice were given azoxymethane (AOM) and dextran sodium sulfate (DSS) to induce colitis-associated colon cancer, and doenjang were orally administered during a certain period. As a result, doenjang especially mixed starter used doenjang treatment decreased the loss of crypt in the colonic mucosa and relieved the inflammation of colonic mucosa. It also significantly reduced total number of colonic neoplasia. Doenjang administration also prevented significant increases in serum pro-inflammatory cytokine levels such as tumor necrosis factor (TNF)-α and interferon (IFN)-γ (Figure 2.24). These cytokines are important mediators of inflammation and its related cancer and elevated production of pro-inflammatory cytokines is found in the inflamed colons. This finding suggests that doenjang-treated mice were protected from inflammation and its

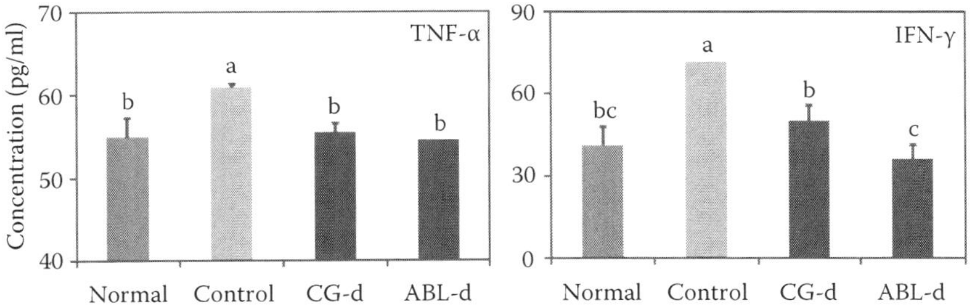

Figure 2.24 Effects of doenjang treatment on serum pro-inflammatory cytokines in AOM and DSS-induced colitis associated cancer (CAC) mice. Normal, group received PBS without AOM and DSS; control, group received PBS and induced CAC by AOM and DSS; CG-d, group received commercial doenjang and induced CAC by AOM and DSS; ABL-d, group received mixed starter used doenjang and induced CAC by AOM and DSS; a–c, means with the different letters on bars representing SD are significantly different ($p < 0.05$) by Duncan's multiple range test.

associated cancer. Treatment with doenjang caused a marked decrease in the expression of iNOS and COX-2, which are important factors related to inflammation and its associated cancer, in colonic tissue of the mice. In addition, pro-apoptotic Bax significantly decreased and anti-apoptotic Bcl-2 increased in DSS and AOM induced CAC mice and they showed a contrary tendency in doenjang-treated group. The expression of p53 tumor suppressor gene and p21 Cdk inhibitor increased in doenjang-fed group compared with the control group (Figure 2.25). Collectively, these results confirm that doenjang has protective effect in the progression of colitis-associated cancer, and it is correlated with a reduction of iNOS and COX-2 expression in colon tissues and an induction of cancer cell apoptosis by regulating Bax, Bcl-2, p21, and p53.

Doenjang has been documented as having antioxidative properties. Many studies have reported that doenjang inhibited lipid oxidation, lipoxygenase activity, and increased metal chelation and free radical scavenging activities in in vitro model system (Cheigh et al., 1990; Lee and Cheigh, 1997). Doenjang can also reduce cardiovascular disease by having fibrinolytic, antihypertensive, and serum cholesterol lowering effects (Park and Jung, 2005). Recently, anti-obesity effect of doenjang and genistein has also been observed in vitro and in vivo (Kwak et al., 2012).

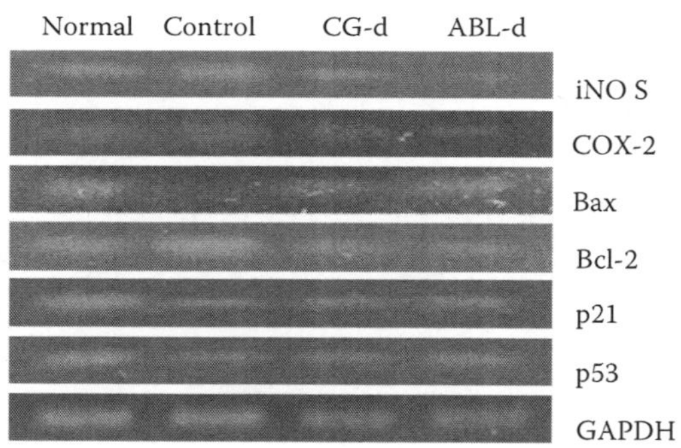

Figure 2.25 Effects of doenjang treatment on mRNA expressions of iNOS, COX-2, Bax, Bcl-2, p21, and p53 in the colon tissue of AOM- and DSS-induced CAC mice. Normal, group received PBS without AOM and DSS; control, group received PBS and induced CAC by AOM and DSS; CG-d, group received commercial doenjang and induced CAC by AOM and DSS; ABL-d, group received mixed starter used doenjang and induced CAC by AOM and DSS.

2.1.1.4.4 Conclusion Doenjang has long been traditionally consumed and manufactured in Korea. Recently, traditional manufacturing method has been replaced by a more convenient and hygienic method. However, maintaining the consistent quality, and manipulating increased quality and functionality were still difficult problems in Korean soybean paste industry. Therefore, many attempts have been made and continue be made to develop the additional standardized manufacturing, and one of such methods is the use of effective and beneficial microorganisms as starter cultures.

2.1.1.5 Chongkukjang

Jang-Eun Lee, Ggot-Im Lee, and Cherl-Ho Lee

2.1.1.5.1 Origin of Chongkukjang Chongkukjang is a type of Korean soybean-fermented product. It is a bacterial-fermented soybean food, similar to natto of Japan, but different in the usage. It is made by easy and rapid fermentation process using cooked soybeans. It can be eaten within 2–3 days, unlike other soybean products with a long fermentation period. It is also used as one of the fermented soybean condiments stemmed from the long history of soybean fermentation in Korea, as classified in Figure 2.26.

Shi is prepared by fermenting cooked soybean whole grain with bacteria, mainly *Bacillus subtilis*, while maljang is made by meju, a fermented soybean loaf where molds and bacteria are grown. *Aspergillus oryzae* is the major mold outside of meju, and *Bacillus* spp. are the main bacteria inside of meju. The term "shi," the Chinese word indicating fermented soybean, first appeared in *Jijiupian*, written in the Han period (206 BC to 208 AD) of China. *Bowuzhi* of Jin (265–420 AD) of China described that shi originated in a foreign country as the word "shi" is a dialect. *Xintangshu* (618–807 AD) of Tang Dynasty

Figure 2.26 The position of chongkukjang in Korean fermented soybean products. (Modified from Lee, C.H., *Fermentation Technology in Korea*, Korea University Press, Seoul, 2001.)

of China named shi as a special product of Balhai or Bohai (688–826 AD), a nation founded by the refugees from defeated Kokuryo Dynasty, the ancient Korea (37 BC–668 AD). *Chi-Min-Yao-Shu*, written by Jia-Si-Xie of Late-Weiin in the 6th century, described several methods of shi preparation (Lee, 2001).

It is generally recognized that Koreans are the first people who conducted soybean fermentation, sparking the soy sauce culture in the world. The history book of Wei Dynasty praised the fermentation skill of Korean people. Their techniques were transferred to Japan during 200–300 AD from Kokuryo. The Japanese term of soybean paste, "miso," was originated from the Korean language maljang, an old term referring to fermented soybean products (Lee, 1990). *Masakurainbunsyo*, written in Japan in 752, described miso as a dialect from Koryo and called as Koryojang (Lee, 1990). In the classification of Figure 2.26, "jang" is the soybean product fermented with other cereals such as rice, wheat, and barley to make koji, mostly developed in Japan, although the term "jang" is generally used for the fermented condiments in Asia. Chongkukjang is called as yumshi in the early Korean literatures like *Sam-kuk-sa-ki*, written by Kim Bu-Shik (1075–1151 AD). The 18th century literatures, like *San-lim-kyung-je* (1715), described it as "jeonkukjang," whose name implies the need for rapid processing of soybean paste during the war time of Byungjahoran in 1636, as "jeon" means war and "kukjang" means soup in Korean. Food historians believe that the phonetic change from "jeonkuk" to "chongkuk" took place at that time. Another story is that the Qing army used to carry the instant fermented soybean as military ration, and it was called chongkukjang (Chong/Qing – kukjang). What all these names imply is that this product was made for extraordinary situation such as war or famine conditions for the urgent supply of nutritious savory food ingredient.

2.1.1.5.2 Methods of Preparation The manufacturing process of chongkukjang is unique compared with the other Korean soybean foods, doenjang and gochujang, which are mainly prepared by mixing and fermenting meju with salt and water or red pepper powder, glutinous rice, salt, and water, respectively. While the other two pastes need at least 6 months of fermentation time for ripening, chongkukjang is made by fermenting microorganisms naturally transferred from the air or rice straw on cooked soybeans.

The first written record on the preparation of jeonkukjang appeared in *Jeung-bo-san-lim-kyung-je* written by Jung-Im Yoo in 1765. Newly harvested soybean is cooked and covered with straw mat, and placed on the warm "ondol" (typical Korean stone floor heated by fire underneath) for 3 days until the mucous string (Figure 2.27) is formed with a strong fermented smell. It is mixed with dehulled and roasted soybean powder, pounded in a stone mortar with the addition of salt, and sun-dried. The mixing ratio of fermented soybean and roasted soybean powder is 2 to 1, and the amount of salt addition is determined by its taste. The product appears to be dried form, easy for storage and transport, and suitable for the military use. This process is slightly different from the current method.

The current process is very simple as shown in Figure 2.28. Soybean is cooked and covered with straw mat or cloth and placed on the warm ondol for 3–4 days until the mucous string is formed. It is mixed with chopped ginger, chopped garlic, and salt, and pounded slightly until the bean kernels are divided into halves, and stored in an earthen jar (Figure 2.29a). Industrial fermentation is made by adding pure culture of *B. subtilis* or placing pieces of rice straw to cooked soybean. Rice straw is the source of *B. subtilis*, which grows fast at 40°C and becomes naturally dominating flora in the system (Lee, 1995). Chongkukjang has its characteristic flavor and smell that are partially masked by ginger and garlic. The spicy seasoning is thus prepared in 3–4 days, and nowadays, the product is marketed in attractive packages (Figure 2.29b).

2.1.1.5.3 Biochemical Composition During the production of chongkukjang, proteases and lipases produced by microorganisms can generate amino acids, organic acids, and fatty acids. Analysis of the

Figure 2.27 Chongkukjang with mucous strings.

Soybean

↓

Soaked in water overnight

↓

Cooked for 3–4 h

↓

Covered with rice straw and put on warm (~40°C) floor for 2 days

↓

Mucous string formed

↓

Crushed by mixing salt and spices

↓

Packed in an earthen jar

↓

Chongkukjang

Figure 2.28 Chongkukjang making process. (From Lee, C.H., *Fermentation Technology in Korea*, Korea University Press, Seoul, 2001.)

Figure 2.29 Chongkukjang packages: traditional earthen jars (a) and a modern package (b).

chemical composition of samples of chongkukjang collected from various regions in Korea revealed that chongkukjang contains many biochemical components including fatty acids, amino acids, carbohydrates, and organic acids. Through chongkukjang analysis, various amino acids such as alanine, glycine, valine, leucine, isoleucine, γ-aminobutyric acid, serine, threonine, methionine, proline, aspartic acid, arginine, phenylalanine, glutamic acid, asparagine, lysine, histidine, tyrosine, tryptophan, and β-alanine are found. Also, lactic acid, oxalic acid, malonic acid, succinic acid, fumaric acid, malic acid, tartaric acid, citric acid, itaconic acid, 2-hydroxyglutaric acid, gluconic acid, and galactaric acid are the major organic acids of chongkukjang. For fatty acids, myristic acid (14:0), palmitic acid (16:0), palmitoleic acid (16:1), stearic acid (18:0), oleic acid (18:1), linoleic acid (18:2), linolenic acid (18:3), arachidic acid (20:0), and behenic acid (22:0) are detected in chongkukjang. Especially, tryptophan, citric acid, β-alanine, itaconic acid, 2-hydroxyglutaric acid, γ-aminobutyric acid, leucine, malic acid, and tartaric acid are the major components of chongkukjang (Park et al., 2010). The biochemical compositions of chongkukjang and other soybean products are shown in Table 2.7.

The pH value of chongkukjang is increased from about pH 6.0 to 8.0 during fermentation (Baek et al., 2008; Jeong et al., 2009). Figure 2.30 shows the pH change of chongkukjang during fermentation. The initial pH was 6.3, which gradually increased as fermentation proceeded, reaching 8.1 in chongkukjang produced by *B. subtilis* at 40°C for 72 h. Different pH values reflected different capabilities of strains to produce metabolites such as ammonia, responsible for the

Table 2.7 Chemical Composition of Soybean and Chongkukjang

PARAMETER	SOYBEAN (YELLOW)	CHONGKUKJANG
Proximate (g/100 g)		
Energy (kcal/100 g)	420	108
Water	9.7	70.7
Protein	36.2	10.2
Fat	17.8	0.8
Ash	5.6	3.4
Carbohydrate	30.7	14.9
Mineral (mg/100 g)		
Calcium	245	96
Phosphorus	620	177
Iron	6.5	3.8
Potassium	1340	602
Sodium	2	961
Vitamin (mg/100 g)		
Vitamin B_1	0.53	0.15
Vitamin B_2	0.28	0.29
Niacin	2.2	1.5
Vitamin C	0	0

Source: RDA (Rural Development Administration), *The Korean Food Composition Table.* 8th revision, Rural Development Administration (National Academy of Agricultural Science), Suwon, 2011.

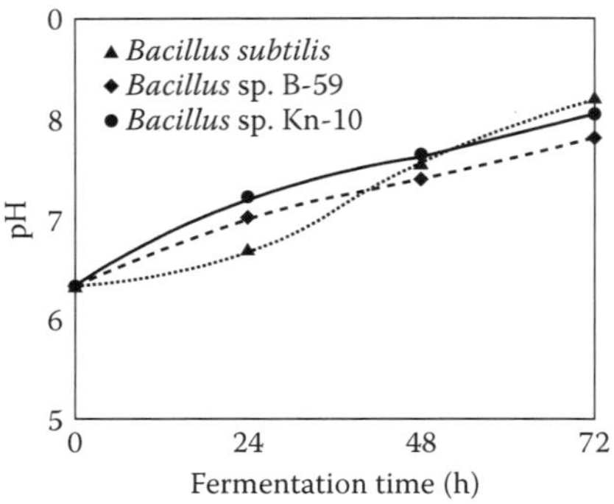

Figure 2.30 Changes in pH during chongkukjang preparation using different strains of *Bacillus* for 72 h at 40°C. (From Baek, L.M. et al., *Korean Journal of Food Science and Technology*, 40, 400–405, 2008.)

pH increase in chongkukjang. The pH values of chongkukjang are in the same range as reported by others (Baek et al., 2008).

The maximum proteolytic activity reaches after 12–24 h, and it remains fairly constant during the remaining period of fermentation. Degradation of carbohydrate to low-molecular-weight compounds also occurs due to the fermentation. The major metabolic activity of *B. subtilis* is the proteolysis of proteins and utilization of released amino acids, and thus, ammonia is produced. It is responsible for the pH increase and the unpleasant strong ammonia flavor. The production of ammonia is a consequence of the utilization of amino acids by the bacteria as sources of carbon and energy. The typical smell can be controlled as regulating ammonia production by bacteria (Jeong et al., 2009).

2.1.1.5.4 Usage and Function *Bacillus subtilis* produces strong proteolytic enzymes, and soybean proteins rapidly undergo partial hydrolysis into peptides and even to amino acids and ammonia during the short fermentation period. This process produces the typical chongkukjang flavor, and a strong meaty and sharp smell. It is used for making chigae, a stew of vegetables, meats, fish, and shellfish, same as doenjang. Chongkukjang chigae is one of the most favorite menus in Korea, especially for the aged group. The amount of chongkukjang added to the stew is large enough to supplement protein to the diet. The strong desire for chongkukjang of elderly Koreans may be not only the sensory nostalgia, but also the physiological demands for their better health.

Recently, scientists have discovered the beneficial physiological functions of some components produced in chongkukjang. The soybean peptides are inhibiting angiotensin converting enzyme (ACE), and thus expected to suppress the incidence of high blood pressure. The mucous substance in chongkukjang is peptido-polysaccharides produced by *B. subtilis* and contains enzyme kinase, which has fibrinolytic and immunomodulating activities. The isoflavonoids in soybean support estrogen hormone function and relieve the malfunction of human menopause (Lee, 2004).

It is not surprising to find that fermented soybean products have been used as medicine in many Asian medicine subscriptions, such as *Dong-eui-bo-gam*, a classic of eastern medicine written by Hur Jun in 1611. Recently, the dried forms of the bacterial fermented soybean, either in granules, pills, or powder form, are sold as a healthy food in Korea.

2.1.1.5.5 Safety Aspects Chongkukjang is generally considered as a safe food, which includes sufficient thermal processing for cooking soybean and relatively high temperature (40°C) fermentation. During the production of chongkukjang the content of free amino acids increases, and it has the potential to produce biogenic amines. For biogenic amines to be produced, free amino acids and bacteria containing decarboxylase enzymes need to be present together with suitable temperature and pH. As the microbial spoilage of food may be accompanied by the increased production of decarboxylases, the presence of biogenic amines might serve as a useful indicator of food spoilage. For these reasons, it is important to monitor biogenic amine levels in foods. Biogenic amines in commercial chongkukjang sold in Korea were analyzed, and the results are given in Table 2.8. Currently, histamine is the only biogenic amine for which maximum limits have been set in the European Union and the United States. The U.S. Food and Drug Administration (FDA) considers histamine level of ≥500 mg/kg in food to be a danger to health (Shukla et al., 2011). The levels found in the commercial products do not seem to be hazardous for human health.

The possible biological and chemical hazards in chongkukjang manufacturing process are presented in Table 2.9. Ten samples of commercial products were analyzed to detect pathogenic bacteria (*Escherichia coli*, *Bacillus cereus*, *Salmonella* spp., *Staphylococcus aureus*, *Yersinia enterocolitica*, and *Listeria monocytogenes*) and aflatoxin B_1 (Lee, 2010). Potential

Table 2.8 Levels of Biogenic Amines in Commercial Chongkukjang in Korea

SAMPLE	BIOGENIC AMINE (mg/kg)										
	PUT	CAD	SPD	SPM	TRY	HIS	TYR	PHE	SER	NOR	DOP
1	8.3 ± 1.2	4.7 ± 0.3	54.3 ± 5.7	9.4 ± 101	76.9 ± 15.2	3.6 ± 0.9	483.1 ± 28.4	32.6 ± 5.4	8.7 ± 2.7	ND	1.5 ± 0.2
2	121.3 ± 8.4	14.6 ± 2.9	44.5 ± 8.9	13.5 ± 5.2	19.2 ± 1.9	54.3 ± 2.7	164.9 ± 23.4	12.2 ± 3.3	5.2 ± 0.9	1.4 ± 0.2	2.5 ± 0.6
3	8.0 ± 1.2	9.3 ± 4.8	67.7 ± 9.2	22.8 ± 4.4	31.1 ± 6.3	0.4 ± 0.1	92.7 ± 18.5	17.1 ± 1.9	8.9 ± 1.0	ND	1.5 ± 0.2
4	4.8 ± 0.9	4.6 ± 0.5	65.9 ± 11.7	23.1 ± 5.4	22.4 ± 3.8	0.2 ± 0.1	5.9 ± 1.5	8.4 ± 1.1	9.8 ± 2.8	2.2 ± 0.4	16.5 ± 3.4
5	4.7 ± 1.1	5.3 ± 1.9	57.5 ± 3.7	12.5 ± 2.4	18.0 ± 5.1	1.3 ± 0.1	4.2 ± 0.9	6.4 ± 0.1	5.7 ± 1.9	ND	6.2 ± 1.3
6	112.4 ± 9.2	9.6 ± 2.2	60.7 ± 3.6	22.4 ± 5.6	6.7 ± 0.9	1.6 ± 0.2	354.2 ± 18.9	0.3 ± 0.1	10.1 ± 2.7	1.9 ± 0.5	0.8 ± 0.2
7	21.9 ± 4.2	13.3 ± 0.9	39.6 ± 7.6	8.4 ± 0.7	30.8 ± 7.4	5.7 ± 0.5	22.7 ± 6.4	26.3 ± 4.4	7.4 ± 2.2	1.2 ± 0.1	2.0 ± 0.1
8	78.8 ± 6.9	7.0 ± 0.4	40.4 ± 7.5	14.2 ± 1.5	17.6 ± 4.9	38.2 ± 8.7	153.1 ± 16.4	ND	3.2 ± 0.6	0.4 ± 0.1	1.2 ± 0.4
9	148.5 ± 0.7	70.6 ± 4.4	33.4 ± 0.8	15.7 ± 6.8	66.3 ± 9.7	70.3 ± 12.6	457.4 ± 15.2	23.8 ± 9.2	11.3 ± 0.6	ND	11.3 ± 1.3
10	5.4 ± 1.2	8.9 ± 1.8	57.4 ± 4.9	7.2 ± 1.9	16.8 ± 0.5	1.0 ± 0.1	105.0 ± 3.4	9.0 ± 0.2	2.7 ± 0.5	0.4 ± 0.1	ND

Source: Han, G.H. et al., *Korean Journal of Food Science and Technology*, 39, 541–545, 2007.

Note: PUT, putrescine; CAD, cadaverine; SPD, spermidine; SPM, spermine; TRY, tryptamine; HIS, histamine; TYR, tyramine; PHE, 2-phenylethylamine; SER, serotonin; NOR, noradrenaline; DOP, dopamine; ND, not detected.

Table 2.9 Possible Hazard Components in Chongkukjang

COMPONENT	HAZARD
Biological hazard	Phathogenic bacteria (*Escherichia coli, Bacillus cereus, Bacillus anthracis,* etc.), yeasts, molds
Chemical hazard	Aflatoxin, Pesticide residue
Physical hazard	Extraneous material (weed, seed, stem, piece of wood, straw, etc.)

Source: Lee, S.O., A study on the hazard analysis and HACCP system for Chongkookjang manufacturing, M.Sc. thesis, Hoseo University, Asan, Korea, 2010.

hazards associated with raw materials, manufacturing process, and work environment were also analyzed to establish HACCP model. The results showed that the pathogenic bacteria, except for *B. cereus*, *S. aureus*, and aflatoxin B_1, were not detected in the samples. Both pathogenic bacteria and aflatoxin B_1 were not detected in raw materials, manufacturing facility, and work environment; however, general bacteria levels were highly detected. Therefore, three CCPs (boiling, sterilization, and metal detection) were selected to secure the safety. As chongkukjang is mainly eaten boiled rather than raw in Korea, it is likely free of microbial hazards.

2.1.1.6 Thua Nao

Ekachai Chukeatirote

Thua nao (in Thai, "thua" means soybean and "nao" means spoiled) is an ethnic, fermented soy product of Thailand. In general, thua nao has been produced on a small-scale basis using traditional method widely practiced among local communities of Northern Thailand (i.e., Chiang Rai, Chiang Mai, and Mae Hong Son). This product, exhibiting a distinct smell of ammonia, can be eaten directly or used as an important ingredient in various local dishes. Thua nao (Figure 2.31a)

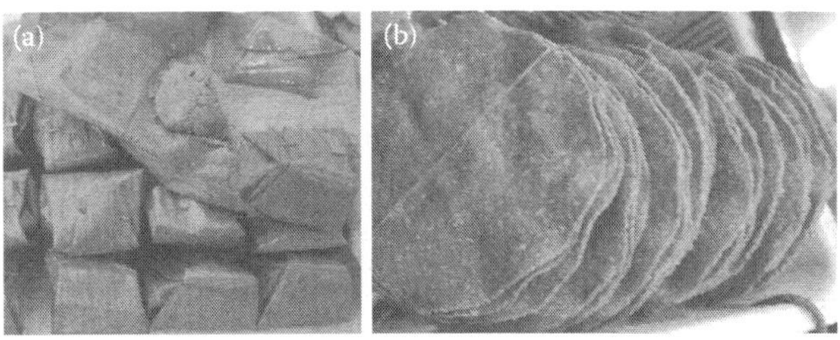

Figure 2.31 Appearance of roasted thua nao covered in banana leaves (a) and its dried form, called thua nao kab (b), as marketed in Chiang Rai. (Courtesy of Pichitchai Nantarat.)

can be consumed directly as a staple food and/or as a flavor enhancer in many local dishes. Traditional thua nao has an unpleasant odor, grayish or brownish color, and slightly sticky cover. In contrast, natto produced by a selective starter culture of *Bacillus subtilis* does not (or slightly) have an ammoniacal smell and is covered with a white massive mucous substance (Hosoi and Kiuchi, 2003).

Like most other alkaline-fermented foods, traditional thua nao preparation relies upon natural microbial activity, especially that of *Bacillus* group (Chantawannakul et al., 2002). Proteolytic activity is a key step in the fermenting process. During the fermentation, soy proteins are hydrolyzed into peptides, amino acids, and ammonia resulting in an alkaline condition (pH 8–9) that would help in preventing a contamination of spoilage microorganisms (Steinkraus, 1996a).

2.1.1.6.1 Conventional Preparation of Thua Nao The major raw materials of thua nao are soybean, clean water, and naturally occurring bacteria. The production is presently performed largely on a small-scale basis. Traditional preparation of thua nao has been passed from generation to generation, and thus the technique used can be different depending on the community. However, overall procedures are very similar including four major steps of soaking, boiling, fermenting, and post-fermenting of soybeans. For detailed practice (Figure 2.32), it typically starts with washing and soaking the soybean seeds with tap water overnight. The soaked seeds are then boiled in a pot under normal atmospheric pressure for 3–4 h (or until soft so that the seeds can be easily crushed by two fingers). The cooked soybeans are collected and transferred to bamboo baskets lined and covered with banana leaves. The baskets are then covered with cheesecloth (or plastic bag) and allowed to undergo spontaneous fermentation at ambient temperature for 2–3 days. However, there are variations in the procedure and equipment used. For example, at Fang District in Chiang Mai, soybeans are cooked by boiling up to 7 h. A specific type of fern leaves is also used to wrap the soybeans during fermentation. As a result, the appearance and attributes of thua nao can vary considerably depending on several factors, including soybean varieties, natural bacterial flora, fermenting time, and so on.

Typical appearance of mature thua nao is indicated by grayish to brownish color of soybean seeds with the presence of a slight viscous substance. Another distinct attribute is a strong ammoniacal smell.

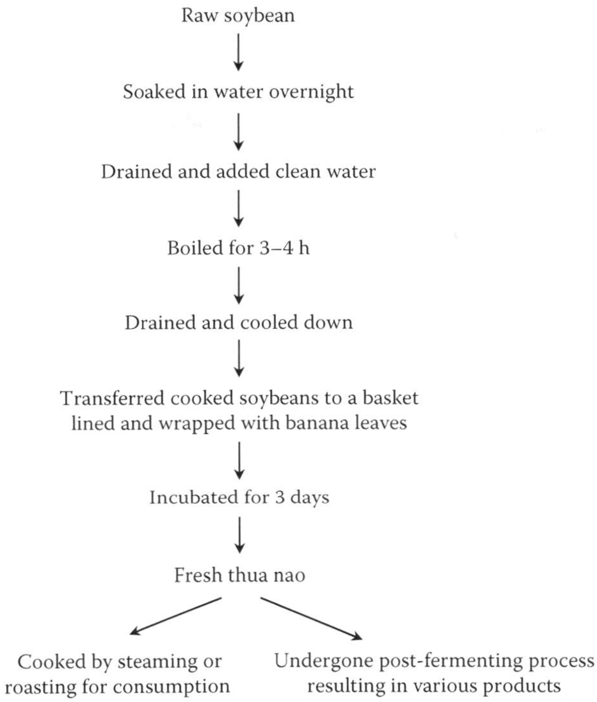

Figure 2.32 Production process of thua nao.

Thua nao is considered as spoiled if the product has a rancid odor or is covered by molds. Once the fermentation is complete, the so-called fresh thua nao can be processed in various ways. For post-fermenting process, fresh thua nao can be cooked by steaming or roasting. Some spices such as chillies, onion, and salt can be added to enhance the taste. Alternatively, fresh thua nao is formed into a circular flat disk, known as "thua nao kab" (Figure 2.31b). This is the dried form of thua nao, which can be achieved by grinding fresh thua nao into a paste, molding into a ball and pressing into a thin disk. Typically, thua nao kab is exposed to sunlight for a day and can be kept for a long period (~6 months). At present, thua nao is available in powder form too.

2.1.1.6.2 Biochemistry and Microbiology Enzymatic degradation of soy proteins is definitely a key process in thua nao production. Soy proteins are sequentially hydrolyzed into peptides, amino acids, and ammonia. This proteolysis and other enzymatic activities contribute to distinct attributes of thua nao, including texture, appearance, flavor, and aroma. Due to release of ammonia, the pH of thua nao rapidly reaches as high as 8 (or even higher). This alkaline nature not only provides a selective condition for certain bacteria (i.e., *Bacillus* spp.) but also causes difficulty for other microbes to grow. Ammonia is considered as a major chemical causing a strong smell to the product. An attempt has been made to control ammonia formation using buffering system in the fermentation that can be achieved by either adding phosphate buffer in the fermenting soybeans or fermenting the soybeans under carbon dioxide atmosphere (Leejeerajumnean, 2000). The changes in enzymatic profiles of proteases, amylases, phytases, and lipases have also been reported during the production of thua nao (Chukeatirote et al., 2006).

The essential bacteria involved in the production of thua nao belong to *Bacillus* species. *Bacillus* spp., especially *B. subtilis*, are predominant from the start to the end of the fermentation (Leejeerajumnean, 2000; Chantawannakul et al., 2002; Chukeatirote et al., 2006; Inatsu et al., 2006). These *Bacillus* spp. are strongly proteolytic, and are thus a key target for strain screening in the research aiming to improvise the production of thua nao. *Bacillus subtilis* strains 38 and TN51 exhibiting a strong proteolytic activity are used as pure starter cultures in thua nao making (Chantawannakul et al., 2002; Dajanta et al., 2009a). An increased proteolysis of thua nao indicating a high rate of fermentation is also described when using *B. subtilis* BIOTEC 7123 (Visessanguan et al., 2005). Apart from *Bacillus* spp., a few bacterial groups (i.e., lactic acid bacteria and Gram-positive cocci) are also present but their role in fermentation may not be relevant (Chukeatirote et al., 2006).

2.1.1.6.3 Nutritional Quality Soybean and its related products are well known as a good source of protein containing all essential amino acids, except methionine (Imsande, 2003). Fermented soybeans in particular are also regarded as functional foods due to their several health benefits including reducing cancer risk (Jung et al., 2006), antioxidation (Kim et al., 2008), thrombolysis (Mine et al., 2005), and prevention of osteoporosis (Jung et al., 2010).

During soybean fermentation, natural microbial starter cultures propagate and are responsible for all biochemical changes. Proteolysis is widely accepted as one of the major and complex events occurring during the fermentation. Other biochemical changes include a bioconversion of lipids and isoflavones. Such changes not only affect the nutritional values, but also have an influence on the taste characteristics. In this review, the detailed characteristics of thua nao have been provided with an expectation that these data can be used as a reference (Table 2.10). For example, these values can be used as a guideline for food regulatory standards. They can also be used in comparison with those of other similar fermented soybeans (including those fermented by pure culture starter).

2.1.1.6.4 Conclusion Thua nao is an indigenous fermented soybean food widely consumed in Northern Thailand. Due to several drawbacks (pungent odor and inconsistent quality), the product does not gain much attention. This is in contrast to natto, which is unarguably one of the best characterized alkaline-fermented soybean foods. There is an attempt to improvise the quality of thua nao product. Several

Table 2.10 Chemical and Microbiological Qualities of Thua Nao

PARAMETER	VALUES
Chemical	
pH[a]	6.7–8.6
Moisture (g/100 g)[a]	56.40–64.91
Fat (g/100 g)[a]	16.97–25.22
Ash (g/100 g)[a]	4.70–5.73
Fiber (g/100 g)[a]	11.93–28.06
Total sugar (g/100 g)[a]	4.79–7.38
Reducing sugar (g/100 g)[a]	2.70–7.74
Ammonia (mg/ml)[a]	0–0.03
Free amino acids (g/kg dry sample)[b]	11.03–61.23
Isoflavone compounds (μg/g dry sample)[c]	320
Phenolic compounds (mg GAE/g extract)[d]	30.46–44.58
Microbiological[e]	
Viable count (log cfu/g)	9.57–10.59
Spore count (log cfu/g)	9.42–9.84
Yeast and mold (log cfu/g)	<1

Note: Values obtained were determined from traditionally prepared thua nao.

[a] Data from Chukeatirote, E., and Thakang, P., *Chiang Mai Journal of Science*, 33, 243–245, 2006; Dajanta, K. et al., *Chiang Mai Journal of Science*, 39, 562–574, 2012.

[b] Data from Dajanta, K. et al., *International Food Research Journal*, 18, 588–592, 2011.

[c] Data from Dajanta, K. et al., *Acta Biologica Szegediensis*, 53, 93–98, 2009.

[d] Data from Dajanta, K., Production of high nutritional fermented soybean (thua nao) by *Bacillus subtilis*. Ph.D. thesis, Chiang Mai University, Chiang Mai, Thailand, 2010.

[e] Data from Dajanta, K. et al., *Chiang Mai Journal of Science*, 39, 562–574, 2012.

researchers have demonstrated that the use of pure bacterial starter culture can help in improving the organoleptic quality of the product. This appropriate means is promising to ensure the product quality in terms of nutritional and safety issues.

2.1.1.7 Meitauza

Li-Te Li and Yan-Li Ma

2.1.1.7.1 Okara: The Raw Material of Meitauza Production Okara is a by-product from the production of soybean foods such as soymilk and tofu. It is also called soybean residue, bean curd residue, douzha, or tofuzha in China, tofukasu in Japan, or bejee in Korea. At least 1.1 kg of fresh okara is produced from every kilogram of soybeans processed for soymilk production. Approximately 20 million tons of fresh okara are produced annually in China, while 0.7 million tons are produced from tofu production industry every year in Japan. Some of the okara is utilized as feed for domestic animals, but most is discarded as industrial waste, which causes social and developmental problems. Although it is "only" soy residue, okara is rich in nutrients. A typical dry okara sample generally contains 25.4%–28.4% protein with high nutritive quality and superior protein efficiency ratio, suggesting that it is a potential source of low-cost vegetable protein for human consumption. It also contains 52%–58% dietary fiber composed of cellulose, hemicellulose, and lignin (O'Toole, 1999).

Okara is a health food material because it reduces calorie intake and increases dietary fiber. It would be ideal if okara could be used as a new type of nutritional food. However, okara undergoes natural spoilage quickly because of its high content of moisture and nutritive protein. In addition, some components in okara have anti-nutritional or adverse effects when consumed in large quantities. The phytic acid in okara can reduce calcium balance and availability of some metal ions (Matsuo, 1996). The soluble carbohydrates, including raffinose and stachyose, can cause meteorism (flatulence). Besides, okara has an undesirable "beany" flavor, which is a significant problem to the development of an okara- or soy-based food product. This undesirable flavor or odor is a result of the oxidation of unsaturated fatty acids by lipoxygenase enzymes during the processing of soy protein products.

Fermentation is an effective method to improve sensory quality, preservation, and nutritional value of food. Okara is rich in carbohydrate, protein, and other nutrients, making it a potentially useful substrate for microbial fermentation. Furthermore, fermentation has been an effective method for eliminating flatulence factors and increasing the edibility of soybean. Therefore, the production of fermented okara would be a potential method to reuse okara. Fermented okara is a kind of traditional food, and there are different kinds of fermented okara in China, Japan, Indonesia, and other places.

In Indonesia, okara-ontjom can be obtained using *Neurospora intermedia* fermentation (Matsuo, 1997a). In Japan, Matsuo (1989a, 1989b) produced okara-tempeh and okara-koji using *Rhizopus oligosporus* and *Aspergillus oryzae*, respectively. The typical traditional fermented okara in China called meitauza, also known as mold okara or fermented okara, is an indigenous alkaline-fermented food. It is a traditional fermented soybean food in Hubei province, especially in Wuhan region. The historic background of when the production of Wuhan meitauza began is not available for reference. The production process of meitauza also lacks written records and has been handed down by generations in words and deeds. The fermentation process of meitauza is similar with the pre-fermentation process of sufu (fermented soybean curd). Based on this, it can be speculated that production of meitauza may have started later than that of sufu.

2.1.1.7.2 Production Methods of Meitauza Meitauza is produced from okara using *Mucor*, naturally obtained from rice straw, during traditional family production process. The traditional production process is shown in Figure 2.33. The main steps are as follows: (1) Soaking: Fresh okara is mixed with water (1: 2, w/v) and certain amount of soy whey left over from making tofu. The mixture is stirred evenly and soaked for about 24 h at room temperature (about 20°C). When there is shimizu on okara paste surface and extrusion water is no longer cloudy, the soaking process is completed. (2) Pressing: Soaked okara is filtered and squeezed to remove the excess moisture. (3) Steaming: The pressed okara is rubbed to be scattered and steamed for about 30 min. (4) Cooling: Steamed okara is spread on a clean mat to cool. When the okara temperature drops to room temperature, the cooling process is completed. (5) Shaping: Okara can be made into different

Fresh okara
↓
Soaked
↓
Pressed
↓
Steamed
↓
Cooled
↓
Shaped
↓
Covered with rice straw
↓
Fermented at room temperature
↓
Meitauza

Figure 2.33 Traditional production process of meitauza.

shapes according to demands. (6) Fermentation: Shaped okara curds, spacing about 2 cm apart from each other, are arranged on straw. The surface of okara curds is also covered with straw. Okara is naturally inoculated by molds on the straw. Then, okara is fermented at 10°C–20°C.

After about 3 days of fermentation, white hair (mycelium) can be vaguely observed on okara surface and the internal temperature of okara rises to above 20°C. At this moment, okara curds should be placed upside down to cool down. As okara continues to be fermented, pure white hair would then densely cover the okara. One day later, the color of hairs changes from pure white to pink or yellow. At this time, the production process of meitauza is completed. According to traditional methods, meitauza is produced below 20°C for about 8–14 days depending on the environment and the activity of microorganisms on rice straw. Generally, producers make a judgment of whether meitauza is matured based on changes in its texture and smell.

In addition to traditional natural fermentation, some microorganisms isolated from meitauza, such as *Actinomucor elegans* and *Zymomonas mobilis*, are also inoculated onto the okara before shaping process to produce meitauza. Specific production process can be simply described as the following: Okara containing about 60% moisture is autoclaved for 20 min under steam pressure at 0.18 MPa and pressed at 1.1 g/cm^2 to a thickness of 2.5 cm. Starter, containing spores, is then sprinkled onto the okara surface to give a final concentration of 0.3%. The inoculated okara, tightly covered with plastic sheeting, is incubated at 15°C for 80 h. During fermentation, acid proteases are produced and there is significant liberation of ammonia and an 8-fold increase in non-protein nitrogen, accompanied by an increase in pH from 5.48 to 7.50. In general, the termination of fermentation is set at 6 days when most parameters are equal or higher than the natural samples (10–14 days of fermentation).

2.1.1.7.3 Microbiology of Meitauza Production *Actinomucor elegans* is the most commonly used microorganism in the production of meitauza. Kronenberg and Hang (1984) isolated a strain of *A. elegans* from meitauza and tested the first pure culture in okara under 15°C for 80 h. Some other suitable microorganisms were also used to produce meitauza with good flavor and functionality. *Rhizopus oligosporus* and *A. oryzae* were used to ferment okara to improve its nutritional qualities as a high-fiber, low-energy foodstuff suitable for human consumption. *Bacillus subtilis* (*natto*), isolated from commercial natto, was tried for producing fermented okara (O'Toole, 1999). *Bacillus subtilis* is an important starter culture for fermented soybean foods like Japanese natto, Thai thua nao, Indian kinema, and Chinese douchi (Steinkraus, 1997). Zhu et al. (2008) applied *B. subtilis* B2 isolated from meitauza to a pure culture fermentation of okara at 40°C for 24 h.

2.1.1.7.4 Chemical Changes Accompanying Meitauza Production There are some changes in chemical composition during okara fermentation process. Krishna (2005) investigated the physical and chemical properties of meitauza produced by *Mucor*. After fermentation, the activity of protein enzyme in okara was 600 U/g, the content of crude protein rose from 19.76% to 22.96%, and the content of amino acid nitrogen increased from 0.26% to 1.45%. With the action of microorganisms, the coarse fiber of okara was degraded and the taste of okara was milder without the fishy smell of soybean. All of these greatly improved the sensory quality and processing properties of okara. Xu et al. (2012) produced meitauza using *A. elegans* and *Z. mobilis* and investigated the changes in some chemical characteristics during fermentation. As shown in Figure 2.34, during the meitauza production process, the pH value rose from 6.3 to 8.1 during the first 4 days of fermentation and had a narrow change to around 7.9. The moisture content dropped steadily from 85.8% to 74.5%. The protease activity was low in the first 36 h and subsequently increased rapidly to 1175 U/g during 2 to 4 days. The main production of α-amino nitrogen content took place after 2 days, and then rose to 6.52 mg/g after 5 days and reached the maximum to 6.78 mg/g after 6 days. As a result

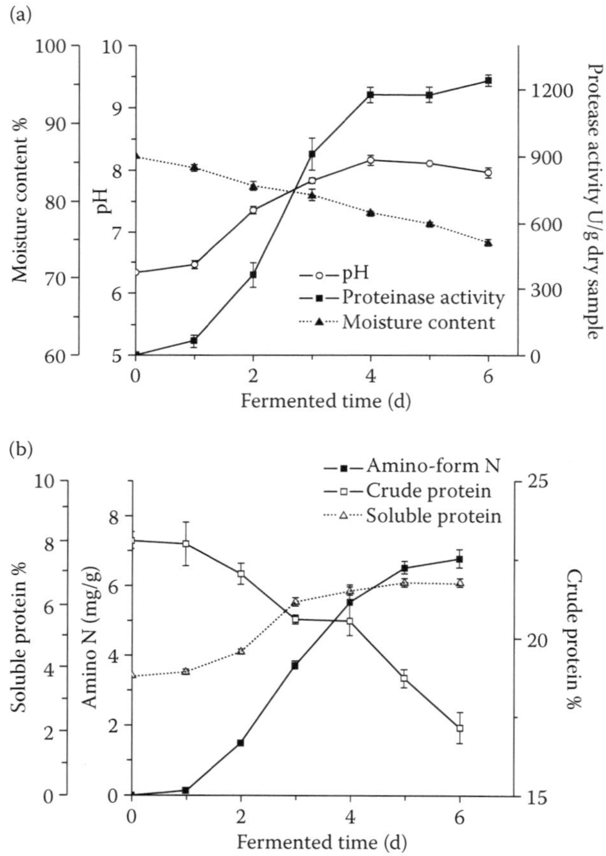

Figure 2.34 Changes in moisture content, pH, and protease activity (a) and contents of crude protein, soluble protein, and amino nitrogen (b) during meitauza production by *Actinomucor elegans* and *Zymomonas mobilis*. (From Xu, X. et al., Study on the Meitauza production from Okara by *Actinomucor elegans* and *Zymomonas mobilis*, in *Information Technology and Agricultural Engineering*, eds. E. Zhu, and S. Sambath, pp. 329–336, Springer, Berlin, 2012.)

of microbial metabolism, the amino acid production and consumption gradually reached a balance in the last days of fermentation. The soluble protein had a steep change after 3 days and remained stable at 6.78%. Under the interaction between microbes and open fermentation, the crude protein content dropped rapidly from 23.1% to 17.15% in the form of ammonia (Xu et al., 2012).

As described in the literature reported by Kronenberg and Hang (1984), a 70 h-incubation with only *A. elegans* will increase the pH from 5.48 to 7.50. In addition, an 8-fold increase in non-protein nitrogen was observed and an acid protease was produced during this co-cultivation study. The pH changed from 6.50 to 7.95, a 4.5-fold increase in α-amino nitrogen and double in soluble nitrogen. This suggests that the fungus had broken down some of the insoluble fiber and protein. The resulting product had good organoleptic properties without the presence of soybean flavor and had a smooth mouthfeel.

A significant increase in total free amino acids was observed during meitauza production process (Table 2.11). Compared with fresh

Table 2.11 Changes in Free Amino Acid Content during the Fermentation of Okara to Produce Meitauza

	FERMENTATION PERIOD (DAYS)		
AMINO ACID (mg/g)	0	3	6
Asp	0.76	0.77	7.90
Glu	0.69	4.36	10.48
Ser	0.34	1.54	1.82
Gly	0.19	0.42	1.64
His	0.59	1.48	0.85
Arg	0.73	0.47	1.35
Thr	1.24	0.50	1.67
Ala	0.71	1.49	4.50
Pro	0.29	1.57	1.41
Tyr	0.35	0.56	1.18
Val	0.36	0.57	2.00
Met	0.04	0.19	0.43
Cys	–	0.03	0.06
Ile	0.26	0.30	1.33
Leu	0.68	0.44	2.23
Phe	0.38	0.49	1.23
Lys	1.03	1.25	1.68
Total	8.37	16.41	36.67

okara, the strong-tasting amino acids, such as glutamic acid and glycine, increased greatly to over 10-fold. Aspartic acid, alanine, and arginine also increased several folds, and sweetening amino acids, such as lecine, valine, serine, and methionine, also increased several folds. These total free amino acids spell abundant fresh sweet flavor, which greatly increased the taste and nutrition of meitauza (Xu et al., 2012).

Some additives are occasionally added during okara fermentation. Matsuo (1990) used *R. oligosporus* to ferment okara with rice bran (10% level). Addition of rice bran resulted in several changes in enzyme activity, vitamin content, and physicochemical properties. Fungus quantity, protease activity, and thiamine and riboflavin contents increased, whereas cellulase and phytase activities and inorganic phosphate quantity decreased. An increase in emulsifying capacity and an improvement in taste were noted in this preparation. These results indicated that the addition of rice bran to okara increased its value as a foodstuff.

2.1.1.7.5 Nutrition and Functionality of Meitauza Microbial fermentation can improve protein digestibility, and the nutrients in meitauza are increased as it contains more free amino acids and free sugars but has lesser fiber content than okara. Meitauza is widely consumed by the Chinese people as a health food due to its high content of dietary fiber. The dietary fiber content of meitauza is greater than 50%, but the energy it provides is only half of that of wheat flour. Fermentation can increase the content of soluble fiber in okara, which improves its nutritional quality and processing properties.

Fermentation induces a significant decrease in lipid and lignin contents in okara. Being rich in hemicellulose with a low lignin and phytic acid content, it causes rats in feeding experiments to show lower weight gain and a significant decrease in the protein efficiency ratio while having no influence on food intake. It is a good source of dietary fiber and is also suitable as another ingredient for the manufacture of low-energy foods. Matsuo (1989a) investigated the composition and properties of okara fermented by *A. oryzae* and *R. oligosporus* and found that fermentation can greatly improve the nutritional qualities of okara, making meitauza a low-energy and high-fiber food. Rashad et al. (2011) evaluated the nutritional quality improvement of okara fermented by different yeast strains. It was found that some levels of value addition occurred as a result of the fermentation, in comparison with non-fermented okara. The protein contents increased by 20%–54%, while crude fiber decreased by 7%–45% with different strains. On the other hand, while the ash content increased, the carbohydrate and lipid contents were reduced in all the samples.

Okara itself contains some anti-nutritional factors, while fermented okara may have more significant advantages in a diet. It can act as a suitable replacement for digestible food in a prepared food to reduce calorie intake, reduce cholesterol levels in the blood stream, and as a food that contains antioxidant activity. Similar to that of vitamin E, it can reduce the level of free radicals in the body. The aldose reductase inhibitor and 8-hydroxydaidzein (Figure 2.35) were isolated and identified from a methanolic extract of okara fermented with *Aspergillus* sp. HK-388 (Fujita et al., 2004). Other soy components that are also likely present in okara include isoflavones (genistein and daidzein), lignans, phytosterols, coumestans, and saponins. These compounds have various physiological and therapeutic functions, including antioxidant activity, prevention of cardiovascular diseases, and effective chemopreventive agents for certain types of cancer (Qitain et al., 2006).

Studies have shown that meitauza exhibits cholesterol-lowering activity. The protein and fiber components in fermented okara affected the lipid metabolism in rats. Animals fed with okara fermented by *R. oligosporus* had significantly lower plasma levels of cholesterol and bile acids as compared with a casein-fed control group (Matsuo and Hitomi, 1993). It was suggested that there was a complementary action between the water-soluble dietary fiber and the protein levels, which led to a decrease of plasma cholesterol levels. Matsuo (1996) also looked at sterol secretion, flatulence, and iron absorption by rats fed with either okara fermented with *A. oryzae* or unfermented okara as the dietary fiber source. The excreted sterol found in the feces of

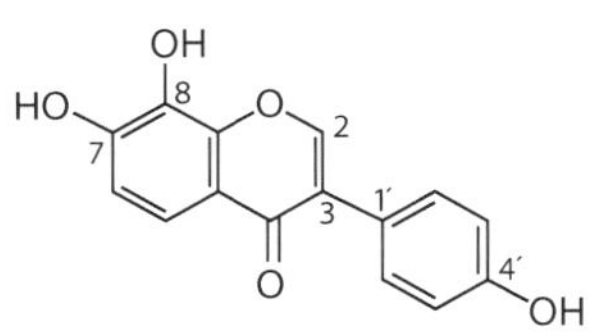

Figure 2.35 Structure of 8-hydroxydaidzein.

rats fed with fermented okara was greater than that in the feces of rats fed with unfermented okara. The residual insoluble dietary fiber in the feces showed more porosity in the fermented okara than the unfermented okara, which suggested it could, along with the hemicellulose, account for the lowered cholesterol level. Lower levels of raffinose and stachyose were present in the fermented okara diet, so it was suggested they should not cause meteorism. Greater iron absorption by rats on the fermented okara diet suggested a reduced effect of phytic acid on its absorption when compared with unfermented okara. The amount of excreted sterol in the feces from rats fed with okara fermented by *A. oryzae* was greater than that from rats fed with okara.

Antioxidant activities of fermented okara were significantly enhanced. Studies have shown that meitauza with high antioxidant activity can be produced under appropriate process conditions. To improve the antioxidant activity of meitauza, *B. subtilis* B2 was used to ferment okara to obtain meitauza (Zhu et al., 2008). Results showed that the antioxidant activity of okara could be substantially improved after 24 h of fermentation. Fermented okara apparently contains an increased amount of antioxidants. Yokota et al. (1996) extracted a crude antioxidant preparation from fermented okara and compared its action against the pathogenesis of gastric ulcer in stressed rats with *R*-tocopherol. Factors in the gastric mucosa, such as the areas of gastric mucosal lesions, levels of thiobarbituric acid reactive substances, prostaglandin E2, and hyaluronic acid, were compared. The work revealed that fermented okara exerted an anti-inflammatory effect on gastric injury in the gastric mucosa and can scavenge the superoxide anion both in vitro and in vivo (Yokota et al., 1996). In further experiments, Matsuo (1997b) found that the antioxidants in okara fermented with *A. oryzae* were γ- and α-tocopherol, isoflavones genistin, daidzein, genistein, and 3-hydroxyanthranilic acid. An 80% methanol extract of the fermented okara showed stronger inhibition of linoleate peroxidation than hexane or hot water extract and accelerated formation of 12-hydroxy-eicosatetraenoic acid in membrane lipids in vitro. In vivo effects were tested by feeding rats on a vitamin E deficient diet with fermented okara or okara supplemented with oxidized oil. In rats fed with oxidized oil-supplemented okara, the body weight increased, while the thiobarbituric acid value in the plasma and glutathione peroxidase activities in plasma and liver were lower than that of rats fed with fermented okara. This suggested that the antioxidants in the fermented okara scavenged lipid peroxides in vivo. Matsuo (1997b) found that okara fermented by *A. oryzae* would scavenge lipid peroxides in vivo. The activity could be caused by antioxidants in fermented okara such as γ-tocopherol, δ-tocopherol, genistin, daidzein, genistein, and 3-hydroxyanthranilic acid. Rashad et al. (2011) reported that the total phenolic content and all the parameters of antioxidative activities were increased in okara fermented by different yeast strains.

2.1.1.7.6 Problems about Meitauza Production Currently, there is little research about microorganisms for meitauza production. Natural inoculation is usually adopted during meitauza production, which is not conducive for large-scale production and quality stability. In fact, okara usually contains 80% moisture, adequate protein, fat, carbohydrates, and other nutrients, which prove suitable for the growth of mold, bacteria, and other microorganisms. Therefore, special microorganisms can be screened out in okara based on different targets. For example, microorganisms can produce antioxidant substances, hypoglycemic components, and so on. *Bacillus subtilis* B2, isolated from meitauza with high antioxidant activity, was used to produce meitauza. It was found that antioxidant activity of okara could be substantially improved after 24 h of fermentation using *B. subtilis* B2. Fermentation with *B. subtilis* B2 offers a novel strategy to enhance the value of okara so as to produce functional meitauza (Zhu et al., 2008).

Okara contains more than 50% crude fiber, among which there includes high content of dietary fiber, which has a very important role on human health. However, okara is not widely accepted as an edible soy product mainly because of its poor mouthfeel. Therefore, further work is necessary to improve the rough taste of meitauza. *Aspergillus oryzae* and *Aspergillus niger* were used to ferment okara to reduce the particle size distribution and improve the mouthfeel of fermented okara. The results showed that both the strains improved the mouthfeel of okara through fermentation at 28°C, with relative humidity of 95%. After fermentation with *A. oryzae* and *A. niger*, the particle size of okara decreased. The morphological analysis also showed the reduced particle size of fermented okara (Figure 2.36).

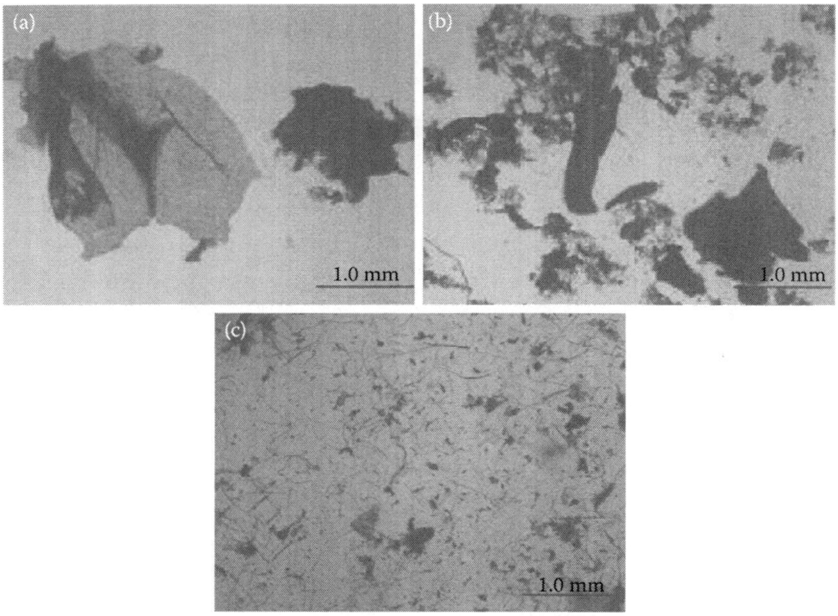

Figure 2.36 Microphotographs of okara fermented by *Aspergillus oryzae* and *Aspergillus niger* for 10 days (4×10): (a) unfermented okara; (b) *A. oryzae*–fermented okara; (c) *A. niger*–fermented okara. (From Li, Y. et al., *Nongye Gongcheng Xuebao/Transactions of the Chinese Society of Agricultural Engineering*, 28, 248–253, 2012.)

The protein, fat, and fiber in okara will undergo some changes during fermentation and at the same time, some new functional ingredients may generate in the role of microorganisms. However, for a long time, the functionality of meitauza had not been given sufficient attention. In the future, the functionality research should be strengthened to enrich functional factors and promote human health.

2.1.1.7.7 Consumption and Prospects of Meitauza The worldwide production and intake of soybean seeds and derived products (mainly soymilk and tofu) are on the increase. Thus, with the consequent rise and accumulation of okara, the utilization and disposal of okara could well become an increasing problem. The huge production and rich nutritional value make okara an ideal feedstock for fermentation. For its physicochemical properties, meitauza could be considered as a good fiber-rich alternative in the development of new functional foods with prebiotic effect. At the same time, meitauza product will add value to this soybean residue and avoid environmental accumulation.

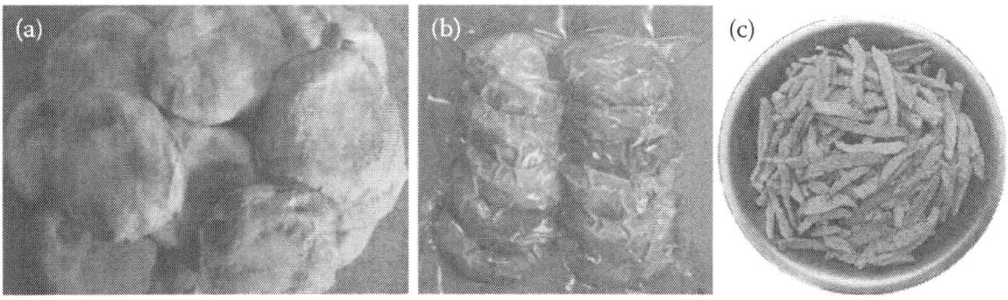

Figure 2.37 Different types of meitauza in China. (a) Fresh meitauza; (b) vacuum-packaged meitauza; (c) homemade strip meitauza.

Currently, vacuum packaged meitauza is available in the Chinese market while homemade meitauza is more common (Figure 2.37). The future research of meitauza can be attributed to screen suitable microbial strains, improving the taste and deep processing for meitauza. In all, meitauza is a healthy food that has not yet been fully developed. It should be paid more attention to increase its value.

2.1.1.8 Yandou

Hongjiang Yang

Yandou is a Chinese traditional soybean-fermented product. In Chinese, "yandou" means salty beans because salt is added to the soybeans after fermentation. As a popular snack food, yandou is widely consumed in eastern China mainly located in Jiangsu province and surrounding areas.

It is well known that soybeans originated in China. In ancient times, Chinese emperor Shennong cultivated and assigned rice, soybean, wheat, millet, and barley as five major food crops for the Chinese people. According to the previous studies, soybean was first domesticated about 4000 years ago in North China plain, located in the downstream of Yellow River (Risal et al., 2010). For thousands of years, many soybean foods have been developed in China using various techniques. The foods fall into two categories, fermented and non-fermented soybean foods. While tofu is the representative of the latter, yandou is unique among the traditional fermented soybean foods.

2.1.1.8.1 History Yandou has a long history. There is a legendary story about its origin that has been very popular for many years. After the downfall of the Qin Dynasty (221–207 BC), the Chu-Han war took place during 206–202 BC between Xiang Yu, the king of Chu, and Liu Bang, the king of Han and who became the emperor of Xi Han dynasty (202 BC to 9 AD). Due to shortage of vegetables, soybean was boiled and mixed with salt as a main course for the Chu soldiers. One time, they were attacked suddenly just after the soybeans had been cooked. The soldiers had to collect the cooked soybeans and package them into cattail bags for retreat. Several days later, the cooked soybeans were found covered with white sticky material with a strange smell and foul taste. At that time, they were short of food and there was no way for them to dump the spoiled soybeans. They added salt to the soybeans and ate them no matter how disgusted they were. However, no one got ill after eating the soybeans.

After the appearance of yandou in the Chu-Han war, thousands years later, yandou has become very popular in Jiangsu province and around. During the Ming Dynasty (1368–1644 AD), yandou became so popular that it was found in almost every local restaurant and household. According to the description in the book *Yue Shan Cong Tan*, written by Li Wenfeng (1510–1552 AD), the Ming Dynasty had 13 provinces and every province had its own nickname based on their native specialties. Yandou was used to describe Jiang-Zhe province at that time.

2.1.1.8.2 Manufacturing Procedure According to the traditional process of yandou preparation, soybeans (Figure 2.38a) are washed to remove dust first and soaked in water for hours to soften the material. Soybeans are then heated with rapid fire to boiling point and simmered with slow fire for 1 h until these become tender for easy chewing. Boiled soybeans are transferred to a clean container, which

Figure 2.38 Raw soybeans (a), and yandou (b), made by using *Bacillus subtilis* HB-1 as starter strain.

is sealed and stored inside haystacks to retain the warm temperature. The storage time is around 2–5 days depending on the environmental temperature when yandou is prepared. Fermentation of the boiled soybeans takes place during storage time. Longer fermentation makes the product sour. Fungi contamination occurs occasionally and should be removed. Salt is added to the fermented soybeans (Figure 2.38b) to taste and become ready to eat. However, if fresh yandou is not to be consumed quickly, sun-drying is required to preserve yandou for a longer time. Whenever dry yandou is ready to eat, it is usually soaked in clean water to soften.

To mask the objectionable smell and bad taste of yandou, it is usually further processed after fermentation. In addition to salt, shredded ginger slices, red pepper powder, and radish slices are the common ingredients added to the fermented soybeans with thorough mixing. After waiting for a few days, yandou becomes ready to eat. Sometimes, yandou with special aromatic flavor is prepared by adding various seasonings including cassia, aniseed, and pricklyash peel. Yandou is a snack food and can be eaten directly. Also, yandou can be cooked with some vegetables such as Chinese cabbage. Some families add yandou when they make scrambled eggs, pancakes, and so on.

Nowadays, yandou and a variety of related products are found in supermarkets (Figure 2.39). However, all commercial products are made by following the traditional protocols. No pure culture starter is used for yandou production. The quality of yandou cannot be assured at this point.

2.1.1.8.3 Starter Culture By reviewing the procedure of yandou production, it is expected that all the microorganisms, except the spores, in soybeans should be eradicated by the boiling step. Recently, the microbial community of yandou and soybean samples was investigated. The samples were subjected to different treatments for microbial colony enumeration. More than one type of colonies was observed in samples without boiling treatment. On L-agar plates, one type of colonies was relatively large, dull, or jagged in shape, showing the classical colony morphology of *Bacillus* species. They were further proved to be the starter microorganisms for soybean fermentation. Colonies with different morphologies were also observed on L-agar plates in comparable numbers. They were believed to be contaminated microbes during

Figure 2.39 Commercial yandou products with outer and inner packages.

manipulations following the fermentation procedure since only one type of colony with large dull shape existed on the L-agar plates from the samples exposed to boiling treatment prior to spreading on L-agar plates. The heat-resistant strain was called HB-1 and determined to be a member of *Bacillus* after Gram-staining (Qin et al., 2013).

The 16S rRNA gene of the isolate HB-1 was sequenced and analyzed. With Classifier of RDP II, HB-1 was found to belong to the genus *Bacillus* with 100% confidence. By Biolog system, the metabolism of HB-1 on a panel of substrates was tested, and HB-1 was defined to be *Bacillus licheniformis*. However, BLAST searching result shows that 16S rRNA gene sequence of HB-1 shares 99% homology to that of *Bacillus subtilis*, *B. licheniformis*, and *Bacillus amyloliquefaciens* strains; all of them belong to the *B. subtilis* group. The members of the *B. subtilis* group are very close to each other, and it is difficult to discriminate between them based on 16S rRNA gene sequence algorism (Wang et al., 2007c). To determine HB-1 to the species level, the 16S–23S internal transcribed spacer (ITS) regions of HB-1 was further amplified, cloned, and sequenced. All the cloned sequences of HB-1 showed 99% identity to those of the reference strain *B. subtilis* subsp. *subtilis* strain 168, indicating that HB-1 is one *B. subtilis* strain responsible for yandou production (Qin et al., 2013).

2.1.1.8.4 Nutritional and Physicochemical Characteristics Yandou, like other fermented soybean foods, is rich in various nutrient components. Among them, peptides and free amino acids bring great benefits to consumers. Their presence is due to the secretion of an array of proteases produced by the fermentative microorganisms. *Bacillus subtilis* is well known for the production of large amounts of multiple proteases. The extracellular proteolytic activity of HB-1 was evaluated by spotting HB-1 culture on the casein agar plate. The formation of hydrolytic clearing zone showed the presence of proteolytic activity of HB-1 culture. Proteolytic activity was also quantitatively measured and 43.4 U protease per milliliter was reached after 48 h of growth (Figure 2.40) (Qin et al., 2013).

Fresh yandou is always covered with sticky material, indicative of the production of some type of extracellular polymer. When the strain HB-1 was inoculated on the medium supplemented with maltose or L-glutamic acid, circular and sticky colonies were formed, different from its morphology on L-agar plates (Figure 2.41). These results suggested that the sticky material might be composed of γ-polyglutamic acid (γ-PGA). The levels of γ-PGA was assessed in strain HB-1 cultures, and the concentration of γ-PGA was 9.0 g/l at the time of fermentation

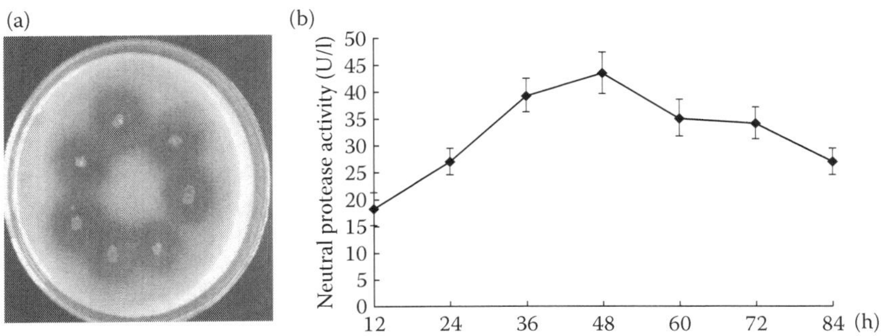

Figure 2.40 Proteolytic activity of *Bacillus subtilis* strain HB-1: clear zones of casein hydrolysis around colonies inoculated on casein agar plate (a), and neutral protease activity measurements at different time intervals (b). (From Qin, H. et al., *Food Control*, 31, 22–27, 2013.)

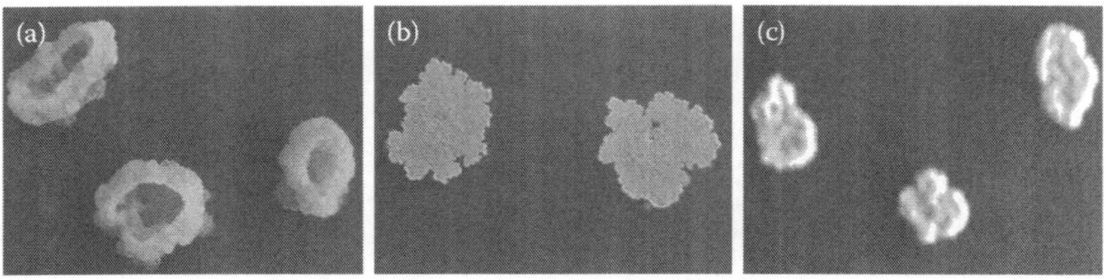

Figure 2.41 Morphology of *Bacillus subtilis* strain HB-1 on three different agar plates: enzyme production medium (a), Luria Bertani (LB) medium (b), and LB medium supplemented with 3% L-glutamic acid (c). (From Qin, H. et al., *Food Control*, 31, 22–27, 2013.)

under shaking. The polymer on soybean surface mainly contributes to the texture and slime attributes of yandou (Qin et al., 2013).

To test if strain HB-1 was responsible for the majority, if not all, of the fermentation, yandou was prepared in laboratory using strain HB-1 as starter to mimic the native fermentation process. After fermentation, the laboratory-made yandou was evaluated by sensory tests. No significant difference was observed between the laboratory-made and homemade yandou in all sensory variables, including appearance, color, smell, texture, and slime (Qin et al., 2013).

Right now, yandou production still depends on native fermentation processes even in the small businesses, let alone homemade yandou. To scale up yandou production, a pure culture starter for fermentation is absolutely required to comply with food safety regulations, to ensure consistent food quality, and to achieve reproducibility of the product. *Bacillus subtilis* HB-1 is the first *B. subtilis* strain identified and characterized in yandou. It could be a good candidate for commercial production of yandou in the future.

2.1.2 Non-Soybean Products

In West and Central Africa, indigenous fermented vegetable proteins are an important cheap source of dietary protein. They contribute significantly to the dietary intake of the people, and are used as flavorsome seasoning agents in soups and sauces. Although the seeds used for the production of these condiments differ from one region to the other, the production and processing conditions are generally similar. Seeds of non-soybean leguminous plants, including locust bean, oil bean, yam bean, sickle pod, monkey pod, mesquite, and *Cathormion*, have been traditionally used in the production of alkaline-fermented foods.

2.1.2.1 Dawadawa (Iru) and Similar Products

Olusola Bandele Oyewole, Adewale Olusegun Obadina, and Paulin Azokpota

Dawadawa or iru is the most important food condiment in many countries of West and Central Africa. The condiment is used to enhance the flavor of many dishes including soups and sauces. For the last 10 years, these are considered as the most important food condiments consumed in Nigeria and Benin by the rural poor people as well as high-income urban families. It is produced by fermenting the seeds of African locust bean tree (*Parkia biglobosa* (Jacq.) G.Don), and the production for use as food condiments is craft-based. Remarkably, in many areas today it is still made in traditional ways with success depending upon the observance of good manufacturing practices and control of environmental conditions during the manufacturing phase (Achi, 2005).

In Benin, the Fons and Baribas produce similar products, called afitin and sonru, respectively, from the locust bean (Gutierrez et al., 2000; Azokpota et al., 2006a). In Togo and some other countries of West Africa, a similar food, called "tonou," is traditionally prepared from the locust bean. While dawadawa (iru), afitin, and sonru are made exclusively of locust bean, tonou is prepared from locust bean and peanut (Konlani et al., 1999).

2.1.2.1.1 Antiquity Dawadawa (also known as irú, carob, daddawa, dawa dawa, and ugba in Nigeria, khinda in Sierra Leone, kpalugu in Ghana), locust bean cakes, néré, and nététu in Senegal and The Gambia, iru, sonru, and afitin in Benin, and soumbara or soumbala in other Francophone West African countries (Niger, Burkina Faso, Côte d'Ivoire, French Guinea, Mali, Senegal, and Mauritania) represents the fermented oily seeds of the locust bean tree, a member of the Leguminosae family. The tree is native to Central and West Africa, and can now be found in Benin, Burkina Faso, Cameroon, Central African Republic, Chad, Cote d'Ivoire, Congo, Gambia, Ghana, Guinea-Bissau, Mali, Niger, Nigeria, Sao Tome, Senegal, Sierra Leone, Togo, and Uganda. The fruit pulp and seed extracts provide nutritious ingredients for traditional soups, sweetmeats, and condiments across West Africa.

2.1.2.1.2 Traditional Method of Production The methods employed in the manufacture of fermented condiments differ from one region to another because these processes are based on traditional systems, some of which are summarized in Figures 2.42 and 2.43. Fermentation takes place under conditions that the producers have found to be favorable for the appropriate growth and activity of the microorganisms. Traditionally, raw beans are cooked for at least 12 h in excess water until they are very soft to allow dehulling using hands. The length of cooking time depends on the strength of the testa and on the dehulling efficiency of the seeds. Boiling in 0.1 M Na_2CO_3 solution reduces the cooking time to 4 h and increases the dehulling efficiency to 89% (Achi and Okereka, 1999). While *Telfairia* Hook.

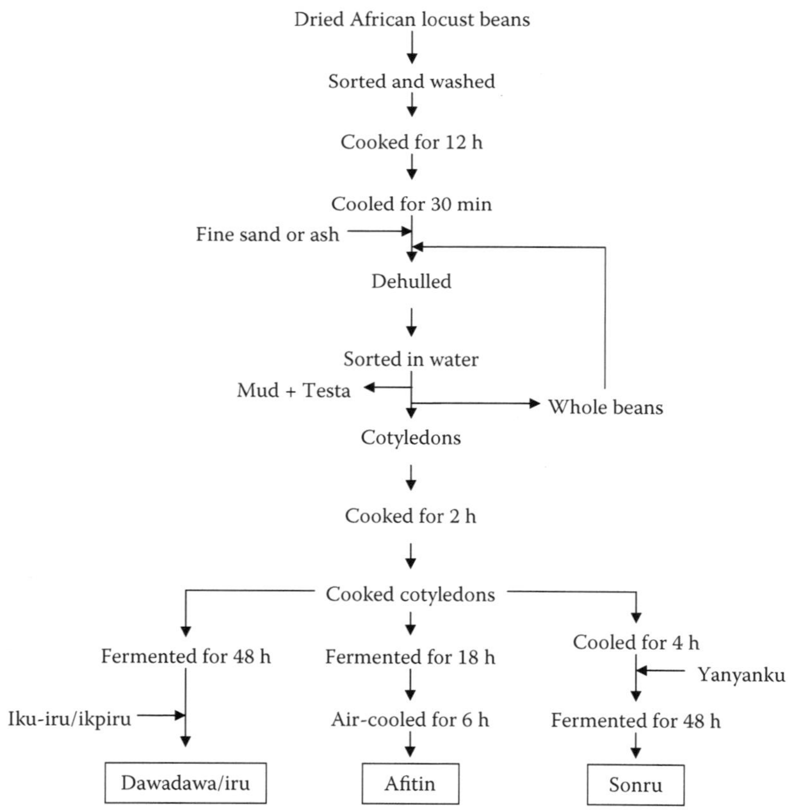

Figure 2.42 Traditional process of dawadawa (iru), afitin, and sonru production. (From Azokpota, P. et al., *International Journal of Food Microbiology*, 107, 304–309, 2006a.)

Figure 2.43 Main steps of dawadawa (iru) production.

seeds require a cooking for 30 to 60 min, melon (*Citrullus lanatus* (Thunb.) Matsum. & Nakai) and castor (*Ricinus communis* L.) seeds are deshelled before cooking.

At present, most dehulling processes rely on the tedious method of hand separation. Abrasive dehulling of dry beans of *Parkia* or *Prosopis* is not possible because of the strong adherence of hulls to the seed. Nevertheless, roasting prior to boiling in water has been identified as a technique for achieving a reduction in cooking time (Ijabo et al., 2000). Boiling in excess water serves not only the purpose of partial cooking but also facilitates microbial penetration and human digestion. The bitter and beany taste of soybean disappears within 15 min at 95°C (Nout et al., 1985). After the cooking and dehulling processes have been completed, the cotyledons are prepared for fermentation. For some seeds, such as the African oil bean (*Pentaclethra macrophylla* Benth.), the cotyledons may be boiled again for 1–2 h. A softening agent "kaun" (potassium carbonate) may be added to aid softening of the cotyledons (Odunfa, 1985a; Ikenebomeh et al., 1986; Achi, 2005).

2.1.2.1.3 Starters As shown in Figure 2.42, no additive is used in the processing steps to produce afitin. However, iru and sonru productions involve the use of two additives, which are the results of the know-how of rural Beninese food condiment processors. Ikpiru (iku-iru) and yanyanku are used as inoculums or starter culture-like additives for the fermentation of locust beans to produce iru and sonru, respectively (Azokpota et al., 2006a, 2011; Agbobatinkpo et al., 2011).

The production of iru and sonru has increased significantly in recent years, requiring an increase in the daily production of ikpiru and yanyanku, which are also produced commercially by traditional processors. Thus, ikpiru and yanyanku are two products with potential for scaling up and industrialization; they could be an alternative to standard starter culture utilization by traditional way. The products are traditionally processed at the household level and have a shelf life about 10 years (Agbobatinkpo et al., 2011; Azokpota et al., 2011). The ikpiru and yanyanku production processes are shown in Figures 2.44 and 2.45, respectively.

In Nagot language, ikpiru means "that helps to rise," and in Bariba language, yanyanku means "product having ability to degrade" (Agbobatinkpo et al., 2011; Azokpota et al., 2011). According to the traditional producers, ikpiru and yanyanku accelerate the fermentation of locust beans to produce iru and sonru, respectively, because untreated beans are too hard to be easily fermented spontaneously (Agbobatinkpo et al., 2011; Azokpota et al., 2011). Ikpiru and yanyanku could be considered as softening agents that probably play similar role as a standard starter culture during the fermentation.

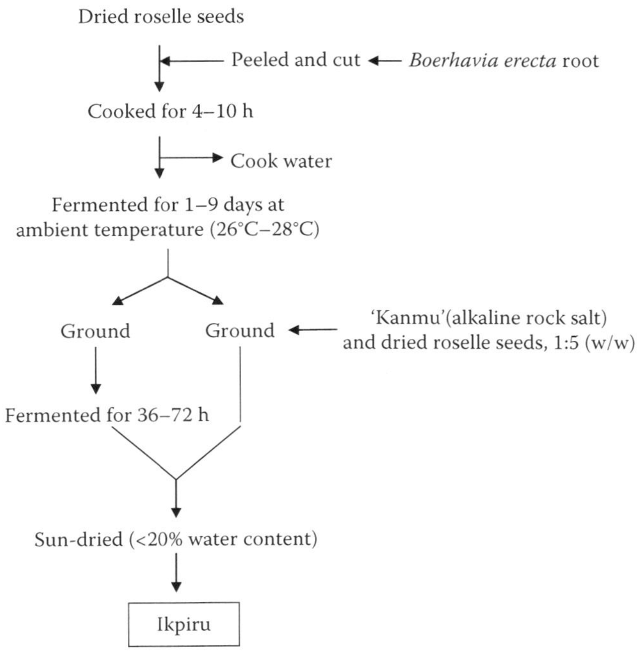

Figure 2.44 Flowchart of traditional process of ikpiru production. (From Agbobatinkpo, P.B. et al., *Ecology of Food and Nutrition*, 50, 101–114, 2011.)

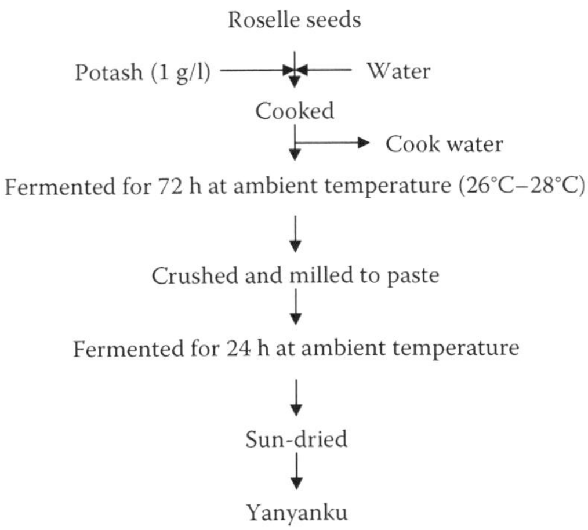

Figure 2.45 Flowchart of traditional process of yanyanku production. (From Azokpota, P. et al., *Journal of Food Processing and Preservation*, 35, 813–821, 2011.)

2.1.2.1.4 Microbial Succession The dynamics of fermentation in any food matrix is a complex microbiological process involving interactions between different microorganisms (Daeschel et al., 1987). The contribution of the accompanying flora of fermenting substrates is determined by the substrate composition and hygiene during production. At the time of fermentation, the microorganisms use the nutritional components of the seeds, converting them into products that contribute to the chemical composition and taste of the condiment. *Bacillus* spp. were found to be the dominant microorganisms in dawadawa (iru), afitin, sonru, and tonou, suggesting that they could be targeted as possible starters for the manufacture of safe and reproducible quality products (Odunfa, 1981; Azokpota, 2005). Furthermore, species of the *Bacillus subtilis* group have been reported to be generally regarded as safe (GRAS) by the U.S. Food and Drug Administration, and their role in the fermentation of locust bean to produce different condiments has already been investigated (Allagheny et al., 1996; Beaumont, 2002).

An investigation on 200 presumptive *Bacillus* isolates collected from afitin, iru, and sonru was performed. They were initially identified by their phenotypic characteristics as *Bacillus* spp. ITS-PCR-RFLP and 16S rDNA sequencing revealed that 79% of the isolates were *Bacillus* spp., 5% were *Staphylococcus*, and 16% were ungrouped bacteria. About 75% of the *Bacillus* spp. identified by ITS-PCR-RFLP were shown to belong to the *B. subtilis* group, and 25% to the *Bacillus cereus* group (Azokpota et al., 2007). Additional biochemical and physiological tests applied to 50 isolates randomly selected from the *B. subtilis* group divided the latter into 80% *B. subtilis* and 20% *Bacillus licheniformis*, some of which showed different PFGE-RFLP bandpatterns. Strains belonging to the *B. cereus* group were differentiated by PCR with primers BCW1F and BCW1R, and the PCR products were then digested with *Eco*RI, *Sau*3AI, and R*sa*I. All the strains but one were found to be *B. cereus.* Enterotoxin testing showed that all the *B. cereus* isolates were negative for nonhemolytic enterotoxin complex NHE, but that 17.9% (7/39) were positive for BceT, and 12.8% (5/39) were positive for the hemolytic enterotoxin complex HBL. Iru, afitin, and sonru have not been associated with cases of any food-borne disease (Azokpota et al., 2007). So the predominance of *B. subtilis* in iru, afitin, and sonru with a high diversity among the strains was reported. However, minor groups of *B. licheniformis* and

B. cereus were also found. The *Bacillus* strains isolated from iru, afitin, and sonru opened new perspectives on the diversity of the principal characteristics of products obtained by controlled fermentation.

A study on 50 strains of *Bacillus* spp. including *B. subtilis* and *B. licheniformis* previously isolated from iru, afitin, and sonru were examined for esterase and protease activities. The electrophoretic profiles of fermented African locust bean protein (ALBP), using strains presenting the highest protease activities in casein agar, were analyzed by SDS-PAGE (sodium dodecyl sulfate–polyacrylmide gel electrophoresis) to select strains with good ability to be used as starter cultures (Azokpota et al., 2006b). All the *Bacillus* strains tested showed esterase activity against tributyrin with high variability among them. The strains showing the highest esterase activities were *B. subtilis*, primarily isolated from iru, sonru, and lastly in afitin. Only 62% (31/50) of the *Bacillus* strains tested showed perceptible, but highly variable protease activity in casein agar. *Bacillus* strains showing the highest protease activities comprised strains of *B. subtilis* isolated from afitin and iru, and strains of *B. licheniformis* isolated from iru and sonru. A strain of *B. subtilis* isolated from afitin showed high esterase as well as high protease activity. The electrophoretic profile by SDS-PAGE of ALBP fermented by the *Bacillus* spp. having the highest protease activities showed degradation products with a wide range of molecular size between 4 and 250 kDa. Variability of the characteristics of the *Bacillus* spp. tested gives new opportunities for their use as starter culture for product development (Azokpota et al., 2006b).

Yeasts and other bacteria are also "seen"; however, only a part of them can be considered to play a substantial role in the fermentation processes. For example, non-fermenting species may just be ubiquitous contaminants although they may affect the flavor of the final product when occurring in high numbers. Other bacteria present include *Staphylococcus* spp. Proteolytic and salt-tolerant microorganisms have been detected in appreciable numbers (10^4–10^6 cfu/g) in samples of fermenting *Parkia* seeds after 36 h (Odunfa, 1981). Members of the Enterobacteriacceae also contributed to the ecology of dawadawa, especially at the early stages. These species do not survive until the end of the fermentation, presumably because of the modified environment, which are developed at later stages (Mulyowidarso et al., 1989; Achi, 1992, 2005).

2.1.2.1.5 Chemical Changes Since the major constituents of legumes and oilseeds used in dawadawa production are proteins, fats, and fewer carbohydrates, the organisms responsible for the hydrolysis must be capable of utilizing these constituents.

Bacillus spp. have been reported as producers of certain hydrolases, such as amylase, galactanase, galactosidase, glucosidase, and fructofuranosidase, which are involved in the degradation of carbohydrates (Aderibigbe and Odunfa, 1990; Sarkar et al., 1997a; Kiers et al., 2000; Omafuvbe et al., 2000; Achi, 2005). Microbial activities hydrolyze carbohydrates into sugars, which are then readily digestible by humans. Similarly, galactanases soften the texture of the seeds and liberate sugars for digestion.

Dawadawa is a high-protein product; the extent of protein hydrolysis is one of the most important factors in the changes in texture and flavor (Whitaker, 1978). Soluble low-molecular-weight peptides and amino acids that contribute to flavor are produced through the enzymatic breakdown of proteins (Odunfa, 1985a; Njoku and Okemadu, 1989; Ogbonna et al., 2001; Ouoba et al., 2003b; Achi, 2005). Dawadawa of good quality has a characteristic strong smell of ammonia and a dark color and is semi-hard. Ammonia produced because of extensive protein hydrolysis is responsible for the gradual increase in pH up to 7.5–8.0 (Achi, 1992; Barber and Achinewhu, 1992; Sarkar et al., 1993; Barimalaa et al., 1994). The increase in pH into the alkaline range may be physiologically important for tolerance and adaptation of fermenting microorganisms in the environment. However, the pH increase due to the formation of ammonia from amino acids may encourage the growth of spoilage organisms such as lactobacilli and pediococci (Barber and Achinewhu, 1992). The effect of vegetable protein fermentation on the total nitrogen is negligible. Free amino acids increase, but a longer fermentation results in losses of lysine or other essential amino acids (Winarno and Reddy, 1986). Although oils constitute up to 40% of the legumes used in the fermentations, extensive lipolysis does not take place. Odunfa (1985a) reported only a low level of lipase activity in *P. biglobosa* during dawadawa production.

The fermentation process dramatically improves the food quality of the beans. Principally, this makes the proteins in the beans much more digestible. Although dawadawa is used principally as a flavor enhancer, particularly in Nigerian cuisine, it adds thiamine and riboflavin to diets and acts as a source of protein. Fermentation also makes free lysine available in the fermented bean mix, which is a boom for diets traditionally rich in maize (as this is deficient in the plant).

2.1.2.1.6 Volatile Compounds Profile Traditional iru, afitin, and sonru are characterized by a strong odor, which is the main criterion used for their acceptance. Volatile compounds in these products were identified and quantified, using the Likens-Nickerson simultaneous distillation–extraction techniques, followed by GC-MS (gas chromatography–mass spectrometry) analysis (Azokpota et al., 2008, 2010). Thirteen chemical groups of volatile compounds were found from which six major groups were distinguished: pyrazines, aldehydes, ketones, alkanes, esters, and alcohols. From these groups, compounds such as tetramethyl pyrazine, trimethyl pyrazine, 2,5-dimethyl pyrazine, 3-methyl butanal, 2-decanone, 3-hydroxy-2-butanone, heptadecane, linoleic acid ethyl ester, and 3,5-dimethyl benzene methanol were particularly found in higher concentrations (Azokpota et al., 2008). Afitin was characterized by high concentrations of 2-decanone and 3,5-dimethyl benzene methanol, whereas 3-methyl butanal, tetramethyl pyrazine, and 2,5-dimethyl pyrazine were found specifically in higher concentration, in both iru and sonru. The concentrations of heptadecane and linoleic acid ethyl ester were higher in sonru (Azokpota et al., 2008).

In the condiments obtained from the controlled fermentation using pure strains of *B. subtilis*, a total of 94 volatile compounds have been found including 53 compounds identified in relatively high concentrations and were subdivided into 7 main groups with the predominance of four major groups: pyrazines, aldehydes, ketones, and alcohols (Azokpota et al., 2010). Compared with the spontaneously fermented condiments, volatile compounds identified in controlled fermented condiments have been found in high number and in concentrations that varied according to the inoculums of *B. subtilis* used. The condiments produced with starter cultures scored significantly higher for odor than the spontaneously fermented condiments (Azokpota et al., 2010). However, the overall acceptability of the two types of condiments was similar. The investigated *B. subtilis* strains, BS1, BS2, and BS3 can be considered as potential starter cultures for the fermentation of locust bean to produce good quality food condiments (Azokpota et al., 2008, 2010).

2.1.2.1.7 Culinary Practice and Economics Dawadawa is widely eaten throughout West Africa as a staple diet and is used in one daily meal for up to 90% of the year in some areas. Offering gifts is an important social practice, and dawadawa is one of the most appreciated culinary gifts in West Africa. The seeds and processed products are frequently traded in local markets. The widespread popularity of dawadawa makes the locust seed an important commercial item across West Africa. It is estimated that 200,000 tons of locust seeds are collected annually for dawadawa, just in Northern Nigeria. The production and sale of dawadawa constitute an important economic activity for women.

Thousands of West African women earn cash income by making dawadawa in low-capital input, small-scale enterprises at home and selling it at local markets. Each locust bean tree is individually owned, usually by men who hold the rights to the plot in which the tree is standing. The women must either buy the beans or buy bean-harvesting rights from the tree owner. The seeds commonly fetch two to four times as much as major staples such as maize, sorghum, and millet on the market. Purchase of locust bean accounts for 10%–20% of regular weekly expenditures of most rural women in the Bassila region of Benin. In Burkina Faso, the seed sale accounts for up to 25% of household income.

2.1.2.2 Ugba

Abiodun Isiaka Sanni and Folarin Anthony Oguntoyinbo

Ugba is a traditional name for fermented seeds of African oil bean tree (*Pentaclethra macrophylla* Benth.; family Leguminosae), and it originated from the Igbo speaking area of southeastern Nigeria (Odunfa and Oyeyiola, 1985; Sanni et al., 2002). Although the African oil bean tree is abundant in the rainforest areas of West and Central Africa, food prepared from it is unique and distinct to the different communities (Enujiugha and Akanbi, 2005) and is defined by the traditional culinary practice in the different geographical locations.

Ugba is a product that is prepared in homes as a family business based on tradition art handed over from generation to generation. Thus, the products from different sources are not uniform and taste may also differ (Sanni et al., 2002; Ogueke et al., 2010). Ugba is an important and cheap source of protein for people whose staple foods are mainly carbohydrates. The condiment can be consumed as a native salad when fermented for 2–5 days or used as soup flavoring after 6–10 days of fermentation (Obeta, 1983; Isu and Ofuya, 2000; Sanni et al., 2002; Ogueke et al., 2010). Ugba has also been reported to be eaten with bitter kola (*Garcinia kola* Heckel) or kolanuts (*Cola nitida* (Vent.) Schott & Endl.), and when cooked with garden egg (*Solanum melongena* L.) leaves it can be used as a sauce to eat fried or boiled yam and cocoyam (Mbajunwa et al., 1998; Ogueke et al., 2010).

The production of ugba (Figure 2.46) is by alkaline fermentation of the seeds of African oil bean tree. The seeds (Figure 2.47a) are boiled in water for 12–15 h to soften the hard seed coat. The boiled seeds are dehulled, and the cotyledons are cut into long narrow slices (about 0.2 cm thick; Figure 2.47b), washed, boiled for additional 1–2 h to remove the bitter substances, and the water is discarded. This is followed by soaking the boiled cotyledon slices overnight (12 h) in an earthen pot covered with water, to remove most of the bitter

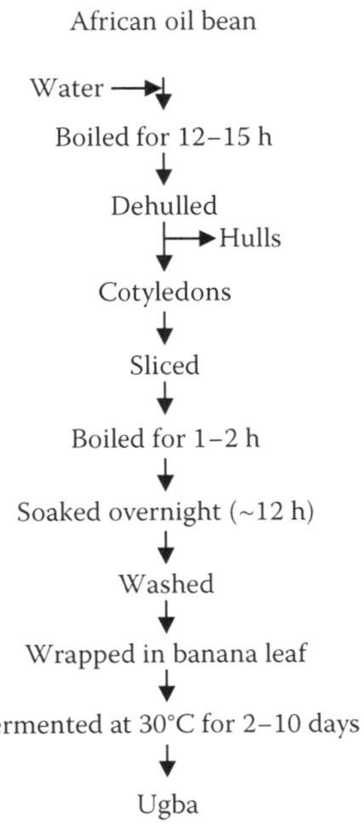

Figure 2.46 Traditional process of ugba production.

Figure 2.47 African oil beans (a), sliced oil bean cotyledons (b), and aerial view of ugba (c).

substances. The water is discarded, and the sliced cotyledons are washed severally in clean water to completely remove the bitter components and then placed in a basket lined with banana leaves to allow the water to drain out completely. The sliced cotyledons are divided into small portions and placed in blanched banana leaves, wrapped tightly, and left in a basket to undergo natural fermentation for 2–5 days at ambient temperature (28°C–30°C). The fermented bean slices are kept near smoldering firewood to develop the unique ugba flavor and the product is consumed as native salad as previously stated. However, fermentation for a longer period of time (6–10 days) produces very soft ugba (Figure 2.47c), which is used as soup flavorings (Odunfa and Oyeyiola, 1985; Sanni et al., 2002).

The fermentation process is spontaneous and the main microorganisms identified are the proteolytic *Bacillus* spp., which include *B. subtilis* (most predominant), *B. licheniformis*, *B. megaterium*, *B. coagulans*, *B. macerans*, and *B. circulans* (Obeta, 1983; Sanni, 1993b; Isu and Ofuya, 2000; Sanni et al., 2002). Other opportunistic organisms reported to be associated with the oil bean fermentation are coagulase negative *Staphylococcus* sp., *Micrococcus* sp., *Lactobacillus plantarum*, *Enterobacter* sp., and *Escherichia coli*. These other microorganisms may have been introduced through air, water, or leaves used in wrapping the fermenting cotyledons or through handling during the preparatory stages. However, *Bacillus* spp. were observed to persist until the end of fermentation period with the increase in population, while other associated microorganisms decreased after 24 h of fermentation (Oyeyiola, 1981; Obeta, 1983; Sanni et al., 2002).

Several studies had reported degradation of proteins as the most important metabolic activity, which leads to an increase in protease activity during alkaline fermentation (Odunfa, 1986; Wang and Fung, 1996; Oguntoyinbo et al., 2007). The softening of the fermenting cotyledons had been suggested to be due to the pectinolytic and proteolytic activities of *B. subtilis*, which hydrolyze the pectin and protein components of the seeds (Mbajunwa et al., 1998; Sanni et al., 2002; Parkouda et al., 2009). Njoku and Okemadu (1989) had earlier reported proteinase as the most important enzyme that complements other bacterial enzymes during ugba production.

Mbajunwa et al. (1998) further suggested that the color of ugba developed during the boiling stage is due to non-enzymatic browning reaction, which involves the amino acids and non-reducing sugar constituents of oil bean and microbial activity. The presence of essential amino acids makes ugba a potential source of protein; thus, its utilization as a flavoring component in soups in southeastern Nigeria (Ogueke et al., 2010).

Studies have also been carried out on the modification of some parameters that affect the fermentation process during ugba production. Such parameters are temperature, relative humidity (RH), and the use of starter cultures (Isu and Njoku, 1997; Sanni et al., 2002; Ogueke et al., 2010). Isu and Njoku (1997) suggested that an RH of 74.5%, which is atmospheric RH, is more favorable for the fermentation process than the humid environment (RH = 80%), created by wrapping of the sliced cotyledons in plantain/banana leaves, which limit the accessibility of air to the fermenting substrate. A temperature of 35°C was found to be most suitable for the fermentation by generating a peak amino-nitrogen content of 19.6 mg/100 g dry matter within 48 h, thus reducing the fermentation time by 24 h in the traditional fermentation process.

Sanni et al. (2002) also screened some *B. subtilis* strains isolated from ugba for proteolytic and amylolytic activities, and stickiness. Although all the selected strains exhibited a high activity for the enzymes, the strain MM-4:B12 demonstrated the highest proteolytic and amylolytic activities and production of whitish mucilaginous polymer, which is responsible for the stickiness of ugba (Table 2.12).

Table 2.12 Screening of Selected *Bacillus subtilis* Strains Isolated from Ugba for Proteolytic and Amylolytic Activities, and Stickiness

STRAINS	PROTEOLYTIC ACTIVITY (mm)	AMYLOLYTIC ACTIVITY (mm)	STICKINESS (cm)
BM-5:B7	6.4	5.1	15.8
BM-1:B24	6.8	5.2	16.3
BM-7:B6	5.3	4.4	15.2
MM-1:B47	5.7	4.4	15.6
MM-4:B12	7.4	5.6	17.3
MM-6:B31	7.0	5.2	16.4
SM-3:B21	5.3	4.4	15.1
SM-7:B23	6.7	5.1	15.5
SM-2:B11	5.2	4.5	15.3

Source: Sanni, A.I. et al., *European Food Research and Technology*, 215, 176–180, 2002.
Note: Values represent means of three independent determinations.

The use of MM-4:B12 strain as a starter culture was observed to accelerate the fermentation, which was completed in 48 h relative to the traditional fermentation period, and pH of 8.1, 46.5% moisture content, and viable count of 1.4×10^{10} cfu/g was obtained at the end of the fermentation period (Table 2.13).

Ugba has unique aroma and flavor, which are due to the production of volatile compounds by the fermenting microorganisms (Ogueke et al., 2010), but there is dearth of scientific information in this direction. However, Kabuo et al. (2007) reported the aroma and flavor compounds of ugba produced by using *B. subtilis* and *B. licheniformis* as starter cultures. The compounds identified were ethyl linoleate (14.14%), ethyl phenol (6.94%), ethyl oleate (4.70%), ethyl acetate (3.6%), and ethyl octanoate (2.69%). The control (spontaneously fermented) ugba sample contained ethyl benzoate (18.60%), ethyl carbonate (5.56%), ethyl octanoate (4.72%), and methyl pentanone (1.67%).

Table 2.13 Changes in pH, Moisture Content, and Viable Count of Fermenting Oil Bean Seeds Using *Bacillus subtilis* MM-4:B12

FERMENTATION PERIOD (h)	pH	MOISTURE (%)	*B. subtilis* CELLS (cfu/g)
0	6.8	38.4	4.1×10^4
12	7.0	40.6	2.5×10^5
24	7.4	43.3	1.9×10^7
36	7.7	44.2	6.1×10^9
48	8.1	46.5	1.4×10^{10}

Source: Sanni, A.I. et al., *European Food Research and Technology*, 215, 176–180, 2002.
Note: Values represent means of three independent determinations.

Although there have been no reports of pathogens such as *Clostridium perfringens*, *Clostridium botulinum*, *Salmonella*, *Shigella*, and *Vibrio* species, *E. coli*, and *Staphylococcus aureus* have been isolated and these are capable of causing food infection and poisoning. Since the culinary practice of utilizing the condiment in soups and sauces involves cooking, these organisms are likely to be eliminated during the process (Obeta, 1983; Odunfa and Oyeyiola, 1985; Sanni et al., 2002; Ogueke et al., 2010).

2.1.2.3 Kawal

Abiodun Isiaka .Sanni and Folarin Anthony Oguntoyinbo

Kawal is an indigenous fermented food that serves as a favorite meat substitute or meat extender in Kordofan and Darfur provinces of Western Sudan and eastern part of Chad. It is produced by solid-state fermentation of the leaves of sicklepod (*Senna obtusifolia* (L.) H.S.Irwin & Barneby; family Leguminosae), which is readily available and of little economic importance. It is a wild African plant found in wasteland during the rainy season (Mbaiguinam et al., 2005). The seeds are toxic and unpalatable in their natural state, but when fermented yield an organoleptically acceptable and highly nutritional product, called kawal (Harper and Collins, 1992).

For its production (Figure 2.48), sicklepod leaves are pounded into a paste and packed into an earthenware "zeer," buried to the neck in ground under a shade. A layer of sorghum leaves are placed on

Sicklepod leaves
↓
Pounded into paste
↓
Packed in an earthen vessel
↓
Vessel covered and left buried under soil for 11–15 days
(Paste mixed at every 3 days interval)
↓
Molded into balls
↓
Sun-dried for 5 days
↓
Kawal

Figure 2.48 Traditional preparation of kawal. (From Mbaiguinam, M. et al., *African Journal of Biotechnology*, 4, 1080–1083, 2005.)

the surface of the paste, and the zeer is covered with a lid that is sealed with mud. At 3-days interval, the zeer is opened, the sorghum leaves are removed, and the paste is remixed thoroughly by hand. The re-mixed paste is again covered with sorghum leaves and the zeer is resealed. After 11–15 days, the strong smelled black mass is removed, molded into small balls, and sun-dried for 5 days (Figure 2.49). In the traditional preparation of kawal, two or three indigenous food processing techniques are a common practice. This involves fermentation, drying, and salting (not in all cases) (Mbaiguinam et al., 2005).

The microorganisms that have been reported to be involved in the fermentation process include *Bacillus subtilis*, *Lactobacillus plantarum*, *Propionibacterium* sp., *Staphylococcus sciuri*, and two yeasts, *Candida krusei* and *Saccharomyces* sp. Although the organisms occurred throughout the fermentation period at varying population, the predominant fermenting microorganisms during kawal production are *B. subtilis* and *Propionibacterium* sp. (Dirar, 1984; Harper and Collins, 1992).

Fermentation causes a significant change in the major nutrients of sicklepod leaves, viz. increase in the contents of protein (24.8%–35.2%), fat (3.5%–4.5%), and ash (13.7%–18.0%), but a decrease in fiber content (Dirar, 1984; Nuha et al., 2010; Ibnouf, 2012). On the other hand, the contents of anti-nutritional factors such as tannin, phytate, and total polyphenols were significantly reduced at the end of fermentation and finally the cooking stage (Nuha et al., 2010). The fermented leaves of sicklepod plant contain the major elements needed by the human body. In addition to the amino acid composition (Table 2.14), the nutritional value of kawal as a meat substitute for many poor and rural people of Sudan is justified and especially when taken with cereals (Mbaiguinam et al., 2005).

Figure 2.49 The appearance of kawal. (Courtesy of Warda Abdelgadir.)

Table 2.14 Amino Acid Composition of Unfermented and Fermented Sicklepod Leaves

AMINO ACID	UNFERMENTED LEAVES (mg/100 g)	FERMENTED LEAVES (KAWAL) (mg/100 g)	RECOMMENDED DAILY ALLOWANCE (mg/50 kg ADULT)[a]
Threonine	980	522	350
Valine	1606	1282	500
Cystine	13	4	650 (Cys + Met)
Methionine	11	18	
Isoleucine	1181	987	500
Leucine	2234	1803	700
Tyrosine	435	239	700 (Tyr + Phe)
Phenylalanine	1562	1022	
Lysine	1477	790	600
Histidine	664	356	500
Arginine	1314	601	
Aspartic acid	3786	1345	
Serine	863	505	
Proline	1769	858	
Glycine	1428	1074	
Alanine	1595	1638	

Source: Mbaiguinam, M. et al., *African Journal of Biotechnology*, 4, 1080–1083, 2005.

[a] Data from *Recommended Dietary Allowances*, 10th ed., National Academy Press, Washington, DC, p. 17, 1989.

Although Mbaiguinam et al. (2005) reported increase in the methionine and alanine content at the end of fermentation, Harper and Collins (1992) observed that alanine composition of kawal did not increase during the fermentation, and the lactic acid production was negligible. Volatile fatty acids, mainly *n*-butyric (8%), acetic acid (5%), and *n*-propionic (9%) formed 15% of the fermented product (Dirar, 1984; Harper and Collins, 1992). The fermentation converts inedible leaves of no economic value to acceptable meaty food that is reach in sulfur and amino acids and also safe for a population that often does not have sufficient nutrient intake (Dirar, 1984; Harper and Collins, 1992).

Considering the importance of indigenous knowledge that is used for food processing and preservation, kawal plays a key role in maintaining food security in the communities where it is consumed. The product can be preserved and consumed years after production and is readily available in the rural areas where it represents important alternative food source during food shortage and famine (Dirar, 1984;

Ibnouf, 2012). During the famine experienced in southern Sudan in 1998, kawal contributed significantly in saving the lives of a large number of famine victims because of its unique characteristics, affordability, and a long shelf life (Guvele et al., 2003; Ibnouf, 2012).

2.1.2.4 Okpehe

Abiodun Isiaka Sanni and Folarin Anthony Oguntoyinbo

Okpehe is a traditional fermented African mesquite (*Prosopis africana* (Guill. & Perr.) Taub.) beans. Mesquite is a sub-tropical or Guinea savanna leguminous tree plant that belongs to family Leguminosae, which spreads across the countries in West Africa. The seed's cotyledon is fermented to produce condiments in many African countries. Okpehe is very popular among the people of mid-western, eastern, and middle belt of Nigeria. In Nigeria, it is called by different names: kiriya (Hausa), kpaaye (Tiv), ayan (Yoruba), and okpehe or okpiye (Idoma). Taxonomy of the genus *Prosopis* includes species such as *P. pubescens* Benth., *P. juliflora* (Sw.) DC., *P. caldenia* Burkart, *P. glandulosa* Torr., *P. laevigata* (Willd.) M.C.Johnst., and *P. reptans* Benth. in South and North America, while *P. cineraria* (L.) Druce and *P. farcta* (Banks & Sol.) J.F.Macbr. were species identified in Asia. It appears that *P. africana*, *P. cineraria*, and *P. pallida* are indigenous to Africa (Ffolliot and Thames, 1986; Agboola, 2004; Vautier and Sacandé, 2007).

2.1.2.4.1 Traditional Method of Preparation The traditional preparation of okpehe (Figure 2.50) generally follows the procedures stated by Sanni (1993a, 1993b), Obeta and Ugwuanyi (1996), and Oguntoyinbo et al. (2007). The mesquite seeds (Figure 2.51a) are boiled for 15–24 h until they become softened to manually dehull by pressing them between the palms of the hands, after which the cotyledons are washed and boiled again for 3–5 h. These are then spread in wide calabash trays or earthen pots already lined with pawpaw leaves, and the trays are then tightly wrapped with jute bags. Fermentation, which takes place for 3–5 days at 30°C–35°C, results in a sticky, dark brown mash (Figure 2.51b) with a pungent smell. Although the condiment is added generously to soups in many homes as a flavoring agent, many rural and urban people also use it as a meat substitute.

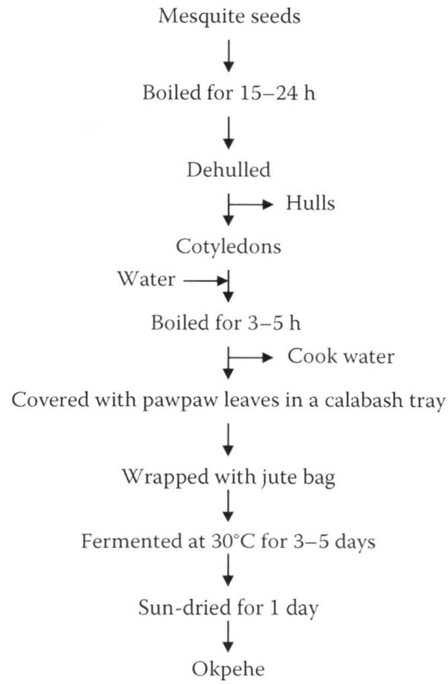

Figure 2.50 Traditional preparation of okpehe. (From Oguntoyinbo, F.A. et al., *International Journal of Food Microbiology*, 113, 208–218, 2007.)

Figure 2.51 Mesquite seeds (a) and okpehe (b).

2.1.2.4.2 Microbiology It appeared that only bacteria were involved during the process, and different studies reported species of *Bacillus* as the dominant microorganism responsible for the fermentation of mesquite beans producing okpehe (Omafuvbe et al., 1999; Oguntoyinbo et al., 2010). Detailed genomic characterization revealed *Bacillus subtilis*, *Bacillus amyloliquefaciens*, *Bacillus licheniformis*, and *Bacillus cereus* as the dominant bacilli group responsible for the traditional fermented condiments in West Africa (Oguntoyinbo et al., 2010). Extracellular enzyme hydrolysis of the beans by the bacilli was observed as significant to product development. However, there is variation in the ability of amylase and protease production among the *Bacillus* spp. The population of *B. subtilis* and *B. amyloliquefaciens* was higher; their enzymatic and hydrolytic potentials must have facilitated dominance. A population count of 10^{10} colony-forming units (cfu) per milliliter has been reported after 72 h of fermentation. In Nigeria, *Staphylococcus*, *Micrococcus*, and *Enterococcus* species were reported to have also been isolated from other traditional fermented condiments. The relevance of these bacteria to product development may, however, be insignificant. Oguntoyinbo et al. (2007) observed that the involvement of enterococci during okpehe production produced an undesirable cheese-like aroma that made the product unacceptable to consumers. Similarly, toxigenic potential of *B. cereus* in the traditionally produced okpehe has been documented (Oguntoyinbo and Sanni, 2007). Development of starter cultures for the production of okpehe with consistent quality attributes has been shown in laboratory-scale fermentation to be promising. Bacteriocin-producing *B. subtilis* BFE 5301 strain in co-culture with proteolytic and amylolytic *B. amyloliquefaciens* BFE 5372 strain was used to produce okpehe in the laboratory with improved shelf life and good organoleptic characteristics. The product was found to be free of toxigenic *B. cereus* and enterococci (Oguntoyinbo and Sanni, 2007).

During fermentation of the beans, pH increases progressively with time, reaching pH 8 after 72 h, which was suggested to be due to consistent hydrolysis of protein to amino acids and formation of ammonia. Proteolysis appeared to be the most important biochemical change during the fermentation. Also, the fermentation resulted in a net increase in concentrations of total soluble sugars and free amino acids, both reaching a peak after 72 h followed by decline thereafter. Corresponding increases were observed in amylase and protease activities, respectively. Similarly, lipase activity was observed, and this increased gradually throughout the fermentation period (Odibo et al., 2008). Enzyme activities of microorganisms isolated from okpehe showed that α-chymotrypsin, leucine arylamidase, α-galactosidase, and α-amylase were produced. In addition, a weak lipase activity was exhibited by *B. licheniformis* and *Enterococcus* species (Oguntoyinbo et al., 2007). The fatty acids identified in both unfermented and fermented seeds were linoleic, oleic, stearic, and palmitic acids. Oleic acid was present in large proportion as free fatty acid, and linoleic acid in appreciable amounts in the fermented sample (Sanni et al., 1993; Odibo et al., 2008).

2.1.2.4.3 Nutritional Value The nutritional composition of okpehe is shown in Table 2.15. It was reported to be comparable to fermented products of *Parkia biglobosa* (Barminas et al., 1998; Odibo et al., 2008; Gberikon et al., 2010). During fermentation, the protease activity and amino acid increased in the fermented beans with a corresponding decrease in total nitrogen. Amino acid has been shown to fluctuate during fermentation, but at the end of 72-h fermentation, higher

Table 2.15 Nutritional Composition of Unfermented and Fermented Mesquite Seeds

PARAMETER	UNFERMENTED COTYLEDON	OKPEHE
Proximate (g/100 g)		
Crude protein	20.54	38.70
Fat	12.93	11.96
Soluble carbohydrate	nd	0.15
Crude fiber	6.51	0.51
Ash	5.67	2.30
Mineral (mg/100 g)		
Calcium	24.1	58.0
Potassium	144	225.8
Magnesium	60.2	63.0
Manganese	14.4	25.9
Iron	16.8	58.8
Zinc	4.6	29.0
Copper	1.5	1.8
Sodium	36.0	42.0
Vitamin (mg/100 g)		
Vitamin A	0.89	nd
Vitamin C	0.92	nd

Source: Barminas, J.T. et al., *Plant Foods for Human Nutrition*, 52, 325–328, 1998; Odibo, F.J.C. et al., *Journal of Industrial Microbiology and Biotechnology*, 35, 947–952, 2008; Gberikon, G.M. et al., *Continental Journal of Food Science and Technology*, 4, 60–64, 2010.

Note: nd, No data.

Table 2.16 Amino Acid Profile of Unfermented and Fermented Mesquite Seeds

AMINO ACID (g/100 g PROTEIN)	UNFERMENTED COTYLEDON	OKPEHE
Lysine	6.60	13.43
Histidine	2.30	2.55
Arginine	12.15	11.33
Asparagine	6.53	10.04
Threonine	2.57	3.13
Serine	3.82	3.22
Glutamine	19.25	19.62
Proline	5.20	5.04
Glycine	3.13	3.44
Alanine	2.53	3.68
Cystine	13.60	20.23
Valine	5.45	4.15
Methionine	3.93	5.98
Isoleucine	3.01	4.04
Leucine	6.89	7.15
Tyrosine	1.46	2.01
Phenylalanine	2.07	1.45

Source: Odibo, F.J.C. et al., *Journal of Industrial Microbiology and Biotechnology*, 35, 947–952, 2008.

concentration of glutamine followed by aspartic acid, cysteine, and lysine in decreasing order was observed (Table 2.16) (Odibo et al., 2008). Flavor characteristics of condiments are mainly due to accumulation of glutamic acids. The mineral composition after fermentation was observed to include calcium, phosphorus, potassium, manganese, and zinc. Increase in calcium and iron was consistently observed in samples studied by Balogun and Oyeyiola (2012). However, the vitamin and anti-nutritional composition of okpehe is poorly reported in literature. It could be assumed that increased availability of phosphorus in the fermented product must have been due to hydrolysis of phytic acid in the cotyledon.

2.1.2.5 Otiru

Yemisi Adefunke Jeff-Agboola

The three major raw ingredients for otiru production are the seeds of African yam bean (AYB) plant (*Sphenostylis stenocarpa* (A.Rich.) Harms), water, and naturally occurring bacterial starter culture. AYB plant is grown in West Africa, particularly in Cameroon, Côte d'Ivoire, Ghana, Nigeria, and Togo. In Nigeria it is found in localized areas of the Volta region, where peasant farmers grow it as a security crop for both its edible seeds and its tubers. Indeed, there AYB is preferred to other legume seeds (Njoku et al., 1989). This leguminous plant is an annual vigorous vine, which twines and climbs to a height of about 3 m and requires staking. It flowers profusely in 100 to 150 days, producing brightly colored flowers, which may be pink, purple, or greenish white. The slightly woody pods contain 20 to 30 seeds, are up to 30 cm long and mature within 170 days. The plant produces underground tubers that are used as food in some parts of Africa and that serve as organs of presentation in the wild (Porter, 1992; Ogbonna et al., 2001).

2.1.2.5.1 Traditional Process of Otiru Production The raw AYB (Figure 2.52a) must first be hand-sorted and cleaned, and then washed, soaked in water for 12 h, dehulled, washed, and boiled for about 1 h until soft. The cooked beans are wrapped in plantain leaves and left to ferment at ambient temperature for 5 days (Figure 2.53). However, there are differences in the procedures, and equipment employed in different areas and communities (Jeff-Agboola, 2007).

The formation of otiru is indicated by the appearance of a grayish brown color (Figure 2.52b), a mucilaginous substance, and a strong ammoniacal odor. Otiru is regarded as spoiled when it liberates a putrid or rancid smell or appears contaminated with mold or yellow-pigmented slimy material on the beans if stored under ambient conditions. Hence, to extend shelf life, the product is finely ground, added with salt, molded into balls or flat discs, and sun-dried, when it can be stored for several months.

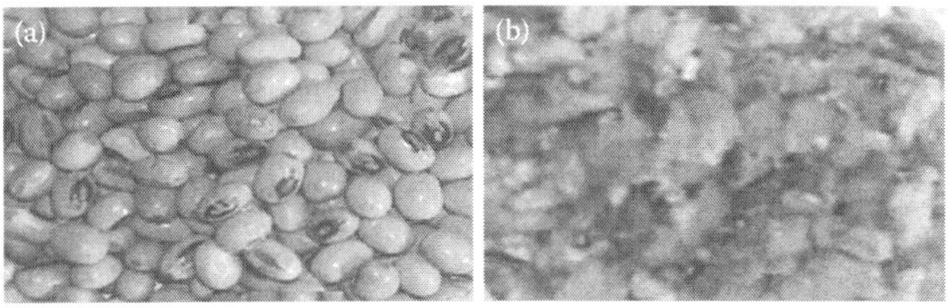

Figure 2.52 Raw seeds of (a) yeam bean tree and (b) otiru. (Modified from Jeff-Agboola, Y.A., *Research Journal of Microbiology*, 2, 816–823, 2007.)

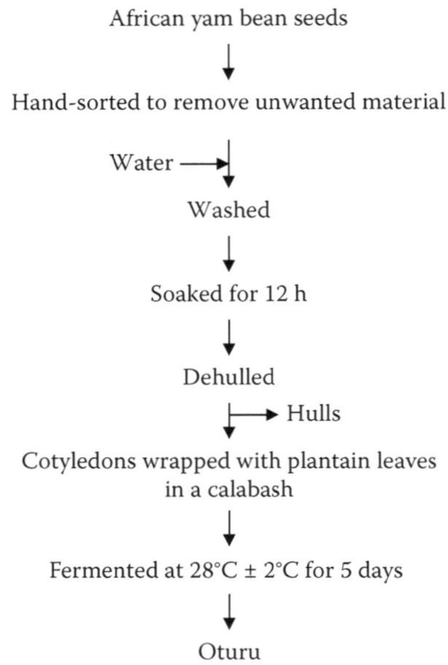

Figure 2.53 Processing of African yam bean seeds into otiru.

2.1.2.5.2 Microorganisms Traditionally fermented AYB were found to contain bacteria and yeasts, while raw beans contained bacteria, yeasts, and a mold, and boiled seeds contained bacteria only (Jeff-Agboola, 2007). The various types of bacteria and yeasts isolated from the fermenting seeds were *Lactobacillus jensenii*, *Bacillus coagulans*, *Aerococcus viridans*, *Pediococcus pentosaceus*, *Saccharomyces cerevisiae*, and *Candida vini*. *Aspergillus niger* was the only mold isolated along with bacteria (*L. jensenii*, *B. coagulans*, and *A. viridans*) *and* yeast (*C. vini*) from raw seeds. Seeds boiled for 1–2 h were found to contain *B. coagulans* only at a load of 2×10^2 cfu/g.

Of the four species of bacteria isolated from fermenting seeds, *L. jensenii* and *P. pentosaceus* were predominant (Table 2.17). They occurred even until 120 h of fermentation. While *S. cerevisiae* was present until the end of fermentation, *C. vini* could not be detected from the final stage of fermentation. In all the cases, there was a decrease in the cell count after 72 h of fermentation.

2.1.2.5.3 Changes in Proximate Composition During the production of otiru, the proximate composition of AYB seeds changed significantly (Table 2.18). After soaking, there was a 5 times increase in the

Table 2.17 Changes in the Microbial Load during Fermentation of African Yam Bean Seeds to Produce Otiru

	LOG cfu/g AT FERMENTATION PERIOD (h)					
MICROORGANISM	0	24	48	72	96	120
BACTERIA						
Lactobacillus jensenii	2.30	3.30	3.49	3.54	3.32	3.04
Bacillus coagulans	2.48	3.30	3.56	3.54	ND	ND
Aerococcus viridans	2.00	3.30	3.65	3.68	ND	ND
Pediococcus pentosaceus	ND	3.00	3.48	3.51	3.32	3.04
YEASTS						
Saccharomyces cerevisiae	ND	3.60	3.69	3.80	3.48	2.48
Candida vini	2.30	3.30	3.30	3.30	ND	ND

Note: ND, not detected.

Table 2.18 Changes in Proximate Composition during the Production of Otiru from African Yam Bean Seeds

PARAMETER	RAW SEEDS	SOAKED AND DEHULLED SEEDS	OTIRU
Moisture (g/100 g)	11.03 ± 0.06	50.05 ± 0.02	32.01 ± 0.50
Crude protein (g/100 g)	22.17 ± 0.24	22.12 ± 0.02	44.22 ± 0.07
Lipid (g/100 g)	4.17 ± 0.03	4.14 ± 0.03	0.99 ± 0.03
Carbohydrates (g/100 g)	54.53 ± 1.22	15.58 ± 0.09	13.85 ± 0.40
Crude fiber (g/100 g)	5.06 ± 0.04	4.10 ± 0.02	1.07 ± 0.01
Ash (g/100 g)	4.14 ± 0.06	4.02 ± 0	7.86 ± 0.01
pH	6.06 ± 0	6.47 ± 0.02	6.55 ± 0
Titratable acidity (mg/100 g)	0.28 ± 0.01	0.26 ± 0.01	0.35 ± 0.01

Note: Values are mean ± SD of three determinations.

moisture content of seeds; however, during fermentation there was 36% decrease in the moisture content. The crude protein content of the raw seeds increased 100% after fermentation (Jeff-Agboola, 2007). The result obtained on the crude protein of raw AYB is contrary to the observation of Njoku et al. (1989) (15.31g/100 g) but similar to that of Aletor and Aladetimi (1989) (21.91 g/100 g). The percent increase in crude protein content of AYB during fermentation to produce otiru compared favorably with the results of Njoku et al. (1989). On the other hand, with the increase in the period of fermentation there was reduction in the lipid, carbohydrates, and crude fiber contents of the sample (Jeff-Agboola, 2007). The reduction in the crude lipid content of the boiled seeds may be due to breaking down and leaching of the fat content and other metabolic activities taking place during the fermentation (Kylen and McCready, 1975). Reduction in the fiber content of dehulled, boiled, and fermented seeds may be due to the dissolution effect on the fiber as well as enzymatic degradation of the fibrous material during fermentation (Ikenebomeh et al., 1986). The pH of naturally fermented samples increased with the increase in fermentation time. Ogbonna et al. (2001) recorded the pH of 5-day fermented AYB as 7.52. There was 35% increase in the content of titratable acidity during fermentation (Jeff-Agboola, 2007).

2.1.2.6 Oso

Olusola Bandele Oyewole and Adewale Olusegun Obadina

Oso is a fermented product of the seeds of *Cathormion altissimum* Hutch. (family Leguminosae). Oso is used as a condiment by incorporating it into soups and foods or eaten as a main course meal by the people of Yewa (Egbado) in Ogun State and those in parts of the Old Oyo province in southwestern Nigeria. The production of oso is similar to most other flavoring agents used in soups in Nigeria. It is usually prepared by the local communities using rudimentary utensils and techniques. However, the food is safe enough for consumption as there is no report of food poisoning or toxicity resulting from consumption of oso (Popoola et al., 2005).

2.1.2.6.1 Traditional Method of Production Traditionally, oso is produced (Figure 2.54) by washing a particular quantity of the *C.*

Cathormion altissimum seeds
↓
Washed with potable water
↓
Boiled for 4–5 h
↓
Cooled
↓
Fermented for 4 days
↓
Oso

Figure 2.54 Traditional process of oso production. (From Popoola, T.O.S. et al., *Tropical Science*, 44, 187–189, 2004.)

Figure 2.55 Oso in a calabash lined with banana leaves.

altissimum seeds several times with potable water and boiled in excess water for 4–5 h so as to soften the hard seed. Thereafter, the boiled seeds are allowed to cool to room temperature. These seeds are then transferred into a clean calabash lined with washed banana (*Musa x paradisiaca* L.) leaves (Figure 2.55) and left to ferment for at least 4 days.

2.1.2.6.2 Chemical and Microbial Succession and Composition During the fermentation period for the production of oso, the dominant microorganisms have been identified to be bacterial species (Popoola et al., 2004, 2005, 2006). *Bacillus subtilis* and *Bacillus licheniformis* are dominant and persistent throughout the period of fermentation. The population of *Bacillus* spp. increased steadily until the fourth day of fermentation. *Leuconostoc mesenteroides* is also at the maximum on the third day of fermentation but decreased in number on the fourth day. *Staphylococcus* sp. was also present at the onset of the fermentation although low initially, but later increased steadily and dominated the fermentation.

The temperature and pH of the medium increased steadily throughout the fermentation period, which, as a result of microbial activities, particularly the breakdown of amino acids and vitamins, provides the optimum condition for proliferation of the microorganisms (Popoola et al., 2006). The pH of the medium started as slightly above neutral and stayed above neutral throughout the experiment, to a maximum of 8.5 on the second day of fermentation (Popoola et al., 2005). This may be attributed to proteolytic activities (Ikenebomeh et al., 1998), hence the characteristic ammonia smell that is associated with vegetable protein fermentation.

It has been discovered that bacterial fermentation has varying effects on the chemical composition of the seeds. After fermentation, oso was found to contain a lower amount of crude protein, carbohydrates, and crude fiber but an unchanged ($p < 0.05$) content of lipid as compared with substrate (Table 2.19). The fermentation of the seeds brought about decrease in the values of protein, fiber, and carbohydrate components. Reduction in the value recorded for carbohydrate is probably due to partial bioconversion of the substrate by the microorganisms involved in the fermentation process, as they derive their energy from carbohydrate metabolism. The long period of boiling also can denature some of the protein content of the seeds (Popoola et al., 2005). During soaking and blanching, about 1%–3% of the protein in some cereals and legumes is lost (Anon., 1973). Popoola et al. (2005) recorded a significant increase in the content of zinc, phosphorus, and calcium, while the values for copper and iron decreased slightly (Table 2.19). The substantial amount of phosphorus and calcium in oso is important in diets for healthy living. The fermented seeds of *C. altissimum* contain many essential amino acids, and fermentation

Table 2.19 Changes in Proximate and Mineral Composition after Fermentation of *Cathormion altissimum* Seeds

PARAMETER	UNFERMENTED SEEDS	OSO
Proximate (g/100 g)		
Moisture	24.3	40.3
Crude protein	29.5	25.3
True protein	20.9	17.0
Carbohydrates	15.6	10.0
Crude fiber	10.3	6.4
Lipid	16.5	16.9
Ash	3.8	1.0
Mineral (mg/100 g)		
P	560.0	720.0
Ca	127.0	129.0
Zn	5.8	6.4
Cu	2.6	1.4
Fe	2.0	1.8

Source: Popoola, T.O.S. et al., *Journal of Food Technology*, 3, 149–151, 2005.

Table 2.20 Amino Acid Content of *Cathormion altissimum* Fresh Seeds and Oso

AMINO ACID (g/16 g N)	UNFERMENTED SEEDS	OSO
ESSENTIAL AMINO ACID		
Isoleucine	2.82	2.77
Leucine	5.10	4.89
Lysine	3.78	4.82
Methionine	0.96	0.83
Cystine	1.56	1.44
Tyrosine	2.19	2.83
Phenylalanine	3.13	3.06
Threonine	2.37	2.03
Valine	2.83	3.54
NON-ESSSENTIAL AMINO ACID		
Aspartic acid	9.25	9.90
Glutamic acid	13.25	15.85
Alanine	3.65	3.22
Arginine	4.04	3.80
Glycine	3.12	2.60
Histidine	1.36	1.24
Proline	3.06	2.85
Serine	2.73	2.65

Source: Popoola, T.O.S. et al., *Journal of Food Technology*, 3, 149–151, 2005.

has a positive role in such an increase of amino acids, particularly lysine, leucine, tyrosine, phenylalanine, glutamic acid, and valine (Table 2.20). The proteolytic activities of the bacterial isolates involved in the fermentation improve the digestibility of oso.

Hence, oso is rich in protein, lipid, carbohydrates, fiber, and minerals. It is also a good source of amino acids.

2.2 Non-Legume Products

Mainly in African countries, seeds of non-leguminous plants like castor, watermelon, fluted pumpkin, roselle, baobab, and cotton, and also leaves of cassava have been traditionally used in the production of alkaline-fermented foods. In Indonesia, coconut residue, after extraction of milk or cream, is used for the production of a delicacy. An interesting type of bread is a precious tradition in certain areas of the United States.

2.2.1 Ogiri

Maureen-Theodore Chinwe Ojinnaka

In recent years, the use of fermented food condiments and flavoring agents is becoming popular in the diets of many nations. Apart from the fact that condiments improve sensory properties of foods, they add nutritional values by providing dietary fiber, energy, minerals, and vitamins. Some of them contain antioxidants and nutraceuticals that provide health benefits. Fermented food condiments have constituted a significant proportion of the diet of many people as they give pleasant aroma to soups and sauces in many countries, especially in Africa and India where protein calorie malnutrition is a major problem (Sarkar et al., 1993). They also have a great potential as protein and fatty acid sources, and are good sources of gross energy (Adewusi et al., 1991; Omafuvbe et al., 2004). In Africa, many proteinaceous oily seeds are fermented to produce food condiments (Sanni and Ogbonna, 1991) that serve as basic ingredients for food supplementation and are named according to the substrate or raw materials employed (Odunfa, 1985b).

Ogiri is a traditionally fermented castor oil bean (*Ricinus communis* L.) condiment that is consumed in the eastern part of Nigeria as a protein-rich meat substitute. With respect to sensory attributes this alkaline fermentation product is similar to dawadawa (see Section 2.1.2.1). Ogiri can also be produced from seeds of melon (*Citrullus lanata* (Thunb.) Matsum. & Nakai) (Ogueke and Nwagwu, 2007; David and Aderibigbe, 2010) and fluted pumpkin (*Telfairia occidentalis* Hook.f.) (Omafuvbe and Oyedapo, 2000). Ogiri serves as an inexpensive soup condiment, particularly among the poor rural dwellers. Ogiri is characterized by a very strong pungent odor. Among consumers, there are preferences for ogiri produced from a specific locality. The production process, being a local art, makes the quality of product vary. The fermented products are also stored at ambient temperature (28°C ± 2°C) for varied length of time (days or weeks). The population and types of microorganisms involved during fermentation and storage affect the quality of the product (Falegan, 2011).

According to Ojimelukwe et al. (2011), castor oil seeds are first dehulled, wrapped in plantain leaves and boiled for 6 h. The boiled seeds are then ground into paste using a clean mortar and wrapped into plantain leaves and allowed to ferment for four days (Figure 2.56). Fermentation is usually carried out in a moist solid state involving contact with appropriate microorganism(s) at ambient temperature. The completion of fermentation is indicated by the formation of mucilage and overtones of ammonia produced as a result of breakdown of amino acids during fermentation (Onyenekwe et al., 2012). The fermented product (Figure 2.57) can be salted and dried (Aworh, 2008; Oyewole and Isah, 2012). Ogiri has a characteristic ammoniacal odor and flavor that enhance the taste of traditional soups and sauces. It also contributes protein and essential fatty acid intake in West Africa (Oke and Umoh, 1978). Achinewhu (1983) reported that ogiri has a

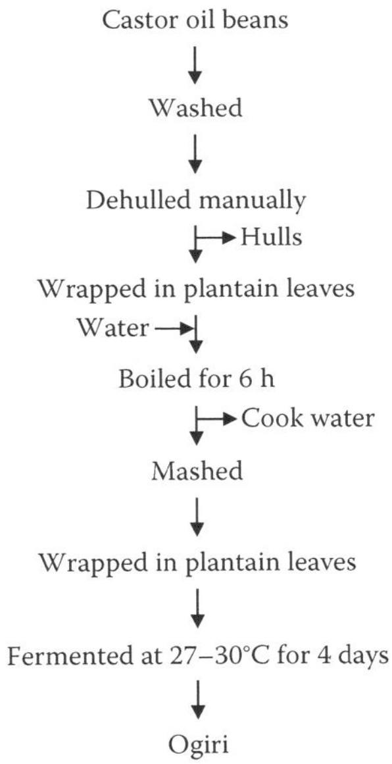

Figure 2.56 Traditional production of ogiri from castor oil bean.

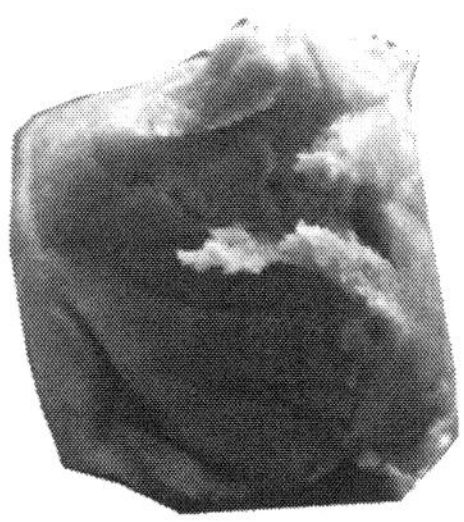

Figure 2.57 Appearance of ogiri.

high percentage of fat, reasonable amounts of maltose, fructose, glucose, and other identified sugars. It has emulsifying and thickening properties when used in preparing soup. The presence of minerals in the form of calcium, magnesium, potassium, and fiber has also been reported (Achinewhu, 1983).

2.2.1.1 Microorganisms Involved in Traditional Fermentation The fermentation of castor oil seeds into ogiri, has been found to be caused mainly by *Bacillus* spp., although some lactic acid bacteria (LAB) have been isolated from the ogiri condiment. Some species of *Bacillus* and LAB, isolated from the traditionally fermented castor oil bean mash and identified using Analytical Profile Index (API) kit software, include *Bacillus subtilis*, *Bacillus circulans*, *Bacillus stearothermophilus*, *Brevibacillus brevis*, *Bacillus megaterium*, *Lactobacillus pentosus*, and *Lactobacillus plantarum* (Ojinnaka et al., 2013b). However, *B. subtilis* was found to be predominant and has been used as a monoculture starter for the production of three different varieties of fermented castor oil bean products. Alkaline fermentations are carried out by *Bacillus* spp., particularly *B. subtilis*, which produce a sticky mucilage on the surface of the beans and strong proteolytic enzymes that hydrolyze the proteins to peptides and amino acids. The species of *Bacillus*, *Pseudomonas*, *Micrococcus*, and *Streptococcus* have also been reported to be isolated from spontaneously fermented castor oil bean (Barber et al., 1988). A key reaction is the release of ammonia, which results in an increase in pH from neutral to pH 8.0 and higher. The relatively high pH in combination with natural toxicity of ammonia inhibits the contaminants (Njoku and Okemadu, 1989; Sarkar and Tamang, 1994). Fermentation of some plant materials is brought about by a consortium of microbes and in most cases initiated by *Bacillus* species with the assistance of LAB (Antai and Ibrahim, 1986). *Leuconostoc mesenteriodes* has been isolated from fermented dawadawa (Uaboi-Egbenni et al., 2009) and they also reported that LAB from dawadawa may play preservative role in the final product from African locust bean fermentation.

Fermentation of castor oil bean into ogiri is usually marked by temperature and pH increase, which is characteristic of alkaline fermentation (Ojinnaka and Ojimelukwe, 2012). An increase in temperature from 27°C up to 30.5°C and an increase in pH values from

6.74 up to 7.15 have been reported during the fermentation of castor oil bean into ogiri using *B. subtilis* as monoculture starter (Ojinnaka and Ojimelukwe, 2012). Thus, castor oil bean fermentationi can be said to be exothermic. Enzyme studies have revealed that α-amylase, proteolytic, and lipolytic enzyme activities attained their maximum level at 24–36 h of fermentation (Njoku and Okemadu, 1989). Thus, it could be the enzymes already produced rather than the presence of microorganisms that continued the fermentation later. The rise in temperature indicates that the fermentation is exothermic with the changes being due to the metabolic activities of the microorganisms (Achi, 1992). The heat generated in the fermenting mash possibly provided the ideal temperature conditions for the optimal activity of the proteolytic enzymes (Odunfa, 1985b).

Ojinnaka and Ojimelukwe (2012) reported favorable microbial growth in 2%-salted ogiri from castor oil seeds, and this is in agreement with Omafuvbe (2006) who also reported that viable cell count was significantly higher in 1%-salted dawadawa, implying that the increase in viable cell count in 1%-salted dawadawa is an indication that low-salt concentration creates a favorable medium for the growth of *B. subtilis*.

2.2.1.2 Nutritional Value Anosike and Egwuatu (1980) stated that fermented castor oil seeds contain a higher level of unsaturated fatty acids than the unfermented seeds, and that these fatty acids produce a characteristic flavor in the fermented product. The castor oil seeds compare with legume seeds in amino acid content. Satar et al. (1979) reported that the tryptophan content of castor oil cake is 3.8% and methionine 1.3%. These values show similar profile with those of legumes. However, unlike legumes, the protein from castor oil seeds showed deficiency in lysine (Polit and Sgarbieri, 1976). Anosike and Egwuatu (1980) observed the presence of some free amino acids like tryptophan, tyrosine, serine, glutamic acid, and cysteine in ogiri, but none of these amino acids were found in the unfermented castor oil seeds. Glutamine, which is the amine of glutamic acid, was found to have contributed to the characteristic flavor of ogiri (Anosike and Egwuatu, 1980). Odunfa (1983) and Sarkar et al. (1993) also reported that during fermentation, the inedible legumes are made edible by extensive hydrolysis of nitrogenous compounds into amino acids.

Thus, the increase in percentage protein could be due to the release of proteins from carbohydrate matrix as well as the generation of free amino acids by proteolytic enzymes. which increase the free measurable nitrogen in the system.

2.2.1.3 Enzymes Involved during Fermentation The fermentation of castor oil seeds for ogiri production is a process where proteolysis and lipolysis as well as degradation of carbohydrates seem to be equally important, each metabolism being predominant at a specific time. For most alkaline fermentations caused by *Bacillus*, proteolysis is the most important metabolism throughout the fermentation (Diawara et al., 1998; Chukwu et al., 2010). Ojinnaka et al. (2013b), in their study of enzymes using high-performance liquid chromatography (HPLC) during fermentation of castor oil seeds into ogiri, reported that proteinase and lipase enzyme levels remained unchanged; the mean proteinase enzyme values ranged from 1.79 to 2.55 μg/ml, while the mean lipase enzyme values were in the range of 3.63 to 5.57 μg/ml. On the other hand, there was significant difference in the amylase activity during the fermentation of castor oil seeds into ogiri. Amylase activity, which is indicative of carbohydrate utilization, is reduced in the presence of 2% NaCl as well as lime, suggesting that carbohydrates are the major nutrients utilized in the absence of these additives (Ojinnaka et al., 2013b).

2.2.1.4 Ammonia and Ricin Contents in Ogiri During the fermentation of castor oil seeds into ogiri, the ammonia content increases as the fermentation period progresses. Ojinnaka and Ojimelukwe (2012) reported a steady increase in the ammonia content of the fermented castor oil seed samples as fermentation time increased. Ogiri containing 2% NaCl has been reported to have the highest value of ammonia content of 0.655 μg/ml (Ojinnaka and Ojimelukwe, 2012). This is in agreement with Omafuvbe (2006) who reported that ammonia concentrations in soy-dawadawa increased significantly during the first 48 h of fermentation, after which the value remained stable. The increase in pH coincides with the period of increased production of ammonia in the fermenting seeds (Visessanguan et al., 2005; Omafuvbe, 2006). Ammonia formed also makes the substrate unsatisfactory for contamination that might spoil the products

(Steinkraus, 1996b; Visessanguan et al., 2005). Omafuvbe (2006) also reported that ammonia concentration was significantly higher in 1%-salted dawadawa and could be a reflection of the enhanced proteolytic activity and release of ammonia following the utilization of amino acids by the increased population of *B. subtilis*. The gradual development of unmistakable ammoniacal odor (Beaumont, 2002) is in agreement with the observed pH changes (alkaline pH ranges), and this agrees with the reported alkaline pH range by Barimalaa et al. (1994). Chukeatirote et al. (2010) also reported increased ammoniacal odor resulting in the distinct feature of thua nao products. They also observed that the increase in pH was due to proteolysis and the release of ammonia following the utilization of amino acids by the fermenting microorganisms.

Fermentation decreases and/or eliminates the toxic components of castor oil seeds, ricin. Toxicological evaluation of ogiri using chicken embryo showed that the toxicity of the seeds, possibly due to toxic proteins, decreased significantly in the fermented product (Odunfa, 1985b). Anosike and Egwuatu (1980) reported that the toxicity of ricin is destroyed when a water solution of the ricin is boiled and that given enough time, the toxicity diminishes with an increase in the extent of denaturation. One of the physical effects of thermal denaturation of protein is that its solubility is decreased. The oil in the product after boiling and fermenting could coat the protein molecules, preventing the hydrophobic group of the protein from interacting with the water thereby resulting in decreased solubility of the protein (Ihekoronye and Ngoddy, 1985).

2.2.1.5 Organic Acid Contents and Volatile Compounds in Fermenting Mash Organic acids have been identified in ogiri using HPLC. Oxalic, citric, tartaric, malic, succinic, lactic, formic, acetic, propionic, and butyric acids were identified in ogiri from castor oil seeds (Ojinnaka et al., 2013a). They reported that there were fluctuations in most of the organic acids present in the samples monitored from 0 to 96 h of fermentation period. The use of 2% NaCl has been reported to suppress the production of organic acids during the fermentation of castor oil seeds into ogiri (Ojinnaka et al., 2013a). Organic acids occur in fermented products as a result of hydrolysis, biochemical metabolism, and microbial activity (Andersson and Hedlund, 1983).

In almost all fermentations involving *Bacillus* species, there is production of primary organic acids, especially butyric acid (Moat, 1979). The production of organic acids is undoubtedly the determining factor on which the shelf life and the safety of the final product depend, while the inhibition of pathogenic and spoilage flora is also dependent on a rapid and adequate formation of these organic acids (Lindgren and Dobrogosz, 1990; Olaoye et al., 2008). The antimicrobial effect of organic acids lies in the reduction of pH as well as the undissociated form of the molecules (Podolak et al., 1996).

The volatile compounds of fermented castor oil seed condiment analyzed using gas chromatography–mass spectrometry (GC-MS) have been reported by Ojinnaka and Ojimelukwe (2013). The volatiles were identified quantitatively by comparing their retention time and mass spectra with the known components stored in the database of National Institute Standard and Technology (NIST) having more than 62,000 patterns. A total of 17 volatile constituents were identified in the fermented castor oil seed condiment samples. Those include acids, esters, alcohol, furan, ketone, and others. Acids were reported as the dominant constituent group and constitute over 50% of total volatiles in the fermented castor oil seed condiment samples. This result is contrary to the findings of Azokpota et al. (2008) who found pyrazines as the dominant constituent group in afitin, iru, and sonru, the three fermented soybean food condiments from Benin and Dajanta et al. (2011) in thua nao, a Thai fermented soy product. Onyenekwe et al. (2012) reported aldehydes as the dominant volatile compound in three fermented food flavor enhancers; melon seed ogiri, soybean dawadawa, and locust bean dawadawa. The use of 3% lime in the fermentation of castor oil seeds into ogiri has led to the production of the highest quantity of esters in the fermented products, while the use of additives (2% NaCl and 3% lime) eliminated alcohols in the fermented product, ogiri. However, the use of 2% NaCl promoted the production of acids in the fermented product, suggesting a superior preservative effect (Ojinnaka and Ojimelukwe, 2013).

2.2.1.6 Conclusion It could be speculated that the subtle taste and aroma of ogiri from castor oil seeds seem to depend not only on particular key components but rather on a critical balance or a weighted concentration ratio of all the components present. However, based on the nutrients and flavoring property of ogiri, the product can serve as a suitable substitute for most of the imported flavoring agents in Nigeria. Some of the biochemical indices examined in the fermentation of castor oil seeds into ogiri can serve as a reference material to food processors. It has also been observed that the addition of 1%–3% NaCl and 1%–3% lime to the fermenting product could be a simple and practical way of extending the shelf life of ogiri from castor oil seeds, add variety with respect to product, and market differentiation as well as improve the aroma of food.

2.2.2 Bikalga/Furundu/Mbuja

Labia Irène Ivette Ouoba

Hibiscus sabdariffa L. (family Malvaceae) is a herbaceous plant found in tropical areas in Africa, Asia, and America. It is called by various names such as karkade, roselle, oseille, guinean sorrel, red sorrel, Jamaica sorrel, Indian sorrel, lemon bush, jelly ocra, or Florida cranberry (Mahadevan and Shivali, 2009; Parkouda et al., 2009). Different parts of the plant are edible: the leaves are used to prepare sauces and soups, the flowers are used for the production of delicious juices, and the seeds are fermented to use as a condiment. The fermentation of the seeds for human consumption is popular mainly in Africa. The end product of the fermentation is known as bikalga in Burkina Faso, dawadawa-botso in Niger, datou in Mali, furundu in Sudan, mbuja in Cameroon, and mungza-ntusa in Nigeria. The technology of production, which varies according to the country and region, is still traditional, uncontrolled, and involves an alkaline/acidic fermentation, with various species of *Bacillus* as the main microorganism responsible for the fermentation (Mohamadou et al., 2008; Ouoba et al., 2008a; Savadogo et al., 2011). In some areas of Burkina Faso, seeds of *H. cannabinus* L., *H. asper* Hook.f., *H. esculentus* L., *Adansonia digitata* L., and *Ceiba pentandra* Gaertn. are sometimes used along with *H. sabdariffa* seeds to produce bikalga. Bikalga and other alkaline-fermented condiments, such as soumbala, are important for the diet of the African population, mainly those of the low-income group. These are an important source of nutrients, including proteins, lipids, sugars, essential amino and fatty acids, minerals, and vitamins, and are commonly used as a meat substitute (Abu-Tarboush et al.,

1997; Ouoba et al., 2003a; Yagoub et al., 2004; Bengaly et al., 2006a; Parkouda et al., 2009).

2.2.2.1 Traditional Technology for Production and Product Utilization Bikalga (Figure 2.58) is mainly produced by women in households. It can be purchased at the production sites or local markets, and constitutes a source of income for the producers. In Burkina Faso, the traditional technology of bikalga production involves cleaning and washing of the crude seeds and cooking (12–24 h) followed by a first step of fermentation (2–3 days) in a closed container (clay pot/jar or metallic vessel). The fermented seeds are then pounded, molded into big balls, and left to ferment a second time under mild steaming for 1–2 days. They are then pounded again, if necessary, remolded into smaller balls, and properly steamed overnight. After steaming, the balls are manually split into small parts or crushed and sun-dried (Figure 2.59) to obtain the final product. In some cases, dried bikalga

Figure 2.58 The appearance of bikalga.

Figure 2.59 Drying of fermented roselle seeds.

is roasted conferring to the final product-specific organoleptic characteristics. During cooking and pounding, a traditional form of potash (locally called "zinm" by the Mossi tribe) is sometimes added. In some localities of Burkina Faso, the production of bikalga involves only one step of fermentation for 3–4 days. In Cameroon, the process involves the following steps (Mohamadou et al., 2009): washing of crude seeds, boiling for more than 3 h, cooling, first fermentation for 7 days in a closed earthen pot, pounding, second fermentation for 3 days, and drying to obtain the final product called mbuja. Unlike bikalga, the mbuja making process does not involve steaming steps and the time for both the fermentation steps is longer. Addition of alkalizing substances seems uncommon in mbuja preparation. For the production of furundu in Sudan, the process involves washing of the crude seeds, cooking, cooling followed by fermentation for 9 days in a closed earthen pot, crushing, and sun-drying to obtain the final product. Unlike bikalga and mbuja, furundu production process involves only one long step (9 days) of fermentation.

Fermented roselle seeds have a dark brown color and a strong ammoniacal smell and taste. The product is used mainly as a condiment to flavor soups, stews, sauces, vegetables, and other foods. In Burkina Faso, the common method is to first soak the fermented seeds in water for a few minutes, then use the steeping water for the preparation of desired dishes.

2.2.2.2 Physicochemical and Nutritional Characteristics Depending on the country of origin and the process, fermented roselle seeds exhibit alkaline, neutral, or acidic pH (4.7–8.5). Parkouda et al. (2008) reported that during the production of bikalga, a first increase in pH was observed after the first fermentation (from pH 6 up to 8). After the second fermentation, the pH rather stayed stable but decreased after the second steaming and drying (down to pH 7.7–6.1). During the production of furundu, a continuous decrease in pH (from 6.1 down to 5.6; 6.3 to 5.8) has been mostly reported (Yagoub et al., 2004; Mohammed and Yagoub, 2007). A slight increase in pH (5.93 up to 6.00) was reported by Mohammed and Yagoub (2007), and this happened when the seeds were sprouted for 24 h before fermentation. For mbuja, both a decrease (pH 6.53 down to 4.73) and an increase (pH 6 up to 8.5) in pH were reported (Mohamadou et al., 2008, 2009).

These variations in pH values observed may be related to variations in processes as well as various physicochemical and microbiological factors. An alkaline fermentation happens, but the results obtained for the various samples suggest that this is associated with an acid fermentation. Parkouda et al. (2009) reported that the fermentation of roselle seeds for bikalga production is a process where proteolysis, lipolysis, and degradation of carbohydrates seem to be equally important; each metabolism becomes predominant at a specific time.

Unfermented and fermented roselle seeds are rich in proteins, lipids, and carbohydrates (Table 2.21). The protein fractions include glutelin, albumin, globulin, and prolamin (Bengaly et al., 2006a; Yagoub and Mohammed, 2008). The main protein fraction is globulin (36%–42%) in bikalga and glutelin (32.2%) in furundu. Investigation on the quality of oil from crude seeds revealed that total lipids include neutral lipids (94.1%), glycolipids (2.6%), and phospholipids (2.1%). Further,

Table 2.21 Chemical Composition of Unfermented and Fermented Roselle Seeds of the Two Countries in Africa

	BURKINA FASO[a]		SUDAN[b]	
PARAMETER	RAW SEED	BIKALGA	RAW SEED	FURUNDU
Moisture (g/100 g)	7.05 ± 1.50	8.24 ± 2.70	ND	ND
Ash (g/100 g)	6.32 ± 1.44	9.03 ± 1.90	5.13 ± 0.12	5.52 ± 0.05
Total proteins (g/100 g)	28.69 ± 1.15	26.47 ± 1.50	28.59 ± 0.01	28.60 ± 0.03
Crude lipid (g/100 g)	21.93 ± 0.74	23.19 ± 1.25	22.15 ± 0.05	24.60 ± 0.06
Total carbohydrates (g/100 g)	26.39 ± 1.03	13.70 ± 0.62	27.79 ± 0.05	22.77 ± 0.42
Total soluble sugars (g/100 g)	5.84 ± 0.69	8.57 ± 0.60	11.07 ± 0.40	5.58 ± 0.47
Crude fiber (g/100 g)	11.22 ± 0.78	12.22 ± 0.76	12.29 ± 0.18	14.48 ± 0.09
Essential amino acids (g/100 g protein)				
Threonine	3.61 ± 0.24	3.38 ± 0.04	3.34	3.42
Valine	6.09 ± 0.06	5.10 ± 0.07	5.83	5.66
Methionine	1.90 ± 0.01	1.86 ± 0	1.11	0.93
Isoleucine	4.51 ± 0	3.70 ± 0.04	4.24	4.13
Leucine	8.59 ± 0.66	7.48 ± 0.04	7.99	7.73
Phenylalanine	5.09 ± 0	5.08 ± 0	5.35	5.19
Lysine	4.57 ± 0.13	4.83 ± 0.04	4.84	4.95
Histidine	1.31 ± 0	1.51 ± 0.01	2.22	3.99
Tryptophan	0.37 ± 0.05	0.48 ± 0.01	ND	ND

Note: Values are mean ± SD; ND, not determined.

[a] Data from Bengaly, M. et al., *Electronic Journal of Food and Plants Chemistry*, 1, 1–6, 2006a; Parkouda, C. et al., *African Journal of Biotechnology*, 7, 916–922, 2008.

[b] Data from Yagoub, A.E.G.A. et al., *Journal of Agriculture and Food Chemistry*, 52, 6143–6150, 2004; Yagoub, A.E.G.A., and Mohammed, M.A., *Pakistan Journal of Nutrition*, 7, 352–358, 2008.

carotenoids (120 mg β-carotene/100 g) and phenols (6.6 mg gallic acid/100 g) are present (Yagoub et al., 2004; Bengaly et al., 2006a; Mahadevan and Shivali, 2009; Zoué et al., 2012). Carbohydrates of the crude and fermented seeds include starch, cellulose, pentosans, and soluble sugars (Yagoub et al., 2004; Bengaly et al., 2006a; Mahadevan and Shivali, 2009). Although there are no changes in the content of total proteins and lipids, a very significant decrease (up to 50%) of total carbohydrates, soluble sugars, and starch has often been observed in the final product (Harper and Collins, 1992; Abu-Tarboush et al., 1997; Yagoub et al., 2004; Bengaly et al., 2006a; Parkouda et al., 2009).

Unfermented and fermented seeds are a great source of amino acids. In addition to the essential amino acids mentioned in Table 2.21, non-essential amino acids, including aspartic acid, serine, glutamic acid, proline, glycine, alanine, arginine, cystine, and tyrosine are also found in the seeds (Yagoub et al., 2004; Bengaly et al., 2006a). Glutamic acid is the most abundant amino acid, and leucine is the predominant essential amino acid in the final product. The fatty acid composition of the fermented seeds has not been documented, but studies on nutritional quality of the oil extracted from the unfermented seeds revealed that the seeds are a great source of unsaturated fatty acids (UFA), with linoleic (39%) and oleic (32%) as the major UFA (Nzikou et al., 2007; Cissé et al., 2009). The major saturated fatty acids are palmitic (22%) and stearic acids (6%). Other fatty acids detected in low quantity in the oil include arachidic, linolenic, myristic, palmitoleic, eicosenoic, eicosatrienoic, and behenic acids. Volatile free fatty acids in the fermented seeds from Sudan (furundu) include isobutyric, *n*-butyric, isovaleric, and *n*-valeric acids (Harper and Collins, 1992). Other acidic components, such as lactic, acetic, and propionic acids are also detected in the product contributing to the decline in pH observed.

Fermented roselle seeds are rich in various minerals, including calcium, magnesium, phosphorus, iron, manganese, potassium, sodium, zinc, sulfur, copper, and aluminium (Harper and Collins, 1992; Bengaly et al., 2006b; Mohamadou et al., 2008; Parkouda et al., 2009). The change in the mineral content during the process is variable, and the increase observed for some minerals, such as potassium, calcium, sodium, aluminium, zinc, and iron has been attributed to addition of alkalinizing substances, such as ash leachate, which contains high levels of potassium and other minerals, and also to the diffusion of the minerals from the container (clay or metallic vessel) used to ferment the seeds. Components such as polyphenols and phytic acid present in the raw seeds are degraded during cooking and fermentation, which contributes to increasing the level of minerals such as iron and zinc in the final product (Bengaly et al., 2006b). The vitamin composition of the fermented seeds has not been documented, but the unfermented seeds have been shown to contain thiamine (vitamin B_1), riboflavin (vitamin B_2), and niacin (vitamin B_3).

2.2.2.3 Microbiology of Fermentation The main microorganisms responsible for the fermentation of roselle seeds are not only *Bacillus* spp. and *Brevibacillus borstelensis* but also lactic acid bacteria (LAB). The *Bacillus* spp., isolated from bikalga, include *B. subtilis*, *B. licheniformis*, *B. amyloliquefaciens*, *B. pumilus*, *B. cereus*, *B. badius*, *B. sphaericus*, *B. fusiformis*, *B. coagulans*, and *B. firmus* (Ouoba et al., 2008a; Savadogo et al., 2011). *Bacillus subtilis* and closely related species, such as *B. amyloliquefaciens*, are the predominant species in the final product. From mbuja, *B. subtilis*, *B. pumilus*, *B. licheniformis*, *B. firmus*, *B. cereus*, *B. polymyxa*, *B. circulans*, *Brevibacillus brevis*, and *Brevibacillus laterosporus* have been isolated (Mohamadou et al., 2007, 2009). *Bacillus subtilis* dominates the early stage of the fermentation, and *B. brevis* and/or *B. laterosporus* the end. In furundu, the presence of only *Bacillus* spp. has been reported, but no clear species has been identified.

The dominance of *Bacillus* spp. during the fermentation of roselle seeds is related to various factors, such as the high protein content of the seeds, the long cooking time, which selects spore-forming bacteria, and the addition of alkalinizing substance during the process. They are associated with the degradation of proteins, lipids, and carbohydrates, with a release of amino acids, fatty acids, and reducing sugars. Further degradation of amino acids leads to the production of alkalinizing substances, such as ammonia, which is responsible for the increase in pH. Protein degradation contributes to the development of the texture, flavor, and organoleptic quality of the product. It also increases the digestibility of the final product compared with the raw material. The enzymatic activity of the *Bacillus* species contributes to the production of various acids such as lactic and acetic acids, which are responsible for the buffering and/or pH decrease observed during the process.

The second most important group of bacteria involved in the fermentation of roselle seeds is LAB. In bikalga, the LAB include *Pediococcus acidilactici*, *Enterococcus faecium*, and *Enterococcus hirae* (Ouoba et al., 2010). *Pediococcus acidilactici* (63%) dominates throughout the fermentation. In mbuja, LAB include *Pediococcus pentosaceus*, *Lactobacillus brevis*, *Leuconostoc mesenteroides* subsp. *dextranicum*, and *Lactobacillus bulgaricus* (Mohamadou et al., 2007, 2009). *Pediococcus pentosaceus* (91%) dominates the early stage of the fermentation, and *Leuc. mesenteroides* the end. The decline in pH observed during the fermentation, which is associated with production of lactic, acetic, and other volatile fatty acids, can be attributed to the metabolic activities of LAB in addition to the amylolytic activity of the *Bacillus* spp. (Harper and Collins, 1992; Parkouda et al., 2009).

In addition to *Bacillus* and LAB, *Staphylococcus* spp., yeasts, and molds have been reported in the fermented seeds (Mohammed and Yagoub, 2007; Yagoub and Mohammed, 2008).

2.2.2.4 Safety Regarding microbial safety of the fermented seeds, Mohamadou et al. (2009) analyzed pathogenic bacteria and quality indicators in the product and reported the absence of total coliforms, spore of sulfite-reducing anaerobe, and coagulase-positive staphylococci. The potential antimicrobial activity of the *Bacillus* spp. and LAB found in the product as well as the long cooking time and low pH observed in some cases probably contribute to the safety of the product eliminating some pathogenic bacteria. However, *B. cereus* and enterococci of possible fecal origin were detected. *Bacillus cereus* has also been isolated from bikalga, and shown to produce heat-labile enterotoxins. Genes encoding production of the heat-labile enterotoxins (NheA, NheB, NheC, HblA, HblC, and HblD) and cytotoxin K (CytK) have been found in the isolates, but no genes encoding production of heat-stable emetic toxins were detected (Ouoba et al., 2008b). Bikalga is usually cooked for long time in stews/sauces before consumption, and it is anticipated that the heat-labile toxins are degraded during cooking. In addition, no report of food poisoning and/or intoxication related to consumption of bikalga has been reported. However, although *B. cereus* is not the dominant species in bikalga and not always present in the product, care should be taken to avoid the presence of this species in the product by selecting non-toxigenic *Bacillus* starter cultures for controlled fermentation.

2.2.2.5 Functional Properties Fermented roselle seeds have been claimed by the local population to cure high blood pressure, diarrhea, rubella, and eye infection. The product is also believed to be diuretic, antiscorbutic, and laxative and to enhance health and strength. It is also used as an antiseptic. These attributes can be related to potential probiotic properties of the *Bacillus* and LAB isolates (Mohamadou et al., 2007) in addition to their potential antimicrobial activity. Oil extracted from the seeds also demonstrated antimicrobial activity against fungi such as *Aspergillus fumigatus* and *Trichophyton mentagrophytes* (Zoué et al., 2012). Mohamadou et al. (2007) reported an increased antioxidant activity in the fermented product compared with the crude seeds. Antioxidants are considered to be important in the prevention of cardiovascular diseases. Ismail et al. (2008) suggested that inclusion of roselle seeds in the diet might potentially prevent risk of atherosclerosis and coronary heart diseases. They also reported that the high content of protein and dietary fiber in the seeds might contribute to hypocholesterolemic and hypoglycemic effects. The dietary fiber may also decrease colon cancer and obesity.

2.2.2.6 Conclusion Fermented roselle seeds constitute a very nutritious condiment, rich in proteins, lipids, carbohydrates, essential amino and fatty acids, minerals, and vitamins, and can be used to fight nutritional deficiencies and to improve the diet of the people. It also exhibits medicinal and functional properties, and consumption of the product is anticipated to enhance health of the people. Due to detection of toxigenic bacteria in the product, a selection of multifunctional starter cultures for controlled fermentation is necessary to improve the safety, nutritional quality, stability, and marketability of the product.

2.2.3 Maari/Dikouanyouri/Tayohounta

Abiodun Isiaka Sanni and Folarin Anthony Oguntoyinbo

Maari is a traditional food condiment produced from the alkaline fermentation of seeds of baobab tree (*Adansonia digitata* L.). The baobab plant, a member of the family Malvaceae, is characteristic of the Sahelian region with a very massive large trunk (up to 10 m diameter) that grows more than 25 m in height and may live for 100 years (Diop et al., 2006; Parkouda et al., 2009). Baobab fruit is ovoid and contains about a hundred seeds, which contains (per 100 g) 21.75 g protein, 5.01 g ash, and 6.71 g fiber (Diop et al., 2006; Nkafamiya et al., 2007).

Maari produced from the baobab seed is found in different regions of West Africa under different names depending on the locality and ethnicity. It is called "dadawa higgi" or "issai" in Nigeria, "dikouanyouri" in Benin republic, "n'gono" in Mali, and "maari," "mackaari," "kaando," "thyou," or "teed bikalgo" in Burkina Faso (Nkafamiya et al., 2007; Parkouda et al., 2009, 2010; Kaboré et al., 2012).

Maari is used as a flavor enhancer in sauces, soups, and other dishes in some African countries (Chadare et al., 2008; Parkouda et al., 2010, 2011). Traditional maari is characterized by a strong pungent aroma, which is the main standard that determines its acceptability and quality by consumers. The flavor has been attributed to various volatile compounds mediated by metabolic activities of microorganisms during the fermentation or processing conditions (Ouoba et al., 2005; Parkouda et al., 2011). Like all other African fermented condiments, traditional processing varies among the tribes and countries, and this affects the quality of the final products (Parkouda et al., 2011).

Figure 2.60 shows the traditional production of maari. The baobab seeds are cleaned and cooked with ash lye solution (traditional potash) (Figure 2.61a) until soft. Then, a two-stage spontaneous fermentation process takes place. First, the cooked seeds are left to ferment for 48–72 h where the pH increases from 6.7 up to 7.3. At the end of the first fermentation step, the seeds are pounded and an alkalizing agent, obtained by leaching ashes with water, is added resulting in an increase in pH up to 9. The second fermentation lasts for 40 h, and the final product having a pH slightly above 7 is molded into small balls, which are steamed cooked and sun-dried for 2–5 days. The resulting product is a dark brown (Figure 2.61b) and pungent smelling condiment (Kaboré et al., 2012).

The fermentation leading to the production of maari is spontaneous and uncontrolled with inherent hygienic, nutritional, and organoleptic defects, resulting in inconsistency of product quality. The dominant

Baobab seeds

↓

Winnowed and wet cleaned

↓

Cooked with ash lye solution

↓

Sieved

↓

Fermented for 48–72 h

↓

Pounded

↓

Molded with ash lye

↓

Fermented again for 40 h

↓

Fresh maari

↓

Molded

↓

Steam cooked

Figure 2.60 Traditional production of maari. (From Kaboré, D. et al., *African Journal of Biotechnology*, 11, 1197–1205, 2012.)

microorganisms were identified as *Bacillus subtilis* strains, which also initiated the fermentation process. However, after 24 h, lactic acid bacteria (LAB), which include *Enterococcus faecium*, *Enterococcus casseliflavus*, and *Pediococcus acidilactici* appeared and remained until the end of fermentation. Other microorganisms implicated in the fermentation included *Bacillus cereus*, *Staphylococcus sciuri*, *Staphylococcus gallinarum*, and *Corynebacterium* sp. (Parkouda et al., 2010; Kaboré et al., 2012).

The changes in chemical composition (Table 2.22) involved a decrease in carbohydrate, which is mainly due to cooking of the seeds

Figure 2.61 Boabab seeds being boiled (a), and dried maari (b). (Courtesy of Charles Parkouda.)

Table 2.22 Chemical Composition of Unfermented Baobab Seeds and Maari

PARAMETER	UNFERMENTED SEEDS	DRIED READY-TO-EAT MAARI SAMPLE		
		ZB2	ZC2	ZV2
Moisture (g/kg)	59 ± 0.5d	134 ± 0.5a	103 ± 0.6c	131 ± 0.2b
Protein (g/kg dw)	156 ± 4.1a	147 ± 0.7b	142 ± 4.1c	134 ± 0.5d
Lipid (g/kg dw)	122 ± 5.3c	181 ± 1.5a	177 ± 3.0a	151 ± 3.8b
Ash (g/kg dw)	45 ± 0.4a	73 ± 3.1b	68 ± 1.3c	88 ± 3.6a
Carbohydrate (g/kg dw)	619	465	511	491
Titratable acidity (% OAE dw)	0.17 ± 0.03d	1.47 ± 0.07b	1.87 ± 0.08a	0.77 ± 0.01c
Tannin (mg/g CE dw)	0.29 ± 0.09a	0.10 ± 0.06b	0.25 ± 0.08a	0.09 ± 0.01b

Source: Parkouda, C. et al., *African Journal of Biotechnology*, 10, 4197–4206, 2011.
Note: dw, dry weight; OAE, oleic acid equivalent; CE, catechin equivalent. Data are expressed as mean ± standard deviation (n = 3). Mean values in a row followed by the same letter are not significantly different at 95% confidence level.

(Parkouda et al., 2011), but the formation of some volatile aroma compounds belonging to ketone and aldehyde groups had earlier been suggested to play a role in the process (Ledauphin et al., 2003). The observed increase in titratable acidity in the fermented product may be due to the accumulation of acid from hydrolytic activities of the amylolytic enzymes as well as the role of microbial metabolism such as that of LAB and *Bacillus* spp. during the fermentation (Schaffner and Beuchat, 1986; Parkouda et al., 2011). Just like other traditional fermented condiments, volatile aroma compounds are important for quality and sensorial attributes of maari. The fermentation leads to an increase in the concentration of volatile compounds (Table 2.23), which is due to the microorganisms involved in the process and processing factors such as boiling and roasting (Azokpota et al., 2010).

To improve and optimize the production of maari, it is necessary to control the fermentation by development of starter cultures, which should be able to inhibit the growth of pathogenic and spoilage microorganisms thereby leading to a product of consistent quality (Ouoba et al., 2008b; Kaboré et al., 2012). The use of baobab seeds for maari production as one of the alternative sources of food despite the presence of anti-nutritional components such as phytate, oxalate, saponin, and tannin (Nkafamiya et al., 2007) is a form of biological ennoblement for the rural consumers.

According to Parkouda et al. (2009), maari is one of the lesser known traditional fermented condiments that have received little

Table 2.23 Volatile Compounds in Unfermented Baobab Seeds and Maari Samples

		RELATIVE AMOUNT OF THE COMPOUND GROUPS (%)						
		MAARI SAMPLE						
AROMA COMPOUNDS	UNFERMENTED COOKED SEEDS	ZV BD	ZV AD	ZC BD	ZC AD	ZB BD	ZB AD	DTA
Acids	4.8	5.5	3.7	6.3	9.0	7.3	1.3	6.6
Alcohols	17.3	24.1	12.0	26.0	17.9	11.9	24.1	3.5
Aldehydes	0.0	0.0	9.3	16.8	5.8	6.0	14.3	17.7
Alkanes	0.7	0.0	0.4	4.2	1.1	0.4	1.2	0.4
Alkenes	0.0	0.0	3.3	1.7	2.5	0.0	3.0	1.9
Esters	16.8	33.1	16.4	20.9	23.6	65.4	9.4	23.9
Ketones	16.5	37.3	31.4	17.4	23.0	5.1	32.3	22.4
Pyrazines	0.0	0.0	0.0	0.0	0.0	0.7	1.6	20.0
Sulfurs	26.0	0.0	17.9	1.3	5.6	1.8	3.6	0.5
Others	17.8	0.0	5.5	5.4	11.5	1.4	9.2	3.2

Source: Parkouda, C. et al., *African Journal of Biotechnology*, 10, 4197–4206, 2011.
Note: BD, before drying; AD, after drying; DTA, commercial sample.

scientific investigation, and therefore, further studies on the product should be carried out.

2.2.4 Ntoba Mbodi

Abiodun Isiaka Sanni and Folarin Anthony Oguntoyinbo

Traditional fermentation processes have been developed in Central Africa to eliminate cyanhydric acid in cassava leaves and roots, which makes them less toxic to humans (Kobawila et al., 2005). One of the products of such fermentation is ntoba mbodi, a popular condiment in the Republic of Congo. It is a product of alkaline semi-solid fermentation of cassava (*Manihot esculenta* Crantz; family Euphorbiaceae) leaves. The leaves (Figure 2.62) are harvested and allowed to wilt for 2–3 days at room temperature. The stalks and petioles are removed, and the laminas cut into small pieces and washed. The washed leaves are distributed into smaller portions and wrapped with pawpaw (*Carica papaya* L.) leaves followed by fermentation in baskets for 2–4 days. The product, ntoba mbodi (Figure 2.63), is used to prepare sauces that are eaten with starch-based foods made from cassava tubers (Kobawila et al., 2005; Ouoba et al., 2010).

Cassava leaves

↓

Sun-dried for 2–3 h

↓

Stalks and petioles removed

↓

Leaves washed and cut

↓

Wrapped in pawpaw leaves

↓

Fermented for 2–4 days at ambient temperature

↓

Ntoba mbodi

Figure 2.62 Traditional preparation of ntoba mbodi. (From Kobawila, S.C. et al., *African Journal of Biotechnology*, 4, 689–696, 2005.)

The softening of the leaves is the main characteristic indicating the end of fermentation. The fermentation process is caused mainly by the degradation of cell wall by enzymes secreted by the associated microorganisms among which are *Bacillus macerans*, *Bacillus subtilis*, *Bacillus pumilus*, *Staphylococcus xylosus*, and *Erwinia* sp. These microorganisms play an important role by their pectinolytic activities, which improve the digestibility of the resulting product (Louembé et al., 2003; Jayani et al., 2005; Mokemiabeka et al., 2011). Lactic acid bacteria (LAB) found to be implicated in the fermentation were identified using genotypic technique. The LAB associated were *Enterococcus faecium*, *Enterococcus hirae*, *Enterococcus casseliflavus*, *Weisella confusa*, *Weisella cibaria*, and *Pediococcus* sp. Although there was a decrease in the population of *W. confusa* during the fermentation that could be linked to

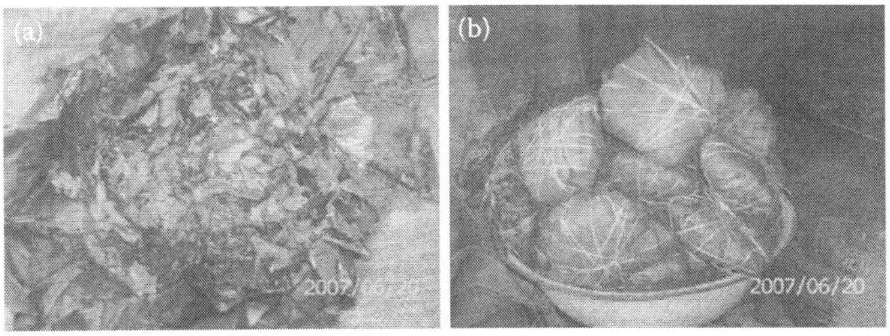

Figure 2.63 Ntoba mbodi: appearance (a) and packed as sold at retail outlets (b). (Courtesy of Simon C. Kobawila.)

the increase in pH (8.5), the organism was still detected at the end of the fermentation. On the other hand, *Enterococcus* and *Pediococcus* spp. known to thrive at pH 9.0 persisted throughout the fermentation period with the relative increase in growth (McHugh et al., 2004). Despite the report of occurrence of LAB in African traditional alkaline fermentations, their role in such fermentation is yet to be clarified (Ouoba et al., 2010; Mokemiabeka et al., 2011).

During the fermentation, the pH at the beginning of the process (6.5) rose sharply to 8.1 within the first 12 h and finally reached pH 8.5–8.96 at the end. Cyanogenic acid was found to reduce significantly by 75% during the process, thereby confirming that fermentation is effective in the reduction of endogenous cyanic compounds from cassava leaves (Agbor-Egbe et al., 1995; Louembé et al., 1997; Kobawila et al., 2005). Alkaline pH favors the decrease of cyanogenic acid because cyanohydrin acetone produced by the hydrolysis of linamarin is cleaved spontaneously when the pH is above 5.0 or by the action of hydroxynitrile lyase to give acetone and cyanhydric acid (Conn, 1969; Cooke et al., 1978; Formunyan et al., 1985; Kobawila et al., 2005). During the fermentation process, amines are produced by *Bacillus* spp., especially *B. pumilus*, which have the capacity to utilize cyanhydric acid for nutrition (Knowles, 1976). This invariably reduces the cyanide content of the fermenting cassava leaves during the fermentation. The decrease in the nitrogen content from 5.1% to 4.5% obtained during the fermentation was suggested to be due to the solubilization and volatilization of hydrogen cyanide (Kobawila

Table 2.24 Mineral Composition in Unfermented Cassava Leaves and Ntoba Mbodi

MINERAL	UNFERMENTED LEAVES	NTOBA MBODI
Calcium	0.980	1.376
Phosphorus	4.270	4.520
Potassium	1.826	2.363
Sodium	0	0.013
Magnesium	0.676	0.913
Nitrogen	5.063	4.446
Ash	9.826	13.290

Source: Mokemiabeka, S. et al., *Advance Journal of Food Science and Technology*, 3, 418–423, 2011.
Note: Data represent g/100g.

et al., 2005) and also linked to the release of some proteins into the water that accompanies softening of the cassava leaves. There was an increase in the ash content (from 9.83 up to 13.29%) and an added advantage of low sodium content (0.013%) (Table 2.24). Ntoba mbodi can therefore be considered a fermented vegetable rich in protein and minerals (Mokemiabeka et al., 2011).

2.2.5 Cabuk

Kris Herawan Timotius

Indonesia is rich in diversity, not only in terms of races and ecosystems, but also in traditional fermented foods or condiments. Cabuk, one of them, is almost neglected by the society, the scientific community, and the government. This fermented food condiment is still found only in certain areas in Central Java. It is not produced widely anymore. Many people, especially the young generation in those areas, are not even aware of cabuk. It is also no longer available in most of the traditional markets. This may be due to its very low economic value, presence of other modern seasoning materials in the market, and lack of care to honor indigenous food culture.

Cabuk is a defatted sesame seed-based fermented food. In Indonesia, many species of sesame are cultivated; however, the most popular one is *Sesamum indicum* L. Sesame seed oil is needed as a raw material for pharmaceutical and cosmetic uses. Sesame oil is well known for its essential unsaturated fatty acids, mainly oleic and linoleic acids content. Its saturated fatty acid content is relatively low, but it is rich in vitamin E (Adriani et al., 1998).

Hence, defatted sesame seeds (sesame cakes) constitute an important by-product of sesame oil industry and are used as a source of protein in cattle feed in Indonesia. It is also used as a substrate of cabuk production, which is an alkaline solid-state fermentation. Scientific information on alkaline fermentation, especially of the traditional fermented foods of Indonesia, is very limited. It is likely that a long time ago the people were very creative in using material that had no economic value to create new useful products. It is not known when the production and use of cabuk began. Cabuk is usually processed with garlic, onion, chili, tomatoes, and salt into a delicious side dish.

Here, the fermentation substrate and process, microbiology, nutritional value, and role of cabuk in the diet will be described. Since scientific publications are limited, visits to local producers were made in order to enrich this article.

2.2.5.1 Substrates of Cabuk Two main substrates for cabuk fermentation are defatted sesame seeds (Indonesian "bungkil minyak wijen") and rice straw ash water (Javanese "londho"). Although sesame seeds are rich in oil (48%), after defatting, the lipid content becomes subdominant (Table 2.25). Sesame seeds contain carotene, tocopherol, antioxidants, minerals, and secondary metabolites (Handajani, 2007).

Rice straw ash water is made from straw ash of rice (*Oryza sativa* L.). Rice straw is first burnt, and the black ash obtained is ground to obtain homogeneous fine particles. If necessary, the ash is sieved. About 100 g of ash is mixed with 250 ml of water, stirred, and left until the ash settles. The clear brown-black supernatant is filtered and ready for use. This rice straw ash water (Indonesian "air abu merang") is also referred to as "ki" water (Indonesian "air ki"). Its pH is 8–9. Alternatively, limestone water (Indonesian "kapur sirih," $Ca(OH)_2$) is used instead of rice straw ash water. This rice straw ash water is well known as a traditional shampoo. It is used as a natural food preservative, such as for noodles, and as a natural black colorant for various traditional cakes. It is probably not only necessary to create

Table 2.25 Proximate Composition (%) of Sesame Seeds and Cabuk

PARAMETER	RAW SEEDS	DEFATTED SEEDS	CABUK
Lipid	48.4	19.6–28.8	25.8–28.3
Protein	17.8	37.1–40.8	36.0–40.0
Carbohydrate	15.5	NR	NR
Crude fiber	8.5	4.6–6.0	10.9–12.0
Ash	1.4	7.8–10.5	11.5–11.8

Source: Handajani, S. et al., *The queen of oil seeds, potensi agribisnis komoditas Wijen,* Andi Offset, Yogyakarta, 2006 (in Indonesian); Handajani, S., *Prospek pengembangan industri berbahan baku wijen,* Presentasi dosen berprestasi, Jakarta, Indonesia, 2007 (in Indonesian); Handajani, S. et al., *Kajian mutu dan* in vivo *virgin sesame oil dan bumbu masak berbahan baku wijen dengan variasi proses produksi. Laporan akhir, Hibah kompetensi angkatan I. Pusat Penelitian dan Pengembangan Pangan Maret,* Gizi dan Kesehatan Masyarakat, Universitas Sebelas, Surakarta, 2009 (in Indonesian).

Note: NR, not reported.

an alkaline environment for cabuk production but also to provide additional minerals, and other compounds that allow a supportive environment for the growth and activity of certain bacteria during cabuk production.

2.2.5.2 Preparation of Cabuk Defatted sesame seeds are steamed and then rice straw ash water is added to make an alkaline environment. Alternatively, the defatted sesame seeds are mixed with ash or lime solution from rice straw and steamed for 30 min. The mixture is cooled, transferred into a bamboo tray or basket, and left to ferment for 36–48 h at room temperature. It is also possible to incubate at a higher temperature (40°C–50°C), and the quality of cabuk is better at a higher temperature; the generation of free fatty acids is higher at higher temperature (Utama, 2010). The process is done under aerobic or microaerobic conditions. The end products are brown to black small balls (diameter, 1–1.5 cm) or clay-like cakes (Javanese "lempung"; Indonesian "tanah liat"). The variety appearance of cabuk is dependent on whether the product is dried or not. A mixture of garlic, chili, and salt is blended with the fermented seeds prior to consumption (Figure 2.64).

Cabuk can be processed to a sauce (like soy sauce). First, cabuk is kept in a closed container for a month to allow optimum fermentation. The fermentation product is then boiled until viscous and filtered to make a delicious sesame sauce. Scientific information on cabuk sauce is limited (Handajani, 2007).

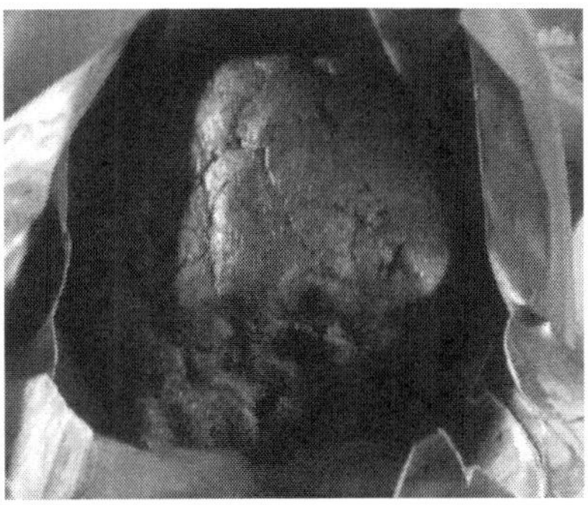

Figure 2.64 Appearance of ready-to-eat cabuk.

2.2.5.3 Microorganisms Involved So far, there is no report on the use of bacterial starter or inoculum for cabuk production. The fermentation is spontaneous. The existence and growth of bacteria in fermenting defatted sesame seeds are determined by several factors created by the characteristics of the substrate and the environment. The steaming treatment prior to the fermentation kills unwanted non-spore-forming bacteria. Since the substrate is rich in protein and lipid, the bacteria should have the ability to produce extracellular enzymes needed for hydrolysis of these macromolecules. The bacteria should also be heat-resistant but tolerant of the alkaline environment created by the addition of rice straw ash water in seeds.

Even while there is no 100% guarantee that cabuk is free of pathogenic microbes, it is assumed that there is no pathogenic potential for creating disease to the consumers. So far, there is no report on the toxicity or pathogenesis of cabuk. Cabuk is assumed as a safe food condiment. The alkaline pH may inhibit the growth of pathogenic and spoilage bacteria. The presence of pathogenic bacteria is limited, but these may still be present since no standard control is applied and most of the producers have never received proper training on the hygienic aspects.

Alkaline environment at least at the beginning of the fermentation elects the growth of certain bacteria, probably members of *Bacillus*, which can grow well in alkaline environments and compete with other bacteria, including spoilage and pathogenic ones. The bacteria grown are probably able to produce antibacterial compounds that suppress the growth of other bacteria. The fermentation microbiota may also produce enzymes, in this case, alkaline protease or lipase, which may produce amino acids and fatty acids that help to decrease the pH marginally during the fermentation.

The predominant bacteria in cabuk are aerobic spore-forming bacteria, notably *Bacillus* species. Many species of *Bacillus* are able to grow in an alkaline environment. Other microbes, like yeasts, *Micrococcus* and *Staphylococcus* are associated with the spoilage rather than with beneficial effects. Like other alkaline fermentations, this fermentation is spontaneously carried out by a mixed bacterial cultures, dominated by *Bacillus subtilis*. *Bacillus* is known to have a strong antimicrobial activity and to inhibit Gram-positive as well as Gram-negative pathogenic bacteria and molds (Deshpande et al., 2000).

2.2.5.4 Changes Accompanying Cabuk Production Protein represents a major macromolecule of defatted sesame seeds, and these are broken down into peptides and amino acids. During fermentation, there is no smell of ammonia. It may be different from other alkaline fermentations that are associated with the production of ammonia so that the pH increases markedly (Parkouda et al., 2009). Glutamic acid is increased during the production of cabuk (Utama, 2010).

After protein, lipid is an important component in defatted sesame seeds. The main free fatty acids are oleic and linoleic acids in sesame seed, defatted sesame seed, and cabuk. The total free fatty acid content is lower in cabuk than in defatted sesame seeds (Ajinugroho, 2010). This means that the consumption of fatty acids is higher than the production of fatty acids by the lipolytic activity of the bacteria during the fermentation.

Carbohydrate is not much reported and may have a less significant role than lipid and protein, at least in terms of the total amount. As the third macrocomponent, carbohydrate, especially simple oligosaccharides or monosaccharides, may have a role for the growth and activity of microorganisms.

It is reported that the antioxidant activity of defatted sesame seeds is increased during cabuk production. This is probably due to the production of antioxidative substances by the bacteria in cabuk. As happens in many traditional fermented foods, some anti-nutritional factors are decreased in cabuk (Ajinugroho, 2010; Utama, 2010).

At the beginning, the pH is around 9 and then it decreases to 7.2. This is due to the production of amino acids and fatty acids during the fermentation as a result of extracellular enzymes produced by the proteolytic and lipolytic bacteria, which then consume the products of hydrolysis for their growth. However, the proteolysis is not so extensive that it would give rise to the formation of ammonia.

2.2.5.5 Prospects of Cabuk The smell and taste of cabuk are very specific, unique, and strong. Cabuk can be introduced as a potential popular and instant or ready-to-eat condiment for modern society. Its role as a food condiment for various recipes should be improved and popularized. It is interesting not only for cooking purpose, but also from the scientific point of view, such as technology for cabuk production with emphasis on hygienic and safety aspects, microbiological aspects, biochemical aspects, enzymatic aspects, bioactive compounds, flavor formation, shelf life, and nutritive value of cabuk.

Since cabuk is usually made from sesame seeds, it is interesting to examine whether a comparable product can be produced using other defatted seeds, like basil, soybean, corn, or nuts after extraction of oil (Indonesian "kelapa sawit"). Future research on cabuk is challenging, and cooperative efforts from various institutions are needed.

2.2.6 Semayi

Kris Herawan Timotius

Indonesia is a "coconut country." This name is not only due to the fact that Indonesia is the world's largest producer of coconut but also because coconut trees are found in most parts of the country and each part of a coconut tree has its specific local term, especially in Java. Additionally, there are many coconut-based staple dishes and desserts, including semayi. Semayi (or semaji) has been well known for a long time in Java. At present, not many people know about semayi. Now it is found only in limited areas or cities in Java. Semayi is no longer commonly sold in traditional markets.

The objectives of this section are to discuss the technological aspects of semayi production starting from the substrate, the microbiological aspects, the fermentation process and its environmental factors, and the use of the products. There are limited scientific articles that can be searched. This article was made through a simple survey of the producers or old people who have experience in making semayi and the ability to make it even if they have not made semayi for a long time.

2.2.6.1 Ampas Kelapa – The Substrate of Semayi The main substrate of semayi production is the ampas kelapa (waste of white flesh/solid endosperm) of coconut fruit, obtained after extraction of coconut milk (Indonesian "santan") and cream (Figure 2.65). The use of coconut milk is not exclusive to Indonesian cuisine; it is also practiced in India, Thailand, Malaysia, Philippines, and Brazil. Its use is quite extensive in many areas of Indonesia. Besides using for semayi production, the ampas kelapa is used in various non-fermented traditional staples by seasoning and adding spices to the shredded coconut

Figure 2.65 Ampas kelapa.

meat or vegetables. Ampas kelapa can also be cooked and seasoned to make a traditional food called "serundeng," an almost powdery sweet and spicy finely shredded coconut.

Ampas kelapa is rich in various nutrients. The main components of ampas kepala are fat, crude fiber, carbohydrate, protein, and ash (Kuswanto, 1988). In coconut meat, the main two amino acids are glutamic acid and arginine; the main fatty acids are lauric acid, myristic acid, and linoleic acid. Lauric acid is dominant. Polysaccharides are galactomannan and glucomannan. Ampas kelapa has pH around 7, and its water content is around 70%.

2.2.6.2 How to Make Semayi Ampas kelapa is steamed for 15 min, cooled, pressed to squeeze out free water, and then left to ferment in a closed place for 3–4 days at room temperature. Spices, namely aromatic ginger (*Kaempferia galanga* L.; Indonesian "kencur"), onion (Indonesian "bawang merah"), garlic (Indonesian "bawang putih"), chilies, Indonesian bay leaves or Indonesian laurel (*Syzygium polyanthum* (Wight) Walp.; Indonesian "daun salam"), and kaffir lime (*Citrus hystrix* DC.; Indonesian "daun jeruk purut") are finely ground and blended with the fermented ampas kelapa along with over-ripe tempe (Indonesian "tempe busuk"), some coconut sugar (Indonesian "gula merah"), and a small amount of salt. The mixture is wrapped in several layers of banana leaf, steamed for 30 min, and usually roasted until a good smell is detected. Semayi is then ready to be eaten (Figure 2.66). The taste and flavor are unique and stimulate an appetite.

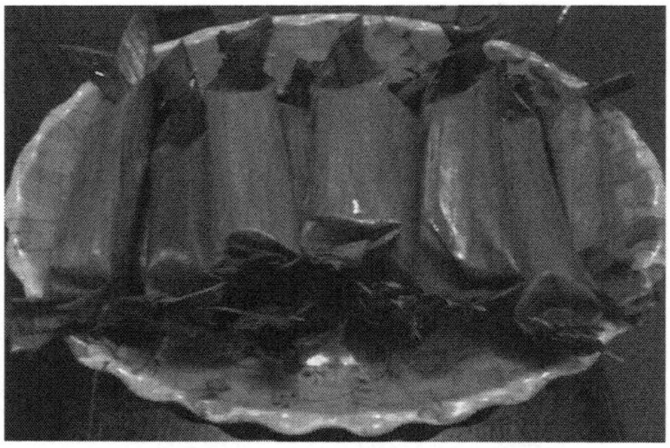

Figure 2.66 Semayi, ready to be eaten.

2.2.6.3 Microbiology of Semayi The fermentation of ampas kelapa to semayi is natural, and no intentional inoculation is made. The steaming treatment selects the existence and the growth of the appropriate bacteria for fermentation. Most of the bacteria are aerobes or facultative anaerobes, motile, and spore-forming. In this case, various *Bacillus* spp., particularly *B. subtilis*, *B. licheniformis*, and *B. polymyxa* were reported (Kuswanto, 1987; Kuswanto et al., 1987) and able to produce extracellular enzymes such as lipase, protease, and mannanase. These enzymes are able to increase the hydrolysis of macromolecules. These bacteria are responsible for producing a degree of sliminess and a unique semayi flavor (Kuswanto, 1988; Steinkraus, 1996b).

Tempe bongkrek, made from bungkil kelapa, can cause a fatal type of food poisoning when *Pseudomonas cocovenanans* grows in the substrate and produces a yellow toxoflavin and/or colorless bongkrek acid (Van Veen, 1967). However, no *Pseudomonas* could be isolated from semayi. Even if any toxigenic strain of *Pseudomonas* exists after fermentation, the subsequent steaming treatment makes semayi a safe food.

2.2.6.4 Changes in Proximate Composition during Fermentation The contents of moisture, fat, protein, carbohydrate, crude fiber, and ash are changed slightly. Soluble solids, free fatty acids, free amino acids, and simple sugars are increased. Interesting is the formation of a unique flavor and taste. Research on semayi, such as analysis of the flavor compounds and bioactive compounds, is warranted.

2.2.7 *Owoh*

Abiodun Isiaka Sanni and Folarin Anthony Oguntoyinbo

Owoh is a traditional fermented soup condiment produced from seeds of cotton plant (*Gossypium hirsutum* L.; family Malvaceae). The condiment serves as a tasty, low cost protein substitute for meat or fish among the people of Bini and Ekiti in the western part of Nigeria. The seed is normally not used in its natural state because it contains some anti-nutritional substances especially gossypol, which inhibits digestion (Sanni and Ogbonna, 1992; Sanni et al., 1998).

The traditional preparation of owoh (Figure 2.67) involves removal of the seeds from the lint followed by boiling in water for about 2 h until soft. The boiled water is discarded, and the seeds (Figure 2.68a) are kept soaked overnight in tepid water. The seeds are dehulled to obtain cotyledons, which are then subjected to washing, allowed to drain, and tightly wrapped with banana leaves (*Musa x paradisiaca* L.) in small packets. The packets are boiled again for 1–2 h, removed from water, and allowed to drain off completely. They are placed in earthen pots or on calabash trays and covered with jute bags. After fermentation

Figure 2.67 Production process of owoh. (Modified from Sanni, A.I. et al., *Advances in Food Science*, 20, 163–167, 1998.)

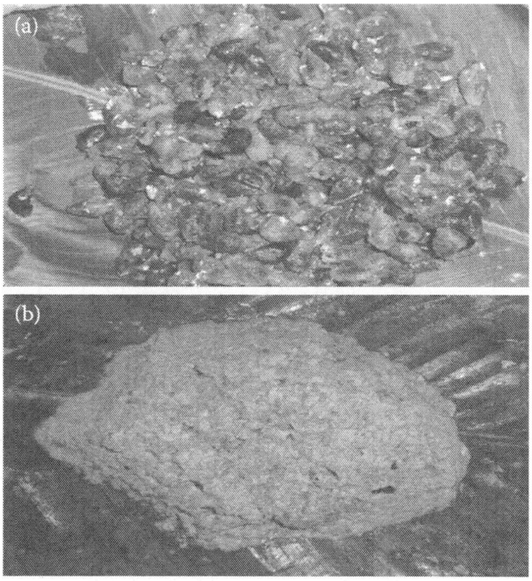

Figure 2.68 Boiled cotton seeds (a) and owoh (b).

for 2–3 days, the resulting mash, called owoh, is a light to dark brown semi-moist solid (Figure 2.68b) with pungent smell. The mash may be ground in a mortar and rolled into balls. For shelf-life extension, the packets may be placed on wire mesh and smoked over charcoal heat for a period of time (Sanni and Ogbonna, 1991; Sanni, 1993b).

Four bacterial species, namely *Bacillus subtilis*, *Bacillus licheniformis*, *Bacillus pumilus*, and *Staphylococcus* sp., were identified during the fermentation process. Yeasts and fungi were, however, not isolated (Sanni and Ogbonna, 1991; Sanni et al., 1998). The bacterial species must have originated from the materials used in the preparation of the condiments such as the banana leaves, water, and spores from the air. The low oxygen tension of the fermenting mash could not favor mold growth (Sanni and Ogbonna, 1992). The dominance of *Bacillus* spp. and occurrence of coagulase-negative *Staphylococcus* sp. during traditional fermentation of vegetable condiments in Nigeria is well documented (Odunfa, 1981; Odunfa and Oyeyiola, 1985; Oguntoyinbo et al., 2007). *Bacillus* spp. are known specifically for their ability to initiate the fermentation of both nitrogenous and carbohydrate products (Sanni et al., 2000).

Increase in pH observed during the fermentation of cotton seeds is due to the production of amines and ammonia by the hydrolytic activity of the associated microorganisms (Table 2.26), while the

Table 2.26 Physical and Microbiological Changes during the Fermentation of Cotton Seeds to Produce Owoh

TIME (h)	TEMPERATURE (°C)	pH	MOISTURE (%)	AEROBIC PLATE COUNT (cfu/g)	ANAEROBIC PLATE COUNT (cfu/g)
0	33	7.8	40.9	4.2×10^6	0.8×10^6
12	37	8.4	42.0	7.5×10^6	1.3×10^7
24	42	8.6	43.6	2.5×10^7	1.7×10^7
36	39	8.9	44.5	2.0×10^7	0.9×10^7
48	34	9.1	46.1	2.1×10^7	0.8×10^7

Source: Sanni, A.I., and Ogbonna, D.N., *Food Microbiology*, 8, 223–229, 1991.
Note: Values represent means of three independent determinations.

decrease in total nitrogen and crude protein (Table 2.27) is due to the loss of ammonia or hydrolysis of the protein in the seeds, which provides excellent sources of nitrogen for microbial growth (Sanni and Ogbonna, 1992). One of the most beneficial effects of this fermentation is the decrease in the gossypol level (a toxic and anti-nutritional component of the seeds). If free gossypol in the cotton seed can be completely eliminated during fermentation, cotton seeds could be a source of dietary protein for poultry and fish in developing countries (Arowosoge, 1987; Sanni and Ogbonna, 1992). Cotton seed is also an excellent source of edible oil and other major nutrients (Oke and Umoh, 1978; Sanni and Ogbonna, 1992).

Like other alkaline-fermented foods in Africa, the production of owoh has remained a traditional art carried out in homes in a crude manner with rudimentary utensils. Longer fermentation period observed in traditional process is due to delay in growth and metabolic activity of the fermenting microorganisms and the competition for nutrients with other opportunistic microorganisms (Sanni and Ogbonna, 1992; Sanni et al., 1998). However, efforts at optimizing some processing conditions for owoh were reported (Sanni, 1993b; Sanni et al., 1998). It is generally agreed that organoleptic quality of traditional fermented soup condiments in Africa is judged by stickiness and pungent ammoniacal smell of the product (Sanni et al., 1998). These are also the attributes for judging similar products such as natto and kinema in Asia (Sarkar et al., 1994). *Bacillus subtilis* EK-4-B1, selected as starter organism for owoh production, was found to elaborate the above attributes by demonstrating better stickiness formation and higher proteolytic activity. Optimal temperature

Table 2.27 Proximate and Mineral Composition of Unfermented and Fermented Cotton Seeds to Produce Owoh

PARAMETER	UNFERMENTED SEEDS	FERMENTED SEEDS (OWOH)
Proximate (g/100 g)		
Total nitrogen	5.08a	4.54a
Crude protein	36.25a	28.57b
Crude fiber	19.40a	14.79b
Lipids	61.08a	42.79b
Ash	8.77a	6.11b
Mineral		
Phosphorus (g/100 g)	0.24a	0.31a
Calcium (g/100 g)	0.29a	1.51a
Magnesium (g/100 g)	0.30a	0.80a
Potassium (g/100 g)	1.50a	0.72a
Sodium (g/100 g)	0.05	0.07
Manganese (μg/g)	47.84b	73.69a
Iron (μg/g)	68.98b	246.76a
Copper (μg/g)	24.39	31.26a
Zinc (μg/g)	68.87b	123.94a

Source: Sanni, A.I., and Ogbonna, D.N., *Food Microbiology*, 9, 177–183, 1992.
Note: Data represent means of three independent determinations. The means followed by the same letters in each row are not significantly different at $p < 0.05$.

and relative humidity, which are important in condiments production because of the direct effect on the microbiota and quality of the product, were determined for the strain. The lengthy traditional fermentation period was also found to be drastically reduced using the starter culture (Sanni et al., 1998). However, the high moisture content of owoh, just like other traditional fermented condiments in Africa, predisposes it to rapid spoilage. The age-long practices of direct heating over a charcoal fire, sun-drying, and salting are adopted to prolong the shelf life and ensure preservation of the product quality.

2.2.8 Salt Rising Bread

Susan Ray Brown and Genevieve Bardwell

Salt rising bread (SRB; Figure 2.69) is a dense, white bread with a fine crumb and a unique, cheese-like flavor (Brown, 2013). Known in the United States since the 1800s, this early American bread is traditionally prepared using a cornmeal–milk–sodium bicarbonate primary starter or a potato–water–salt primary starter. This is followed by an

Figure 2.69 SRB loaves.

adaptive starter with the addition of wheat flour and water, then a dough stage with the addition of wheat flour and salt. SRB's legacy can be traced to women in the Appalachian regions (Bryan, 1839) and to women living on the frontier as they crossed the country in wagon trains (Miles, 1936). Some common ways of eating SRB include a slice with sugared coffee poured over it, a grilled cheese sandwich, and, the most popular preference, as buttered toast.

2.2.8.1 History Contrary to its name, SRB does not contain high levels of salt (nutritional analysis reveals only 20 mg per slice). Rather, the salt in the name of this bread is thought to refer to the early methodology of fermentation. The authors have found two diverging theories that could explain why SRB has held onto its name.

The first theory involves the use of saleratus in the early recipes of SRB (Heinemann and Hefferan, 1909). The use of this chemical as a leaven was first documented in recipes for cookies and biscuits in the late 1700s, in the United States (Simmons, 1958). Saleratus leaches out on the surface of certain types of mineral soils when these soils dry through natural evaporation. It is mainly composed of potassium carbonate. It may be that the use of saleratus in food was knowledge shared with the pioneers by the Native Americans, but it was the pioneer women who first used saleratus in SRB primary starters (Kreidberg, 1975). It was not until the next century that this leavening agent was introduced to Europe and became part of the European cuisine (Simmons, 1958).

The second theory postulates that the name "salt-rising" was derived from the use of a bed of salt kept near hearth fires to coddle the primary starter and thereby hold the heat required for fermentation.

Diaries of pioneer women traveling west in wagon trains describe placing the SRB starter in barrels of salt that were warmed by the sun as they hung against the wagons (Miles, 1936). Once women settled and began cooking on wood burning stoves, they put their starters in the stove's warming ovens, where the temperature was suitable for fermentation. In later years, when cooking gas was available in homes, women found that they could use the heat from the pilot lights on furnaces or hot water tanks to keep their starters warm. Modern bakers of SRB have many options for keeping their starters warm, including the use of crock pots, electric skillets, yogurt makers, and electric hot water baths.

During the late 1800s to the mid-1900s in the United States, SRB's popularity reached its zenith, whereupon many local bakeries routinely made SRB, oftentimes replacing the baking that had been conducted in households. Data gathered in surveys conducted by the authors show that the states of California, Florida, Michigan, New York, Ohio, Pennsylvania, and West Virginia are where SRB is most likely to be found today. These are some of the areas of the United States where people migrated after leaving Appalachia, taking with them the SRB tradition.

In the United States today, only a few individuals and a handful of bakeries make SRB, as the process requires 18 h or longer and strict attendance and observation at specific intervals. Individuals and professional bakers often encounter failure of the primary starter to develop. Such failures historically resulted in lost revenues for bakeries (The Ohio Farmer, 1895; Sun Times, 1909). In the early 1900s, Kohman (1915) addressed the high failure rate of SRB by developing and patenting SRB "yeast." Through the early 1900s, Kohman marketed his SRB "yeast," and by the mid-1900s, virtually all the commercial bakeries that produced SRB across the United States purchased the "yeast" and were baking successful SRB more reliably.

Kohman's SRB "yeast" recipe entails making a successful SRB sponge, dehydrating it, adding appropriate amounts of baking soda and calcium carbonate, and then amplifying the volume with cornmeal (Kohman, 1915). While Kohman's SRB "yeast" gave bakers a higher rate of success with their SRB, failures and losses continued to plague home bakers and bakeries. Thus, by the mid-1900s, bakeries increasingly began to add bread baker's yeast (*Saccharomyces cerevisiae*) to the bread, along with Kohman's SRB "yeast." Although traditional

SRB recipes never incorporate baker's yeast, the adoption of *S. cerevisiae* in SRB recipes became a common practice in bakeries, making failures with SRB far less frequent. The Kohman SRB "yeast" is no longer commercially available. The authors have formulated one similar SRB "yeast" at Rising Creek Bakery, Mt. Morris, PA, which is currently the only known source in the United States.

2.2.8.2 Science SRB utilizes a wild mixture of bacteria containing *Clostridium perfringens* and *Lactobacillus* spp., as leavening agents (Kohman, 1912; Koser, 1923; Clark, 1961). These bacteria seem to originate from the cornmeal or potatoes used in the recipes.

As professional bakers, the authors have been making SRB successfully for years, yet due to the periodic failure of starters to develop, investigations into the science of the bread were undertaken. In 2000, the dry ingredients for a SRB primary starter (flour, cornmeal, and sodium bicarbonate) were sent to RTech Laboratories, in Arden Hills, MN, for confirmation of the microbial content. In their laboratory, scalded milk was added to these ingredients, and this primary starter was kept at 37°C for 22 h. At the end of this period, bubbles and foam began to develop on the surface and a strong cheese-like odor was noted. Next, a sample of this primary starter was analyzed, indicating high levels of *C. perfringens* (1.4×10^8 cfu/g) and a number of *Lactobacillus* species, totaling 5.5×10^8 cfu/g. Counts for coliforms, yeasts, and molds were <10, 10, and 120 cfu/g, respectively.

The preparation of primary starters used in chickpea-fermented Greek bread and lentil-fermented Sudanese bread appears to be somewhat similar to that of SRB (Sherfi and Hamad, 2001; Hatzikamari et al., 2007; Thorsen et al., 2011). Given that *Bacillus* spp. play a key role in the early fermentation of the primary starters of these breads, the authors speculate that the same microbiota may also play a role in the early stage of SRB primary starters. Analysis of SRB primary starters for the presence of *Bacillus* spp. has currently not been carried out to our knowledge.

While the cornmeal recipe for the primary starter typically includes wheat flour, experimentation conducted by another U.S. baker concluded that primary SRB starters as well as the risen bread could be achieved using other grains (Nielsen, 2002). Those grains successfully tested in the primary starter included kamut flour, buckwheat flour, triticale, spelt flakes, wheat flakes, oatmeal, oat bran, wheat bran, wheat gluten, and shredded wheat biscuits.

The authors also collaborated with Bruce McClane, a professor in the Department of Microbiology and Molecular Genetics at the University of Pittsburgh, Pennsylvania, School of Medicine, to test a dozen samples of SRB primary starters and bread. *Clostridium perfringens* was found in high numbers in all of the samples (Juckett et al., 2008). A few colonies were found in the baked bread, indicating that some of the bacteria can survive the high heat during baking, perhaps in a spore stage. Analysis of the *C. perfringens* DNA isolated from the SRB primary starter was conducted using polymerase chain reaction (PCR). All the samples were positive for the *C. perfringens* alpha toxin genes (*C. perfringens* type A) and negative for the beta, epsilon, or iota toxin genes. This confirmed that the bread samples contained *C. perfringens* type A. Additional PCR testing for the *C. perfringens* enterotoxin gene proved negative, indicating that the bread samples contained a non-enteropathogenic strain and were not capable of causing food poisoning (Juckett et al., 2008). It is possible that other bread samples could contain enterotoxin-positive strains of *C. perfringens*, yet consumption of this bread is still unlikely to cause disease. The *C. perfringens* enterotoxin is inactivated at temperatures above 70°C, well below the internal temperatures found in bread baking (87°C and above). Additionally, the infective dose of bacteria is greater than 10^5 cfu/g, and because most bacteria are killed by the heat, dosage levels upon consumption would be much lower. Most significantly, there are no recorded reports of illness from SRB consumption (Dack, 1961; Juckett et al., 2008).

Changes in pH of the primary starter and sponge were measured for the two most common recipes (Table 2.28). The potato primary starter generally has a lower pH than the cornmeal primary starter. The pH of the potato primary starter begins at 6.3–6.9, and after 10–12 h, the pH ranges from 4.9 to 5.7. After the addition of flour to

Table 2.28 Ranges of pH for Various Stages of Primary Starters and Adaptive Starters

	PRIMARY STARTER		ADAPTIVE STARTER	
RECIPE	BEGIN	END	BEGIN	END
Potato	6.3–6.9	4.9–5.7	5.0–6.0	5.3–5.7
Cornmeal	~7.5	6.7–6.8		

the adaptive starter, the pH remained constant at 5.0–6.0. The pH of the initial cornmeal starter was higher, 7.5, then dropped to 6.7–6.8.

2.2.8.3 Baking Methods It is possible that SRB was made by women long before a recipe was ever recorded. Written SRB recipes, however, can be found from as far back as the late 1830s (Bryan, 1839). Although numerous SRB recipes exist, their starter ingredients vary only slightly. Nearly all recipes contain cornmeal, flour, or potatoes, and either salt or sodium bicarbonate.

Two SRB primary starters appear to be the most commonly used. One involves cornmeal, milk, and sodium bicarbonate, while the second involves potatoes, cornmeal, sugar, salt, and water. Both begin by mixing the dry ingredients with heated milk or boiled water, respectively. The primary starter is kept warm for approximately 10–12 h at a temperature ranging 37°C–45°C. When the primary starter is foamy, has risen, and emits a distinctive cheese-like odor (often likened to that of old sneakers or dirty socks), it is used to make an adaptive starter or "sponge" by adding water and flour. Again, it is crucial to keep the adaptive starter at 40°C for a period of 1–2 h. Once this has risen to almost double its original size, it is ready to be made into loaves after mixing in water, wheat flour, and salt. In the authors' experience, it is very important to use the adaptive starter close to the point at which it "drops." This may be the time of the highest levels of *C. perfringens*, the organism responsible for producing the gas that leavens the bread. The loaves are then placed in a warm location until the dough reaches the top of the pan. Finally, the bread is baked at 191°C for approximately 30 mins. Following are two of the most common SRB recipes:

1. Salt Rising Bread: Cornmeal-Milk Starter

   Primary Starter: In a small bowl, combine 1 tablespoon of cornmeal, 1 teaspoon of flour, and 1/8 teaspoon of baking soda. Pour ½ cup of scalded milk over this mixture and stir. Cover the container with a cloth and keep in a warm place overnight until bubbly and foamy (Figure 2.70). In the morning, if the mixture is foamy and has a "stinky" odor, the starter is successful and is ready to be made into the sponge.

   Adaptive Starter or Sponge: Add 2 cups of very warm water and enough flour (1–2 cups) to the foamy mixture to make a

Figure 2.70 SRB primary starter at 0 h (a) and 11 h (b).

slurry having the consistency of pancake batter. Cover and leave this in the same warm place until the starter doubles in volume. This can take 1–2 h.

Dough: Place the risen sponge into a large bowl to make the dough. Add approximately 1 cup of warm water and 1 cup of flour for each loaf of bread you want to make (up to 10 loaves). A teaspoon of salt for each loaf may be added, if desired. Knead just long enough to form the dough. Divide the dough into 1-lb, 12-oz mounds, form into loaves, and place in greased bread pans. Set the loaves in a warm place to rise, making sure that the dough reaches the top edge of the bread pan. A diluted egg wash may be brushed on the tops of the loaves before baking. Bake at 191°C for 25–30 mins. Remove loaves from pans to cool. Yield: 6–10 loaves.

2. Salt Rising Bread: Potato-Water Starter

   Another common SRB starter uses potatoes: Slice 3 medium-sized unpeeled potatoes and place into a medium-sized bowl. Add 3 tablespoons of cornmeal, 2 tablespoons of sugar, and 1 teaspoon of salt. Cover the potatoes with boiling water (Figure 2.71). Proceed as in above recipe. In the morning, remove potatoes from the primary starter and add enough flour (1–2 cups) to make slurry having the consistency of pancake batter, for the adaptive starter. Remaining directions are the same as the above recipe.

Figure 2.71 SRB potato starter at 0 h (a) and 11 h (b).

During the course of baking SRB over many years, two methods for increasing yields of loaves in a timely and reliable manner have been utilized at Rising Creek Bakery. One such method involves reserving two cups of a successful adaptive starter, after it has risen and reached the point when the dough is to be made. With the addition of hot water and flour to this adaptive starter sample, it will rise a second time and support another full batch of loaves. Removing this amount does not significantly reduce the number of loaves that can be made into bread because more water and flour can be added to the dough, proportional to the number of loaves desired.

A second method of increasing the yield of loaves is to take approximately 113 g of adaptive starter or a similar piece of dough to "inoculate" a fresh primary starter. This inoculated primary starter will then be ready in approximately 4–6 h, as opposed to the 10–12 h that is usually required to complete fermentation. Both of these steps can be replicated indefinitely to make as many batches of loaves as time allows.

2.2.8.4 Conclusion SRB is an enigmatic food for which many questions remain unanswered. The succession of microorganisms that populate the primary starter is only partially understood. Most importantly, the reasons for the failures of the primary and adaptive starters are a mystery, and these failures limit the production of this bread on a commercial scale. Nevertheless, it is apparently the rich microbiota that gives the SRB its unique flavor and odor. This flavor is both elusive and highly sought after by generations of Americans. As long as skilled bakers continue to make this bread, others will continue to enjoy and make precious memories of this nearly lost tradition.

References

Abu-Tarboush, H.M., Ahmed, S.A.B., and Al Kahtani, H.A. 1997. Some nutritional and functional properties of karkade (*Hibiscus sabdariffa*) seed products. *Cereal Chemistry* 74: 352–355.

Achi, O.K. 1992. Microorganisms associated with natural fermentation of *Prosopis africana* seeds for the production of okpiye. *Plant Foods for Human Nutrition* 42: 297–304.

Achi, O.K. 2005. Traditional fermented protein condiments in Nigeria. *African Journal of Biotechnology* 4: 1612–1621.

Achi, O.K., and Okereka, E.G. 1999. Proximate composition and functional properties of *Prosopis africana* seed flour. *Journal of Management and Technology* 1: 7–13.

Achinewhu, S.C. 1983. Chemical and nutritional composition of fermented products from plant foods. *Nigerian Food Journal* 1: 115–117.

Aderibigbe, E.Y., and Odunfa, S.A. 1990. Growth and extracellular enzyme production by strains of *Bacillus* species isolated from fermenting African locust bean, iru. *Journal of Applied Bacteriology* 69: 662–671.

Adewusi, S.R.A., Orisadare, B.O., and Oke, O.L. 1991. Studies on weaning diets in Nigeria. 1. Carbohydrate sources. *Cereal Chemistry* 68: 165–169.

Adriani, M., Ramelan, A.H., and Lestari, E. 1998. Teknik pengolahan wijen produk samping dari industry cabuk sebagai minyak pangan dan penerapan teknologinya untuk industry kecil. Laporan Penelitian. Surakarta, Indonesia: Universitas Sebelas Maret, Fakultas Pertanian (in Indonesian).

Agbobatinkpo, P.B., Azokpota, P., Akissoe, N., Kayodé, P., Da Gbadji, R., and Hounhouigan, D.J. 2011. Indigenous perception and characterization of yanyanku and ikpiru: Two functional additives for the fermentation of African locust bean. *Ecology of Food and Nutrition* 50: 101–114.

Agboola, D.A. 2004. *Prosopis africana* (Mimosaceae): Stem, roots, and seeds in the economy of the savanna area of Nigeria. *Economic Botany* 5: S34–S42.

Agbor-Egbe, T., Mbome, I.L., and Trèche, S. 1995. The effectiveness of cyanogens reduction during cassava processing into miondo. In *Transformation Alimentaire du Manioc*, eds. T. Agbor-Egbe, D. Griffon, and S. Trèche, pp. 307–318. France: Orstom.

Ajinugroho, H. 2010. Pengaruh lama fermentasi terhadap karakteristik. Jur. Teknologi Hasil Pertanian (in Indonesian).

Aletor, V.A., and Aladetimi, O.O. 1989. Compositional evaluation of some cowpea varieties and some underutilized edible legumes in Nigeria. *Die Nahrung* 33: 999–1007.

Allagheny, N., Obanu, Z.A., Campbell-Platt, G., and Owens, J.D. 1996. Control of ammonia formation during *Bacillus subtilis* fermentation of legumes. *International Journal of Food Microbiology* 29: 321–333.

Andersson, R., and Hedlund, B. 1983. HPLC analysis of organic acids in lactic acid fermented vegetables. *Zeitschrift für Lebensmittel-Untersuchung und-Forschung* 176: 440–443.

Anon. 1973. *Summary of report of activities and findings: Development and food utilizations of soyabeans.* Agency for International Development (AID) and University of Illinois.

Anosike, E.O., and Egwuatu, C.K. 1980. Biochemical changes during the fermentation of castor oil (*Ricinus communis*) seeds for use as a seasoning agent. *Plant Foods for Human Nutrition* 30: 181–185.

Antai, S.P., and Ibrahim, M.H. 1986. Micro-organisms associated with African locust bean (*Parkia filicoidea* Welw) fermentation for 'dawadawa' production. *Journal of Applied Bacteriology* 61: 145–148.

Arowosoge, I.A. 1987. Nutritional implication of cotton seed meal in diets of *Clarias lazera*. Ph.D. thesis. Ibadan, Nigeria: University of Ibadan.

Ashenafi, M., and Busse, M. 1991. The microflora of soak water during tempeh production from various beans. *Journal of Applied Bacteriology* 70: 334–338.

Aworh, O.C. 2008. The role of traditional food processing technologies in national development: The West African experience. In *Using Food Science and Technology to Improve Nutrition and Promote National Development*, eds. G.L. Robertson, and J.R. Lupien, pp. 1–18. Ontario, Canada: International Union of Food Science & Technology.

Azokpota, P. 2005. Étude de la dynamique physico-chimique et microbiologique de la fermentation des graines de néré pour la production de l'afitin, de l'iru et du sonru. Ph.D. thesis. Cotonou: Université d'Abomey-Calavi; Université Royale Agricole et Vétérinaire de Copenhague au Danemark (in French).

Azokpota, P., Houndenoukon, M.S., Hounhouigan, J.D., Nago, M.C., and Jakobsen, M. 2011. Evaluation of yanyanku processing, an additive used as starter culture to produce fermented food condiments in Benin. *Journal of Food Processing and Preservation* 35: 813–821.

Azokpota, P., Hounhouigan, D.J., and Nago, M.C. 2006a. Microbiological and chemical changes during the fermentation of African locust bean (*Parkia biglobosa*) to produce afitin, iru and sonru, three traditional condiments produced in Benin. *International Journal of Food Microbiology* 107: 304–309.

Azokpota, P., Hounhouigan, D.J., Nago, M.C., and Jakobsen, M. 2006b. Esterase and protease activities of *Bacillus* spp. from afitin, iru and sonru; three African locust bean (*Parkia biglobosa*) condiments from Benin. *African Journal of Biotechnology* 5: 265–272.

Azokpota, P., Hounhouigan, J.D., Annan, N.T., Nago, M.C., and Jakobsen, M. 2008. Diversity of volatile compounds of afitin, iru and sonru, three fermented food condiments from Benin. *World Journal of Microbiology and Biotechnology* 24: 879–885.

Azokpota, P., Hounhouigan, J.D., Annan, N.T., Odjo, T., Nago, M.C., and Jakobsen, M. 2010. Volatile compounds profile and sensory evaluation of Beninese condiments produced by inocula of *Bacillus subtilis*. *Journal of the Science of Food and Agriculture* 90: 438–444.

Azokpota, P., Moller, P.L., Houghouigan, J.D., and Jakobsen, M. 2007. Biodiversity of predominant *Bacillus* isolated from afitin, iru and sonru at different fermentation time. *International Journal of Biological and Chemical Sciences* 1: 211–222.

Baek, L.M., Park, L.Y., Park, K.S., and Lee, S.H. 2008. Effect of starter cultures on the fermentative characteristics of Cheonggukjang. *Korean Journal of Food Science and Technology* 40: 400–405.

Balogun, M.A., and Oyeyiola, G.P. 2012. Changes in the nutrient composition of okpehe during fermentation. *Pakistan Journal of Nutrition* 11: 270–275.

Barber, L., Achinewhu, S.C., and Ibiama, E.A. 1988. The microbiology of ogiri production from castor seed (*Ricinus communis*). *Food Microbiology* 5: 177–183.

Barber, L.A., and Achinewhu, S.C. 1992. Microbiology of ogiri production from melon seeds (*Citrullus vulgaris*). *Nigerian Food Journal* 10: 129–135.

Barimalaa, I.S., Achinewhu, S.C., Yibatama, I., and Amadi, E.N. 1994. Studies on the solid substrate fermentation of bambara groundnut (*Vigna subterranean* (L.) Verdc). *Journal of the Science of Food and Agriculture* 66: 443–446.

Barminas, J.T., Maina, H.M., and Ali, J. 1998. Nutrient content of *Prosopis africana* seeds. *Plant Foods for Human Nutrition* 52: 325–328.

Beaumont, M. 2002. Flavouring composition prepared by fermentation with *Bacillus* spp. *International Journal of Food Microbiology* 75: 189–196.

Bengaly, M., Béré, A., and Traoré, A.S. 2006a. The chemical composition of bikalga, a traditional fermented roselle (*Hibiscus sabdariffa* L.) seeds condiment. Part I: Proximate analysis and protein quality evaluation. *Electronic Journal of Food and Plants Chemistry* 1: 1–6.

Bengaly, M., Béré, A., and Traoré, A.S. 2006b. The chemical composition of bikalga, a traditional fermented roselle (*Hibiscus sabdariffa* L.) seeds condiment. Part II: Evaluation of mineral, total polyphenols and phytic acid content, predicting the iron bioavailability. *Electronic Journal of Food and Plants Chemistry* 1: 7–11.

Brown, S.R. 2013. Susan R. Brown's salt rising bread project. http://home.comcast.net/~petsonk/. Accessed May 14, 2013.

Bryan, L. 1839. *The Kentucky Housewife*. Cincinnati, OH: Shepard and Stearns.

Chadare, F.J., Hounhouigan, J.D., Linnemann, A.R., Nout, M.J.R., and van Boekel, M.A.J.S. 2008. Indigenous knowledge and processing of *Adansonia digitata* L. food products in Benin. *Ecology of Food and Nutrition* 47: 338–362.

Chantawannakul, P., Oncharoen, A., Klanbut, K., Chukeatirote, E., and Lumyong, S. 2002. Characterization of proteases of *Bacillus subtilis* strain 38 isolated from traditionally fermented soybean in Northern Thailand. *Science Asia* 28: 241–245.

Cheigh, H.S., Park, K.S., Moon, G.S., and Park, K.Y. 1990. Antioxidative characteristics of fermented soybean paste and its extracts on the lipid oxidation. *Journal of the Korean Society of Food Science and Nutrition* 19: 163–167.

Chen, J., Cheng, Y.Q., Yamaki, K., and Li, L.T. 2007. Anti-a-glucosidase activity of Chinese traditionally fermented soybean (douchi). *Food Chemistry* 103: 1091–1096.

Cho, K.M., and Seo, W.T. 2007. Bacterial diversity in a Korean traditional soybean fermented foods (Doenjang and Ganjang) by 16S rRNA gene sequence analysis. *Food Science and Biotechnology* 16: 320–324.

Choi, K.S., Lee, H.J., and Kwo, D.J. 2009. Physicochemical and microbiological properties of Korean traditional Meju. *Korean Journal of Food Preservation* 16: 217–222.

Choi, S.Y., Cheigh, M.J., Lee, J.J., Kim, H.J., Hong, S.S., Chung, K.S., and Lee, B.K. 1999. Growth suppression effect of traditional fermented soybean paste (doenjang) on the various tumor cells. *Journal of the Korean Society of Food Science and Nutrition* 28: 458–463.

Choi, Y.H., Choi, B.T., Lee, W.H., Rhee, S.H., and Park, K.Y. 2001. Doenjang hexane fraction-induced G1 arrest is associated with the inhibition of pRB phosphorylation and induction of Cdk inhibitor p21 in human breast carcinoma MCF-7 cells. *Oncology Reports* 8: 1091–1096.

Chukeatirote, E., Chainun, C., Siengsubchart, A., Moukamnerd, C., Chantawannakul, P., Lumyong, S., and Thakang, P. 2006. Microbiological and biochemical changes in Thua nao fermentation. *Research Journal of Microbiology* 1: 38–44.

Chukeatirote, E., Dajanta, K., and Apichartsrangkoon, A. 2010. Thua nao, indigenous Thai fermented soybean: A review. *Journal of Biological Sciences* 10: 581–583.

Chukeatirote, E., and Thakang, P. 2006. Chemical composition of thua nao – A fermented soybean food of Northern Thailand. *Chiang Mai Journal of Science* 33: 243–245.

Chukwu, O., Orhevba, B.A., and Mahmood, B.I. 2010. Influence of hydrothermal treatment on proximate compositions of fermented locust bean (dawadawa). *Journal of Food Technology* 8: 99–101.

Cissé, M., Dornier, M., Sakho, M., N'Diaye, A., Reynes, M., and Sock, O. 2009. Le bissap (*Hibiscus sabdariffa* L.) composition et principales utilisations. *Fruits* 64: 179–193.

Clark, D. 1961. The effect on salt-rising bread of different temperature conditions, loaf sizes, and baking procedures. M.Sc. Thesis, Cornell University.

Conn, E.E. 1969. Cyanogenic glycosides. *Journal of Agricultural and Food Chemistry* 17: 519–526.

Cook, N.R., Kumanyika, S.K., and Cutler, J.A. 1998. Effect of change in sodium excretion on change in blood pressure corrected for measurement error. The Trials of Hypertension Prevention, Phase I. *American Journal of Epidemiology* 148: 431–444.

Cooke, R.D., Blake, G.G., and Battershill, J.M. 1978. Purification of cassava linamarase. *Phytochemistry* 17: 381–383.

Dack, G.M. 1961. Public health significance of flour bacteriology. *Cereal Science Today* 6: 9–10.

Daeschel, M.A., Andersson, R.E., and Fleming, H.P. 1987. Microbial ecology of fermenting plant materials. *FEMS Microbiology Reviews* 46: 357–367.

Dajanta, K. 2010. Production of high nutritional fermented soybean (thua nao) by *Bacillus subtilis*. Ph.D. thesis. Chiang Mai, Thailand: Chiang Mai University.

Dajanta, K., Chukeatirote, E., and Apichartsrangkoon, A. 2011. Analysis and characterisation of amino acid contents of thua nao, a traditionally fermented soybean food of Northern Thailand. *International Food Research Journal* 18: 588–592.

Dajanta, K., Chukeatirote, E., and Apichartsrangkoon, A. 2012. Nutritional and physicochemical qualities of Thua Nao (Thai traditional fermented soybean). *Chiang Mai Journal of Science* 39: 562–574.

Dajanta, K., Chukeatirote, E., Apichartsrangkoon, A., and Frazier, R.A. 2009a. Enhanced aglycone production of fermented soybean products by *Bacillus* species. *Acta Biologica Szegediensis* 53: 93–98.

Dajanta, K., Wongkham, S., Thirach, P., Baophoeng, P., Apichartsrangkoon, A., Santithum, P., and Chukeatirote, E. 2009a. Comparative study of proteolytic activity of protease-producing bacteria isolated from thua nao. *Maejo International Journal of Science and Technology* 3: 269–276.

David, O.M., and Aderibigbe, E.Y. 2010. Microbiology and proximate composition of 'ogiri', a pastry produced from different melon seeds. *New York Science Journal* 3: 18–27.

Deshpande, S.S., Salunkhe, D.K., Oyewole, O.B., Azam-Ali, S., Battcock, M., and Bressani, R. 2000. Fermented grain legumes, seeds and nuts. A global perspective. *FAO Agricultural Services Bulletin* 142.

Diawara, B., Sawadogo, L., Amoa Amwa, W.K.A., and Jakobsen, M. 1998. *HACCP system for traditional African fermented foods: Soumbala*. Taastrup: WAITRO Danish Technological Institute.

Diop, A.G., Sakho, M., Dornier, M., Cisse, M., and Roynes, M. 2006. Le baobab Africain (*Adansonia digitata* L.): Principales caractéristiques et utilisations. *Fruits* 61: 55–69 (in French).

Dirar, H.A. 1984. Kawal, meat substitute from fermented *Cassia obtusifolia* leaves. *Economic Botany* 38: 342–349.

Enujiugha, V.N., and Akanbi, C.T. 2005. Compositional changes in African oil bean (*Pentaclethra macrophylla* Benth) seeds during thermal processing. *Pakistan Journal of Nutrition* 4: 27–31.

Falegan, C. 2011. Microbiology profile and biochemical characteristics of commercial 'ogiri' samples from south-western Nigeria. *Journal of Microbiology, Biotechnology and Food Sciences* 1: 187–203.

Fan, J., Zhang, Y., Chang, X., Saito, M., and Li, Z. 2009. Changes in the radical scavenging activity of bacterial-type douchi, a traditional fermented soybean product, during the primary fermentation process. *Bioscience, Biotechnology, and Biochemistry* 73: 2749–2753.

FAO/WHO (Expert Consultation) 1990. *Protein Quality Evaluation*. Rome: Food and Agriculture Organization/World Health Organization.

FAO/WHO/UNU (Expert Consultation) 1985. *Energy and Protein Requirements*. Geneva: Food and Agriculture Organization/World Health Organization.

Ffolliot, P.F., and Thames, J.L. 1986. *Collection, handling, Storage and Pretreatment of Prosopis seeds in Latin America*. Rome: Food and Agriculture Organization of the United Nations, pp. 2–33.

Formunyan, R.T., Adegbola, A.A., and Oke, O.I. 1985. Technical note: The stability of cyanidrins. *Food Chemistry* 17: 221–225.

Fujita, H., Yamagami, T., and Ohshima, K. 2001. Fermented soybean–derived water-soluble touchi extract inhibits alpha-glucosidase and is antiglycemic in rats and humans after single oral treatments. *Journal of Nutrition* 131: 1211–1213.

Fujita, T., Funako, T., and Hayashi, H. 2004. 8-Hydroxydaidzein, an aldose reductase inhibitor from okara fermented with *Aspergillus* sp. HK-388. *Bioscience, Biotechnology, and Biochemistry* 68: 1588–1590.

Fukushima, D. 1979. Fermented vegetable (soybean) protein and related foods of Japan and China. *Journal of the American Oil Chemists' Society* 56: 357–362.

Gberikon, G.M., Ameh, J.B., Ado, S.A., and Umoh, V.J. 2010. Comparative studies of the nutritional qualities of three fermented African legume seeds using *Bacillus subtilis* and *Bacillus pumilus* as starters. *Continental Journal of Food Science and Technology* 4: 60–64.

Gibson, T., and Gordon, R. 1974. Genus I. *Bacillus* Cohn 1872. In *Bergey's Manual of Determinative Bacteriology*, 8th ed., eds. R. Buchanan, and N. Gibbons, pp. 529–550. Baltimore, MD: Williams & Wilkins.

Grun, I.U., Adhikari, K., Li, C., Li, Y., Lim, B., Zhang, J., and Fernando, L.N. 2001. Changes in the profile of genestein, daidzein, and their conjugates during thermal processing of tofu. *Journal of Agricultural and Food Chemistry* 49: 2839–2843.

Gumbmann, M.R., Dugan, G.M., Spangler, W.L., Baker, E.C., and Rackis, J.J. 1989. Pancreatic response in rats and mice to trypsin inhibitor from soybean and potato after short and long term dietary exposure. *The Journal of Nutrition* 119: 1598–1609.

Gutierrez, M.L., Maizi, P., Nago, C.M., and Hounhouigan, D.J. 2000. *Production et commercialisation de l'Afitin dans la région d'Abomey-Bohicon au Bénin*. Montpellier, France: CERNA, CNEARC, CIRAD: Librairie de CIRAD (in French).

Guvele, C., Deng, L.B., Itto, A., and D'Silva, B. 2003. Food Security Analysis. Proceedings, *The Rumbek Strategy Meeting*, ed. Institute of Development Studies (IDS); University of Sussex. Brighton, UK.

Han, G.H., Cho, T.Y., Yoo, M.S., Kim, C.S., Kim, J.M., Kim, H.A., Kim, M.O., Kim, S.C., Lee, S.A., Ko, Y.S., Kim, S.H., and Kim, D.B. 2007. Biogenic amines formation and content in fermented soybean paste (Cheonggukjang). *Korean Journal of Food Science and Technology* 39: 541–545.

Handajani, S. 2007. *Prospek pengembangan industri berbahan baku wijen*. Jakarta, Indonesia: Presentasi dosen berprestasi (in Indonesian).

Handajani, S., Choirul Anam, J.M., Godras, J.M., and Baskara, K.A. 2009. *Kajian mutu dan* in vivo *virgin sesame oil dan bumbu masak berbahan baku wijen dengan variasi proses produksi. Laporan akhir, Hibah kompetensi angkatan I. Pusat Penelitian dan Pengembangan Pangan Maret*. Surakarta: Gizi dan Kesehatan Masyarakat. Universitas Sebelas (in Indonesian).

Handajani, S., Erlyna, W.R., and Anantanyu, S. 2006. *The Queen of Oil Seeds, Potensi Agribisnis Komoditas Wijen*. Yogyakarta, Indonesia: Andi Offset (in Indonesian).

Hara, T. 1990. Natto no ruhtsu wo motomete (Quest of a root of Natto). *Kagaku to Seibutsu* 28: 676–681 (in Japanese).

Hara, T., Saito, H., Iwamoto, N., and Kaneko, S. 1995. Plasmid analysis in polyglutamate-producing *Bacillus* strain isolated from non-salty fermented soybean food, "kinema," in Nepal. *Journal of General and Applied Microbiology* 41: 3–9.

Harper, D.B., and Collins, M.A. 1992. Leaf and seed fermentations of Western Sudan. In *Applications of Biotechnology in Traditional Fermented Foods*, eds. E.L. Gaden, M. Bokanga, S. Harlander, C.W. Hesseltine, and K.H. Steinkraus, pp. 105–113. Washington, DC: National Academic Press.

Hatzikamari, M., Yiangou, M., Tzanetakis, N., and Litopoulou Tzanetaki, E. 2007. Changes in numbers and kinds of bacteria during a chickpea submerged fermentation used as a leavening agent for bread production. *International Journal of Food Microbiology* 116: 37–43.

Heinemann, P.G., and Hefferan, M. 1909. A study of *Bacillus bulgaricus*. *Journal of Infectious Diseases* 6: 304–318.

Hesseltine, C.W. 1965. A millennium of fungi, food and fermentation. *Mycologia* 57: 149–197.

Hosoi, T., and Kiuchi, K. 2003. Natto – A food made by fermenting cooked soybeans with *Bacillus subtilis* (*natto*). In *Handbook of Fermented Functional Foods*, ed. E.R. Farnworth, pp. 227–245. Boca Raton, FL: CRC Press.

Hu, H., Hao, J., Cheng, Y., Yin, L., Ma, Y., Yu, Z., and Li, L. 2011. Effect of fermented rice culture on the microbiological, biochemical and sensory characteristics of low-salt douchi, a traditional Chinese fermented soybean condiment. *International Journal of Food Science and Technology* 47: 689–695.

Ibnouf, F.O. 2012. The value of women's indigenous knowledge in food processing and preservation for achieving household food security in rural Sudan. *Journal of Food Research* 1: 238–253.

Ihekoronye, A.I., and Ngoddy, P.O. 1985. *Integrated Food Science and Technology for the Tropics*. London: Macmillan Publishers.

Ijabo, O.J., Satirehin, A.A., and Iwe, M.O. 2000. The effect of heat treatments of the cooking time of locust bean. *Nigerian Food Journal* 18: 49–56.

Ikeda, H. 1992. A water soluble vitamin K_2 complex in natto. *Journal of Home Economics of Japan* 43: 643–648 (in Japanese).

Ikeda, H., and Doi, Y. 1990. A vitamin-K_2-binding factor secreted from *Bacillus subtilis*. *European Journal of Biochemistry* 192: 219–224.

Ikenebomeh, M.J., Kok, R., and Ingram, J.M. 1986. Processing and fermentation of the African locust bean (*Parkia folicoidea* Welw.) to produce dawadawa. *Journal of the Science of Food and Agriculture* 37: 273–282.

Ikenebomeh, M.J., Kok, R., and Ingran, J.M. 1998. Proteolytic bacteria in the natural fermentation of African locust bean tree (*Parkia filicoidea* Welw) seeds to produce dawadawa. *Nigerian Journal of Microbiology* 12: 83–87.

Imsande, J. 2003. Sulfur nutrition and legume seed quality. In *Sulfur in Plants*, eds. Y.P. Abrol, and A. Ahmad, pp. 295–304. Dordrecht, The Netherlands: Kluwer Academic Publishers.

Inatsu, Y., Nakamura, N., Yuriko, Y., Fushimi, T., Watanasiritum, L., and Kawamoto, S. 2006. Characterization of *Bacillus subtilis* strains in Thua nao, a traditional fermented soybean food in northern Thailand. *Letters in Applied Microbiology* 43: 237–242.

Ismail, A., Ikra, E.H.K., and Nazri, H.S.M. 2008. Roselle (*Hibiscus sabdariffa* L) seeds-nutritional composition, protein quality and health benefits. *Food* 2: 1–16.

Isu, N.R., and Njoku, H.O. 1997. An evaluation of the microflora associated with fermented African oil bean (*Pentaclethra macrophylla* Bentham) seeds during ugba production. *Plant Foods for Human Nutrition* 51: 145–157.

Isu, N.R., and Ofuya, C.O. 2000. Improvement of the traditional processing and fermentation of African oil bean (*Pentaclethra macrophylla* Bentham) into a food snack – "ugba." *International Journal of Food Microbiology* 59: 235–239.

Japan Natto Cooperative Society Federation 2011. The research report on natto. http://www710orjp/topics/pdf/110913pdf (in Japanese, translated title).

Jayani, R.S., Saxena, S., and Gupta, R. 2005. Microbial pectinolytic enzymes: A review. *Process Biochemistry* 40: 2931–2944.

Jeff-Agboola, Y.A. 2007. Microorganisms associated with natural fermentation of African yam bean (*Sphenostylis sternocarpa* Harms) seeds for the production of Otiru. *Research Journal of Microbiology* 2: 816–823.

Jeong, J.K. 2012. Improvement of quality and probiotic effect of meju and doenjang prepared with mixed starter cultures. Doctoral thesis. Busan, Korea: Pusan National University.

Jeong, W.J., Lee, A.R., Chun, J., Cha, J., Song, Y.S., and J.H., K. 2009. Properties of cheonggukjang fermented with *Bacillus* strains with high fibrinolytic activities. *Journal of Food Science and Nutrition* 14: 252–259.

Juckett, G., Bardwell, G., McClane, B., and Brown, S. 2008. The microbiology of salt rising bread. *West Virginia Medical Journal* 104: 26–27.

Jung, K.O., Park, S.Y., and Park, K.Y. 2006. Longer aging time increases the anticancer and antimetastatic properties of doenjang. *Nutrition* 22: 539–545.

Jung, W.K., Choi, I.W., Hong, G.E., Pyun, C.W., Park, K., Park, P.J., Seo, S.K., Choi, Y.H., and Lee, C.H. 2010. Effects of isoflavone aglycone-rich fermented soybean paste extracts on osteoblastic differentiation of MG-63 cells. *Journal of Korean Society of Applied and Biological Chemistry* 53: 803–809.

Kaboré, D., Sawadogo-Lingani, H., Dicko, M.H., Diawara, B., and Jakobsen, M. 2012. Acid resistance, bile tolerance and antimicrobial properties of dominant lactic acid bacteria isolated from traditional "maari" baobab seeds fermented condiment. *African Journal of Biotechnology* 11: 1197–1205.

Kabuo, N.O., Uzuegbu, J.O., Ubbaonu, C.N., and Onyeka, E.U. 2007. Effect of chemical compounds produced by microorganisms on organoleptic properties of ugba (*Pentaclethra macrophylla* Benth) during fermentation. Proceedings, *Annual Conference of Nigerian Institute of Food Science and Technology*, pp.175–176. Abuja, Nigeria: Oct. 22–25.

Kang, M.J., Kim, S.H., Joo, H.K., Lee, G.S., and Yim, M.H. 2000. Isolation and identification of microorganisms producing the soy protein-hydrolyzing enzyme from traditional mejus. *Journal of the Korean Society of Agricultural Chemistry and Biotechnology* 43: 86–94.

Kanno, A., and Takamatsu, H. 1987. Changes in the volatile components of "natto" during manufacturing and storage (Studies on "natto," Part IV). *Nippon Shokuhin Kogyo Gakkaishi* 34: 330–335.

Kiers, J.L., Van Laeken, A.E.A., Rombouts, F.M., and Nout, M.J.R. 2000. In vitro digestibility of *Bacillus* fermented soya bean. *International Journal of Food Microbiology* 60: 163–169.

Kim, D.H., Lee, K.H., Yook, H.S., Kim, J.H., Shin, M.G., and Byun, M.W. 2000. Quality characteristics of gamma irradiated grain shape improved meju. *Korean Journal of Food Science and Technology* 32: 640–645.

Kim, H.G., Hong, J.H., Song, C.K., Shin, W.H., and Kim, K.O. 2010. Sensory characteristics and consumer acceptability of fermented soybean paste (Deonjang). *Journal of Food Science* 75: S375–S383.

Kim, J.H., Yoo, J.S., Kim, S.Y., and Lee, S.K. 2006. Quality properties of soybean pastes made from meju with mold producing protease isolated from traditional meju. *Journal of the Korean Society for Applied Biological Chemistry* 49: 7–14.

Kim, J.K. 2002. Biological active compounds of soy-fermented foods. *Korea Soybean Digest* 19: 1–7.

Kim, J.K., Chang, H.G., Seo, J.S., and Lee, S.J. 1993. Character impact compounds in flavors of Korean soy sauce manufactured with the traditional and improved meju. *Journal of Microbiology and Biotechnology* 3: 270–276.

Kim, N.Y., Song, E.J., Kwon, D.Y., Kim, H.P., and Heo, M.Y. 2008. Antioxidant and antigenotoxic activities of Korean fermented soybean. *Food and Chemical Toxicology* 46: 1184–1189.

Kim, T.W., Lee, J.H., Kim, S.E., Park, M.H., Chang, H.C., and Kim, H.Y. 2009. Analysis of microbial communities in doenjang, a Korean fermented soybean paste, using nested PCR-denaturing gradient gel electrophoresis. *International Journal of Food Microbiology* 131: 265–271.

Kim, Y.S., Kim, M.C., Kwon, S.W., Kim, S.J., Park, I.C., Ka, J.O., and Weon, H.Y. 2011. Analyses of bacterial communities in Meju, a Korean traditional fermented soybean bricks, by cultivation-based and pyrosequencing methods. *Journal of Microbiology* 49: 340–348.

Kiuchi, K. 2000. Processing suitability of soybeans for natto fermentation and technology of natto fermentation. *New Food Industry* 42: 9–19 (in Japanese, translated title).

Kiuchi, K., and Suzuki, H. 1991. The current state on development of natto containing a high activity of nattokinase. *Food Science* 33: 85–89 (in Japanese, translated title).

Kiuchi, K., Taya, N., Sulistyo, J., and Funane, K. 1987. Isolation and identification of natto bacteria from market-sold natto starters. *Report of National Food Research Institute* 50: 18–21 (in Japanese).

Kiuchi, K., and Watanabe, S. 2004. Industrialization of Japanese natto. In *Industrialization of Indigenous Fermented Foods*, 2nd ed., ed. K.H. Steinkraus, pp. 193–246. New York: Marcel Dekker.

Knowles, C.J. 1976. Microorganisms and cyanide. *Bacteriological Reviews* 40: 652–680.

Ko, B.K., Kim, K.M., Hong, Y.S., and Lee, C.H. 2010. Metabolomic assessment of fermentative capability of soybean starter treated with high pressure. *Journal of Agricultural and Food Chemistry* 58: 8738–8747.

Kobawila, S.C., Louembé, D., Keleke, S., Hounhouigan, D.J., and Gamba, C. 2005. Reduction of the cyanide content during fermentation of cassava roots and leaves to produce bikedi and ntoba mbodi, two food products from Congo. *African Journal of Biotechnology* 4: 689–696.

Kohman, H.A. 1912. Salt-rising bread and some comparisons with bread made with yeast. *Journal of Industrial and Engineering Chemistry* 4: 20–30.

Kohman, H.A. 1915. Ferment and process for producing and preserving the same. U.S. Patent: 1,149,839.

Konlani, S., Lamboni, C., Ameyapoh, Y., and De Souza, C. 1999. Microbiology and biochemical composition of "Tonou", a food condiment from fermented seeds of *Parkia biglobosa* (locust bean) and soybean. *Food Biotechnology* 13: 217–225.

Koser, S.A. 1923. *Bacillus welchii* in bread. *Journal of Infectious Diseases* 32: 208–219.

Kreidberg, M. 1975. *Food on the Frontier: Minnesota Cooking from 1850 to 1900 with Selected Recipes*. St. Paul, MN: Minnesota Historical Society Press.

Krishna, C. 2005. Solid-state fermentation systems-an overview. *Critical Reviews in Biotechnology* 25: 1–30.

Kronenberg, H.J., and Hang, Y.D. 1984. Biochemical changes in Okara during Meitauza fermentation. *Nutrition Reports International* 30: 439–443.

Kubo, Y., Rooney, A.P., Tsukakoshi, Y., Nakagawa, R., Hasegawa, H., and Kimura, K. 2011. Phylogenetic analysis of *Bacillus subtilis* strains applicable to natto (fermented soybean) production. *Applied and Environmental Microbiology* 77: 6463–6469.

Kudo, T. 1990. Warfarin antagonism of natto and increase in serum vitamin K by intake of natto. *Artery* 17: 189–201.

Kuswanto, K.R. 1987. Mikroflora fermentasi semayi. Proceedings, *Simposium Bioproses dalam Industri Pangan*. Yogyakarta, Indonesia (in Indonesian).

Kuswanto, K.R. 1988. Studies on semayi: An Indonesian traditional fermented food. Ph.D. thesis. Tokyo, Japan: Tokyo University of Agriculture.

Kuswanto, K.R., Soedarsono, J., and Kozaki, M. 1987. Microflora of semayi, an Indonesian traditional food. Proceedings, *7th International Congress of Food Science*. Singapore, Sept. 28–Oct. 2.

Kwak, C.S., Park, S.C., and Song, K.Y. 2012. Doenjang, a fermented soybean paste, decreased visceral fat accumulation and adipocyte size in rats fed with high fat diet more effectively than non-fermented soybeans. *Journal of Medicinal Food* 15: 1–9.

Kylen, A.M., and McCready, R.M. 1975. Nutrient in seeds and sprouts of alfalfa, lentils, mung beans and soybeans. *Journal of Food Science* 40: 1008–1009.

Ledauphin, J., Guichard, H., Saint Clair, J.F., Picoche, B., and Barillier, D. 2003. Chemical and sensorial aroma characterization of freshly distilled Calvados. 2. Identification of volatile compounds and key odorants. *Journal of Agricultural and Food Chemistry* 51: 433–442.

Lee, C.H. 1995. *The Mystery of Chongkukjang*. Seoul, Korea: Shinkwang Publishing Co.

Lee, C.H. 2001. *Fermentation Technology in Korea*. Seoul, Korea: Korea University Press.

Lee, C.H. 2004. The role of biotechnology in modern food production. *Journal of Food Science* 69: R92–R95.

Lee, J.H., Kim, T.W., Lee, H., Chang, H.C., and Kim, H.Y. 2010. Determination of microbial diversity in meju, fermented cooked soya beans, using nested PCR-denaturing gradient gel electrophoresis. *Letters in Applied Microbiology* 51: 388–394.

Lee, J.S., and Cheigh, H.S. 1997. Antioxidative characteristics of isolated crude phenolic from soybean fermented foods (doenjang). *Journal of the Korean Society of Food Science and Nutrition* 26: 376–382.

Lee, S.O. 2010. A study on the hazard analysis and HACCP system for Chongkookjang manufacturing. M.Sc. thesis. Asan, Korea: Hoseo University.

Lee, S.W. 1990. Study on the origin and interchange of Dujang (also known as soybean sauce) in ancient Asia. *Korean Journal of Dietary Culture* 5: 313–318 (in Korean).

Leejeerajumnean, A. 2000. *Bacillus* fermentation of soybeans: Characterisation of traditional thua nao manufacture. Ph.D. thesis. Reading, UK: University of Reading.

Leejeerajumnean, A., Ames, J.M., and Owens, J.D. 2000. Effect of ammonia on the growth of *Bacillus* species and some other bacteria. *Letters in Applied Microbiology* 5: 385–389.

Li, F.-J., Yin, L.-J., Cheng, Y.-Q., Yamaki, K., Fan, J.-F., and Li, L.-T. 2009. Comparison of angiotensin-I-converting enzyme inhibitor activities of pre-fermented douchi (a Chinese traditional fermented soybean food) started with various cultures. *International Journal of Food Engineering* 5: 1–10.

Li, Y., Hao, J., Cheng, Y., Zhao, R., Yin, L., and Li, L. 2012. Improvement of okara mouthfeel by *Aspergillus niger* and *Aspergillus oryzae* fermentation. *Nongye Gongcheng Xuebao/Transactions of the Chinese Society of Agricultural Engineering* 28: 248–253.

Liener, I.E. 1994. Implications of anti-nutritional components in soybean foods. *Critical Reviews in Food Science and Nutrition* 34: 31–67.

Lim, S.Y., Park, K.Y., and Rhee, S.H. 1999. Anticancer effect of doenjang in *in vitro* dulforhodamine b (SRB) assay. *Journal of the Korean Society of Food Science and Nutrition* 28: 240–245.

Lindgren, S.E., and Dobrogosz, W.J. 1990. Antagonistic activities of lactic acid bacteria in food and feed fermentations. *FEMS Microbiology Reviews* 87: 149–164.

Liu, Y., Wang, L., Cheng, Y., Saito, M., Yamaki, K., Qiao, Z., and Li, L. 2009. Isoflavone content and anti-acetylcholinesterase activity in commercial douchi (a traditional Chinese salt-fermented soybean food). *Japan Agricultural Research Quarterly* 43: 301–307.

Louembé, D., Kobawila, S.C., Kalou, G.B., and Kéléké, S. 2003. Étude microbiologique des feuilles fermentées de manioc: "Ntoba Mbodi". *Tropicultura* 121: 106–111.

Louembé, D., Malonga, M., Kobawila, S.C., and Mavoungou, O. 1997. Evolution de la teneur en composes cyanés des tubercules de manioc au cours du rouissage – Activité linamarasique de bactéries lactiques. *Microbiologie, Aliments, Nutrition* 15: 53–60.

Ma, Y., and Wang, T. 2010. Deactivation of soybean agglutinin using nonthermal treatments. *Journal of Agricultural and Food Chemistry* 58: 11413–11419.

Mahadevan, N., and Shivali, K.P. 2009. *Hibiscus sabdariffa* Linn: An overview. *Natural Product Radiance* 8: 77–83.

Matsuo, M. 1989a. Morphological and physicochemical properties and composition of "okara" fermented with *Rhizopus oligosporus*. *Journal of the Japanese Society of Nutrition and Food Science* 42: 173–178.

Matsuo, M. 1989b. Morphological, compositional and physicochemical properties of "okara" fermented by *Aspergillus oryzae*. *Journal of the Agricultural Chemical Society of Japan* 63: 1765–1770.

Matsuo, M. 1990. Effects of rice bran on properties of "okara tempe". *Nippon Nogeikagaku Kaishi* 64: 1237–1239.

Matsuo, M. 1996. Acceleration of sterol excretion, little meteorism, and less inhibition of iron absorption by okara koji, a new foodstuff, in rats. *Bioscience, Biotechnology, and Biochemistry* 60: 571–574.

Matsuo, M. 1997a. In vivo antioxidant activity of okara koji, a fermented okara, by *Aspergillus oryzae*. *Bioscience, Biotechnology, and Biochemistry* 61: 1968–1972.

Matsuo, M. 1997b. Preparation and components of okara-ontjom, a traditional Indonesian fermented food. *Journal of the Japanese Society for Food Science and Technology* 44: 632–639.

Matsuo, M., and Hitomi, E. 1993. Suppression of plasma cholesterol elevation by okara-tempe in rats. *Bioscience, Biotechnology, and Biochemistry* 57: 1188–1190.

Mattson, F.H., Volpenhein, R.A., and Erickson, B.A. 1977. Effect of plant sterol esters on the adsorption of dietary cholesterol. *The Journal of Nutrition* 107: 1139–1146.

Mbaiguinam, M., Mahmout, Y., Tarkodjie, M., Delobe, B., and Bessiere, J.M. 2005. Constituents of Kawal, fermented *Cassia obtusifolia* leaves, a traditional food from Chad. *African Journal of Biotechnology* 4: 1080–1083.

Mbajunwa, O.K., Akingbala, J.O., Mulongoy, K., and Oguntimein, G. 1998. Starter culture evaluation for the production of ugba from African oil bean seed (*Pentaclethra macrophylla*). *Journal of the Science of Food and Agriculture* 77: 127–132.

McCue, P., and Shetty, K. 2003. Role of carbohydrate-cleaving enzymes in phenolic antioxidant mobilization from whole soybean fermented with *Rhizopus oligosporus*. *Food Biotechnology* 17: 27–37.

McHugh, C.P., Zhang, P., Michalek, S., and Eleazer, P.D. 2004. pH required to kill *Enterococcus faecalis* in vitro. *Journal of Endocrinology* 30: 218–219.

Miles, Z.W. 1936. Pioneer women of Dixie. Original paper, given by the author at the dedication of the Daughters of the Pioneers Monument in St. George, Utah, September 2nd. Library of Congress, Manuscript Division, WPA Federal Writers' Project Collection. http://www.lofthouse.com/USA/Utah/washington/wpa/women.html.

Mine, Y., Wong, A.H.K., and Jiang, B. 2005. Fibrinolytic enzymes in Asian traditional fermented foods. *Food Research International* 38: 243–250.

Ministry of Education–Culture–Sports–Science and Technology 2010. *Standard Tables of Food Composition in Japan: Food Composition Database.* http://fooddbjp/.

Mitsuboshi, S., and Kiuchi, K. 2007. Development of new itohiki-natto and odor of natto type foods. *Journal of Japan Association on Odor Environment* 38: 151–162 (in Japanese).

Moat, A.G. 1979. *Microbial Physiology*. New York: John Wiley & Sons.

Mohamadou, B.-A., Mbofung, C.M., and Thouvenot, D. 2007. Functional potential of a product from traditional biotechnology: Antioxidant and probiotic potential of Mbuja, produced by fermentation of *Hibiscus sabdariffa* seeds in Cameroon. *Journal of Food Technology* 5: 164–168.

Mohamadou, B.-A., Mbofung, C.M., and Thouvenot, D. 2008. Bacterial fermentation induced mineral dynamics during the production of Mbuja from *Hibiscus sabdariffa* seeds in earthen-ware pots. *Annals of Microbiology* 58: 447–451.

Mohamadou, B.-A., Mbofung, C.M., and Thouvenot, D. 2009. Microbiological and organoleptic profiles of mbuja: A traditional condiment produced by fermentation of *Hibiscus sabdariffa* seeds in Cameroon. *Journal of Food Technology* 7: 84–91.

Mohammed, A.M., and Yagoub, A.E.G.A. 2007. Furundu, from fermented/sprouted roselle (*Hibiscus sabdariffa* L.) seed: Investigation on chemical composition, antinutritional factors, HCl-extractability of minerals, amino acids composition, *in vitro* protein digestibility and microbial growth. *Research Journal of Agriculture and Biological Sciences* 3: 876–885.

Mokemiabeka, S., Dhellot, J., Kobawila, S.C., Diakabana, P., Ntietie Loukombo, R.N., Nyanga-Koumou, A.G., and Louembe, D. 2011. Softening and mineral content of cassava (*Manihot esculenta* Crantz) leaves during the fermentation to produce Ntoba Mbodi. *Advance Journal of Food Science and Technology* 3: 418–423.

Moktan, B., Saha, J., and Sarkar, P.K. 2008. Antioxidant activities of soybean as affected by *Bacillus*-fermentation to kinema. *Food Research International* 41: 586–593.

Mulyowidarso, R.K., Fleet, G.H., and Buckle, K.A. 1989. The microbial ecology of soybean soaking for tempe production. *International Journal of Food Microbiology* 8: 35–46.

Muramatsu, K., Kanai, Y., Kimura, N., Miura, N., Yoshida, K., and Kiuchi, K. 1995. Production of natto with high elastase activity. *Nippon Shokuhin Kagaku Kogaku Kaishi* 42: 575–582 (in Japanese).

Muramatsu, K., Muraoka, Y., Yasui, A., Suzuki, T., and Kiuchi, K. 1997. Study on deterioration of itohiki-natto (formation of struvite). *Nippon Shokuhin Kagaku Kogaku Kaishi* 44: 285–289 (in Japanese).

Muramatsu, K., Yasui, A., Suzuki, T., and Kiuchi, K. 2000. Factors involved in struvite formation by natto bacilli. *Biocontrol Science* 5: 57–60.

Nagai, T., Nishimura, K., Suzuki, H., Banba, Y., Sasaki, H., and Kiuchi, K. 1994. Isolation and characterization of a *Bacillus subtilis* strain producing natto with strong umami-taste and high viscosity. *Nippon Shokuhin Kogyo Gakkaishi* 41: 123–128 (in Japanese).

Nagai, T., Phan Tran, L.-S., Inatsu, Y., and Itoh, Y. 2000. A new IS*4* family insertion sequence, IS*4Bsu*1, responsible for genetic instability of poly-γ-glutamic acid production in *Bacillus subtilis*. *Journal of Bacteriology* 182: 2387–2392.

Nagai, T., and Tamang, J.P. 2010. Fermented legumes: Soybean and nonsoybean products. In *Fermented Foods and Beverages of the World*, eds. J.P. Tamang, and K. Kailasapathy, pp. 191–224. Boca Raton, FL: CRC Press.

Nagai, T., and Yamasaki, F. 2009. *Bacillus subtilis* (natto) bacteriophages isolated in Japan. *Food Science and Technology Research* 15: 293–298.

Nakao, S. 1972. Mame no ryori (Cooking beans). In *Ryori no kigen (The origins of cooking)*, pp. 115–126. Tokyo, Japan: Japan Broadcast Publishing (in Japanese).

Nielsen, R. 2002. Salt rising bread: A continuing conundrum. *Petits Propos Culinaires* 70: 67–80.

Nikkuni, S., Karki, T.B., Vilkhu, K.S., Suzuki, T., Shindoh, K., Suzki, C., and Okada, N. 1995. Mineral and amino acid contents of kinema, a fermented soybean food prepared in Nepal. *Food Science and Technology International* 1: 107–111.

Njoku, H.O., Eli, I., and Ofuya, C.O. 1989. Effect of pretreatment on the cooking time of the African yam bean (*Sphenostylis stenocarpa*). *Journal of Food Science* 54: 758–759.

Njoku, H.O., and Okemadu, C.P. 1989. Biochemical changes during the natural fermentation of the African oil bean seed (*Pentaclethra macrophylla*) for the production of ugba. *Journal of the Science of Food and Agriculture* 49: 457–465.

Nkafamiya, I.I., Osemeahon, S.A., Dahiru, D., and Umaru, H.A. 2007. Studies on the chemical composition and physicochemical properties of the seeds of Baobab (*Adansonia digitata*). *African Journal of Biotechnology* 6: 756–759.

Nolan, K.B., and Duffin, P.A. 1987. Effect of phytate on mineral bioavailability. In vitro studies on Mg2+, Ca2+, Fe3+, Cu2+ and Zn2+ solubilities in the presence of phytate. *Journal of the Science of Food and Agriculture* 40: 79–83.

Nout, M.J.R., Bakshi, D., and Sarkar, P.K. 1998. Microbiological safety of kinema, a fermented soya bean food. *Food Control* 9: 357–362.

Nout, M.J.R., Bonants-Van Laarhoven, T.M.G., De Dreu, R., and Gerats, I.A.G.M. 1985. The influence of some process variables and storage conditions on the quality and shelf-life of soybean tempeh. *Antonie Van Leeuwenhoek* 51: 532–534.

Nout, M.J.R., De Dreu, M.A., Zuurbier, A.M., and Bonants-Van Laarhoven, T.M.G. 1987. Ecology of controlled soyabean acidification for tempe manufacture. *Food Microbiology* 4: 165–172.

Nout, M.J.R., and Rombouts, F.M. 1990. A review: Recent developments in tempe research. *Journal of Applied Bacteriology* 69: 609–633.

Nuha, M.O., Isam, A.M.A., and Elfadil, E.B. 2010. Chemical composition, anti-nutrients and extractable minerals of Sicklepod (*Cassia obtusifolia*) leaves as influenced by fermentation and cooking. *International Food Research Journal* 17: 775–785.

Nzikou, J.M., Mvoula-Tsieri, M., Matos, L., Matouba, E., Ngakegni-Limbili, A.C., Linder, M., and Desobry, S. 2007. *Solanum nigrum* L. seeds as an alternative source of edible lipids and nutriment in Congo-Brazzaville. *Journal of Applied Sciences* 7: 1107–1115.

O'Toole, D. 1999. Characteristics and use of okara, the soybean residue from soy milk production – A review. *Journal of Agricultural and Food Chemistry* 47: 363–371.

Obeta, J.A.N. 1983. A note on the microorganisms associated with fermentation of seeds of the African oil bean tree (*Pentaclethra macrophylla*). *Journal of Applied Bacteriology* 54: 433.

Obeta, J.A.N., and Ugwuanyi, M.A.A. 1996. Microbiology of the production of okpeye, a condiment from seeds of *Prosopis africana* Taub. *Tropical Science* 36: 206–210.

Odibo, F.J.C., Ezeaku, E.O., and Ogbo, F.C. 2008. Biochemical changes during the fermentation of *Prosopis africana* seeds for ogiri-okpei production. *Journal of Industrial Microbiology and Biotechnology* 35: 947–952.

Odunfa, S.A. 1981. Microorganisms associated with fermentation of African locust bean (*Parkia filicoidea*) during iru preparation. *Journal of Plant Foods* 3: 245–250.

Odunfa, S.A. 1983. Biochemical changes during production of ogiri, a fermented melon (*Citrullus vulgaris* Schrad) product. *Plant Foods for Human Nutrition* 32: 11–18.

Odunfa, S.A. 1985a. African fermented foods. In *Microbiology of Fermented Foods*, vol. 2, ed. B.J.B. Wood, pp. 155–191. London: Elsevier Applied Science.

Odunfa, S.A. 1985b. Microbiological and toxicological aspects of fermentation of castor oil seeds for ogiri production. *Journal of Food Science* 50: 1758–1759.

Odunfa, S.A. 1986. Other legume-based fermented foods. In *Legume-Based Fermented Foods*, eds. N.R. Reddy, M.D. Pierson, and D.K. Salunkhe, pp. 219–225. Boca Raton, FL: CRC Press.

Odunfa, S.A., and Oyeyiola, G.F. 1985. Microbiological study of the fermentation of ugba, a Nigerian indigenous fermented food flavour. *Journal of Plant Foods* 6: 155–163.

Ogbonna, D.N., Sokari, T.G., and Achinewhu, S.C. 2001. Development of an owoh-type product from African yam beans (*Sphenostylis stenocarpa*) (Hoechst (ex. A. Rich.) Harms.) seeds by solid substrate fermentation. *Plant Foods for Human Nutrition* 56: 183–194.

Ogueke, C.C., and Nwagwu, A. 2007. Comparative study of melon seeds (*Citrullus vulgaris*) fermented with mixed cultures and pure cultures of bacteria isolated from ogiri egusi. *Life Science Journal* 4: 62–67.

Ogueke, C.C., Nwosu, J.N., Owuamanam, C.I., and Iwouno, J.N. 2010. Ugba, the fermented African oilbean seeds: Its production, hemical composition, preservation, safety and health benefits. *Pakistan Journal of Biological Sciences* 13: 489–496.

Oguntoyinbo, F.A., Huch, M., Cho, G.S., Schillinger, U., Holzapfel, W.H., Sanni, A.L., and Franz, C.M.A.P. 2010. Diversity of *Bacillus* species isolated from okpehe, a traditional fermented soup condiment from Nigeria. *Journal of Food Protection* 73: 870–878.

Oguntoyinbo, F.A., and Sanni, A.I. 2007. Determination of toxigenic potentials of *Bacillus* strains isolated from Okpehe, a Nigerian fermented condiment. *World Journal of Microbiology and Biotechnology* 23: 65–70.

Oguntoyinbo, F.A., Sanni, A.I., Franz, C.M.A.P., and Holzapfel, W.H. 2007. In vitro fermentation studies for selection and evaluation of *Bacillus* strains as starter cultures for the production of okpehe, a traditional African fermented condiment. *International Journal of Food Microbiology* 113: 208–218.

Ojimelukwe, P.C., Okechi, A., and Ojinnaka, M.C. 2011. Physicochemical characteristics of fermenting castor seeds containing lime and NaCl as additives. *African Journal of Food Science* 5: 754–760.

Ojinnaka, M.C., and Ojimelukwe, P.C. 2012. Effect of salt and lime on *Bacillus* fermented castor oil bean condiment. *Global Research Journal of Agricultural and Biological Sciences* 3: 381–389.

Ojinnaka, M.C., and Ojimelukwe, P.C. 2013. Study of the volatile compounds and amino acid profile in *Bacillus* fermented castor oil bean condiment. *Journal of Food Research* 2: 191–203.

Ojinnaka, M.C., Ojimelukwe, P.C., and Ezeama, C.F. 2013a. Effect of fermentation period on the organic acid and amino acid contents of ogiri from castor oil bean seeds. *Malaysian Journal of Microbiology* 9: 201–212.

Ojinnaka, M.C., Ojimelukwe, P.C., and Ezeama, C.F. 2013b. Microbial and enzymatic changes associated with the production of ogiri from castor oil bean using *B. subtilis* as starter culture. *Sky Journal of Food Science* 2: 10–18.

Oke, O.I., and Umoh, I.N. 1978. Lesser known oil seeds: Chemical composition. *Nutrition Reports International* 17: 293–297.

Olaoye, O.A., Onilude, A.A., and Dodd, C.E.R. 2008. Identification of *Pediococcus* spp. from beef and evaluation of their lactic acid production in varying concentrations of different carbon sources. *Advances in Natural and Applied Sciences* 2: 197–207.

Omafuvbe, B.O. 2006. Effect of salt on the fermentation of soybean (*Glycine max*) into daddawa using *Bacillus subtilis* as starter culture. *African Journal of Biotechnology* 5: 1001–1005.

Omafuvbe, B.O., Abiose, S.H., and Adaraloye, O. 1999. The production of okpeye a fermented condiment from *Prosopis africana* seed. *International Journal of Food Microbiology* 51: 183–186.

Omafuvbe, B.O., Falade, O.S., Osuntogun, B.A., and Adewusi, S.R.A. 2004. Chemical and biochemical changes in African locust bean (*Parkia biglobosa*) and melon (*Citrullus vulgaricus*) seeds during fermentation to condiments. *Pakistan Journal of Nutrition* 3: 140–145.

Omafuvbe, B.O., and Oyedapo, O.O. 2000. Observed biochemical changes during natural fermentation of African oil bean (*Pentaclethra macrophylla*) seeds. *Nigerian Journal of Nutritional Sciences* 21: 19–23.

Omafuvbe, B.O., Shonukan, O.O., and Abiose, S.H. 2000. Microbiological and biochemical changes in the traditional fermentation of soybean for 'soydaddawa' – Nigerian food condiment. *Food Microbiology* 17: 469–474.

Onyenekwe, P.C., Odeh, C., and Nweze, C.C. 2012. Volatile constituents of ogiri, soybean daddawa and locust bean daddawa three fermented Nigerian food flavour enhancers. *Electronic Journal of Environmental, Agricultural and Food Chemistry* 11: 15–22.

Ouoba, L.I.I., Cantor, M.D., Diawara, B., Traore, A.S., and Jakobsen, M. 2003a. Degradation of African locust bean oil by *Bacillus subtilis* and *Bacillus pumilus* isolated from soumbala, a fermented African locust bean condiment. *Journal of Applied Microbiology* 95: 868–873.

Ouoba, L.I.I., Diawara, B., Annan, N.T., Poll, L., and Jakobsen, M. 2005. Volatile compounds of Soumbala, a fermented African locust bean (*Parkia biglobosa*) food condiment. *Journal of Applied Microbiology* 99: 1413–1421.

Ouoba, L.I.I., Nyanga-Koumou, C.A.G., Parkouda, C., Sawadogo, H., Kobawila, S.C., Keleke, S., Diawara, B., Louembe, D., and Sutherland, J.P. 2010. Genotypic diversity of lactic acid bacteria isolated from African traditional alkaline-fermented foods. *Journal of Applied Microbiology* 108: 2019–2029.

Ouoba, L.I.I., Parkouda, C., Diawara, B., Scotti, C., and Varnam, A.H. 2008a. Identification of *Bacillus* spp from bikalga, fermented seeds of *Hibiscus sabdariffa*: Phenotypic and genotypic characterization. *Journal of Applied Microbiology* 104: 122–131.

Ouoba, L.I.I., Rechinger, K.B., Barkholt, V., Diawara, B., Traore, A.S., and Jakobsen, M. 2003b. Degradation of proteins during the fermentation of African locust bean *Parkia biglobosa*, by strains of *Bacillus subtilis* and *Bacillus pumilus* for production of soumbala. *Journal of Applied Microbiology* 94: 396–402.

Ouoba, L.I.I., Thorsen, L., and Varnam, A.H. 2008b. Enterotoxins and emetic toxins production by *Bacillus cereus* and other species of *Bacillus* isolated from Soumbala and Bikalga, African alkaline fermented food condiments. *International Journal of Food Microbiology* 124: 224–230.

Owens, J.D., Allagheny, N., Kipping, G., and Ames, J.M. 1997. Formation of volatile compounds during *Bacillus subtilis* fermentation of soya beans. *Journal of the Science of Food and Agriculture* 74: 132–140.

Oyewole, O.A., and Isah, P. 2012. Locally fermented foods in Nigeria and their significance to national economy: A review. *Journal of Recent Advances in Agriculture* 1: 92–102.

Oyeyiola, G.F. 1981. Studies on the microorganisms isolated from ugba, a Nigerian indigenous fermented food. M.Sc. thesis. Ibadan, Nigeria: University of Ibadan.

Park, J.S., Park, H.Y., Kim, D.H., Kim, D.H., and Kim, H.K. 2008. ortho-Dihydroxyisoflavone derivatives from aged Doenjang (Korean fermented soypaste) and its radical scavenging activity. *Bioorganic and Medicinal Chemistry Letters* 18: 5006–5009.

Park, K.I., and Kim, K.J. 1970. Studies on manufacturing of the Korean soy sauce (part I). *Research Report, Central Industrial Research Institute* 20: 56–62.

Park, K.Y., and Jung, K.O. 2005. Fermented soybean products as functional foods: Functional properties of doenjang (Fermented soybean paste). In *Asian Functional Foods*, eds. J. Shi, C.T. Ho, and F. F. Shahidi, pp. 555–596. Boca Raton, FL: CRC Press.

Park, K.Y., and Kweon, M.H. 1987. Effect of l-ascorbic acid on the degradation of aflatoxin B1. *Journal of the Korean Society of Food Science and Nutrition* 16: 1–8.

Park, K.Y., and Lee, E.S. 1989. Effect of ammonia and pH on the degradation of aflatoxin B1 during the storage of Korean soy sauce (kanjang). *Journal of the Korean Society of Food Science and Nutrition* 18: 115–122.

Park, K.Y., Lee, E.S., Moon, S.H., and Cheigh, H.S. 1989. Effects of browning products and charcoal on the degradation of aflatoxin B1 in Korean soysauce (kanjang) and its model system. *Korean Journal of Food Science and Technology* 21: 419–424.

Park, K.Y., and Lee, K.B. 1985. Aflatoxin production by *Aspergillus parasiticus* NRRL2999 in various varieties of soybeans. *Journal of the Korean Society of Food Science and Nutrition* 14: 177–181.

Park, K.Y., Lee, K.B., and Bullerman, L.B. 1988. Aflatoxin production by *Aspergillus parasiticus* and its stability during the manufacture of Korean soypaste (doenjang) and soysauce (kanjang) by traditional method. *Journal of Food Protection* 51: 938–945.

Park, K.Y., Moon, S.H., Cheigh, S.H., and Baik, H.S. 1996. Antimutagenic effects of doenjang (Korean soy paste). *Journal of Food Science and Nutrition* 1: 151–158.

Park, K.Y., Son, M.H., Moon, S.H., and Kim, K.H. 1999. Cancer preventive effects of doenjang in vitro and in vivo. 1. Antimutagenic and *in vivo* antitumor effect of doenjang. *Journal of the Korean Association of Cancer Prevention* 4: 68–78.

Park, M.K., Cho, I.H., Lee, S., H.K., C., Kwon, D.Y., and Kim, Y.S. 2010. Metabolite profiling of Cheonggukjang, a fermented soybean paste, during fermentation by gas chromatography–mass spectrometry and principal component analysis *Food Chemistry* 122: 1313–1319.

Park, S.K., Seo, K.I., Choi, S.H., Moon, J.S., and Lee, Y.H. 2000. Quality assessment of commercial doenjang prepared by traditional method *Journal of the Korean Society of Food Science and Nutrition* 29: 211–217.

Parkouda, C., Diawara, B., Lowor, S., Diako, C., Kwesi Saalia, F., Annan, N.T., Jensen, J.S., Tano-Debrah, K., and Jakobsen, M. 2011. Volatile compounds of maari, a fermented product from baobab (*Adansonia digitata* L.) seeds. *African Journal of Biotechnology* 10: 4197–4206.

Parkouda, C., Diawara, B., and Ouoba, L.I.I. 2008. Technology and physicochemical characteristics of bikalga, alkaline fermented seeds of *Hibiscus sabdariffa*. *African Journal of Biotechnology* 7: 916–922.

Parkouda, C., Nielsen, D.S., Azokpota, P., Ouoba, L.I.I., Amoa-Awua, W.K., Thorsen, L., Hounhouigan, J.D., Jensen, J.S., Tano-Debrah, K., Diawara, B., and Jakobsen, M. 2009. The microbiology of alkaline-fermentation of indigenous seeds used as food condiments in Africa and Asia. *Critical Reviews in Microbiology* 35: 139–156.

Parkouda, C., Thorsen, L., Compaore, C.S., Nielsen, D.S., Tano-Debrah, K., Jensen, J.S., Diawara, B., and Jakobsen, M. 2010. Microorganisms associated with Maari, a baobab seed fermented product. *International Journal of Food Microbiology* 142: 292–301.

Peng, Y., Huang, Q., Zhang, R.H., and Zhang, Y.Z. 2003. Purification and characterization of a fibrinolytic enzyme produced by *Bacillus amyloliquefaciens* DC-4 screened from douchi, a traditional Chinese soybean food. *Comparative Biochemistry and Physiology* 134B: 45–52.

Podolak, R.K., Zayas, J.F., Kastner, C.L., and Fung, D.Y.C. 1996. Inhibition of *Listeria monocytogenes* and *Escherichia coli* 0157:H7 on beef by application of organic acids. *Journal of Food Protection* 59: 370–373.

Polit, P.F., and Sgarbieri, V.C. 1976. Some physicochemical and nutritional properties of castor bean (*Ricinus communis*) protein. *Journal of Agricultural and Food Chemistry* 24: 795–798.

Popoola, T.O.S., Jolaoso, A.A., and Afolabi, R.O. 2004. Microbiology of the production of Oso, a condiment made by fermenting seeds of *Cathormion altissimum*. *Tropical Science* 44: 187–189.

Popoola, T.O.S., Jolaoso, A.A., Afolabi, R.O., and Akintokun, A.K. 2006. Microbiological studies on the production of Oso: A condiment made from fermenting seeds of *Cathormion altissimum*. *Journal of Herbs, Spices and Medicinal Plants* 12: 129–138.

Popoola, T.O.S., Jolaoso, A.A., and Akintokun, A.A. 2005. An assessment of the nutritional value of oso – A condiment made by fermenting seeds of *Cathormion altissinum*. *Journal of Food Technology* 3: 149–151.

Porter, D. 1992. Economic botany of *Sphenostylis* (Leguminosae). *Economic Botany* 46: 262–275.
Premarani, T., and Chhetry, G.K.N. 2010. Evaluation of traditional fermentation technology for the preparation of hawaijar in Manipur. *Assam University Journal of Science & Technology: Biological and Environmental Sciences* 6: 82–88.
Qin, H., Yang, H., Qiao, Z., Gao, S., and Liu, Z. 2013. Identification and characterization of a Bacillus subtilis strain HB-1 isolated from Yandou, a fermented soybean food in China. *Food Control* 31: 22–27.
Qitain, A.T., Oro, K., Katoh, S., and Moriyoshi, T. 2006. Recovery of oil components of okara by ethanol-modified supercritical carbon dioxide extraction. *Bioresource Technology* 97: 1509–1514.
Quinn, L., and Tang, H. 1996. Antioxidant properties of phenolic compounds in macadamia nuts. *Journal of the American Oil Chemists' Society* 73: 1385–1388.
Rashad, M.M., Mahmoud, E.A., Abdou, M.H., and Nooman, U.M. 2011. Improvement of nutritional quality and antioxidant activities of yeast fermented soybean curd residue. *African Journal of Biotechnology* 10: 5504–5513.
RDA (Rural Development Administration) 2011. *The Korean Food Composition Table*. 8th revision. Suwon: Rural Development Administration (National Academy of Agricultural Science).
Recommended Dietary Allowances, 10th ed. 1989. Washington, DC: National Academy Press.
Risal, C.P., Yokoyama, T., Ohkama-Ohtsu, N., Djedidi, S., and Sekimoto, H. 2010. Genetic diversity of native soybean bradyrhizobia from different topographical regions along the southern slopes of the Himalayan Mountains in Nepal. *Systematic and Applied Microbiology* 33: 416–425.
Ryo, S., and Baba, K. 1981. Studies on the effect of postpartum administration of natto by mothers on the content of vitamin K in breast milk. *Perinatal Medicine (Shusanki Igaku)* 11: 1191–1195 (in Japanese, translated title).
Sanni, A.I. 1993a. Biochemical changes during production of okpehe, a Nigerian soup condiment. *Chemie, Mikrobiologie, Technologie der Lebensmittel* 15: 97–100.
Sanni, A.I. 1993b. The need for process optimization of African fermented foods and beverages. *International Journal of Food Microbiology* 18: 85–95.
Sanni, A.I., Ayernor, G.S., Sakyi-Dawson, E., and Sefa-Dedeh, S. 2000. Aerobic spore-forming bacteria and chemical composition of some Nigerian fermented soup condiments. *Plant Foods for Human Nutrition* 55: 111–118.
Sanni, A.I., Lie, E., and Lindberg, A.M. 1993. Fatty acid composition of *Prosopis africana* and its fermented product okpehe. *Chemie, Mikrobiologie, Technologie der Lebensmittel* 15: 89–90.
Sanni, A.I., and Ogbonna, D.N. 1991. The production of owoh – A Nigerian fermented seasoning agent from cotton seed (*Gossypium hirsutum* L). *Food Microbiology* 8: 223–229.

Sanni, A.I., and Ogbonna, D.N. 1992. Biochemical studies on owoh – A Nigerian fermented soup condiment from cotton seed. *Food Microbiology* 9: 177–183.

Sanni, A.I., Onilude, A.A., Fadahunsi, I.F., Ogunbanwo, S.T., and Afolabi, R.O. 2002. Selection of starter cultures for the production of ugba, a fermented soup condiment. *European Food Research and Technology* 215: 176–180.

Sanni, A.I., Onilude, A.A., and Oguntoyinbo, F.A. 1998. Optimization of process conditions for owoh, a fermented cotton seed condiment. *Advances in Food Science* 20: 163–167.

Sarkar, P.K. 2000. Ecology of soybean acidifications and some biochemical changes during controlled Kinema production. *Journal of Food Science and Technology* 37: 477–483.

Sarkar, P.K., Cook, P.E., and Owens, J.D. 1993. *Bacillus* fermentation of soybeans. *World Journal of Microbiology and Biotechnology* 9: 295–299.

Sarkar, P.K., Hasenack, B., and Nout, M.J.R. 2002. Diversity and functionality of *Bacillus and* related genera isolated from spontaneously fermented soybeans (Indian kinema) and locust beans (African soumbala). *International Journal of Food Microbiology* 77: 175–186.

Sarkar, P.K., Jones, L.J., Craven, G.S., and Somerset, S.M. 1997a. Oligosaccharide profiles of soybeans during kinema production. *Letters in Applied Microbiology* 24: 337–339.

Sarkar, P.K., Jones, L.J., Craven, G.S., Somerset, S.M., and Palmer, C. 1997b. Amino acid profiles of kinema, a soybean-fermented food. *Food Chemistry* 59: 69–75.

Sarkar, P.K., Jones, L.J., Gore, W., Craven, G.S., and Somerset, S.M. 1996. Changes in soya bean lipid profiles during kinema production. *Journal of the Science of Food and Agriculture* 71: 321–328.

Sarkar, P.K., Morrison, E., Tinggi, U., Somerset, S.M., and Craven, G.S. 1998. B-group vitamin and mineral contents of soybeans during kinema production. *Journal of the Science of Food and Agriculture* 78: 498–502.

Sarkar, P.K., and Tamang, J.P. 1994. The influence of process variables and inoculum composition on the sensory quality of kinema. *Food Microbiology* 11: 317–325.

Sarkar, P.K., and Tamang, J.P. 1995. Changes in the microbial profile and proximate composition during natural and controlled fermentations of soybeans to produce kinema. *Food Microbiology* 12: 317–325.

Sarkar, P.K., Tamang, J.P., Cook, P.E., and Owens, J.D. 1994. Kinema – A traditional soybean fermented food: Proximate composition and microflora. *Food Microbiology* 11: 47–55.

Satar, A., Huq, M.F., Nurul, H., and Sattaw, M.A. 1979. Studies on the utilization of castor cake. *Bangladesh Journal of Scientific Industrial Research* 14: 325–333.

Savadogo, A., Ilboudo, A.J., Gnankiné, O., and Traoré, A.S. 2011. Numeration and Identification of thermotolerant endospore-forming *Bacillus* from two fermented condiments Bikalga and Soumbala. *Advances in Environmental Biology* 5: 2960–2966.

Sawamura, S. 1906. On the micro-organisms of natto. *Bulletin of College of Agriculture, Tokyo Imperial University* 7: 107–110.

Schaffner, D.W., and Beuchat, L.R. 1986. Functional properties of freeze-dried powders of unfermented and fermented aqueous extracts of legume seeds. *Journal of Food Science* 51: 629–633.

Science and Technology Agency of Japan 1979. *Research report on the revision of the standard table of food composition in Japan.* Tokyo: Science and Technology Agency of Japan (in Japanese).

Sherfi, S.A., and Hamad, S.H. 2001. Microbiological and biochemical studies on Gergoush fermentation. *International Journal of Food Microbiology* 67: 247–252.

Shimakage, A., Shinbo, M., Yamada, S., and Ito, H. 2006. Changes in isoflavone content in soybeans during the manufacturing processes of tubunatto and hikiwari-natto. *Nippon Shokuhin Kagaku Kogaku Kaishi* 53: 185–188 (in Japanese).

Shrestha, A.K., and Noomhorm, A. 2002. Comparison of physico-chemical properties of biscuits supplemented with soy and kinema flours. *International Journal of Food Science and Technology* 37: 361–368.

Shukla, S., Kim, J.K., and Kim, M. 2011. Occurrence of biogenic amines in soybean food products. In *Soybean and Health*, ed. H. El-Shemy, pp. 181–206. Rijeka, Croatia: InTech.

Simmons, A. 1958. *The First American Cookbook, a Facsimile of "American Cookery," 1796.* New York: Oxford University Press.

Singh, A., Singh, R.K., and Sureja, A.K. 2007. Cultural significance and diversities of ethnic foods of northeast India. *Indian Journal of Traditional Knowledge* 6: 79–94.

Soda, K., Kano, Y., Sakuragi, M., Takao, K., Lefor, A., and Konishi, F. 2009. Long-term oral polyamine intake increases blood polyamine concentrations. *Journal of Nutritional Science and Vitaminology* 55: 361–366.

Steinkraus, K.H., ed. 1996a. *Handbook of Indigenous Fermented Foods*, 2nd ed. New York: Marcel Dekker.

Steinkraus, K.H. 1996b. Indigenous fermented foods involving an alkaline fermentation. In *Handbook of Indigenous Fermented Foods*, 2nd ed., ed. K.H. Steinkraus, pp. 349–362. New York: Marcel Dekker.

Steinkraus, K.H. 1997. Classification of fermented foods: Worldwide review of household fermentation techniques. *Food Control* 8: 311–318.

Sulistyo, J., Taya, N., Funane, K., and Kiuchi, K. 1988. Production of natto starter. *Nippon Shokuhin Kogyo Gakkaishi* 35: 278–283.

Sumi, H. 1990. The current state on development of natto containing a high activity of nattokinase. *Bio Industry* 7: 724–730 (in Japanese, translated title).

Sun Times 1909. Boom for salt rising bread. *The Sun Printing and Publishing Associates* 2nd Section: 1.

Svanberg, U., Lorri, W., and Sandberg, A.S. 1993. Lactic fermentation of non-tannin and high-tannin cereals: Effects on in vitro estimation of iron availability and phytate hydrolysis. *Journal of Food Science* 58: 408–412.

Takahashi, S., Katsumata, R., Yoshizawa, K., Yamaki, M., Osako, M., Murakami, M., Mouri, S., Sato, M., Meguro, H., Kubokura, H., Hirose, K., Tanaka, T., Watanabe, S., and Kiuchi, K. 2005. Soft natto manufactured with starter of *Bacillus subtilis* KFP 843 isolated from douchi. *Nippon Shokuhin Kagaku Kogaku Kaishi* 52: 454–461 (in Japanese).

Takemura, H., Ando, N., and Tsukamoto, Y. 2000. Breeding of branched short-chain fatty acids non-producing natto bacteria and its application to production of natto with light smells. *Nippon Shokuhin Kagaku Kogaku Kaishi* 47: 773–779 (in Japanese).

Tamang, J.P. 1999. Development of pulverised starter for kinema production. *Journal of Food Science and Technology* 36: 475–478.

Tamang, J.P. 2001. Kinema. *Food Culture* 3: 11–14.

Tamang, J.P. 2010. *Himalayan Fermented Foods: Microbiology, Nutrition, and Ethnic Values*. Boca Raton, FL: CRC Press.

Tamang, J.P., Chettri, R., and Sharma, R.M. 2009. Indigenous knowledge of northeast women on production of ethnic fermented soybean foods. *Indian Journal of Traditional Knowledge* 8: 122–126.

Tamang, J.P., and Nikkuni, S. 1996. Selection of starter cultures for the production of kinema, a fermented soybean food of the Himalaya. *World Journal of Microbiology and Biotechnology* 12: 629–635.

Tamang, J.P., Sarkar, P.K., and Hesseltine, C.W. 1988. Traditional fermented foods and beverages of Darjeeling and Sikkim – A review. *Journal of the Science of Food and Agriculture* 44: 375–385.

Tamang, J.P., Thapa, S., Dewan, S., Jojima, Y., Fudou, R., and Yamanaka, S. 2002. Phylogenetic analysis of *Bacillus* strains isolated from fermented soybean foods of Asia: Kinema, chongkukjang and natto. *Journal of Hill Research* 15: 56–62.

Tamura, M. 1988. Development of stabilized and low smelling natto. *Up-todate Food Processing* 23: 44–47 (in Japanese).

Tanaka, T., Yamauchi, T., R., K., and Kiuchi, K. 2003. Comparison of volatile components in commercial itohiki-natto by solid-phase microextraction and gas chromatography. *Nippon Shokuhin Kagaku Kogaku Kaishi (Journal of the Japanese Society for Food Science and Technology)* 50: 278–285 (in Japanese).

Tanaka, Y., and Tomiyasu, Y. 1973. Chemical components of the white deposits of natto (fermented soybeans). *Journal of Japanese Society of Food and Nutrition* 26: 473–478 (in Japanese).

Ten Brink, B., Damink, C., Joosten, H.M.L.J., and Huis In 't Veld, J.H.J. 1990. Occurrence and formation of biologically active amines in foods. *International Journal of Food Microbiology* 11: 73–84.

Teng, D., Gao, M., Yang, Y., Liu, B., Tian, Z., and Wang, J. 2012. Biomodification of soybean meal with *Bacillus subtilis* or *Aspergillus oryzae*. *Biocatalysis and Agricultural Biotechnology* 1: 32–38.

The Ohio Farmer 1895. Salt rising bread. *American Periodicals* May 23, 37, 21: 419.

Thorsen, L., Abdelgadir, W.S., Ronsbo, M.H., Abban, S., Hamad, S.H., Nielsen, D.S., and Jakobsen, M. 2011. Identification and safety evaluation of *Bacillus* species occurring in high numbers during spontaneous fermentations to produce gergoush, a traditional Sudanese bread snack. *International Journal of Food Microbiology* 146: 244–252.

Toda, T., Uesugi, T., Hirai, K., Nukaya, H., Tsuji, K., and Ishida, H. 1999. New 6-O-acyl isoflavonee glycosides from soybeans fermented with *Bacillus subtilis* (*natto*). I. 6-O-Succinylated isoflavone glycosides and their preventive effects on bone loss in ovariectomized rats fed a calcium-deficient diet. *Biological and Pharmaceutical Bulletin* 22: 1193–1201.

Uaboi-Egbenni, P.O., Okolie, P.N., Sobande, A.O., Alao, O., Teniola, O., and Bessong, P.O. 2009. Identification of subdominant lactic acid bacteria in dawadawa (a soup condiment) and their evolution during laboratoryscale fermentation of *Parkia biglobosa* (African locust beans). *African Journal of Biotechnology* 8: 7241–7248.

Ueda, S. 1989. Industrial application of *B. subtilis*. In *Bacillus subtilis: Molecular Biology and Industrial Application*, eds. B. Maruo, and H. Yoshikawa, pp. 143–161. Tokyo, Japan: Kodansha.

Utama, H.K. 2010. Kajian karakteristik kimia dan sensoris bumbu masak berbahan baku bungkil wijen (*Sesamum indicum*) dengan variasi lama fermentasi serta suhu pengeringan. Surakarta, Indonesia: Digital Library, Universitas Sebelas Maret.

Van Veen, A.G. 1967. The bongkrek toxins. In *Biochemistry of Some Foodborne Microbial Toxins*, eds. R.I. Mateles, and G.N. Wogan, pp. 43–50. Cambridge, MA: MIT Press.

Vautier, H., and Sacandé, M. 2007. *Prosopis africana (Guill. & Perr.) Taub.* Seed Leaflet No. 132, ed. L Schmidt.

Visessanguan, W., Benjakul, S., Potachareon, W., Panya, A., and Riebroy, S. 2005. Accelerated proteolysis of soy proteins during fermentation of Thua-Nao inoculated with *Bacillus subtilis*. *Journal of Food Biochemistry* 29: 349–366.

Wang, H., Li, Y., Cheng, Y., Yin, L., and Li, L. 2011. Effect of the Maillard reaction on Angiotensin I-Converting Enzyme (ACE)-inhibitory activity of Douchi during fermentation. *Food and Bioprocess Technology* 6: 279–301.

Wang, H.L., Swain, E.W., Wallen, L.L., and Hesseltine, C.W. 1975. Free fatty acids identified as antitryptic factor in soybeans fermented by *Rhizopus oligosporus*. *The Journal of Nutrition* 105: 1351–1355.

Wang, J., and Fung, D.Y. 1996. Alkaline-fermented foods: A review with emphasis on pidan fermentation. *Critical Reviews in Microbiology* 22: 101–138.

Wang, L.J., Li, D., Zou, L., Chen, X.D., Cheng, Y.Q., Yamaki, K., and Li, L.T. 2007a. Antioxidative activity of douchi (a Chinese traditional saltfermented soybean food) extracts during its processing. *International Journal of Food Properties* 10: 385–396.

Wang, L.J., Mu, H.L., Liu, H.J., Bhandari, B., Saito, M., and Li, L.T. 2010. Volatile components in three commercial douchies, a chinese traditional salt-fermented soybean food. *International Journal of Food Properties* 13: 1117–1133.

Wang, L.J., Yin, L.J., Li, D., Zou, L., Saito, M., Tatsumi, E., and Li, L.T. 2007b. Influences of processing and NaCl supplementation on isoflavone contents and composition during douchi manufacturing. *Food Chemistry* 101: 1247–1253.

Wang, L.T., Lee, F.L., Tai, C.J., and Kasai, H. 2007c. Comparison of gyrB gene sequences, 16S rRNA gene sequences and DNA-DNA hybridization in the *Bacillus subtilis* group. *International Journal of Systematic and Evolutionary Microbiology* 57: 1846–1850.

Watanabe, H., Kashimoto, N., Kajimura, J., and Kamiya, K. 2006. A miso (Japanese soybean paste) diet conferred greater protection against hypertension than a sodium chloride diet in Dahl salt-sensitive rats. *Hypertension Research* 29: 731–738.

Whitaker, J.R. 1978. Biochemical changes during the fermentation of high protein foods. *Food Technology* 32: 175–190.

Winarno, F.G., and Reddy, N.R. 1986. Tempe. In *Legume-Based Fermented Foods*, eds. N.R. Reddy, M.D. Pierson, and D.K. Salunkhe, pp. 95–117. Boca Raton, FL: CRC Press.

Xu, X., Liu, H., and Zhou, Y. 2012. Study on the Meitauza production from Okara by *Actinomucor elegans* and *Zymomonas mobilis*. In *Information Technology and Agricultural Engineering*, eds. E. Zhu, and S. Sambath, pp. 329–336. Berlin, Germany: Springer.

Yagoub, A.E.G.A., Mohamed, B.E., Ahmed, A.H., and El Tinay, A.H. 2004. Study on Furundu, a traditional Sudanese fermented roselle (*Hibiscus sabdariffa*) seed: Effect on *in vitro* protein digestibility, chemical composition and functional properties of the total proteins. *Journal of Agriculture and Food Chemistry* 52: 6143–6150.

Yagoub, A.E.G.A., and Mohammed, M.A. 2008. Furundu, a meat substitute from fermented roselle (*Hibiscus sabdariffa* L.) seed: Investigation on amino acids composition, protein fractions, minerals content and HCl-extractability and microbial growth. *Pakistan Journal of Nutrition* 7: 352–358.

Yang, Q., Liu, T., Kuklina, E.V., Flanders, W.D., Hong, Y., Gillespie, C., Chang, M.H., Guinn, M., Dowling, N., Khoury, M.J., and Hu, F.B. 2011. Sodium and potassium intake and mortality among US adults: Prospective data from the Third National Health and Nutrition Examination Survey. *Archives of Internal Medicine* 171: 1183–1191.

Yokota, T., Hattori, T., Ohishi, H., Ohami, H., and Watanabe, K. 1996. Effect of oral administration of crude antioxidant preparation from fermented products of okara (bean curd residue) on experimentally induced inflammation. *LWT – Food Science and Technology* 29: 304–309.

Yonzone, H., and Tamang, J.P. 1998. Consumption pattern of traditional fermented foods in the Sikkim Himalaya. *Journal of Hill Research* 11: 112–115.

Yoo, S.K., Cho, W.H., Kang, S.M., and Lee, S.H. 1999. Isolation and identification of microorganisms in Korean traditional soybean paste and soybean sauce. *Korean Journal of Applied Microbiology and Biotechnology* 27: 113–117.

Zhang, J.H., Tatsumi, E., Ding, C.H., and Li, L.T. 2006. Angiotensin I-converting enzyme inhibitory peptides in douchi, a Chinese traditional fermented soybean product. *Food Chemistry* 98: 551–557.

Zhang, J.H., Tatsumi, E., Fan, J.F., and Li, L.T. 2007. Chemical components of *Aspergillus*-type Douchi, a Chinese traditional fermented soybean product, change during the fermentation process. *International Journal of Food Science and Technology* 42: 263–268.

Zhu, Y.P., Fan, J.F., Cheng, Y.Q., and Li, L.T. 2008. Improvement of the antioxidant activity of Chinese traditional fermented okara (Meitauza) using *Bacillus subtilis* B2. *Food Control* 19: 654–661.

Zoué, L.T., Bédikou, M.E., Gonnety, J.T., Faulet, B.M., and Niamké, S.L. 2012. Two novel non-conventional seed oil extracts with antioxidant and antimicrobial activities. *Tropical Journal of Pharmaceutical Research* 11: 469–475.

3

Diversity of Animal-Based Food Products Involving Alkaline Fermentation

Among the indigenous alkaline-fermented foods (AFFs), fermented fishery products have been widely consumed in Southeast Asian countries as main dishes or condiments, due to their delicacy and high nutritional properties. Fish pastes and sauces are the most popular products due to their salty, slightly cheese-like flavor, and their characteristic appetite-stimulating aroma. Fish paste is obtained through the natural fermentation process of whole fish or shrimp in the presence of salt (3:1) under ambient conditions. Enzymatic fermentation of small fish and shrimp mediated by either microorganisms or indigenous proteases yields short-chain peptides, amino acids, and other nitrogenous compounds, rendering the typical flavor and taste of the product. Apart from serving as a good source of proteins, these products possess bioactivities, especially natural antioxidants, essential fatty acids like linoleic and linolenic acids, and polyunsaturated fatty acids (PUFA) such as icosapentaenoic and docosahexaenoic acids, which provide health benefits.

3.1 Fish Sauces and Pastes

Wonnop Visessanguan and Siriporn Chaikaew

Fish sauces and pastes constitute the largest member of traditional alkaline-fermented fishery products. Owing to their flavor-enhancing properties, fish sauces and pastes are important as a condiment and the main source of protein. There are a number of fish sauces and pastes produced throughout the world. Geographically, the fermentation has been developed in coastal countries of Asia (Figure 3.1). It is likely that the fermentation has been developed first where fish and salt were easily obtained. Although such a wide range of products

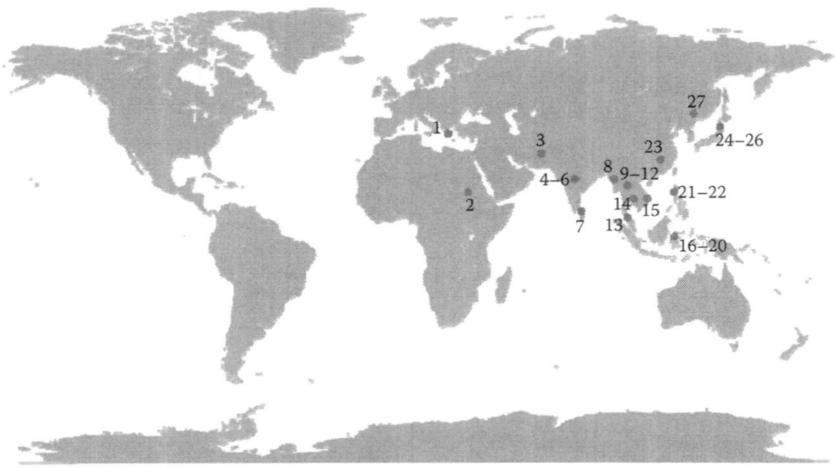

Figure 3.1 The world atlas of fermented fish pastes and sauces. 1, garum (Italy and Greece); 2, terkin (Sudan and Central Africa); 3, mahyaveh (Iran); 4–6, ngari, hentak, and tungtap (India); 7, jaadi (Sri Lanka); 8, ngapi and ngan byar yay (Myanmar); 9–12, nam pla, kapi plaa, tai plaa, and budu (Thailand); 13, budu (Malaysia); 14, prahok (Cambodia); 15, nuoc-mam (Vietnam); 16–20, bakasang, terasi ikan, wadi betok, kecap ikan, and pedah (Indonesia); 21–22, burong isda, bagoong, and patis (Philippines); 23, yu-lu (China); 24–26, shottsuru, ikanago shoyu, and konago (Japan); 27, jeotkal (Korea).

has been classified and the different fish species have been fermented, the main processing techniques and ingredients are mostly similar. The most important features making them distinct from other fermented fish products are that they are usually carried out with high salt addition (15%–40% by fish weight), undergo extensive hydrolysis, exhibit neutral or alkaline pH, and have no carbohydrate substrates added. Not only creating an unfavorable environment for food spoilage microorganisms and pathogens, the addition of salt at 13% w/v or higher also controls protein hydrolysis to a required stage and exerts a positive impact on the nutritional or organoleptic quality of the product (Steinkraus, 1996). In addition, the use of salt allows for a lower preservative content (e.g., sulfur dioxide, benzoic acid), a lower degree of acidity, or a milder physical treatment (e.g., heating). As a consequence of microbial conversion, the original fish gets reduced to the form of paste; otherwise, a complete hydrolysis will change the texture of the end product into fish sauce.

Unlike other fermented fish products, alkaline-fermented foods are those in which alkaline pH in combination with ammonia controls the fermentation. Higher pH allows bacteria to become dominant and also favors the anaerobic breakdown of proteins that releases amine compounds (Steinkraus, 1996). The change in pH toward alkaline side (>7.0) and liberation of ammonia are the consequence of extensive hydrolysis of protein to peptides and amino acids. In most cases the final pH would reach values higher than 7.0, but some foods, for example, fermented fish, have a strong buffering capacity and their pH remains restricted within 7.0 despite extensive hydrolysis of proteins. The combination of high pH and free ammonia along with rapid growth of the essential microorganisms at relatively high temperatures makes it very difficult for other microorganisms that might spoil the product to grow. Thus, the products are quite stable and well preserved especially when dried. Obviously, it is noted that no carbohydrates are added. The addition of carbohydrates is intended for different purposes (Saisithi, 1994). The rice is a source of carbohydrate for the microorganisms involved in the fermentation, carried out by mainly lactic acid bacteria, which give a characteristic flavor and sour taste to the products.

Due to high concentration of salt, microbial conversion is usually influenced by various species of halotolerant and halophilic microorganisms including *Bacillus*, *Pseudomonas*, *Staphylococcus*, *Virgibacillus*, *Halobacillus*, *Filobacillus*, and *Tetragenococcus halophilus* (Chaiyanan et al., 1999; Hiraga et al., 2005; Namwong et al., 2006; Udomsil et al., 2010; Tanasupawat et al., 2011a), which produce a substantial level of extracellular proteolytic activity that was active in extreme conditions.

3.1.1 Garum (Italy and Greece)

Archeological research has indicated that the use of garum might have been extensive already in the ancient Roman Empire (Badham, 1854; Smriga et al., 2010). The production technique of garum involved essentially a dry-salting method, in which Romans placed small, whole fish or the innards of large fish into a vat with salt at a specified ratio perhaps along with other ingredients such as herbs, spices, and the like. They mixed these ingredients together, covered the vat, and allowed the concoction to ferment for up to 3 months (Curtis, 1991). The ancient technique strikingly resembles not only modern fish sauce production in southeast Asia, China, and Japan (Yoshida, 1998; Michihata et al., 2000; Curtis, 2009), but also a process currently utilized in local production sites in southern Italy (Smriga et al., 2010). The use of fish sauces as condiments virtually disappeared from Europe in the early Middle Ages because of sociopolitical chaos following the collapse of the Western Roman Empire and from China and Japan commencing in the 14th century because of an increased use of soy sauces. Analysis of remnants from ancient garum from preserved containers (dolia) located in Pompeii confirmed a pattern of free amino acids that was comparable to the pattern detected in modern Italian and Asian products, with free glutamate being the taste-dominant amino acid, followed by sweet-tasting glycine and alanine (Smriga et al., 2010).

In Greece, the finest quality of garum is reported to have been made from the viscera and blood of mackerel. The fermentation was probably due to the proteolytic enzymes form the fish viscera as they are very halotolerant. It is possible that garum is different from the products made in the present day, as unhydrolyzed fish tissue is decanted to give a clear liquid. Garum is also used to make a stronger, thicker sauce known as alec (Beddows, 1985). The trade in garum was extensive and production of the sauce extended to ancient Greece. Even today, a sauce is made in Greece from fish viscera, particularly the liver, of mackerel (*Scomber colias*). It can be produced very quickly probably because of the high concentration of proteolytic enzymes present; the product is known as garos (Kelaiditis, 1949).

3.1.2 Terkin (Sudan and Central Africa)

Terkin or terkeen is Sudanese fermented fish paste. Terkin production and consumption is centered on Dongola, the ancient town of Nubia region, northern Sudan (Abu-Hassan and Adam Sulieman, 2011). The region has a long experience in fish fermentation (Besyuni, 1979). The fish used are kass (tiger fish; *Hydrocynus* spp.), kawara (*Alestes* spp.), tilapia, and Nile perch (Essuman, 1992; Abu-Hassan and Adam Sulieman, 2011). For the traditional process, the fish are washed and salted by arranging it in alternate layers with salt in an earthen pot or barrel. The salt to fish ratio is 1:5. Pickling and fermentation take place for 9–15 days, after which the fermentation vat or pot is placed in the sun or near a fireplace to speed up fermentation. This process continues for about 4 days, and the fish is stirred daily to break up the fish into a pasty mixture of muscle and bones that has a dark color and a strong odor (Essuman, 1992), having 40% moisture, 23.8% protein, and pH of 6.1 (Mohammed, 2010). However, the nutritive values of commercial terkin paste products could be varied upon region. Significant differences ($p > 0.01$) in crude protein, moisture, and ash were reported between Jebel Al-aulia and Wadi Halfa terkin products. Additionally, the pH of the studied products was at a higher level (7.2) in terkin of Jebel Al-aulia than Wadi Halfa terkin product (6.7).

3.1.3 Mahyaveh (Iran)

Mahyaveh is a traditional fermented fish sauce widely consumed in the southern part of Iran (Zarei et al., 2012). This product is mostly produced according to the family tradition, availability of raw materials, consumer preferences, and climatic conditions of the region. Therefore, wide variations can be seen in production methods, proportions of raw materials, and composition between retail samples from different sources. Mahyaveh is typically composed of sardines (*Sardinella* sp.) or anchovies (*Stelophorus* sp.), salt, mustard (*Brassica juncea* (L.) Czern.), and water. It is traditionally prepared from fresh or dried fish, which are beheaded, washed, and packed along with salt and warm water into earthen or glass jars. The jars are allowed to stand under the sun or at ambient temperature for 25–30 days. The fish–salt mixture is then mashed into slurry and filtered through a stainless steel mesh. The brown liquid portion of fermented fish is then mixed with mustard and other spices. Various spices such as cumin (*Cuminum cyminum* L.), coriander (*Coriandrum sativum* L.), fennel (*Foeniculum vulgare* Mill.) seeds, black pepper (*Piper nigrum* L.), and thyme (*Thymbra capitata* (L.) Cav.) are added according to the consumer preferences. After a thorough mixing, the jars are placed at ambient temperature until the desirable taste and aroma produced, usually after 10–15 days. Zarei et al. (2012) found that the pH of mahyaveh from different locations was in the range of 4.89–7.55, and the concentration of NaCl was in the range of 7.48%–17.1%. Moreover, they observed histamine (2.66 g/kg) was found to be the main biogenic amine in the Iranian fish sauce. The high histamine content can be related to the high levels of bacterial count, especially Enterobacteriaceae and lactic acid bacteria in this product.

3.1.4 Ngari (India)

Ngari is a traditional fermented fish product of the state of Manipur in India. Because of its special flavor, ngari is used as a compulsory item in daily preparation of curry (Sarojnalini and Vishwanath, 1988). It is prepared from small and less expensive sun-dried fishes such as *Puntius sophore* (Ham.) and *Puntius ticto* (Ham.) subjected to fermentation in the absence of salt for 5 to 6 months or more at room temperature. The process of ngari preparation involves a brief washing of the sun-dried fish, followed by draining and drying for 24–48 h. The fish are then pressed hard using stone roller to breakdown head and bones and then mechanically inside the pots using a wooden stick. The pots are then sealed airtight and incubated at room temperature for 6 months. During fermentation, the quantity of most of the free amino acids and peptides increases markedly, which contribute significantly to the characteristics of fermented fish. Total volatile-based nitrogen (TVBN) value and amino nitrogen increase, probably due to the microbiological and biochemical changes in the fish muscle, which impart flavor during fermentation.

To accelerate the fermentation process, the starter culture consisting of three species of *Bacillus* and three species of *Micrococcus* isolated from ngari was used (Sarojnalini and Suchitra, 2009). Proper fermentation was noticed only after 40 days in starter culture-inoculated fish, whereas in naturally fermented fish, the fermentation was complete after 5–6 months. The results confirmed that bacteria are responsible in the ripening process of ngari. The quality of products is acceptable as judged by their microbiological characteristics. The mean values for moisture, total lipid, and total amino acids are 24%, 16%, and 4.4 mg/g, respectively. A pH level of 6.74 was observed for starter-fermented "ngari," while a pH of 6.49 was recorded for the naturally fermented sample after a period of 40 days.

3.1.5 Hentak (India)

Hentak (a fermented fish product of Manipur) is a fermented fish paste prepared from a mixture of sun-dried fish (*Esomus danricus* Ham.) powder and petioles of an aroid plant (*Alocasia macrorrhizos* (L.) G.Don.). A ball of mixture-like thick paste is made, kept in an earthen pot, tightly sealed, and left to ferment for 7–9 days. Naturally fermented sample of hentak has pH value of 6.9 (Sarojnalini and Vishwanath, 1988).

3.1.6 Tungtap (India)

Tungtap, a traditionally fermented fish product of the Khasi and Jaintia tribes in the state of Meghalaya in India, is a fermented fish paste that is made from sun-dried fish (*Pruntius* spp. and/or *Danio* spp.) mixed with salt (1:10). It is kept in an earthen pot, made air-tight, and left to ferment at ambient temperature (18°C–28°C) for 2–6 months. Tungtap is consumed as a pickle or taste enhance (Rapsang and Joshi, 2012).

3.1.7 Jaadi (Sri Lanka)

Jaadi is salt-fermented product, consisting of partially hydrolyzed fish flesh and organs immersed in a liquid exudate from fish. The main fish species used in the process are Indian mackerel (*Rastrelliger kanagurata*), trenched sardine (*Amblygaster sirm*), skipjack tuna (*Katsuwonus pelamis*), and tilapia (*Oreochromis niloticus*) (Lakshmi et al., 2010). The fishes are gutted, cleaned in seawater, and packed in a wooden barrel, an earthen pot, or other suitable vessel with alternate layers of fish and a salt-goraka (*Garcinia gummi-gutta* (L.) Roxb.) mixture. The contents are tightly packed in the vessel or, alternatively, a weight is kept on the contents to ensure good contact of fish with the salt-goraka mixture (at 14%–25% salt) when left to ferment. The fermentation process, terminated by the development of the characteristic aroma and flavor of the product, takes 2–4 months. The product has the pH of 4.9–7.1 and 10.3%–18.3% NaCl (Weerasinghe and Samaradiwakara, 1985). Fermented fish can be kept in good quality for a year if care is taken to prevent contamination (Subasinghe, 1993). *Syncephalastrum racemosum* and *Aspergillus* sp. are found during the fermentation.

3.1.8 Ngapi and Ngan-Byar-Yay (Myanmar)

Ngapi is indigenous food greatly relished by the people of Myanmar and consumed daily along with their staple meal (rice). Ngapi is a protein-rich, nutritious food containing high percentages of calcium, phosphorus, and several essential amino acids; its caloric content is comparable to beef, pork, and mutton (Tyn, 2004). Because of its adoration by the Burmese and its attractive food value, ngapi has attained its respectable status as a national food (Tyn, 1986). Ngapi is generally prepared by pounding or grinding fish with 20%–25% salt and partially drying in the sun for 3–4 days. The product is then stored in earthen jars or concrete ponds for 3–6 months to mature (Tyn, 1986). Nga ngapi, which is prepared from fish, consists of three types including ngapi gaung, yegyo ngapi, and damin ngapi (Tyn, 1993). Ngapi gaung is made from whole fish of fairly large size. After scaling, cleaning, and removing entrails, fins, and kets, the fish are well-weighted and pressed down. After 12–24 h, by which time the superfluous juices, ngan-byar-yay (fish sauce), would have gradually oozed out, the weight is removed, and the fish are gently taken out, rubbed with salt again, and dried in the sun for a day. Next day, they are neatly packed, layer by layer, in a large container with plenty of salt in between and on the uppermost surface. Normally, the fish are ready for consumption after 2–3 months and keep for more than a year. Yegyo ngapi is prepared from fish of medium and small size. The heads, fins, and entrails are removed and the fish cleaned. After mixing with some salt, the fish are spread evenly on bamboo mats and left to dry under the sun for a day. In the evening, the fish are partially pounded and spread on slightly inclined bamboo platforms and left overnight. This procedure is repeated three times. On the fourth day, they are firmly pressed down into a huge container and covered with a thick layer of salt and then by bamboo mats over which heavy weights are placed. The fermentation takes 4–6 months during which a reddish brown liquid, ngan pya ye, is occasionally drawn out from the outlet at the bottom of the containers. Damin ngapi is a fermented paste from various kinds of small fish (<5 cm) caught by a bamboo trap (damin). The fish are washed, mixed with some salt, and spread out on bamboo mats in the sun for 3–4 days until the mass becomes dry. The dried mass is thoroughly pounded or ground for 3 h, adding some salt and occasionally water. The paste obtained is dried for a day. The pounding and sun-drying operations are repeated 4–5 times until a homogeneous and smooth paste is obtained. It is tightly pressed into a large container, the surface is covered with a thick layer of salt, and is left to ferment for 4–6 months. This product contains 70%–85% moisture, 22% NaCl, 15%–25% crude protein, 1%–10% fat (or lipid), 1%–1.5% minerals, and pH of 6.0–7.5 (Tyn, 2004). *Bacillus* spp., *Micrococcus* spp., *Staphylococcus aureus*, and *Pseudomonas* spp. are found during the fermentation (Tyn, 2004).

3.1.9 Nam Pla (Thailand)

Nam pla is a traditional liquid product, having salty taste, and is used as seasoning for cooking and dipping seafood and oriental food. It is preferably made from brackish water, seawater, and freshwater fish *Corica argentius* (siu-kaew), *Corica soborna*, *Stolephorus indicus*, *Stolephorus commersonii* (sai-ton), and *Clupeoides* sp. (ka-tak), or any other kind of fish with the addition of a large amount of salt and is fermented for 12–18 months in an earthen jar or concrete vat. The fish can be found in bountiful supply in the Gulf of Thailand and the South China Sea. Generally, cleaned fish is mixed with salt in a 2:1 or 3:1 ratio (fish/salt, w/w), depending on the area of production. The product is clear, yellow, or brownish yellow-colored, salty, having a fishy aroma. After fermentation, the liquid showing a clear reddish brown color and a fragrant aroma is removed from the tanks, preferably by passing through the layers of fish remains at the bottom of the tanks. Any sediment formed is strained out with a clean cloth. It is then ready for bottling. Second- and third-grade fish sauces are made by adding salt-water to cover the fish remains for 2–3 months, then filtering before bottling. Finally, the fish remains are boiled with salt-water to produce the lowest-grade fish sauce. Because flavor is substantially reduced in each cycle of fermentation, top-grade fish sauce is frequently added to the lower grades to improve their flavor. In fact, many manufacturers do not market top-grade sauce. For this reason, 100% fish sauce is mixed with second- and third-grade sauces to produce larger quantities to sell, which can still be qualified as genuine fish sauce (Saisithi, 1994; Phithakpol et al., 1995; Lopetcharat et al., 2001).

In Thailand, nam pla is divided into three categories by the Thai Ministry of Public Health (2001) as (1) genuine fish sauce, which is obtained from fermentation or digestion of fish or parts of fish or residue of fish remaining from the fermentation, by fish sauce production process; (2) fish sauce made from other animals besides fish, or parts of other animals, or residue of other animal remaining from the fermentation, by fish sauce production process, and shall include fish sauce that is made from other animals mixed with genuine fish sauce; (3) mixed fish sauce means fish sauce as stipulated in (1) or (2), in which some other constituents of non-hazardous to consumption are added, or diluted, or for seasoning. Fish sauce shall be of qualities or standards as follows: color, odor, and flavor inherent of that specific characteristics of it, clear and free of sediment, except natural sediment not more than 0.1 g/l, salt as NaCl not less than 200 g/l fish sauce. In case of usage of KCl as individually used or mixed with NaCl, the quantity of individually or the total quantity of both salts used shall not be less than 200 g. Total nitrogen shall not be less than 9 g/l fish sauce. Nitrogen from amino acids is not less than 40% and not more than 60% of total nitrogen. Glutamic acid per total nitrogen is not less than 0.4% and not more than 0.6%, no color added, except brown sugar or caramel color. Usage of artificial sweetener shall follow the Food Standards of Joint FAO-WHO Codex and may be used in single or in combination with sugar.

The factors that influence nam pla quality are (i) fish species, (ii) type of salt, (iii) ratio of fish and salt, (iv) addition of minor ingredients, and (v) fermentation conditions. Fish species also affects the type of proteins that serve as nutrients for microorganisms and substrates for enzymes that hydrolyze proteins into small peptides, free amino acids, ammonia, and trimethylamine (TMA), which contribute to the specific aroma and flavor in fish sauce.

During the initial fermentation period (1–6 months), strains of *T. halophilus*, *Tetragenococcus muriaticus*, *Halobacterium salinarum*, *Halococcus thailandensis*, *Halococcus saccharolyticus*, *Natrinema gari*, *Leuconsotoc jurispiscarius*, *Leuconostoc halophilus*, *Pisciglobus halotolerans*, *Chromohalobacter salexigens*, *Halobacillus* sp., *Virgibacillus* sp., *Filobacillus* sp., and *Marinococcus* sp. are involved. After 6 months of fermentation, the same species are distributed and *Leuconostoc salicampi* strains are also isolated (Thongthai et al., 1992; Tanasupawat, 1999; Thongsanit et al., 2002; Hiraga et al., 2005; Namwong et al., 2005, 2006, 2007; Tanasupawat et al., 2006, 2009, 2011b; Sinsuwan et al., 2007, 2010; Tapingkae et al., 2008; Phrommao et al., 2010). *Haloarcula salaria* and *Haloarcula tradensis* strains are also found (Namwong et al., 2011).

The release of free amino acids from proteins and large polypeptides causes the pH value (5.3–6.7) to decrease during the process. Nam pla contains indispensable amino acids, namely histidine, isoleucine, leucine, lysine, methionine, phenylalanine, tryptophan, valine, arginine, cysteine, glycine, proline, and tyrosine, and dispensable amino acids, namely alanine, aspartic acid, glutamic acid, and serine (Lopetcharat et al., 2001; Thongthai and Gildberg, 2005). The other components of fish sauce are minerals (potassium, phosphorus, sulfur, sodium, magnesium, calcium, and iron) and vitamins (thiamine, riboflavin, niacin, pyridoxine, and cyanocobalamin), which are essential for healthy diet (Lopetcharat et al., 2001; Park et al., 2001). Often fish sauces are known to contain high levels of histamine, a biogenic amine. These compounds are generally vasoactive and can cause changes in blood pressure, severe headache, hypertension, renal intoxication, intracerebral hemorrhage, and eventually death. The Food and Drug Administration in 1998 established an advisory level of 500 ppm to be hazardous to human health. However, high histamine concentration may not cause severe illness since a consumption of fish sauce is less than 23.5 g/day/person (Yongsawatdigul et al., 2004).

3.1.10 Kapi Plaa (Thailand)

Kapi plaa, a common ingredient used in the cuisine of southern Thailand, is a dark gray-colored fermented fish paste with a salty taste and a characteristic fishy and pungent flavor. The fish used are kowala coval (ka-tak) and *Stolephorus indicus* (ma-li, gluay). For the traditional method, the fish are washed, mixed with salt in the ratio of 4–10:1 by weight, placed in a bamboo basket, and left overnight. The water is drained off. The salted fish on a bamboo mat are spread out and sun-dried for 2–3 days. The mixture is pound to a paste, packed tightly in a container, and left to ferment for 3–4 months until characteristic flavor and color are fully developed. The product contains 35.6%–57.3% moisture, 16.8%–29.9% protein, 1.6%–5.3% fat, 0.1%–0.5% fiber, 12.3%–25.2% ash, 10.9%–19.8% NaCl, 0.1%–0.5% acidity, and pH of 6.2–6.6 (Phithakpol et al., 1995).

3.1.11 Tai Plaa (Thailand)

Tai plaa is a salty and red to dark brown colored fermented fish viscera. The fish used are *Trichogaster trichopterus* (kru-dee), *Lutjanus* sp.

(ka-pong), *Mugil* sp. (ka-bork), *Rastrelliger brachysoma* (too), *Johnius* sp. (chuad), *Channa striatus* (chorn), *Cirrhinus jullieni* (soi), *Scomberoides* sp. (sala and see-siad), and *Megalaspis cordyla* (haang-kaeng). For the traditional method, the fish viscera are cleaned by pressing to remove the stomach and the intestine contents. If this is not well removed, the taste of the product becomes bitter, which is preferred in some places. Salt is added to the washed viscera in the proportion of fish viscera to salt 10:1 to 5:1 by weight. In some places, KNO_3 is added for curing. The ingredients are mixed well and packed in closed containers such as earthen jars, which are left in the sun to ferment for 10–20 days. For eating in the family, the viscera of a big fish such as ppaa-ka-pong and plaa-ka-bork, which have high fat content, is preferable. The product has a relatively short shelf life due to rapid rancidity. It is eaten by adding chopped shallots (*Allium cepa* L. var. aggregatum), chili, lemongrass (*Cymbopogon citratus* (DC.) Stapf), and lemon (*Citrus limon* (L.) Osbeck) juice. The dish is also eaten with fresh vegetables. It is the main ingredient of the famous southern dish gaeng-tai-plaa, which is favorite throughout Thailand. It contains 40.1%–65.8% moisture, 3.3%–16.0% crude protein, 1.3%–28.0% crude fat, and 13.5%–25.3% NaCl, and its pH is 4.0–6.7. *Bacillus subtilis*, *Bacillus polymyxa*, *Micrococcus* sp., and *T. halophilus* were isolated from this product (Phithakpol et al., 1995).

3.1.12 Budu (Thailand and Malaysia)

In Thailand, budu is made from seawater or brackish water fish (*Stolephorus indicus* (sai-ton), *Clupeoides* sp. (ka-tak), *Sardinella* sp. (lang-khieo)), and small fish anchovies (*Anchoviella commersonii*, and *Anchoviella indica*) with high salt (Phithakpol et al., 1995). The product is a viscous brown liquid, with gray colloidal fish flesh separated from the fish bones suspended in it and has a salty taste. The product is rich in salt and soluble nitrogen compounds and has a distinct odor and flavor. In the southern region, it is eaten by adding chili, shallots, sugar, and lime juice. In the central region, it is boiled with lemongrass, galangal (*Alpinia galanga* Willd.), and bitter orange (*Citrus aurantium* L.) leaves, decanted off, and consumed with rice and various vegetables. This is known as "khao-yam," which is a very popular dish in Thailand (Phithakpol et al., 1995). The product contains 45.0%–60.2% moisture, 12.1%–13.2% protein, 1.3%–1.8% fat, 0.2%–0.4%

fiber, 21.6%–23.2% ash, 19.4%–20.6% NaCl, 0.2 mg% total invert sugar, 0.4%–0.9% acidity, and pH of 5.4–6.5 (Phithakpol et al., 1995). *Tetragenococcus halophilus, Staphylococcus piscifermentans, Bacillus subtilis, and Bacillus licheniformis* strains were isolated from this product (Choorit and Prasertsan, 1992; Tanasupawat and Komagata, 1995).

In Malaysia, budu is a fish sauce made from small anchovies (*A. commersonii* and *A. indica*) known collectively as "ikan bilis," which occur in vast shoals at certain seasons off the east coast of Malaysia. The quantity consumed by a rough estimation would be 20–40 ml per person per day. The product is consumed by all socioeconomic groups as a source of protein. It is consumed as condiment on rice, as a dip for certain vegetables, and as a flavoring ingredient in various dishes. In the traditional fermentation, budu is prepared by mixing fish with 25%–33% salt. The mixture is then incubated in a concrete container at ambient temperature for 6–12 months depending on quality. The hydrolyzed mixture is boiled for about 2 h, cooled, decanted, or filtered, and bottled. Mature budu is a dark brown liquid extract containing ground fish residue. The product is rich in salt and soluble nitrogen compounds and has a distinct odor and flavor. Ismail (1977) developed a more rapid process that involves mixing 10% (dry weight) soybean overgrown with *Aspergillus oryzae* (as a source of enzymes) with 90% (w/v) fresh fish. The mixture is incubated at 40°C. The maximum digestion (soluble nitrogen to total nitrogen ratio (SN/TN) of 90%) is achieved in 7 days. The total nitrogen remains constant during the incubation period. The rate of hydrolysis product remaining 5% ground roasted wheat is similar to that without wheat. The pH of product without added wheat is about 6.5, and that with wheat is 5. A good-quality product is obtained in 2 months. The product contains 58%–64% moisture, 7.8%–10.2% protein, 0.5%–0.6% fat, 6.9%–10.3% carbohydrate, 20% NaCl, and pH of 6.5 (Steinkraus, 1996; Ghazali et al., 2011). *Micrococcus* sp. was the predominant bacterium to initiate the fermentation before being replaced by *Staphylococcus* sp. that survived throughout the traditional process. Halophilic bacteria especially *Micrococcus luteus* and *Staphylococcus arlettae* exhibited good proteolytic and lipolytic activities compared with other bacterial strains. *Lactobacillus plantarum* LP1, LP2, *S. arlettae* SA1, *Saccharomyces cerevisiae* SC3, and *Candida glabrata* CG2 strains showed potential probiotic activities. These demonstrate that budu is a mixed culture fermentation involving functional strains in hydrolyzing fish protein into solubilized liquid. Kican ikan is a fish sauce, produced from fish other than anchovies; however, the process is similar to that for budu.

3.1.13 Prahok (Cambodia)

Prahok is a crushed, salted, and fermented fish paste (usually of mud fish) that is used in Cambodian cuisine as a seasoning agent or a condiment (Van Veen, 1965). It originated as a way of preserving fish during the longer months when fresh fish was not available in abundant supply. Because of its saltiness and strong flavor, it was used as an addition to many meals in Cambodian cuisine, such as soups. Prahok has a strong and distinct smell, earning the nickname "Cambodian cheese." Prahok is usually eaten with rice in the countryside or poorer regions. At least two methods of preparation are employed. In the first, cyprinids are used (e.g., *Rasbora* sp. and *Thynnichtys* sp.); in the second, *Ophicephalus* sp. and others are used. The fish are beheaded and trampled by foot to remove scales and press out entrails. The fish are washed and stirred by hand until the scales are completely removed. They are again pressed to remove water. The upper part of the basket is covered with banana leaves and weighted with stones for about 24 h. The following day, the fishes are mixed with coarse salt for about 30 min and spread out on matting in the sun for a day. The salted and dried fish are returned to the basket for packing. The next phase, pounding small quantities of the fish in a wooden mortar for about 20 min takes place in the homes of villagers, and it is salted to taste. The paste is then placed in open earthen jars in the sun. During the evening, the jars are covered to keep out insects. Fermentation or ripening now takes place, and a pickle gradually appears on top of the paste, which is removed every day. The Cambodians use liquid as fish sauce. This phase of process may last a month. When no further pickle forms, the prahok is finished and is ready for immediate consumption (Van Veen, 1965). This product contains 52%–63% moisture and 1.7%–2.7% total nitrogen (Brémond and Rosé, 1919).

3.1.14 Nuoc-Mam (Vietnam)

The Vietnamese fish sauce nuoc-mam is made from fish (or sometimes shrimp). The most important production centers in Vietnam are in the province of Binh-Thuan (South Annam) and on the island of Phu Quoc. The best product has a rather strong cheese-like odor and salty taste (Van Veen, 1965). Traditionally, the fish is ground, pressed by hand, and placed in earthen jars layer by layer with salt in between the layers. The approximate ratio of fish to salt is 3:1 (Salampessy et al., 2010). The jars are then closed tightly and buried almost completely in the ground. In the first stage of fermentation, a bloody liquid (nouc-boi) is decanted carefully, and the remaining fish is allowed to ferment further for 6–18 months depending on the size of fish. Some of this nouc-boi is added back to the jar, and some is added at a later date. The high-quality nuoc-mam is obtained from the first separation of the supernatant (Beddows, 1985). The remaining undigested material is then mixed with brine to get low-quality nuoc-mam (Lopetcharat et al., 2001). The low quality of the nouc-mam can be improved by addition of food additives such as caramel, molasses, roasted maize, or roasted barley to the fish before the hot brine extraction. The sauce is clear brown liquid with rich in salt and soluble nitrogen compounds. It has a distinctive meaty aroma. The taste is predominantly salty, but the contribution of other aroma compounds including the volatiles is very apparent (Beddows, 1985). Nuoc-nam is a valuable supplement to cereal foods and can supply as much as 7.5% of an individual's total nitrogen intake and serve as a rich source of nutrients, vitamins, and minerals in the diet. The chemical composition consists of 12.0% crude fat, 17.5% crude protein, 9.5% carbohydrate, and 1.90%–2.28% total nitrogen (Beddows et al., 1979). Nuoc-nam contains 275 mg organic phosphorus/l. In addition, it contains 350 mg calcium/l, 1.3 g magnesium/l, and approximately 10 g sulfur/l. *Bacillus* sp., *Pseudomonas* sp., *Micrococcus* sp., *Staphylococcus* sp., *Halococcus* sp., *Halobacterium salinarium*, *Halobacterium cutirubrum*, *Lactobacillus* sp., *Streptococcus* sp., and *Pediococcus* sp. were found during the fermentation (Thongthai et al., 1992; Lopetcharat et al., 2001).

3.1.15 Bakasang (Indonesia)

Bakasang, a traditional salt-fermented fish sauce, is light brown colored, salty, and slightly acidic (Ijong and Ohta, 1995, 1996). Bakasang is widely consumed in the eastern part of Indonesia, being particularly popular among the Manadonese people (in the North Celebes island).

Bakasang is traditionally produced by fermenting small whole sardines (*Sardinella* sp. or *Stelophorus* sp.). In the process, about 1.5–3.0 parts of salt are mixed with 5 parts of fish, packed into small bottles, and placed in kitchen near the fire place. The temperature ranges from 30°C to 60°C, and the fermentation takes 3–6 weeks. It is usually used as flavoring in many dishes or mixed with red chili, tomato, red onion, and garlic, and then sautéed with coconut oil. The sautéed sauce is eaten with hot porridge mixtures of rice and vegetables called tinutuan. Traditional bakasang contains 66% moisture, 17% salt, 14.5% protein, 1.4% lipid, and pH value of 5.9. The fermentation process seemed to increase the soluble nitrogen of the fish. This could probably lead to an increase in the biological availability of the protein. With the total essential amino acids content (42–47 mg/ml), bakasang is a good source of essential amino acids such as valine, isoleucine, phenylalanine, and lysine in the diet.

3.1.16 Terasi Ikan (Indonesia)

In Indonesia, especially in Sumatra, a fish paste called terasi ikan is manufactured in large quantities in Bagan Si (Putro, 1993). Terasi is consumed in small quantities as a flavor additive. The product is not only for local consumption but also exported mainly to the Netherlands and Suriname in powder form. Fishes such as *Stelophorus* sp. and *Engraulis* sp. are mixed with 2%–5% salt and left for 1–3 days at ambient temperature. The salted mass is dried in thin layers in the sun, mixed with more salt, and pounded until a paste is obtained with no more water drains out. Terasi ikan is usually mixed with dyestuffs to give it a more attractive appearance. It contains 35%–50% moisture, 20%–45% protein and protein degradation products, 10%–25% minerals (NaCl and calcium salts), and small percentage of fatty substances (Van Veen, 1965). Jansen et al. (1953) found terasi ikan to have a high content of vitamin B_{12}. *Bacillus* sp., *Lactobacillus* sp., *Staphylococcus* sp., and *Proteus* sp. were found during the fermentation (Budhyatni et al., 1982). Terasi ikan usually has a strong smell, keeps for a very long time, and is often mixed with Spanish peppers etc. to give a very spicy product called sambal that is used regularly in Indonesia with every rice meal together with dried and salted fish.

3.1.17 *Wadi Betok (Indonesia)*

Wadi betok is one of the popular traditional fermented fish products in South Kalimantan in Indonesia (Petrus, 2009). This freshwater fish is climbing perch (*Anabas testudineus* Bloch) or "walking fish" bearing local name "ikan betok" or "ikan papuyu." Having 10%–15% protein content, it is used as the raw material in wadi betok production (Basrindu, 1987; Rahayu et al., 1992; Arianti, 2004). Petrus (2009) reported that whole freshwater fish after added with 30%–100% (w/w) salt were layered in plastic jars and left for spontaneous fermentation for 7 days to 4 months at room temperature. Physicochemical characteristics, sensory acceptability, and microbial quality of wadi betok are influenced by the level of salt addition (Petrus et al., 2013). The physicochemical characteristics of wadi betok were pH of 6.74–7.87, a_w of 0.75–0.93; salt content of 2.4%–13.7%, and protein content of 11.5%–16.5% (Petrus et al., 2013).

3.1.18 *Kecap Ikan (Indonesia)*

Kecap ikan is a clear yellow to brown liquid with a salty taste and specific aroma. Rahayu et al. (1992) reported that the raw material for kecap ikan production could be any marine fish, such as anchovy (*Stolephorus* sp.), herring (*Clupea* sp.), or ponyfish (*Leiognathus* sp.). Putro (1993) reported that along the Bali strait, a kecap ikan processor used sardine (*Sardinella* sp.) as the raw material. Some freshwater fish, such as java barb (*Puntius* sp.) and bony-lipped barb (*Osteochilus* sp.), are also suitable raw materials (Rahayu et al., 1992).

The processing of kecap ikan begins by eviscerating, washing, and mincing the fresh fish. The cleaned fish are mixed with 20%–30% salt and arranged in the fermentation tank. The fish are arranged in layers, and each layer is covered with salt. The tank is then closed, and fermentation is allowed to take place for 4–12 months (Huda, 2012). The kecap ikan liquid collected from outlet is then filtered and mixed with brown sugar and spices (Putro, 1993). Because kacap ikan is a liquid product, its moisture content is more than 70%. Poernomo et al. (1984) reported the salt, moisture, protein, and fat contents of kecap ikan of 19.7%–22.4%, 66.7%–76.9%, 10.2%–10.5%, and 0.5%–0.7%, respectively. Halophilic bacteria are the main orgamisms found in kecap ikan. Rahayu et al. (1992) reported that the mold species found in kecap ikan were *Cladosporium herbarum*, *Aspergillus fumigatus*, and *Penicillium notatum*, whereas the yeast found in kecap ikan was *Candida claussenii*.

3.1.19 Pedah (Indonesia)

Pedah is processed using mackerel (*Rastrelliger* sp. and *Scomber kanagurta*) as raw material mixed with 30%–50% salt (Putro, 1993; Desniar et al., 2009). The basic pedah processing involves salting in two steps. The first salting step takes several weeks to develop its characteristic flavor and texture, and this is followed by a maturation phase. Fish and salt are arranged in layers alternately in a salting tank. By the end of the process, the fish are soaked in saturated brine solution with coarse salt remaining at the bottom and top covering the fish. This salting process normally takes 3 days (Huda, 2012). Maturation is performed by mixing the fish and salt thoroughly. During maturation, the fish are kept in bamboo baskets about 1–2 weeks (Rahayu et al., 1992). Generally, pedah is described as being a fatty, partly dried salty fish with reddish brown color and as being moist and slightly pasty with a flabby texture and a specific flavor and is cheesy, tasty, salty, and often mixed with a mild rancid flavor (Putro, 1993). Pedah contains 20.2%–21.5% protein, 1.2%–1.4% fat, 2.9% carbohydrate, 52.7%–54.0% moisture, 11.0%–14.4% salt, and pH of 6.0 (Poernomo et al., 1984; Sarnianto et al., 1984; Desniar et al., 2009). *Lactobacillus* sp. and *Staphylococcus* sp. were found during the fermentation (Sarnianto et al., 1984). They showed that the histamine content of pedah was highest among Indonesian fermented fish products. The histamine content reported was approximately 107–133 mg%, which is over the limit of 50 mg% regulated in Indonesia. Therefore, utilization of fresh fish and hygienic practices during processing is necessary to produce higher quality pedah.

3.1.20 Bagoong and Patis (Philippines)

Bagoong and patis are processed in a similar manner, except that in the latter, the fermentation is allowed to proceed until the flesh of the fish disintegrates into a liquid phase (Sanchez, 1977). Bagoong is characterized by its reddish brown color and is a salty paste with slightly cheese-like odor. Dilis (*Stolephorous* spp.), tamban (*Sardinella fimbriate*), and galunggong (*Decapterus* spp.) are the fish employed. The fish is washed thoroughly and drained. For the traditional method of production, bagoong is prepared from wet treatment of fish with salt over a period of 3–12 months at ambient temperature with occasional stirring to make sure the salt is spread evenly until it develops the characteristic flavor and aroma (Espejo-Hermes, 1998). Salt is mixed with the drained fish at varying proportion from 1:3 to 2:7 (salt:fish), depending on its size. Angkak (or red yeast rice) is added to give the bagoong its characteristic red or pink color. Bagoong can be blended with spices, flavoring, and other approved coloring materials (Beddows, 1985). Some manufacturers grind the fermented product finely and sell the resulting mixture as fish paste. The finished product is consumed as is or used as a flavor enhancer in the preparation of other types of food (Montano et al., 2001). The action occurring during fermentation of bagoong is partly one of liquefaction. Shortly after salting, considerable moisture is extracted from the fish. The nitrogen content of this liquid is low initially, but as the storage period continues, the fish protein is hydrolyzed and dissolves in the salt brine.

Patis is clear straw-yellow to amber in color depending on the type of fish used. The chemical composition of 6-month-old patis showed that it has a pH of 5.85, 26.10% NaCl, 2.26% total nitrogen, 1.31% formol nitrogen, 0.31% ammonia nitrogen, 1.0% amino nitrogen, and 0.84% lactic acid. The dominant amino acids found in patis are lysine (1083 mg%), histidine (976 mg%), and glutamic acid (960 mg%). The enzymes from *Bacillus subtilis* and *B. coagulans* in addition to the endogenous ones in the intestines of the fish cause protein hydrolysis. *Bacillus pumilus* is the dominant species throughout the fermentation process. Other bacteria responsible for the early stage of fermentation were *B. coagulans*, *Bacillus megaterium*, and *B. subtilis*, while *B. licheniformis*, *Micrococcus colpogenes*, *Micrococcus roseus*, *Micrococcus varians*, and *Staphylococcus* species on the later stage of fermentation.

3.1.21 Bagoong Isda (Philippines)

Bagoong isda is a fish paste product in the Philippines, making from different species of fresh fish, *Ophicephalus striatus* (burong dalag),

Tilapia plumbeus (burong tilapia), *Chanos chanos* (burong bangus), and *Therapon plumbeus* (burong ayungin) with or without addition of angkak. The fish is scaled, gills and fins are removed, split into butterfly fillet, washed, packed alternately with a mixture of cooked rice and 2.5%–20% salt at the ratio of 35:65 (salted cooked rice fish), and fermented for 7 days at 28%–30°C. It is a popular traditional food in the central part of Luzon, particularly in Pampanga, Bulacan, Cavite, and Nueva Ecija. It is consumed as a sauce or as main dish after sautéeing in vegetable oil with garlic and onion. Since it contains 8%–25% protein, it is a good source of protein in the Filipino diet. The meat flavor is comparable to the soybean-based sauces, paste, or miso in Japan and China. *Stolephorus commersonii* (dilis) and *Decapterus macrosoma* (galunggong) are the two common fish species for the production of bagoong isda. *Bacillus subtilis, B. cereus, Lactobacillus brevis, Streptococcus faecalis. Leuconostoc mesenteroides, Pediococcus cerevisiae, Lactobacillus fermentum*, and *Lactobacillus plantarum* were isolated during the fermentation.

3.1.22 Yu-Lu (China)

Yu-lu is a traditional fermented fish sauce widely consumed as condiment for cooking in the southern and eastern parts of China, especially in Guangdong and Fujian Provinces. It is produced by fermenting anchovies (or the other small marine fish), which have been preconditioned with salt at the level of 30%–40%, salt/fish (w/w). After fermentation for 12–18 months, the mixture is kept at 40°C–50°C for 1–2 weeks. The obtained liquid is filtered, and the filtrate of fish sauce is called top-grade fish sauce. The fish residue is mixed with saturated brine again. The mixture is left for 1–4 months. The obtained liquid is boiled, filtered, and added with caramel or other additive. The diluted fish sauce is obtained at this step. Fermentation results from the action of enzymes in the fish and some halotolerant and halophilic microorganisms (Jiang et al., 2007). The clear brown liquid portion of the fermented fish is called yu-lu, which has a strong and complicated umami taste and nutritive value. Yu-lu contains all essential amino acids and many vitamins and minerals (Lopetcharat et al., 2001). Present yu-lu processing, like some traditional processes, lacks quality assessment and quality control. Jiang et al. (2007) reported that the total soluble nitrogen, formol nitrogen, TCA (trichloacetic acid) soluble peptides, total titratable acid, and non-enzymatic browning index increased throughout the fermentation period. Moreover, glutamic acid, lysine, leucine, and valine were prominent in yu-lu. The results from quantitative descriptive analysis (QDA) test showed that incubation after 180 days for a week can accelerate the flavor formation of yu-lu. For Taiwan, fresh small fish, such as silver anchovy, are used as the raw material. The main procedures include salting, fermentation, agitating, aging, and separation. The clear supernatant is yu-lu and the thick precipitate is fish paste (Chen, 1993).

3.1.23 Shottsuru (Japan)

The manufacture of shottsuru, the Japanese fish sauce, is a household industry in small fishing villages. It is generally made using species of small fish, particularly the liver and meat of squid. In Japan, other fermented fish products are kusaya which are horse mackerel (*Irachurus japonicus*) pickled in fermented brine and then sun-dried in the Izu islands. Shottsuru may be prepared from sandfish (*Arocoscopus japonicas*), Japanese sardine (*Sardinops melanostictus*), anchovy (*Engraulis japonicas*), horse mackerel, pacific mackerel (*Scomber japonicas*), and mysid (*Neomysis isaza*). Shottsuru is made from fish meat added to 10% mysid. The fish and salt mixture, usually in a ratio of 3:1 or 7:2, is allowed to disintegrate in earthen jars or in wooden or cement tanks for 12–18 months. During fermentation, it is stirred well and more salt is added several times. The liquid is extracted, boiled, and filtered through sea sand. After a few months of settling, the supernatant liquor is decanted off carefully for excepted sediment. This may be improved by the addition of caramel, soy sauce, or seasoning made from chemical hydrolysate of vegetable protein to the fish before subsequent extraction. The shottsuru is used in cooking shottsuru dish. In shottsuru, glutamic acid and lysine predominate, followed by aspartic acid, alanine, glycine, leucine, and valine, while tyrosine, methionine, cystine, and tryptophan are in small quantities. Generally, the amino acid composition in shottsuru shows a pattern similar to that in fish meat used as materials. The flavor of the sauce depends on the quality of the amino acid components of fish protein. Glutamic acid in particular contributes the mild flavor and meaty aroma of the product.

3.1.24 Ikanago Shoyu (Japan)

Ikanago in Takamatu of Kagawa Prefecture is manufactured by mixing two parts of sand eel (*Ammodytes personnatus*) to one part of soy sauce and then allowing it to pickle for over 100 days. After hydrolysis, the salt liquid is made to extract and filter. Ika-shoyu in Hokkaido Prefecture and ishiru in Noto of Ishikawa Prefecture are produced by the process same as ika-shiokara (fish paste) except that the fish flesh is allowed to disintegrate further into liquid state (Itoh et al., 1993).

3.1.25 Konago (Japan)

Konago may be prepared from smaller sand eel instead of konago. Iwashi-shoyu is manufactured by mixing five parts of anchovy to one part of salt and placing the mixture in vats to pickle. After 2 days, the fish are removed and washed with tap water, de-fatted, mixed with salt, put in a barrel, covered, and weighted with stones for 3 years in a cool and dark place. After fermentation, the liquid is filtered (Itoh et al., 1993).

3.1.26 Jeotkal (Korea)

Jeotkal is a generic name for high-salt-fermented fish products, which contains partially hydrolyzed tissues of fish organs immersed in liquid exudates of the fish (Lee et al., 1993). There are more than 30 kinds being processed and consumed in Korea. The specific name of jeotkal is based on the raw materials used such as sae-jeot for fermented small shrimp and myeolchi-jeot for fermented anchovy. Only salt is used as the curing stuff in most fish fermentations found in Korea and various salt concentrations are used depending on raw materials (Mheen, 1993). The ingredients are *Engraulis japonicas* (anchovy), myeolchi-jeot and/or *Ammodytes* sp., and salt at the level of 20%–30% (Mah et al., 2008; Guan et al., 2011). It is made by salting fish, whole or after slicing, followed by aging for a long time (few years) at a proper temperature, and is characterized by its specific flavor, which is formed by enzymatic digestion of the substrate. The type of microorganisms involved in the fermentation is mainly dependent upon the concentration of salt. The minimum amount of salt is determined by the salt tolerance of infectious and putrefactive microorganisms present in the product. Jeotkal, which has the original shape of raw materials, is made after 2–3 months of fermentation for use as sidedishes with rice meals. Jeotkuk, which is liquid and used as fish sauce, is made by holding jeotkal for a longer time (over 6 months) for extensive enzymatic hydrolysis and subsequent crushing, filtering, and pasteurization. The most important conditions for proper aging of jeotkal are temperature and salt concentration. An ageing temperature of 20°C and salt concentration of 20%–21% yield jeotkal of the best quality. It is used as ingredient for kimchi, a Korean fermented vegetable food, and for condiments (Lee et al., 1993). It consists of 67.6% moisture, 1.1% crude fat, 7.4% crude protein, 0.1% carbohydrate, 22.7% salt, and pH of 6.0. *Micrococcus*, *Brevibacterium*, *Sarcina*, *Leuconostoc*, *Bacillus*, *Pseudomonas*, *Flavobacterium*, *Staphylococcus*, *Paenibacillus tyraminigenes*, and yeasts are found initially, while halophilic bacteria such as *B. subtilis*, *L. mesenteroides*, *Pseudomonas halophilus*, and *Sarcina itoralis* appear during the aging period (Lee et al., 1977; Mheen, 1993). *Achromobacter*, *Corynebacterium*, *Micrococcus*, *Pediococcus*, *Alcaligenes*, *Aeromonas*, and *Moraxella* were also isolated from jeotkal (Lee et al., 1993).

3.2 Shellfish Products

Wonnop Visessanguan and Siriporn Chaikaew

Nowadays, shellfish is a common part of indigenous cuisines throughout the globe. Despite the name, shellfish are not a kind of fish but are simply animals that live in water. “Shellfish” is a common term for exoskeleton-bearing aquatic invertebrates used as food, including various species of mollusks, crustaceans, and echinoderms harvested from both marine and freshwater environments. Familiar mollusks enjoyed as a food source by humans include many species of cephalopods (squid, octopus, and cuttlefish) and bivalves (clams, mussels, oysters, winkles, and scallops). Some crustaceans commonly eaten are shrimp, lobster, crayfish, krill (or shrimp-like zooplankton), and crabs. Although echinoderms are not as frequently harvested for food as mollusks and crustaceans, sea urchin roe is quite popular in many parts of the world. Archeological findings provide evidence that humans have been making use of shellfish as a food item for thousands of years. Shellfish provides an important source of dietary amino acids and proteins around the world, especially in countries with coastal areas. In addition, inter-tidal herbivorous shellfish such as mussels and clams can help people reach a healthy balance of ω-3 and ω-6 fats in their diets. For this reason, the eating of shellfish is often encouraged.

Due to their highly perishable nature, shellfish are susceptible to rapid quality changes and spoilage. In order to preserve them, fermentation has been practiced and consumption is widespread mainly in continental Southeast Asian countries. These products are known by diverse names by local population of each nation. Fermented shrimp products of each group in each region are processed with almost the same technologies but the fermentation time and salt contents of the products are somewhat different. These products are usually used as side dishes, condiments, or main dishes in which the primary function is to provide a salty and umami taste to the food (Hajeb and Jinap, 2012). Like alkaline-fermented fish products, they are usually carried out with relatively high salt addition, undergo extensive hydrolysis, exhibit neutral or alkaline pH, and have no carbohydrate substrate additives. As a consequence of microbial conversion, the original form of shellfish is commonly reduced to the forms of paste or sauce. Nevertheless, some fermented shellfish products especially from crab, squid, and mussel are still present as their original forms even after prolonged fermentation. Compared with fish, this may be elaborated by the distinct muscle structure and composition of shellfish that cannot be easily digested by enzymatic hydrolysis through both microbial and endogenous enzymes evolved during fermentation and the fact that no mechanical disruption process is involved in their preparation.

3.2.1 Shrimp/Krill Paste

3.2.1.1 Bagoong (Philippines) Bagoong is shrimp paste made from minute shrimp or krill. It is a popular condiment in the Philippines (Montano et al., 2001). It is also used in Hawaii, and other regions of the Pacific. Bagoong is the undigested residue of partially hydrolyzed shrimp. The characteristics of this product vary depending on the region where it is made and consumed. In Tagalog provinces, the shrimp paste is completely fermented and ground, with or without coloring matter added. In the Ilocos region and Pangasinan Province, the products are either partially or completely fermented. In Visayas and Mindanao, the product is slightly fermented without liquid. As a result, the shrimp is hard and solid salt is present (Olympia, 1992). Bagoong can also be made from shrimp fry. This type of bagoong is known as "bagoong alamang" or "bagoong aramang" in Ilocano or simply "alamang" or "oyap" in the south, in western Visayas, simply "ginamos" or "dayok."

For bagoong production, the shrimps (*Atya* sp.) are cleaned thoroughly and washed in weak (10%) brine solution. The shrimp are then mixed with salt (20%–25%). The mixture is kept inside a large earthen jar, covered, and left to ferment for a period of 3–12 months at ambient temperature with occasional stirring to make sure the salt is spread evenly (Martin and Sulit, 1955; Espejo-Hermes, 1998). The finished product has a paste-like consistency and is consumed as is, or used as a flavor extender in the preparation of other types of indigenous food. Pink and salty bagoong is marketed as "fresh." Some final products are pale gray to white in color. To obtain the characteristic red or pink color of some bagoong, angkak made from rice inoculated with a species of red mold (*Monascus purpureus*) is added (Miller et al., 1946). The smell is extremely pungent and some consider it offensive.

The product contains 27%–40% moisture, 28%–40% protein, 1.4%–2.6% fat, 20%–28% ash, 13%–25% NaCl, and pH of 7.2–7.8. *Bacillus*, *Micrococcus*, and *Moraxella* species are found during the fermentation (Fujii et al., 1980). Moreover, the increase in free glutamic acid during fermentation enhances the desirable umami taste in the product. Prolonged fermentation of shrimp paste for more than 1 year brings about a decrease in free amino acids, especially free glutamic acid (Olympia, 1992). Salting has a significant impact on the rate and quality of fermentation due to their effects on the bacteria involved in the process (Sanchez, 2008). Bagoong is eaten raw or cooked, and used mainly as a flavoring agent or in condiment preparation, sautéed with onion and garlic, and served as an appetizer with tomatoes or green mangoes, eaten with vegetables, or often as a main source of protein (Olympia, 1992).

3.2.1.2 Belacan (Malaysia) Belacan or balachan is a popular fermented shrimp paste of Malaysia. It is made from fresh tiny shrimp (geragau, geragok, *Acetes*, or mysid) and 10%–15% salt (w/w) (Abu Bakar, 2003; Leong et al., 2009; Hajeb and Jinap, 2012). The production begins as soon as the shrimps are harvested. There are various methods of belacan production. In general, the shrimps are mixed with salt and immediately dried under the sun for half a day or one whole day. The semi-dried salted shrimps are pounded; at this point the paste mixture is placed in a tightly sealed container and kept for 7 days at room temperature to allow fermentation to take place. The fermented shrimp paste is dried, mashed, and kept for several days until a desired texture is achieved. Finally, the paste is formed into different shapes and sizes (Adnan, 1984). Various techniques used to produce belacan result in composition and quality differences among products (Hajeb and Jinap, 2012). Belacan is salty in taste and has a strong shrimp odor, mainly due to fermentation (Adnan, 1984). It is a thick paste with a grayish pink to grayish purple color and it has a pungent shrimp taste, i.e., buttery, sting taste with a touch of sweetness. The product contains 37% moisture, 27% ash, 23% crude protein, 1% fat, 5% carbohydrate, 20% NaCl, and pH of 7.9 (Abu Bakar, 2003; Sharif et al., 2008). Many types of bacteria are known to be involved in the fermentation of belacan. Several researchers have identified them as the members of *Bacillus*, *Clostridium*, *Enterobacter*, *Escherichia*, *Micrococcus*, *Proteus*, *Streptococcus*, *Acetobacter*, *Pediococcus*, *Lactobacillus*, *Sarcina*, *Staphylococcus*, *Brevibaterium*, *Flavobacteirium*, and *Corynebacterium* (Merican et al., 1980; Steinkraus, 1996). Belacan contains less than 15% of NaCl, which is prone to the presence of *Salmonella* (Anon., 1985). The detection of *Salmonella enterica* subsp. *enterica* serovar Kentucky and Chincol in belacan indicates either post-processing contamination or the ability of these *Salmonella* serovars to survive the salting process during the manufacture of the shrimp paste. The presence of *Salmonella* also poses a serious health hazard as shrimp paste is served as a sauce and is not subjected to severe heat treatment (Arumugaswamy et al., 1995). Moreover, Adnan (1984) recommended that a high-quality belacan should consist of about 50% moisture and 13%–17% salt. The best-quality belacan will be obtained after 1 or 2 weeks of fermentation. It has a strong umami taste and contains considerable amounts of glutamate and 5′-nucleotides (Khairunnisak et al., 2009; Jinap et al., 2010). The concentration of free glutamic acid found in different brands of

Malaysian belacan was reported to be 180–530 mg/100 g. The total concentration of 5′-nucleotides of belacan ranges from 0.85 to 42.25 mg/g (Jinap et al., 2010). Dishes containing belacan have been shown to have a high intensity of meaty flavor (Jinap et al., 2010). A high content of salt in belacan makes it inconsumable in the raw form, but it is often used as flavor enhancer and condiment in several local dishes such as chili belacan (sambal belacan), spicy noodle soup (laksa), fried rice with belacan (nasi goreng belacan), stir fried chili paste (sambal tumis), sour and spicy fish stew with belacan (asam pedas), stir-fried water convolvulus with belacan (kangkung goreng belacan), and Indian fried noodles (Lee, 2001; Hutton, 2003, 2005). Belacan is commonly sold in dried blocks that range in color from pink to dark brown. Generally, belacan is roasted prior to usage to enhance the flavor as well as killing bacteria (Hutton, 2005).

In Brunei Darussalum, belacan is made from shrimp (*Acetes* spp.) mixed with 15%–20% salt. The mixture is kept overnight. The shrimp and salt mixture is sun-dried for a few hours, and then pounded to form cohesive paste. Sometimes dry shrimp is used, but it is not popular because it does not produce a cohesive paste. The paste is allowed to ferment in a container for 2 months or longer. The end product has been described as a pungent, salty paste with a strong shrimp flavor and sold in ball/block forms.

3.2.1.3 Kapi (Thailand and Cambodia) Kapi is a typical traditional fermented shrimp paste, generally prepared from the planktonous shrimp or krill (*Acetes vulgaris* or *Mesopodopsis orientalis*), mixed with salt (25%–35%), sun-dried to decrease the moisture content, and then blended thoroughly. The paste is compacted and allowed to ferment for at least 2 months until the desired aroma is developed (Phithakpol, 1993). Kapi has a paste-like consistency and a distinctive odor and flavor. The color varies slightly depending on the raw materials used. Kapi from koei-dtaa-daeng has a pinkish purple color and that from koei-dtaa-dam has dark purple color. Kapi possesses a characteristic flavor at the time of ripening, which varies for each batch (Phithakpol et al., 1995).

Kapi has moisture and protein contents of 36.8%–49.9% and 20%–25%, respectively. Kapi contains a high amount of salt, ranging from 20% to 25%. This is mainly due to a large amount of salt added during the processing and the drying during fermentation. Moreover, kapi has 1.5%–2% fat, 2.3%–13.0% carbohydrate, 24%–29% ash, and pH of 7.4–7.7. Kapi is very rich in umami taste and contains a high amount (647 mg/100 g) of free glutamic acid (Mizutani et al., 1987). *Bacillus licheniformis*, *Bacillus sphaericus*, *Bacillus amyloliquefaciens*, *Micrococcus* sp., *Corynebacterium* sp., *Lactobacillus* sp., *Straphylococcus aureus*, *Staphylococcus epidermidis*, *Tetragenococcus halophilus*, *Tetragenococcus muriaticus*, *Virgibacillus halodenitrficans*, *Oceanobacillus kapialis*, *Lentibacillus kapialis*, *Salinicoccus siamensis*, *Alkalibacterium thalassium*, *Alkalibacterium pelagium*, *Alkalibacterium putridalgicola*, and *Alkalibacterium kapii* are found during the fermentation (Kobayashi et al., 2003; Pakdeeto et al., 2007a, 2007b; Ishikawa et al., 2009; Namwong et al., 2009; Wittanalai et al., 2010; Tanasupawat et al., 2011a).

Due to the high content of salt, kapi is used in various spicy soups or for flavoring dishes. Another main use is to make a condiment called "namprik" for which kapi is mixed with garlic, bird chili, and lime juice or tamarind. The ingredients vary according to local practice and taste. Kapi is also cooked with fried rice or coconut milk to make a vegetable dip (Phithakpol et al., 1995). Apart from serving as a good source of proteins, those products might possess bioactivities, especially natural antioxidants, which provide health benefits. Accordingly, these products could be marketed as health foods with high market value. Moreover, kapi exhibits antioxidant activity, which is stable to pH and heating. Therefore, it can be an important source of natural antioxidants as well as nutrients (Faithong et al., 2010).

3.2.1.4 Mam Ruoc and Mam Tom (Vietnam) Mam ruoc and mam tom are Vietnamese fermented shrimp pastes. These are more popular in North and Central Vietnam. Depending on the shrimp used, mam ruoc is used in the central and southern parts of Vietnam and is made from small shrimp caught during the rainy season. On the other hand, mam tom is a northern style and made from larger shrimp. Shrimps are dried in the sun for 3 months, mixed with salt (2%–5%, w/v), ground to a powder, and placed in a jar to pickle in the sun for another 1.5 months. Sugar is then added to that mixture and left to ferment for another month. The mixture is then dried again in the sun for 10 days. Mam ruoc tends to be light pinkish-gray and is a thick paste rather than a block. It is used as a flavor enhancer in bun bo hue (Vietnamese hue-style beef noodle soup) and dipping sauce.

3.2.1.5 Nappi (Bangladesh) Nappi is produced in coastal villages in Cox's Bazaar district of southern Bangldesh by the Rakhiyane people, a minority population of Arakanese origin. The main customers inhabit the Chittagong hill tracts, to which it is sold via Rangamati market. The peak shrimping season is September–January, when paste production also peaks. However, since some shrimp can be caught year-round, some nappi is produced throughout the year. Unlike most other parts of Asia, in this region the shrimp fermented are mostly not planktonic organisms, but juveniles of *Penacus indicus*, *Metapenaeus monoceros*, and *Perapaeniopsis stylifera*. The reported conversion rate of fresh shrimp to paste is 40:15 (Choupaldandi) and 40:20 (Moshiskhali island). Local salt made by solar evaporation is used to prepare fermented shrimp paste in this region (Ruddle, 1990). Only simple tools are needed to make fermented shrimp paste, and they can mostly be fabricated locally from readily available materials. The exceptions are the "behundi" net for catching shrimp, and various metallic containers, where they are used to unload the shrimp from the boats, bamboo mats for sun-drying the shrimp, mortars and pestles for comminution, bamboo packing-transport baskets, and leaves for lining the basket (Ruddle, 1990). For the traditional method, the shrimp (juveniles of *P. indicus*, *M. monoceros*, and *P. stylifera*) is comminuted with salt and kept one day for sun-drying, followed by moistening and comminution, crumbling, and sun-drying (1 day), and sometimes blending to obtain uniform color and finally packing (Ruddle, 1990).

3.2.1.6 Ngapi (Myanmar) Ngapi is a pink to reddish colored shrimp paste made from very small mysis or planktonic shrimp. To produce this type of paste, shrimp is mixed with salt and dried under the sun for 3–4 days. The mass is later pounded with the addition of extra salt and water to obtain paste. The paste is sun-dried again for a day, pounded for 3–4 times to get a homogeneous and smooth mass, which is then fermented for 4–6 months. This shrimp paste is rich in umami taste. Hmyinnga-pi contains considerable amount of free glutamic acid (10.1 mg/100 g). Ngapi, an important element of Myanmar food for a long time, is consumed locally, and there is no export of this commodity.

Small planktonic shrimp from Tanintharyi and Ayeyarwady Delta areas is used to prepare ngapi. Ngapi made from Tanintharyi is called "seinsa ngapi," whereas the one made from Ayeyarwady Delta shrimp is called "hmyin ngapi." Actually, these shrimps in different small pelagic larval stages of a shrimp call mysis are caught from brackish water (Tyn, 2004). For the traditional method, the small planktonic shrimp (*Acetes* sp. and *Mysidacea* sp.) is mixed with salt (20%–25%, w/v) and dried under the sun for 3–4 days. The mass is later pounded with the addition of extra salt and water to obtain paste, which is sun-dried for a day and pounded for 3–4 times to get a homogeneous and smooth paste. The paste is then fermented for 4–6 months (Tyn, 1986). Seinsa ngapi is a pink to reddish colored shrimp paste that is rich in umami taste. The product contains 70%–85% moisture, 15%–25% protein, 1%–10% fat, 1%–1.5% minerals, and pH of 7–8.2 (Myint et al., 1997; Sein, 1999). Hmyin ngapi contains considerable amount (10.08 mg/100 g) of free glutamic acid (Tyn, 1983). *Staphylococcus aureus*, *Bacillus* sp. *Pseudomonas* sp., and *Micrococcus* sp. are found during the fermentation. Food-borne pathogenic bacteria such as total coliforms, *Escherichia coli*, *Salmonella*, *Vibrio*, and *S. aureus* were found to be present in shrimp paste. So, it is necessary to thoroughly clean the raw material and also to take measures to prevent contamination during processing and fermentation because most of the contamination is caused by negligence of elementary rules of hygiene (Tyn, 2004). It is used in a wide array of dishes and is eaten in myriad ways: it can be eaten on its own, such as baked or roasted ngapi, as a watery preparation called "ngapi yay," as a salad, as a pounded mixture with chili, or fried like balachaung. It is also used as a soup base and in main dishes. Raw ngapi is not intended for direct consumption.

3.2.1.7 Terasi Udang (Indonesia) Terasi udang, a fermented shrimp paste, is a traditional condiment used commonly in Indonesia (Hajeb and Jinap, 2012). It is consumed in small quantities as a flavoring agent. The most important center for its manufacture is Began Si-api-api (Sumatra) (Van Veen, 1965). Terasi, a variant of dried shrimp paste, is usually purchased in dark blocks but is also sometimes sold ground in powder form (Ruddle and Ishige, 2010). On Java island, shrimp paste is often prepared from pre-cooked shrimps called "brabon" (literally means mother paste) (Hajeb and Jinap, 2012). Terasi udang is usually made from shrimp *Schizopodes* sp., *Atyas* sp., or *Mytis* sp., and 15% (w/v) salt (Van Veen, 1965; Budhyatni et al., 1982; Hajeb and Jinap, 2012). Fermentation usually takes 6–9 months until the desired terasi aroma develops (Yunizal, 1998; Hajeb and Jinap, 2012). The color and aroma of terasi vary depending on which village produces it. Terasi udang is dark brown, gray, or red paste in color with a distinct taste and a strong aroma (Kobayashi et al., 2003). It has a considerable level (1508 mg/100 g) of free glutamic acid and is rich in umami taste (Mizutani et al., 1987). Terasi udang is very popular as appetizer in the daily diet of Indonesian families. It is usually mixed with chili, garlic, and salt, which they call "sambal-terasi" (Hajeb and Jinap, 2012). Terasi is used as a flavor enhancer and as an ingredient in various Indonesian dishes.

The product contains 40% moisture, 30% protein, 3.5% fat, 0.1% calcium, 0.25% phosporus, 0.0031% ferrous, and pH of 7.5 (Soedarmo and Sediaoetama, 1977; Surano and Hosono, 1994). *Tetragenococcus halophilus*, *T. muriaticus*, *Bacillus* sp., *Pseudomonas* sp., *Micrococcus* sp., and *Proteus* sp. are found during fermentation (Budhyatni et al., 1982; Kobayashi et al., 2003). However, *T. muriaticus* strains were found in this product and reported to produce histamine in fermented fish in Japan and Thailand (Kobayashi et al., 2003). *Staphylococcus* sp., *Bacillus* sp., and *Proteus* sp. were found to be present (Budhyatni et al., 1982).

3.2.2 Shrimp/Krill Sauce

3.2.2.1 Nam-Poo (Thailand) In Thailand, fermented crab paste, nam-poo is very popular in the North. It is made from fresh water crab, *Sesarma mederi* (poo-sa-mer) and/or *Somanniathelpusa* sp. (poo-naa) with lemongrass leaves, galangal leaves, guava (*Psidium guajava* L.) leaves, salt, and water. The live crabs are washed, put in a mortar, added with leaves and a little water, and pounded until finely ground. The mixture is pressed through a sieve to obtain liquid. This is normally done by hand through a woven bamboo sieve with mesh of 2–3 mm. This liquid portion is the first extract. The solid fraction is returned to the mortar and a little water is repeatedly added for pounding and extracting of liquid to get second and third liquid portions or until the color shows the liquid as too diluted. The liquid portions are pulled and left to ferment overnight. The next morning, a small amount of salt is added to the greenish yellow liquid, which is gently cooked at low temperature until it turns into a dark paste. This finished product is then packed in a glass jar and screw-capped. Nam-poo is an ingredient in nam-prik, a condiment used for dipping boiled vegetables such as bamboo (*Bambusa vulgaris* Schrad. or *Phyllostachys edulis* (Carrière) J.Houz.) shoot, eggplant (*Solanum melongena* L.), string bean (*Vigna unguiculata* (L.) Walp.), etc. The product contains 20.8% moisture, 30.0% protein, 1.4% fat, trace to 0.6% fiber, 21.5% ash, 10.3% NaCl, 1.1% acidity, and pH of 6.6. The main microorganism is *Bacillus* sp. The microbial quality of this product depends on the species and load of pathogenic bacteria as well as the number of lactic acid bacteria. However, the presence of pathogenic bacteria in nam poo such as *Bacillus cereus*, *E. coli*, *Salmonella typhimurium*, and *S. aureus* may cause direct negative effects on human health (Phithakpol et al., 1995).

3.2.2.2 Nan Ji Xia Jiang Zhi (Taiwan) Nan ji xia jiang zhi is Taiwanese krill sauce. For the traditional method, the frozen krill is mixed with salt at 30% (w/w) in vat. The krill mixture is thawed at room temperature, mixed, covered, and stored at 5°C for 3 months without stirring. The ferment is squeezed and filtered. The filtrate is boiled and bottled (Chen and Chen, 1978). The product contains 64.7% moisture, 1.56% total nitrogen, 519 mg% amino nitrogen, 24.3% salinity, and pH of 5.97 (Chen, 1993).

3.2.2.3 Ngan Pya Ye (Myanmar) Ngan pya ye is a dark brown sauce obtained as a by-product of nga-pi (fermented shrimp), manufactured in Myanmar. Ngan pya ye is usually aged in concrete storage tanks for one year before being consumed. During this period partial hydrolysis occurs resulting in the development of a unique flavor. After aging, the sauce is boiled for 4–6 h to reduce the moisture content; at the same time, the boiling also contributes to partial sterilization of the sauce. Solar dryer or steam boiling is used to concentrate ngan pya ye. To improve the dark color and strong smell of ngan pya ye, activated carbon together with heating and continuous stirring is used for bleaching ngan pya ye. Using this method, the color of the bleached ngan pya ye turns from dark brown to golden yellow and part of the odor can also be removed simultaneously (Tyn, 1983). Boiled ngan pya ye contains 54 mg/100 ml of free glutamic acid, which is responsible for the umami taste in the product (Sanda, 1973).

3.2.2.4 Xia You (Taiwan) Fresh shrimp meat or whole shrimp are mixed with salt (30%, w/w) and kept at room temperature for about 6 months without stirring. The processing procedure is almost the same as that for yu-lu (described in Section 3.1.22) except that pressure (squeeze) filtration is used for shrimp sauce instead of just separation (pouring) for yu-lu. The product contains 69.1% moisture, 1.5% total nitrogen, 571.6 mg% amino nitrogen, 25.6% salinity, and pH of 5.9 (Chen, 1993).

3.2.3 Other Shellfish Products

3.2.3.1 Ogiri-Nsiko (Nigeria) Ogiri-nsiko is a fermented crab condiment prepared in West Africa. In the traditional method of ogiri-nsiko preparation, freshwater whole crab (*Sudananautes africanus africanus*) is cleaned to remove slime and soil, wrapped in fresh banana leaves, and fermented for 48–72 h. Smoking or sun-drying then dries the ferment for 4–5 days. Fermentation increases the pH of the substrate from 6.2 up to 8.4. Fermented crab, as a rich source of protein, offers potential substitute for meat and cultured dairy products (Achi et al., 2007). *Bacillus subtilis*, *B. pumilis*, *Staphylococcus saprophyticus*, *Micrococcus luteus*, and *Pseudomonas* sp. are found during fermentation. It is used mostly in soup preparation and as meat-like seasoning for vegetable stews (Achi, 2005).

3.2.3.2 Jeotgal (Korea) Jeotgal, or jeot, a traditional Korean salted and fermented food, is used as an important additive for improving the taste of foods or alone as a food (Guan et al., 2011). Based on the major raw materials and the regional fermentation methods, jeotgal is classified into more than 160 different kinds, of which approximately 30 are sold commercially in Korea (Guan et al., 2011). The types of jeotgal vary depending on main ingredients, region, and family and personal preferences. Saeujeot, made by fermentation of tiny, salted shrimps (*Acetes japonicus*), is the most representative and best-selling fermented seafood in Korea. Ojeot and yukjeot are saeujeots made with shrimp harvested in May and that made with shrimp harvested in June, respectively. Tohajeot is made with small freshwater shrimp in Jeolla Province. While jogaejeot is made with shellfish, guljeot is made with oyster, ojingeojeot is made with squid, and ggolddugijeot is made with small squid. For eoriguljeot (salted and spiced oysters), oysters are salted for 3 days and then seasoned with red pepper powder, sugar, garlic, and ginger. They are stored in a jar in a cool place until required. For gejeot, shelled crabs are salted and cut into pieces. The seasoning used is soy sauce, garlic, sesame seeds, ginger, sugar, sesame oil, and red pepper threads, along with monosodium glutamate. It must be added at least 2 h before serving. The crab meat can be returned to its shell for serving. These crabs must be eaten within a couple of days.

The fermentation period of jeotgal depends on the salt concentration and fermentation temperature: 2 months for jeotgal with low salt levels (6%–18%) and a few years for jeotgal with a high salt content (>20%). Salinity and pH are 17.9%–28.7% and 7.82–8.74, respectively. Spontaneous fermentation without the use of starter cultures or sterilization during the seafood fermentation leads to the growth of various microorganisms, which are primarily derived from raw materials, sea animal-associated environments (e.g., seawater or sea-tidal flat), and solar-salts used for fermented seafood preparation. *Bacillus* sp., *Acromobacter* sp., *Brevibacterium*, *Carnobacterium jeotgali*, *Halalkalicoccus jeotgali*, *Natronococcus jeotgali*, *Halomonas jeotgali*, *Nesterenkonia jeotgali*, *Haloterrigena jeotgali*, *Methylobacterium jeotgali*, *Jeotgalicoccus Halotolerans*, *Psychrobacter jeotgali*, *Clostridium*, *Flavobacterium*, *Halobacterium*, *Leuconostoc*, *Micrococcus*, *Lentibacillus*, *Pediococcus*, *Pseudomonas*, *Sarcina*, *Sporosarcina*, and *Serratia*, as well as the yeasts *Saccharomyces* and *Torulopsis* are reported during fermentation (Lee, 1969; Sands and Crisan, 1974; Cha et al., 1983; Chen, 1993; Yoon et al., 2001, 2003a, 2003b, 2006; Aslam et al., 2007; Roh et al., 2007, 2009; Jung et al., 2010; Kim et al., 2010).

The favorable taste and flavor may be developed gradually during fermentation by several enzymatic reactions and the process of microbial degradation. Protein hydrolysis of jeotgal may rely mainly on the action of endogenous proteinases associated with marine animals and only partially on those of halophilic bacteria. As a result, the fermented seafood contains relatively high amounts of amino acids and the degradation products of seafood protein (Kim, 1988; Mok et al., 2000; Jung et al., 2013). Jeotgal is mainly used as a condiment in pickling kimchi and as a dipping sauce for pig's feet (jokbal) and blood/noodle sausage (sundae). Sometimes jeotgal, commonly saeu-jeot, is added to Korean style stews (jjigae) and soups (guk and tang), for flavor instead of using salt or soysauce (ganjang).

3.2.3.3 Shiokara (Japan) Shiokara is a traditional processed seafood and used as a flavor enhancer and ingredient in various Japanese dishes. Squid shiokara has three varieties, known as akazukuri, kurozukuri, and shirozukuri depending on their color (red, black, and white). Akazukuri, the must common type with the largest production volume, is made from sliced squid meat complete with skin. Shirozukuri uses squid mantle meat without the skin, while kurozukuri (a speciality of Toyama Prefecture) is made with the addition of squid ink. The main squid shiokara production areas are Hokkaido, Aomori, Iwate, and Miyagi Prefecture, although it is also made in various other places. At present, consumers prefer fresh shiokara with texture similar to that of raw squid, which is rather hard. The hardness of squid mantle muscle in shiokara was reduced with the degradation of paramyosin and myosin heavy chain. To control the hardness of shiokara, rice seed oryzacystatin was used, which suppresses proteolysis by papain-like cysteine proteinases (Funaki et al., 2010). Shichimi spice or yama wasabi (mountain wasabi) is added to shiokara. Black olives are also a nice pairing to shiokara. Hotaruika no shiokara is prepared in the same manner but using firefly squid. To prepare, slices of squid are mixed with salt (>10%, w/v) and stirred well in a barrel for 10–20 days. It is sealed and allowed to ferment, which can take up to 1 year (Subba Rao, 1967; Beddows, 1985; Okuzumi and Fujii, 2000). Its unique taste and texture arise from the combination of squid liver and tendered muscle. During the process of ripening, muscle proteins were digested by proteinases present in squid liver. The squid muscle becomes soft, and the flavor of the muscle changes during ripening. Many studies have shown that the umami taste, especially that of free amino acids, increases during ripening. This taste was thought to be generated by proteases or microbes present in squid liver (Mori et al., 1979; Takai et al., 1992).

The product contains 74.2% moisture, 7.8% salt, 11.6% protein, 8.68% ash, and pH of 6.8–7.7 (Amano, 1992; Okuzumi and Fujii, 2000). *Staphylococcus xylosus*, *S. saprophyticus*, *Staphylococcus equorum*, and *Staphylococcus wurneri* were found during fermentation (Okuzumi and Fujii, 2000). Whereas *Micrococcus* sp. was responsible for the ripening, halophilic *Vibrio* spp. and *Achromobacter* spp. were responsible for spoilage (Zenitani, 1955). Many food poisoning and putrefying bacteria are unable to grow in shiokara due to its high salt content. For example, *Vibrio parahaemolyticus*, the halophilic bacterium grows well in the conditions of 2%–5% salt, but cannot grow if salt concentration is more than 10%. Of food poisoning bacteria, many staphylococci are present in shiokara but *S. aureus* is not found at all. In traditional shiokara, a salt concentration of 10% or more prevents putrefaction, and the functions of autolytic enzymes proceed to digest the raw material. At the same time, its characteristic flavor is created. However, because the growth of putrefying bacteria is not inhibited in low-salt shiokara, it cannot be laid in for a long period and umami cannot be produced by ripening. Therefore, the product has to be flavored by seasonings and its preservability has to be maintained by means other than adding salt. To this end, it is treated with a combination of low-temperature storage, adjustment of pH and water activity, improved preservability through the use of additives, and so on. The final products are closer to dressed dishes than fermented food (Okuzumi and Fujii, 2000).

3.3 Hongeohoe

Ggot-Im Lee and Cherl-Ho Lee

Hongeohoe is a fermented raw skate that is traditionally eaten in the southern area of South Korea. Hongeo is a generic name of skate in Korea. Skates are cartilaginous fish belonging to the family Rajidae in the superorder Batoidea of rays. Skates live in oceans all over the world from fresh water to the depth of 3000 m. The history of hongeohoe is traced back to the 14th century in Korea. A few documents of the late Koryo Dynasty recorded that the residents of Heuk-San Do, an island located far away in the southeastern part of Korea moved into the mainland (a town named Naju) due to the frequent foreign invasion. They, however, kept on going out fishing to the sea near Heuk-San Do because they knew very well about the currents of the sea and had lots of information of fisheries of that region. Heuk-San Do was so far away that it took several days for them to sail back home, and this caused hongeo to be fermented on board. They learned

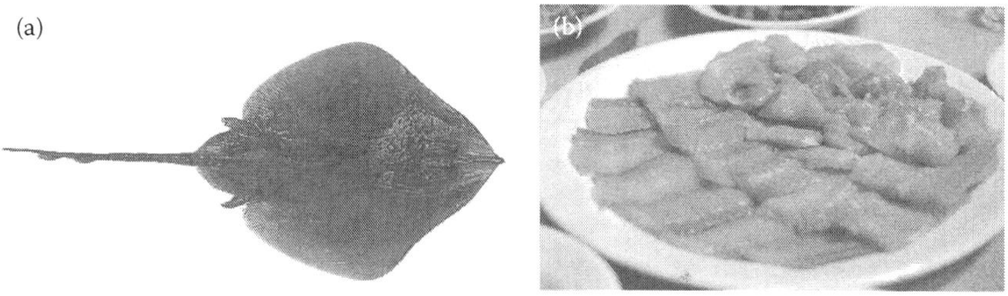

Figure 3.2 Skate (hongeo) (a) and hongeohoe (b).

that eating fermented hongeo not only did no harm to their stomachs even in hot summer but also its taste was even better than fresh one. Fermented hongeo (called hongeohoe) has usually been served uncooked (without any further preparation) with other side dishes like kimchi or rice wine, called makgeolli (Figure 3.2). Hongeohoe is a widely loved food and especially essential for the banquets of wedding and mourning nowadays.

3.3.1 Manufacturing Process of Hongeohoe

The traditional making of hongeohoe differs depending on the climate (Figure 3.3). It is not easy to have hongeo fermented because of extreme cold weather in winter. Therefore, the people keep hongeo in a

Skate (hongeo)

↓

Cleaned with paper

↓

Separated viscera from the body

↓

Cleaned with paper

↙ ↘

Put it within straw (in warm climate) or within compost (in cold climate) | Put it in an earthen jar or refrigerator for 7–8 days

↓

Hongeohoe

Figure 3.3 Manufacturing process of hongeohoe.

compost pile to make it fermented. They, on the other hand, make hongeo fermented in a stack of straws, which maintains warmth. However, for the sake of safety control and constant quality, refrigerators or low-temperature chambers are presently used for the fermentation process.

3.3.2 Chemical Composition and Physiological Function

Proximate composition of hongeohoe according to their fermentation temperature (4°C, 10°C, and 20°C) and fermentation period was analyzed (Choi et al., 2003). The crude protein, crude lipid, and crude ash content using freeze-dried samples were 86%–88%, 2.3%–2.7%, and 9.6%–11.3%, respectively, when it was fermented at 20°C for 6 days, and they did not show any big difference between fermentation period (days). The protein and lipid contents did not show significant change in proximate composition according to fermentation temperature and period. However, ash content was increased from 9.6% up to 11.3% over time, resulting from the loss of extracts (Table 3.1). These results are consistent with those from skate skin. The moisture, protein, lipid, and ash content of fresh and fermented skate skin were studied (Cho et al., 2004). Fresh and fermented skate skin contained moisture content of each 76.0% and 74.5%, protein content of 22.7% and 21.8%, lipid content of 0.5% and 0.7%, and ash content of 0.6% and 0.9%, respectively. The predominant minerals were potassium and phosphorus (53.5 and 33.0 mg/100 g in fresh skin and 10.46 and 10.5 mg/100 g fermented skin, respectively). Amino acid concentrations were lower in the fermented skin compared with the fresh skin. Histidine, glycine, alanine, and glutamic acid were the major free amino acids in both skins. Palmitic acid was the major fatty acid in both fresh (16.68%) and fermented (20.38%) skate skin.

Table 3.1 Changes in Proximate Composition during Fermentation of Hongeo

PARAMETER	BEFORE FERMENTATION	AFTER FERMENTATION (AT 20°C FOR 6 DAYS)
Crude protein (g/100 g)	88.09 ± 2.68	86.04 ± 6.11
Crude lipid (g/100 g)	2.29 ± 0.03	2.72 ± 1.06
Crude ash (g/100 g)	9.55 ± 0.24	11.31 ± 0.80
pH	6.30	8.93

Source: Choi, M.R. et al., *Korean Journal of Life Science* 13: 675–683, 2003.
Note: Values are mean ± SD.

Hongeo has plenty of functional peptides that affect antihypertension (Lim, 2003). Antimicrobial property of the peptide, kenojeinin I, has been proven against *Bacillus subtilis*, *Escherichia coli*, and *Saccharomyces cerevisiae* (Cho et al., 2005). The increased taurine content during fermentation was known to prevent the accumulation of cholesterol. Liver oil, especially, contains a large amount of ω-3 fatty acids such as eicosapentaenoic acid (EPA) and docosahexaenoic acid (DHA), which reduce low-density lipoproteins (LDL)-cholesterol and increase high-density lipoproteins (HDL)-cholesterol (Nam and Lee, 1999). Furthermore, anticancer and antibacterial activities increased with the progress of fermentation (Choi et al., 2003).

3.3.3 Microbial and Chemical Changes

Microbial and chemical properties of hongeo change significantly during fermentation. Especially, hongeohoe, which has finished fermentation, has a strong ammoniacal odor. The unique taste and odor of hongeohoe come from the nitrogenous compounds such as ammonia, urea, and trimethylamine, which are essential for osmoregulation in the deep sea. Ammonium concentration of fermented skate collected from Korean market ranges from 4.4 to 14.1 mg/g (Lee et al., 2008). Another study showed a similar result that ammonia concentration in skate extracting solutions was not detected before fermentation, but increased to 14.2 mg/ml on day 8 at 10°C (Choi et al., 2003). Due to the ammonia, the pH value of hongeohoe, which is less than 7 before fermentation goes up to 9.4 after fermentation, which shows that hongeohoe is one of the AFFs (Lee et al., 2008). The ammonia that is produced during the fermentation process neutralizes the stomach acid and inhibits bacterial growth (Cho and Kim, 2008). The major microflora in the both fresh and fermented skin includes *Photobacterium* sp. (2.4×10^4 cfu/g) and *Vibrio* sp. (7.7×10^7 cfu/g). These bacteria inhabit on the skate skin are known to promote fermentation (Cho et al., 2004).

3.3.4 Safety Aspects and HACCP

The research on safety aspects including a risk factor was carried as an R&D project under Korea Food and Drug Administration in

2006 (Kim, 2006). From the fields such as manufacturing process, on-site supervision, and shelf-life study to laboratory tests, all different aspects using 17 fermented skate samples were tested. *Escherichia coli* and *Staphylococcus aureus* were detected in one and two samples, respectively, out of 17 samples. It occurred in the sample contaminated with *S. aureus* when the catering cook used the same gloves in various jobs. To examine shelf life, samples were stored at 4°C and 10°C, and then changes of total plate counts, volatile basic nitrogen (VBN), pH, and ammonia concentration were examined. The pH ranged from 6.72 to 9.43, and VBN values were significantly different between samples (12.61–621.91 mg/100 g). The ammonia concentration was 5.2–14.8 mg/g. During the storage period, VBN values showed severe changes depending on the temperature. The VBN value increased to 500 mg/100 g at 10°C after 7 days of fermentation, but it increased to 200 mg/100 g at 4°C during the same period. Most of all, storage temperature plays a key role in above the safety factor, and thus it is suggested to be kept below 4°C.

3.3.5 Conclusion

Critical control point (CCP) at which controls can be applied might exist in the storage and distribution process of fermented skates. As hongeohoe is eaten raw without heating process, pathogens can be a major control point in the defrosting, washing, caring for fish offal, fermentation process (container and temperature), packaging, and storage temperature. However, as long as hongeo is properly fermented, it is generally as safe as other AFFs.

3.4 Pidan

Soottawat Benjakul and Palanivel Ganesan

Pidan, or "century egg," also known as "alkaline egg," "preserved egg," "hundred-year egg," and "thousand-year-old egg," is a Chinese cuisine ingredient made by preserving duck, chicken, or quail eggs in a mixture of clay, ash, salt, lime, and rice straw for several weeks to several months depending on the method of processing (Wang and Fung, 1996). The formation of pidan is mainly caused by the penetration of alkali through the egg shell and membrane, leading to several chemical changes in the egg components (Ganasen and Benjakul, 2010). Traditionally, lead has been used as the essential ingredient for making pidan of acceptable quality, since it can yield pidan with a stable and firm albumen gel with acceptable color. This divalent cation can induce gelation of the albumen during pickling or mud coating under the alkaline condition. Without the appropriate cations, the gel of pidan cannot be formed, and it is reliquefied. Ganasen and Benjakul (2010) confirmed that cations in the pickling solution have the pronounced effect in the gelation of pidan white and yolk. However, the effectiveness and influence on pidan formation vary with the types of cations used for the pickling process. It has been believed that the use of tea, wood ash, and local clay gives pidan a more detectable taste and unique character. Apart from gel-like texture, brown, or amber color is desirable for pidan, especially for its albumen. Maillard reaction between the glucose of the albumen and amino acids mainly contributes to browning in pidan (Li and Hsieh, 2004). Additionally, the pigment from oxidized tea phenols used in the pickling media can partially contribute to the development of brown color of pidan. It has been known that pidan has a unique ammonical flavor that is developed during pickling or storage via deamination or decomposition of proteins in egg. Hydrogen sulfide produced from the decomposed protein reacts with the iron in the yolk, which gives pidan its typical dark green-colored yolk. The pine-floral-like structures formed in pidan come from the degraded protein products such as alkalinized amino acids. In general, high-quality pidan has more pine-floral crystals.

Due to the safety concerns, most consumers typically request "lead-free pidan." Therefore, the alternative technology, especially the use of safe cations, has been proposed intensively. Divalent cations, especially zinc, have been used alternatively to produce pidan, instead of lead (Chen and Su, 2004; Ganasen and Benjakul, 2010). Pidan can be free of lead, but still have the acceptability equilavent to that produced with the aid of lead. As a consequence, consumers can have the nutritive pidan, which is rich in protein without risk of the poisonous lead.

3.4.1 Manufacturing of Pidan

Pidan has been produced in the countries of Southeast Asia, like Thailand, Malaysia, and Singapore, and East Asian countries, including China, Taiwan, and South Korea. Recently, small- and medium-sized food companies make pidan in large amounts to sell in the food markets. Due to the various processes and technologies applied to meet the increasing demand for pidan, a variety of pidan with different characteristics, such as color or appearance, have been produced. With a better understanding of chemical reactions involved in pidan formation, the use of the proper and safe ingredients with a shorter production time allows the industries to produce pidan with consistently good quality for the growing market.

Generally, duck egg is popular for pidan production; however, hen or quail eggs can also be used. It was reported that hen egg has smaller shell pores and needs a higher concentration of sodium hydroxide. Furthermore, pidan from hen egg has a poorer taste than that produced from duck egg (Wang and Fung, 1996). The freshness of the egg is another important factor determining the quality of pidan. Freshness has an influence on gelation and reliquefaction time. When older eggs are used for pidan production, the gelation time is longer and the reliquefaction time is shorter (Wang and Fung, 1996).

3.4.1.1 Traditional Processing The principle of pidan-making, which originated in China around 500 years ago, was transferred to other countries when the Chinese migration took place during the 19th century AD. The methods used for pidan-making can be varied or slightly adapted among countries. China and the Southeast Asian countries have produced pidan for many decades.

Typically, there are three types of pidan, viz. pine-floral pidan (songhuadan), soft-yolked pidan (tangsindan), and hard-yolked pidan (Wang and Fung, 1996). These pidans are produced by modifications of the main processing method. The processes used directly determine the characteristics of pidans in different fashions.

The traditional processing of pidan is classified into three types: rolling powder method, coating method, and immersion method (Wang and Fung, 1996).

3.4.1.1.1 Rolling Powder Method Traditionally, fresh duck eggs are coated with a thin layer of mud paste and rolled in the rolling powder, in which all the ingredients have been included, before being packed and sealed in a jar. The powder-rolled eggs are allowed to stand at room temperature for 20–30 days. The ingredients used for the rolling powder method vary slightly according to the season. The process flow diagram is shown in Figure 3.4. The rolling powder method produces hard-yolk pidan. The advantages of this method are its low cost and the ease in handling. However, mud paste cannot be reused and it is difficult to control the ingredients used.

3.4.1.1.2 Coating Method For this method, a muddy paste containing the necessary ingredients is required. The ingredients used for this method are shown in Figure 3.4, as reported by Wang and Fung (1996). Fresh duck eggs are completely coated with the prepared muddy paste and rolled in rice hulls. The hulls are used to prevent the coated eggs from sticking together. They are then placed into jars, sealed with mud, and allowed to stand for 40 and 50 days in summer and winter, respectively.

The sealing step is very important for producing a high-quality product. A well-sealed jar prevents the dehydration of the mud coating. This coating method prevents the formation of soft-yolk pidan. The coating method, as reported by Lin (1983), is described as follows: for 100 eggs, 125 g of sodium carbonate, 625 g of wood ash, 1 kg of calcium oxide,

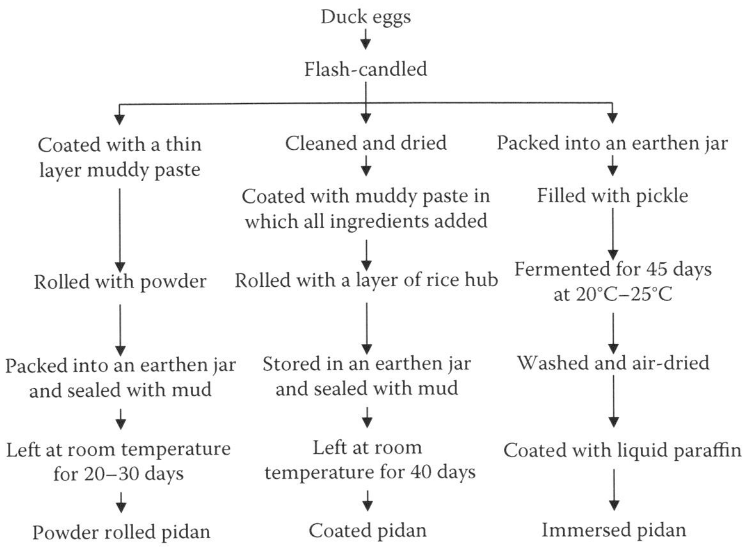

Figure 3.4 Flow diagram of different pidan processing methods. (Modified from Liu, Y.C., and Zhang, S.A., *Egg Processing and Technology*, Agriculture Press, Beijing, China, 1989.)

100 g of salt, and 500 ml of water are mixed to form a paste. The surface of the eggs is coated with the prepared paste to obtain a thickness of 1 cm, followed by rolling the egg in rice hulls. Coated eggs are packed in wooden jars and then sealed. After 5–6 months of pickling, the pidan with desired characteristics is ready for sale.

3.4.1.1.3 Immersion Method First, all ingredients are mixed into a pickling solution. Eggs are immersed in the solution at room tempertaure. After the pickling process is completed, the eggs are removed, washed with water, and allowed to air-dry. The eggs can be further coated with liquid paraffin before packaging and marketing. This can prevent the dehydration of pidan albumen, in which the quality can be maintained. The following pickling composition can vary from place to place. Mainly, the solution contains sodium carbonate, lime (CaO), salt, and Chinese tea. The process for making pidan using immersion method is given by Wang and Fung (1996) (Figure 3.4).

For the immersion method, ginger and cinnamon can be added to the pickling solution to reduce the pungent taste. Zhang (1988) suggested the immersion of 1000 eggs in a mixture of 3.6 kg of sodium carbonate, 2 kg of common salt, 1 kg of wood ash, 1.5 kg of tea, 14 kg of CaO, 0.37 kg of PbO, 0.25 kg of pine needle, 0.5 kg of dry mud, and 50 l of water.

3.4.1.2 Commercial and Novel Processing of Pidan

3.4.1.2.1 Production of Pidan Due to the long time of pidan production by the traditionl process, the manufacturers have paid attention to the use of a rapid method to shorten the production time. In the traditional method, sodium hydroxide is generated from the mixture of sodium carbonate and lime (Wang and Fung, 1996):

$$CaO + H_2O \rightarrow Ca(OH)_2$$

$$Ca(OH)_2 + Na_2CO_3 \rightarrow 2\ NaOH + CaCO_3$$

$$Na_2O_3 + H_2O \rightarrow 2\ NaOH + O_2$$

$$K_2O + H_2O \rightarrow 2\ KOH$$

where Na_2CO_3 and K_2O are from the wood ash.

However, the sediment from the mixture can block the egg shell pores, thereby lowering the penetration of NaOH into the egg. As a consequence, the pidan pickling time becomes longer. To tackle this problem, NaOH has been introduced directly for pidan manufacturing to decrease the pidan pickling time and avoid the sediment from insoluble ingredients used in the pickling medium. Using 4.1% NaOH as the picking solution, pidan could be formed within 14 days. Nevertheless, pidan obtained from this method has a strong alkaline taste. Ganasen and Benjakul (2010) produced pidan by immersing duck eggs in pickling solution containing 4.2% NaOH, 5% NaCl, 2% Chinese tea, and different cations. With this process, pidan could be obtained within 3 weeks. Su and Lin (1993) produced pidan from duck eggs using the pickling solution containing 4.2% NaOH and 5% NaCl.

NaCl is involved in both liquefaction and gelation of pidan. At high level, it may impede the gelation of egg white. However, it may bring about the hardening of egg yolk. From the viewpoint of preservation, NaCl can extend the shelf life of pidan by retardation of microbial growth. It also helps in separation of pidan from shells (Wang and Fung, 1996). Ganasen et al. (2013) found that pidan white and yolk had NaCl contents of 1.10%–1.11% and 0.80%–0.85%, respectively.

To accelerate the pickling process, the moving and closing system, in which the pickling solution is allowed to flow during pickling, has been implemented (Liu and Zhang, 1989). Thus, the pickling solution can contact and penetrate into the egg faster and more uniformly. The solution with alkaline pH directly causes the changes in egg, in the fashion by which pidan can be formed. However, other ingredients such as salt and divalents have been considered to be involved in the formation of pidan during pickling (Ganasen and Benjakul, 2011a, 2011b).

Acid pretreatment of egg shell is another approach for acceleration of pidan production. The reduction of shell thickness was achieved by acetic acid pretreatment. Ganasen and Benjakul (2011c) soaked duck eggs in 5% acetic acid for 30 min at room temperature, followed by washing to remove the acetic acid and drying in air for 1 h. The pretreated eggs were used for pidan production, in which the penetration of NaOH, divalents as well as tea components into the eggs can be enhanced (Ganasen and Benjakul, 2011c).

3.4.1.2.2 Coating of Pidan When the pickling is complete, the coating is introduced to prolong the shelf life of pidan. Coating plays a role in keeping the quality of pidan with the following aspects: (1) prevent the invasion of spoilage microorganisms, (2) prevent moisture loss, and (3) reduce H_2S and NH_3 formation in pidan (Wu, 1992). However, this coating material must be inexpensive, so that the manufacturers can apply this for pidan production. Originally, pidans produced by the traditional process, in which coating with mud or other materials are applied, were considered to have a long shelf life. However, it is not attractive and some of the coating materials are removed or broken during packaging and handling (Wu, 1992).

3.4.2 Characteristics of Pidan

3.4.2.1 Textural Property During the pickling of egg, alkaline substances, mainly NaOH, gradually penetrate through the pores and membrane of shell. The pH of egg white becomes alkaline, around 11–12, as the pickling proceeds (Ganasen and Benjakul, 2010). Under the alkaline condition, proteins in egg white become negatively charged. As a consequence, the repulsion between the protein molecules takes place and the egg white is liquefied.

To induce the gel formation under alkaline condition, divalent or tetravalent cations are required to ensure the development of a firm and elastic gel. Those cations are able to cross-link the negatively charged proteins via “salt bridges” or ionic interaction. Furthermore, those cations are more likely involved in shielding of negative charge of proteins (Ganasen and Benjakul, 2011b). They found that PbO_2, $ZnCl_2$, and $CaCl_2$ could lower the negative charge of egg white proteins when tested by Zeta potential analysis. The decrease in negative charge was in agreement with the increase in turbidity of egg white protein solution. Among all the cations, lead had the higher binding capacity with egg protein at a very alkaline pH and the aggregate was stable, associated with the tris–thiol ligand binding with cysteine residues (Godwin, 2001). Electrostatic attraction between the positively charged Zn–water complex and the carboxylic groups of the negatively charged proteins forms an aggregate (Shi et al., 2008). This phenomenon contributes to the gel formation of egg white. Ganasen and Benjakul (2011a) found that cations had an influence on the texture of pidan white. PbO_2 at a level of 0.2% was found to maintain the hardness of pidan white during pickling of 3 weeks and further aging of 3 weeks more effectively than $ZnCl_2$ or $CaCl_2$ at the same level. Pidan white treated with 0.2% PbO_2 exhibited closer aggregates with the denser network, while pidan white treated with 0.2% $CaCl_2$ or 0.2% $ZnCl_2$ had a looser network with larger voids and irregular shape (Figure 3.5) (Ganasen and Benjakul, 2010).

For pidan production, the appropriate cations play a crucial role in gelation of egg white and maintainance of egg white gel of pidan. The simplified scheme representing the gel formation of egg white induced by cations under the alkaline condition is shown in Figure 3.6. Due to concerns about the hazardous lead poisoning in pidan, zinc has been widely used as a substitute in the manufacturing of unleaded pidan. In the presence of NaOH, Zn^{2+} and ZnO_2^{2-} (zinc oxide anion) are prevalent and can penetrate into the egg, inducing an aggregation of proteins (Wang and Fung, 1996).

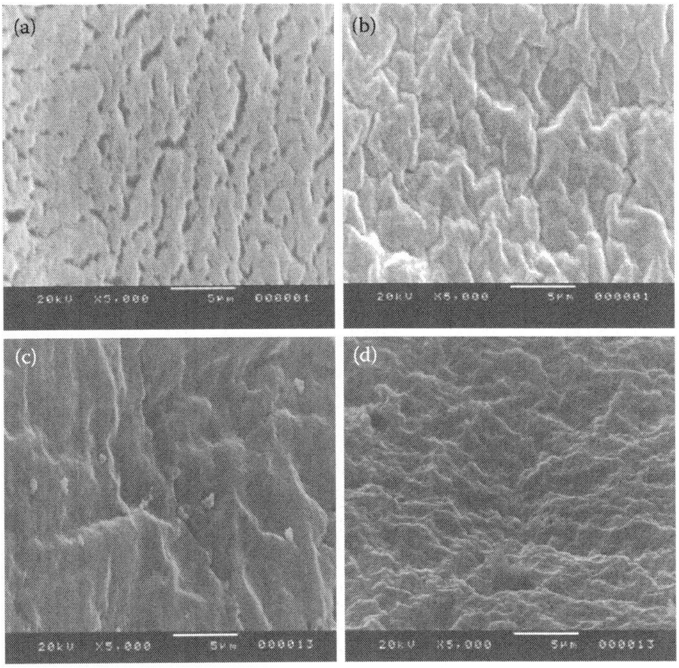

Figure 3.5 Scanning electron microscopic photograph of pidan white after 3 weeks of pickling. Magnification: 5×. (a) No cations; (b) 0.2% PbO_2; (c) 0.2% $ZnCl_2$; (d) 0.2% $CaCl_2$. (From Ganasen, P., and Benjakul, S., *Journal of Food Biochemistry*, 35, 1528–1537, 2011a.)

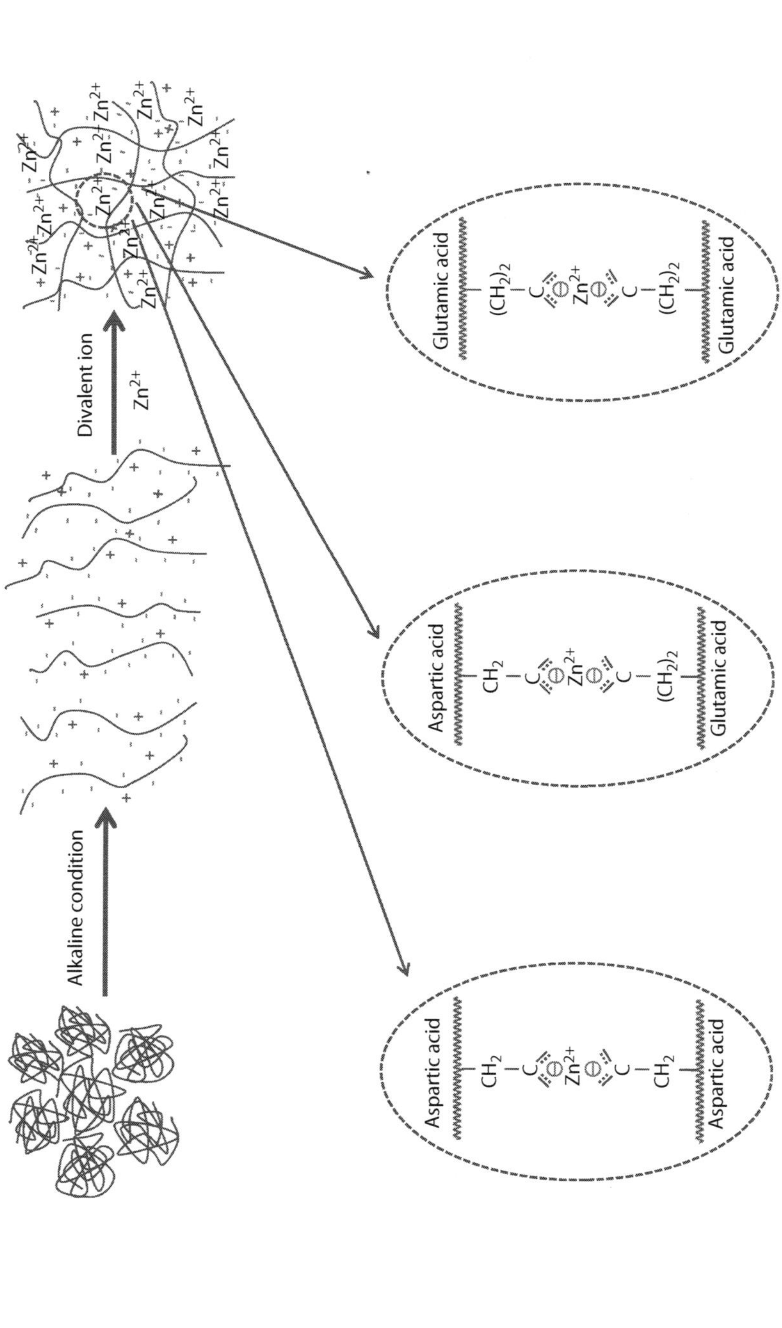

Figure 3.6 Gelation mechanism of pidan white.

Hydrolysis via either enzymatic or non-enzymatic reactions is generally found in pidan white. Ganasen and Benjakul (2011a) also reported that degradation of several proteins in egg white occurred during pickling and aging. However, pidan made from pickling solution containing lead ions had the lower degradation of proteins (Figure 3.7). The degraded proteins with the shorter chain cannot undergo gelation effectively since the junction zone cannot be formed. Li and Lai (1978) studied the proteolytic properties in duck egg white. The optimum temperature for protease activity in egg was 40°C, and the activity was negligible at a temperature below 25°C or above 55°C. The optimum pH for protease activity in egg was 10, and the enzyme activity was inactivated at pH 11.5. The addition of PbO can inhibit egg white reliquefaction in pidan processing. Apart from inducing the salt bridges, lead can lower the proteolytic activity of egg white, which is active at the beginning of the pickling process.

During pickling, the duck egg yolk becomes hardened and no marked changes were found during the aging (Ganasen and Benjakul, 2010). NaCl and NaOH can migrate into the yolk during pickling. In the presence of salt, dehydration of the yolk takes place via the "salting out" effect. This causes the solidification of pidan yolk; however, pidan

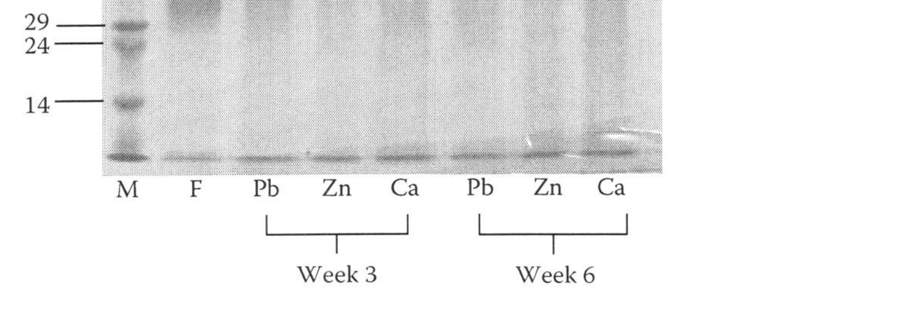

Figure 3.7 Protein patterns of fresh egg white and pidan white treated with different divalent cations after pickling (3 weeks) and ageing (6 weeks). M, molecular weight standard; F, fresh egg white; Pb, 0.2% PbO_2; Zn, 0.2% $ZnCl_2$; Ca, 0.2% $CaCl_2$. (From Ganasen, P., and Benjakul, S., *Journal of Food Biochemistry*, 35, 1528–1537, 2011a.)

treated with PbO_2 had a higher moisture content than that treated with $ZnCl_2$ or $CaCl_2$ (Ganasen and Benjakul, 2010).

At the initial stage, the yolk becomes harden, especially at the surface, named "exterior egg yolk," whole viscous liquid yolk is found inside, named "interior egg yolk." Hardening ratio of pidan yolk increased continuously during 3 weeks of pickling and a slight increase was noticeable during ageing (Ganasen and Benjakul, 2010). A decrease in adhesiveness occurred in pidan yolk, and there was no adhesiveness in pidan yolk after pickling. Alkali penetration was associated with the major release of free lipids, which subsequently underwent saponification (Ganasen and Benjakul, 2010).

When pickling and aging of pidan proceeds, a greater release of lipid occurs and can be associated with the poorer emulsion capacity of proteins in yolk. Irregular shapes of both lipid and protein were found in both the middle and the exterior yolk. The amount of lipids released and alteration of lipid shape were more pronounced in exterior portion. A greater release of free lipids from lipoprotein induced by the pickling process was obtained in the presence of PbO_2. For pidan treated with $ZnCl_2$, the lesser release of free lipids was noticeable and this might be related with harder pidan yolk treated with $ZnCl_2$, compared with those treated with PbO_2. Confocal laser scanning micrographs (CLSM) of duck egg yolk of pidan treated with 0.2% $ZnCl_2$ after pickling and aging for 6 weeks are illustrated in Figure 3.8.

3.4.2.2 Color The color of both egg white and yolk of pidan is another characteristic feature affecting the quality and acceptability of pidan (Figure 3.9). As NaOH penetrates through the egg shell and membrane via osmosis into the egg, a number of reactions take place. Alkaline condition induces the hydrolysis of several proteins of egg white (Ganasen and Benjakul, 2011a). The available free amino aids can be generated and subsequently undergo Maillard reaction with reducing sugar in egg under the alkaline condition. Maillard reactions between the glucose of the albumen and amino acids, combined with the pigment of the tea, contribute to the development of a brown-colored albumen gel (Wang and Fung, 1996). Maillard reaction is a non-enzymatic browning reaction that links the carbonyl group of reducing carbohydrates and the amino group of free amino acids,

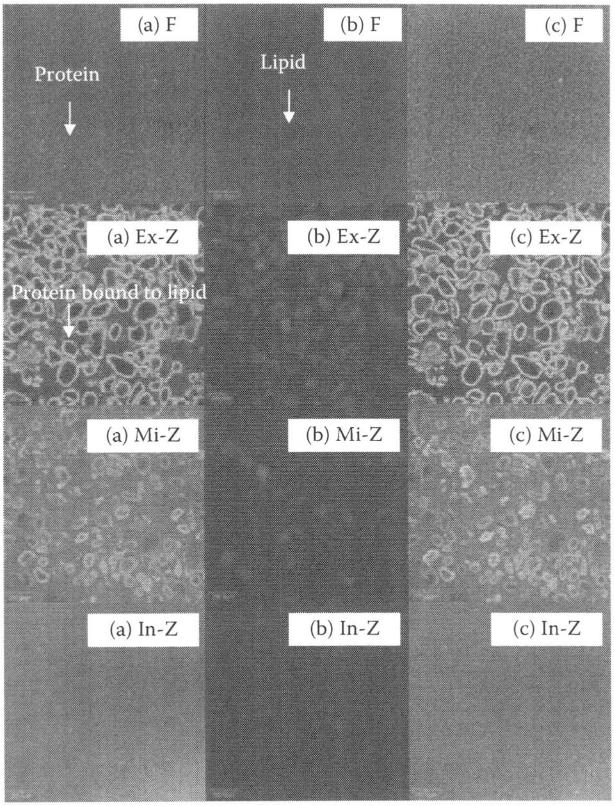

Figure 3.8 Confocal laser scanning micrographs (CLSMs) of duck egg yolks from pidan treated with $ZnCl_2$ at a level of 2 g/kg after pickling and aging for 6 weeks. (a) Protein distribution; (b) lipid distribution; (c) combined image of protein and lipid. F, fresh yolk; Ex-Z, exterior yolk; Mi-Z, middle layer yolk; In-Z, interior yolk. Magnification: 200× (zoom ×2.5). (From Ganasen, P., and Benjakul, S., *LWT– Food Science and Technology*, 43, 77–85, 2010.)

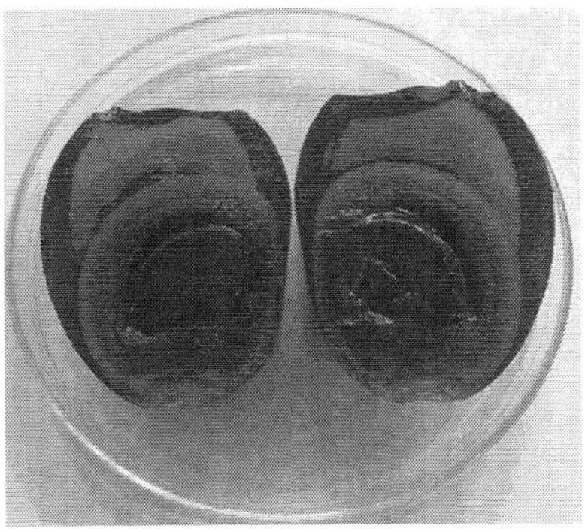

Figure 3.9 Pidan, as cut to halves.

especially lysine residues (Ajandouz et al., 2001). Maillard reaction is considered important in pidan white due to the significant amount of glucose present in egg white (Ganasen and Benjakul, 2011c).

Chinese tea in the pickling solution is believed to contribute to the brown color of pidan. The oxidation of flavonols in tea is governed by pH. Flavonols are colorless at neutral pH; however, they turn to yellow at alkaline pH (Mizooku et al., 2003). During the black tea fermentation, an enzymatic oxidation of tea polyphenols, especially tea catechins, takes place, leading to the formation of a series of colored chemical compounds such as theaflavins (TFs) and thearubigins (TRs), which are responsible for the characteristics of the black tea liquors (Biswas and Biswas, 1971). Green tea contains mainly catechins, and black tea is rich in TFs and TRs, which are dimers of catechins (Majchzrak et al., 2004). Traditionally, people believe that the addition of tea contributes to the color of pidan. However, Chinese tea has the influence of browning of pidan white mainly within the first week of pickling. Thereafter, it showed no impact on browning in pidan white (Ganasen, 2010). The addition of Chinese tea in pickling solutions yielded pidan with loose texture and lighter color, which might be because of the block of tea flavonoids on the pores of eggshell during pickling (Ganasen, 2010). In general, flavonoids in tea are known to interact with proteins to form complexes whose properties depend on the structure of both flavonoids and protein (Yuksel et al., 2010).

When duck egg was pretreated with 5% acetic acid prior to pickling, both Chinese tea and green tea in the pickling solution increased the browning of pidan white after pickling (3 weeks) and aging (6 weeks). Tea components in the pickling solution can penetrate more effectively when the egg shell becomes thinner. The coloring effect of both Chinese tea and green tea was in dose-dependent manner. Therefore, tea components can affect the color of pidan when they are able to migrate into the egg at sufficient concentrations. For the egg without acid pretreatment, the oxidized compounds in tea can be adducted and blocked at the pores of shell, thus lowering their migration. Green tea and Chinese tea in pickling solution at 5% showed the similar effect of color of pidan white prepared from egg pretreated with acetic acid (Ganasen and Benjakul, 2011c). Pickling solution with alkaline pH can favor the oxidation of catechin and other components.

Maillard reaction is the major reaction actively involved in the browning of pidan. This was related with the decrease in both free amino group and glucose contents in pidan white (Ganasen, 2010). This reaction took place continuously during both pickling and aging, thereby leading to continuous increase in the brown color of pidan. To increase the browning intensity, glucose can be added into the pickling solution to enhance the browning in pidan white (Ganasen, 2010).

The color of pidan yolk is dark to greenish, which is the typical characteristic feature of pidan yolk. Cysteine produced by protein hydrolysis can be continuously decomposed into ammonia and hydrogen sulfide. Additionally, hydrogen sulfide produced from the denatured protein reacts with the iron in the yolk, which gives pidan its typical dark green-colored yolk (Wang and Fung, 1996). Ferric ions (Fe^{3+}), which are present only in phosvitin, are reduced by S^{2-} to Fe^{2+}, and the residual S^{2-} combines with Fe^{2+} to form FeS, which gives a typical dark green color to the pidan yolk. When a pidan is cut in half, the yolk is both dark and light colored. This is why pidan is called chaidan (colored egg) in China. Four different color layers occur. From external to internal, the first layer is a dark color; the second is light; the third is the darkest; and the fourth is the lightest. The dark and light colors are due to different iron contents in different layers. Higher iron content results in a darker color. The first change in color occurs at the third layer, as it forms deep green. When pidan is cut and exposed to the air at room temperature, the deep green color of the pidan yolk will turn back to its original yellow color (Wang and Fung, 1996).

In some processes, lead has been used to stabilize egg white gel. Pb^{2+} has been known to affect the color of pidan. Pb^{2+} combines with S^{2-}, generated from protein hydrolysis process, leading to the formation of black-colored lead sulfide that forms sediments inside and outside of the egg shell and reduces the size of the egg shell pores. This negatively affects the penetration of NaOH into pidan. One observation for consumers to select lead-free pidan is the absence of black spot on the egg shell, which is caused by PbS. Due to safety concerns, most consumers typically request "lead-free" pidan. Zn^{2+} has been used to produce pidan with no black spots on the egg shell, and the color of the pidan's albumen and yolk is more stable (Chen and Su, 2004).

3.4.2.3 Odor Pidan has a characteristic pungent ammoniacal plus sulfide smell. Ammonia is formed in the pidan white, and yolk, possibly by deamidation. Deamidation of proteins occurs at pH above 8.0, dependent upon the H^+ or OH^- concentration and adjacent amino acid residues (Riha et al., 1996). The ingredients used in pickling solution have an influence on the formation of ammonia. Higher ammonia content was found in pidan white and yolk treated with 0.2% $ZnCl_2$-treated pidan, compared with that of pidan treated with 0.2% PbO_2 (Ganasen et al., 2013). Several volatile compounds contribute to the aroma of pidan white and yolk. Chau and Chung (2001) reported 60 and 76 compounds in pidan white and yolk, respectively. Four ketones were found in only pidan white, and 3-hydroxy-2-butanone was dominant in pidan white. Amomg the 10 aldehydes found in pidan yolk, decanal was the predominant compound. Hydrogen sulfide and acetaldehyde were the key precursor for sulfur-containing compounds in pidan (Chau and Chung, 2001).

3.4.2.4 Pine-Floral Crystal Pine-floral structure, which is star-like in shape, can be found in pidan white and on the surface of the yolk in good-quality pidan (Wang and Fung, 1996). Previously, it was believed that it was caused by the penetration of pinewood ash into the eggs. Nevertheless, pidan made without pinewood ash can have pine-floral crystals. Pine florals are caused by crystallization of the degraded protein products, such as alkalinized amino acids (Wang and Fung, 1996). NaOH concentration affects the formation of pine-floral crystals. At NaOH concentration below 0.2% in pidan protein, the formation of pine-floral crystals does not take place. Pine-floral crystals reflect the conversion of the protein into amino acids. Therefore, the number of pine-floral crystals in pidan can be used as a quality index of pidan (Wang and Fung, 1996).

3.4.3 Lysinoalanine Formation in Pidan

Food proteins have been alkali-treated to improve the extractability, solubilization, gelation, and dispersability, in which the texturized products can be prepared (Chang et al., 1999a). However, alkaline condition generally causes the formation of lysinoalanine (LAL) and lanthionyl residues, racemization and degradation of amino acid residues, and Maillard reactions, especially when the protein-containing foods are thermally treated during the processing (Liardon and Hurrel, 1983). Moreover, most of these reactions are the main causes of nutritional losses (Friedman et al., 1984). LAL and related cross-linked amino acids may be derived from the reaction of lysine with dehydroalanine (DHA) residues formed from substituted serines, cystine, and cysteine residues in proteins through nucleophilic reactions (Fletcher et al., 1963). The lowered digestibility of proteins and the loss of lysine are the main causes of nutritional losses of proteins.

Several factors affecting LAL formation include pH, alkaline treating time, and heating temperature as well as sources of protein (Friedman et al., 1984). DHA is found in some food proteins, which have been heated or treated under alkaline condition. Under such a condition, serine in the protein chain can be dehydrated. Subsequently, DHA is able to alkylate lysine to yield the crosslinked LAL (Figure 3.10). DHA formed through the β-elimination reaction is unstable at high pHs and it is liable to degradation, in which pyruvic acid and ammonia can be formed. Also, it is susceptible to reaction to form lanthionine and β-aminoalanine (Fletcher et al., 1963). Cysteine was found to be most unstable to alkali and could mostly be destroyed when casein was alkali-treated at pH 9.0 and 75°C for 3 h

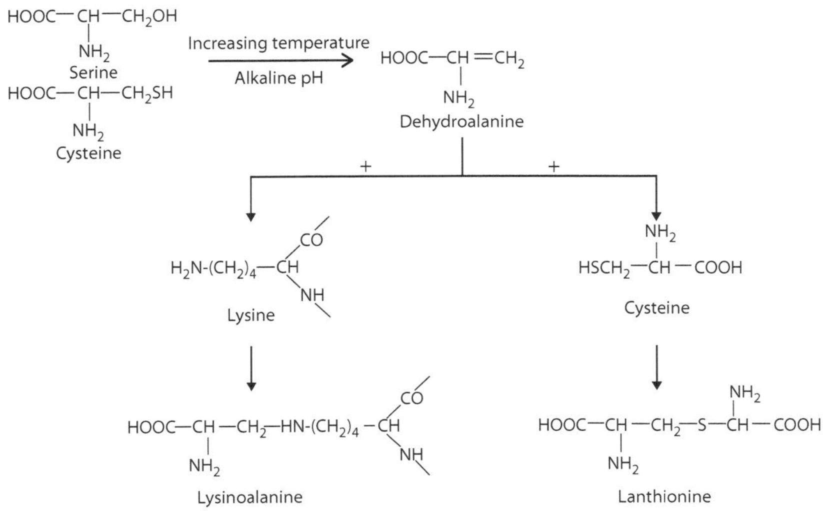

Figure 3.10 Formation of lysinoalanine in pidan.

(Friedman and Masters, 1982). Acylation of the ε-amino group of lysine, complex formation of metal ions with the imidazol group of histidine, and ε-amino group of lysine were effective in reducing the formation of LAL (Henkin, 1974). About 65% and 35% of cysteine in the albumen and yolk, respectively, of alkali-pickled duck eggs were degraded (Chang et al., 1999c). The decline in lysine was enhanced with increasing pH and alkali-treating time, accompanied by LAL formation (Chang et al., 1999a).

To prevent or control the formation of LAL, several approaches have been proposed. The LAL formation in both pidan albumen and yolk was markedly reduced by the addition to the pickling solution of certain additives, such as $ZnSO_4$, PbO, and L-cysteine or reduced glutathione (Chang et al., 1999a). The inhibition of LAL formation by L-cysteine and glutathione are owing to the production of a relatively large quantity of negatively charged sulfur-containing compounds that are able to retard β-elimination and DHA formation (Chang et al., 1999a).

3.4.4 Nutritive Value of Pidan

3.4.4.1 Amino Acids Pidan can be an excellent source of proteins and amino acids. Amino acid composition in pidan changes to some degree, in comparison with fresh egg. Although pidan can serve as the source of nutrients, its nutritive value can be reduced as a result of destruction of essential amino acids, e.g., threonine and the racemization of some amino acids. In pidan, cysteine, serine, and threonine were destroyed by 67%, 27%, and 12%, respectively (Chang et al., 1999c). Serine (42%), aspartic acid (36%), glutamic acid (20%), and phenylalanine (16%) in pidan white were D-enantiomers (Chang et al., 1999b). Ganasen et al. (2013) reported that higher contents of glutamine, aspargine, and leucine were found in pidan white and yolk, in comparison with those found in fresh counterparts (Benjakul and Ganasen, unpublished data) (Table 3.2). On the other hand, amino acids, such as serine and lysine were decreased in pidan white and yolk, in comparison with those of fresh counterparts. Some amino acids underwent degradation during alkaline pickling, but the rate of degradation varied with the type of cations (Ganasen, 2010). Thus,

Table 3.2 Amino Acid Contents of White and Yolk of Fresh Egg and Pidan (Expressed as Residues/1000 Residues)

	WHITE			YOLK		
AMINO ACID	FRESH	PIDAN-Pb	PIDAN-Zn	FRESH	PIDAN-Pb	PIDAN-Zn
Aspartic acid/Asparagine	91	92	91	89	92	93
Threonine	67	67	65	60	62	63
Serine	97	96	90	99	94	86
Glutamic acid/glutamine	123	126	124	110	112	114
Glycine	61	61	62	54	57	59
Alanine	68	70	70	77	77	78
Cysteine	1	1	2	4	3	4
Valine	70	70	74	64	64	67
Methionine	54	55	55	26	30	30
Isoleucine	41	42	44	50	49	49
Leucine	81	82	83	86	88	88
Tyrosine	33	33	33	34	35	35
Phenylalanine	58	59	59	36	40	39
Hydroxylysine	0	7	16	0	4	12
Lysine	61	49	42	70	64	58
Histidine	17	19	21	37	31	29
Arginine	6	6	5	5	6	6
Tryptophan	33	27	24	52	47	45
Hydroxyproline	0	0	0	0	0	0
Proline	37	40	40	45	46	47

Source: Ganasen, P., Chemical compositions and properties if alkali picked egg (Pidan) as affected by cations and selected pickling ingredients. Ph.D. thesis, 2010.

Note: Pidan-Pb, PbO_2-treated pidan; Pidan-Zn, $ZnCl_2$-treated pidan.

cations used in pickling solution had an influence on the amino acid composition of pidan white and yolk.

3.4.4.2 Minerals Pidan contains several kinds of minerals. This is associated with the uses of mud, salt, ash, etc., in the pickling solution or rolling materials. Ganasen et al. (2013) reported that the higher contents of sodium, potassium, and iron were found in pidan white, whereas calcium, magnesium, sodium, and potassium were higher in pidan yolk, in comparison with those found in fresh counterparts (Table 3.3). Lead was found in only white and yolk of pidan treated with PbO_2. A higher content of zinc was obtained in either white or yolk of pidan treated with $ZnCl_2$. The increase in mineral content

Table 3.3 Elemental Contents of White and Yolk of Fresh Egg and Pidan

	WHITE (mg/100 g)			YOLK (mg/100 g)		
PARAMETER	FRESH	PIDAN-Pb	PIDAN-Zn	FRESH	PIDAN-Pb	PIDAN-Zn
Ca	23.12	19.40	29.20	158.22	186.88	307.84
Mg	79.89	19.94	28.89	18.23	27.18	25.22
Na	754.00	2192.90	2250.90	72.30	711.20	1516.81
K	604.40	878.15	748.60	131.42	212.87	246.67
Zn	ND	0.56	2.63	3.95	3.03	3.75
Pb	ND	<0.005	ND	ND	<0.005	ND
Cu	0.79	0.52	0.35	0.14	0.16	0.34
Fe	1.11	2.33	2.19	8.51	7.38	6.63
N	12.86	13.00	12.68	4.91	5.36	5.32
S	1.70	1.40	1.17	0.28	0.29	0.29

Source: Ganasen, P. et al., *Korean Journal of Food Science and Animal Resources*, 33, 214–220, 2013.

Note: Pidan-Pb, PbO_2-treated pidan; Pidan-Zn, $ZnCl_2$-treated pidan; ND, not determined.

of pidan white and yolk indicated the migration of minerals from pickling solution into egg white and yolk, respectively. It was coincidental with the increase in pH of pidan white and yolk. The high content of sodium (2.1%–2.2%) of pidan white and yolk indicated a greater migration of Na^+ from pickling solution through the egg shell. The decrease in some minerals indicated the dilution effect of major minerals as the pickling solution migrated through the shell to the egg white or yolk (Ganasen et al., 2013). No differences in nitrogen content were found between egg white of fresh egg and pidan. For sulfur content of egg white, it was noted that pidan had a lower sulfur content than fresh egg (Ganasen et al., 2013).

3.4.4.3 Lipid and Cholesterol Pidan yolk contains a high content of lipid as well as cholesterol. Pidan made through immersion and coating methods has lipid contents of 16.99% and 17.41%, respectively (Yang and Chen, 2001). Oleic acid (18:1) and palmitic acid (16:0) constituted the major fatty acid in pidan yolk lipids (Liu et al., 2005). Lipid oxidation product was formed to some level in pidan yolk produced by immersion and coating method, which had the thiobarbituric acid reactive substances of 2.34 and 3.53 mg/kg, respectively (Yang and Chen, 2001). Additionally, egg yolk is rich in cholesterol, which can undergo oxidation during processing or pidan pickling. Cholesterol contents in pidan, made through immersion and coating methods, were 786.7 and 768.6 mg/100 g, respectively. Yang and Chen (2001) reported that pidan made from coating method had higher cholesterol oxidation products (COP; 148.8 μg/g) than that prepared through immersion method (98.9 μg/g). 7-β-Hydroxycholesterol and 20-hydroxycholesterol were the major COPs in pidan yolk, regardless of processing methods used.

References

Abu-Hassan, O., and Adam Sulieman, H.M. 2011. Quality and microbial analysis of local salted-fermented paste product (Terkin). *World Veterinary Journal* 1: 10–13.

Abu Bakar, H.J.H. 2003. Food composition of some fermented seafood in Brunei Darussalam. *Scientia Bruneiana Brunei Muara: Faculty of Science Official Journal, Universiti Brunei Darussalam* 4: 5–8.

Achi, O.K. 2005. Traditional fermented protein condiments in Nigeria. *African Journal of Biotechnology* 4: 1612–1621.

Achi, O.K., Anokwuru, I.C., and Ogbo, F.C. 2007. Microbiological and chemical changes during fermentation of crabs for ogiri-nsiko production. *American Journal of Food Technology* 2: 301–306.

Adnan, N.A. 1984. Belacan-kaedah penyediaan dan kawalan. *Teknologi Makanan Jil* 3: 5–8 (in Indonesian).

Ajandouz, E.H., Tchiapka, L.S., Ore, F.D., Benjibas, A., and Puigserver, A. 2001. Effect of pH on caramelisation and Maillard reaction kinetics in fructose-lysine model systems. *Journal of Food Science* 66: 926–932.

Amano, K. 1992. The influence of fermentation on the nutritive value of fish with special reference to fermented fish products in South East Asia. In *Fish in Nutrition*, eds. E. Heen, and R. Kreuzer, pp. 180–197. London: Fishing News (Books).

Anon. 1985. *Food Act 1983 and Food Regulations 1985*. Kuala Lumpur, Malaysia: Maju Daya Cipta.

Arianti, L. 2004. Isolation and identification of halophilic acid bacteria from wadi betok (*Anabas testudineus* Bloch) cultured in agar media with NaCl 10% (Isolasi dan identifikasi bakteri asam halofilik dari wadi ikan betook (*Anabas testudineus* Bloch) yang ditumbuhkan pada media agar dengan konsentrasi NaCl 10%). B.Sc. thesis. Banjar, Indonesia: University of Lambung Mangkurat (in Indonesian).

Arumugaswamy, R.K., Rusul, G., Abdul Hamid, S.N., and Cheah, C.T. 1995. Prevalence of *Salmonella* in raw and cooked foods in Malaysia. *Food Microbiology* 12: 3–8.

Aslam, Z., Lee, C.S., Kim, K.H., Im, W.T., Ten, L.N., and Lee, S.T. 2007. *Methylobacterium jeotgali* sp. nov., a non-pigmented, facultatively methylotrophic bacterium isolated from jeotgal, a traditional Korean fermented seafood. *International Journal of Systematic and Evolutionary Microbiology* 57: 566–571.

Badham, C.D. 1854. *Prose Balicutes, or Ancient and Modern Fish Cattle.* London: Parker & Sons.

Basrindu, A. 1987. The utilization of cooking salt and palm sugar in preservation of ikan betok (*Anabas testudineus* Bloch). M.Sc. thesis. Yogyakarta, Indonesia: Gajah Mada University.

Beddows, C.G. 1985. Fermented fish and fish products. In *Microbiology of Fermented Foods*, vol. 2, ed. B.J.B. Wood, pp. 1–39. London: Elsevier Applied Science.

Beddows, C.G., Ardeshir, A.G., and bin Daud, W.J. 1979. Biochemical changes occurring during the manufacture of Budu. *Journal of the Science of Food and Agriculture* 30: 1097–1103.

Besyuni, M.A. 1979. History of agriculture in the Sudan. Ph.D. thesis. Cairo, Egypt: University of Cairo (in Arabic).

Biswas, A.K., and Biswas, A.K. 1971. Biological and chemical factors affecting the valuations of North-East Indian plain teas. I. Statistical association of liquor characteristics with cash valuations of black tea. *Journal of the Science of Food and Agriculture* 22: 191–195.

Brémond, H.M.L., and Rosé, E. 1919. Condiments azotes solides en Indochine. *Annales de l'Institut Pasteur* 33: 282–291 (in French).

Budhyatni, S., Murtini, J.T., and Peranginangin, R. 1982. The microflora of terasi powder. *Laporan Penelitian Teknologi Perikanan* 16: 25–33.

Cha, Y.J., Chung, S.Y., Ha, J.H., Jeong, I.C., and Lee, E.H. 1983. Studies on the processing of low salt fermented seafoods: Changes of microflora during fermentation of low salted sardine. *Bulletin of the Korean Fisheries Society* 16: 211–215.

Chaiyanan, S., Maugel, T., Huq, A., Robb, F.T., and Colwell, R.R. 1999. Polyphasic taxonomy of a novel *Halobacillus*, *Halobacillus thailandensis* sp. nov. isolated from fish sauce. *Systematic and Applied Microbiology* 22: 360–365.

Chang, H.M., Tsai, C.F., and Li, C.F. 1999a. Inhibition of lysinoalanine formation in alkali-pickled duck egg (Pidan). *Food Research International* 32: 559–563.

Chang, H.M., Tsai, C.F., and Li, C.F. 1999b. Quantification of racemization of amino acids in alkali-treated duck eggs by micellar capillary electrophoresis. *Journal of Agricultural and Food Chemistry* 47: 479–484.

Chang, H.M., Tsai, C.F., and Li, C.F. 1999c. Changes in amino acid composition and lysinoalanine during formation in alkali-pickled duck eggs. *Journal of Agricultural and Food Chemistry* 47: 1495–1500.

Chau, W.S., and Chung, H.Y. 2001. Volatile compounds in thousand-year old eggs. Proceedings, *IFT Annual Meeting, 23–27 June*. New Orleans, LA.

Chen, H.C. 1993. Fish fermentation technology in Taiwan. In *Fish Fermentation Technology*, eds. C.H. Lee, K.H. Steinkraus, and P.J.A. Reilly, pp. 203–205. Tokyo, Japan: United Nations University Press.

Chen, J.W., and Su, H.P. 2004. A new process for preparing spots-free pidan. *Journal of the Chinese Society of Animal Science* 33: 79–88.

Chen, M.S., and Chen, T.S. 1978. Studies on the manufacture and storage of the Antarctic krill sauce. *Bulletin of the Taiwan Fisheries Research Institute* 30: 343–352.

Cho, S.H., Jahncke, M.L., and Eun, J.B. 2004. Nutritional composition and microflora of the fresh and fermented skate (*Raja kenojei*) skins. *International Journal of Food Science and Nutrition* 55: 45–51.

Cho, S.H., and Kim, K.H. 2008. Quality characteristics of commercial slices ofskate *Raja kenojei*. *Journal of the East Asian Society of Dietary Life* 18: 214–220.

Cho, S.H., Lee, B.D., An, H., and Eun, J.B. 2005. Kenojeinin I, antimicrobial peptide isolated from the skin of the fermented skate, *Raja kenojei*. *Food Hydrocolloids* 26: 581–587.

Choi, M.R., Yoo, E.J., Lim, H.S., and Park, J.W. 2003. Biochemical and physiological properties of fermented skate. *Korean Journal of Life Science* 13: 675–683.

Choorit, W., and Prasertsan, P. 1992. Characterization of proteases produced by newly isolated and identified proteolytic microorganisms from fermented fish (Budu). *World Journal of Microbiology and Biotechnology* 8: 284–286.

Curtis, R.I. 1991. Garum and Salsamenta. In *Studies in Ancient Medicine*, No. 3. Leiden, The Netherlands: Brill Academic Publishers.

Curtis, R.I. 2009. Umami and the foods of classical antiquity. *American Journal of Clinical Nutrition* 90S: 712–718.

Desniar, Poernomo, D., and Wijatur, W. 2009. The influence of salt concentration on peda chubb mackerel (*Rastrelliger* sp) with spontaneous fermentation. *Jurnal Pengolahan Perikanan Indonesia* 12: 73–87.

Espejo-Hermes, J. 1998. *Fish Processing Technology in the Tropics*. Quezon City, Philippines: Tawid Publications.

Essuman, K.M. 1992. *Fermented fish in Africa*. FAO Fisheries Technical Paper 329. Rome: Food and Agriculture Organization of the United Nations.

Faithong, N., Benjakul, S., Phatcharat, S., and Bindan, W. 2010. Chemical composition and antioxidative activity of Thai traditional fermented shrimp and krill products. *Food Chemistry* 119: 133–140.

Fletcher, J.C., Robson, A., and Todd, S. 1963. The sulfur balance of wool. *Biochemical Journal* 87: 560–567.

Friedman, M., Levin, C.E., and Noma, A.T. 1984. Factor governing lysinoalanine formation in soy proteins. *Journal of Food Science* 49: 1282–1288.

Friedman, M., and Masters, P.M. 1982. Kinetics of racemization of amino acid residues in casein. *Journal of Food Science* 47: 760–764.

Fujii, T., Basuki, S.D., and Tozawa, H. 1980. Microbiological studies on the ageing of fish sauce; chemical composition and microflora of fish sauce produced in Philippines. *Nippon Suisan Gakkaishi* 46: 1235–1240.

Funaki, J., Tamura, T., Nishinoaki, M., Misaka, T., Eto, W., and Asakura, T. 2010. Utilization of oryzacystatin for regulating the ripening of squid Shiokara, a traditional Japanese salted and fermented seafood. *Journal of Food Science* 75: 527–530.

Ganasen, P. 2010. Chemial compositions and properties of alkali pickled egg (Pidan) as affected by cations and selected pickling ingredients. Ph.D. thesis. Hat Yai, Thailand: Prince of Songkla University.

Ganasen, P., and Benjakul, S. 2010. Physical properties and microstructure of pidan yolk as affected by different divalent and monovalent cations. *LWT – Food Science and Technology* 43: 77–85.

Ganasen, P., and Benjakul, S. 2011a. Chemical compositions, physical properties and microstructure of pidan white as affected by different divalent and monovalent cations. *Journal of Food Biochemistry* 35: 1528–1537.

Ganasen, P., and Benjakul, S. 2011b. Effect of three cations on the stability and microstructure of protein aggregate from duck egg white under alkaline condition. *Food Science and Technology International* 17: 343–349.

Ganasen, P., and Benjakul, S. 2011c. Effects of green tea and Chinese tea on the composition and physical properties of pidan white. *Journal of Food Processing and Preservation* 35: 907–916.

Ganasen, P., Benjakul, S., and Hideki, K. 2013. Effect of different cations on pidan composition and flavor in comparoson to the fresh duck egg. *Korean Journal of Food Science and Animal Resources* 33: 214–220.

Ghazali, A.R., Rajab, N.F., Wen, L.W., Rahmani, A.S., Abdullah, R., Ramli, N.M., Kamarulzaman, F., Harun, Z., and Hasiah, A.H. 2011. Evaluation of the biochemical profile and biological activity of budu (a local fermented fish product) extracts on HepG2 hepatoblastoma cells. *Australian Journal of Basic and Applied Sciences* 5: 2606–2612.

Godwin, H.A. 2001. The biological chemistry of lead. *Current Opion in Chemical Biology* 5: 223–227.

Guan, L., Cho, K.H., and Lee, J.H. 2011. Analysis of the cultivable bacterial community in jeotgal, a Korean salted and fermented seafood, and identification of its dominant bacteria. *Food Microbiology* 28: 101–113.

Hajeb, P., and Jinap, S. 2012. Fermented shrimp products as source of umami in Southeast Asia. *Journal of Nutrition and Food Sciences* S10:006: doi:10.4172/2155-9600.S4110-4006.

Henkin, R.I. 1974. Metal-albumin-amino acid interaction: Chemical and physiological interrelationships. In *Protein-Metal Interactions*, ed. M. Friedman, pp. 299–305. New York: Plenum Press.

Hiraga, K., Nishikata, Y., Namwong, S., Tanasupawat, S., Takada, K., and Oda, K. 2005. Purification and characterization of serine proteinase from a halophilic bacterium, *Filobacillus* sp. RF2–5. *Bioscience, Biotechnology, and Biochemistry* 69: 38–44.

Huda, N. 2012. Indonesian fermented fish products. In *Handbook of Animal-Based Fermented Food and Beverage Technology*, 2nd ed., ed. Y.H. Hui, pp. 717–734. Boca Raton, FL: CRC Press.

Hutton, W. 2003. *Malaysian Favourites.* Hong Kong: Periplus Editions.

Hutton, W. 2005. *Authentic Recipes from Malaysia.* Hong Kong: Periplus Editions.

Ijong, F.G., and Ohta, Y. 1995. Amino acid composition of Bakasang, a traditional fermented fish sauce from Indonesia. *LWT – Food Science and Technology* 28: 236–237.

Ijong, F.G., and Ohta, Y. 1996. Physicochemical and microbiological changes associated with bakasang processing – A traditional Indonesian fermented fish sauce. *Journal of the Science of Food and Agriculture* 71: 69–74.

Ismail, M.S. 1977. Accelerated fermentation of fish-soy paste and fish-soy sauce by using *Aspergillus oryzae* NRRL 1989. Proceedings, *Symposium on Indigenous Fermented Foods*. Bangkok, Thailand.

Itoh, H., Tachi, H., and Kikuchi, S. 1993. Fish fermentation technology in Japan. In *Fish Fermentation Technology*, eds. C.H. Lee, K.H. Steinkraus, and P.J.A. Reilly, pp. 177–186. Tokyo, Japan: United Nations University Press.

Jansen, B.C.P., van der Linden, A.C., Aussen, B.S., and Jensen, J.D. 1953. Le taux de cobalamine, vitamine B12, dans la trassi, un produit de fermentation des poisons et des crevettes. *Gazette Médicale de France* 60: 33–34 (in French).

Jiang, J.J., Zeng, Q.X., Zhu, Z.W., and Zhang, L.Y. 2007. Chemical and sensory changes associated Yu-lu fermentation process – A traditional Chinese fish sauce. *Food Chemistry* 104: 1629–1634.

Jinap, S., Ilya-Nur, A.R., Tang, S.C., Hajeb, P., Shahrim, K., and Khairunnisak, M. 2010. Sensory attributes of dishes containing shrimp paste with different concentrations of glutamate and 50-nucleotides. *Appetite* 55: 238–244.

Jung, J.Y., Lee, S.H., Lee, H.J., and Jeon, C.O. 2013. Microbial succession and metabolite changes during fermentation of saeu-jeot: Traditional Korean salted seafood. *Food Microbiology* 34: 360–368.

Jung, M.J., Roh, S.W., Kim, M.S., and Bae, J.W. 2010. *Lentibacillus jeotgali* sp. nov., a halophilic bacterium isolated from traditional Korean fermented seafood. *International Journal of Systematic and Evolutionary Microbiology* 60: 1017–1022.

Kelaiditis, G.C. 1949. Technological research on salted sardine and anchovy produced in Greece. *Praktika of the Hellenic Hydrobiological Institute* 3: 5.

Khairunnisak, M., Azizah, A.H., Jinap, S., and Nurul Izzah, A. 2009. Monitoring of free glutamic acid in Malaysian processed foods, dishes and condiments. *Food Additives and Contaminants Part A Chemistry, Analysis, Control Exposure and Risk Assessment* 26: 419–426.

Kim, B. 1988. Changes in the properties of protein during the fermentation of salted shrimp. *Korean Journal of Food Science and Technology* 20: 883–889.

Kim, M.S., Roh, S.W., and Bae, J.W. 2010. *Halomonas jeotgali* sp. nov., a new moderate halophilic bacterium isolated from a traditional fermented seafood. *Journal of Microbiology* 48: 404–410.

Kim, Y.J. 2006. *Study on the safety for traditional foods*: Korea Food and Drug Administration research report.

Kobayashi, T., Kajiwara, M., Wahyuni, M., Kitakado, T., Hamada Sato, N., Imada, C., and Watanabe, E. 2003. Isolation and characterization of halophilic lactic acid bacteria isolated from "terasi" shrimp paste: A traditional fermented seafood product in Indonesia. *Journal of General and Applied Microbiology* 49: 279–286.

Lakshmi, W.G.I., Prassanna, P.H.P., and Edirisinghe, U. 2010. Production of Jaadi using Tilapia (*Oreochromis niloticus*) and determination of its physico-chemical and sensory properties. *Sabaramuwa University Journal* 9: 57–63.

Lee, C.-H., Steinkraus, K.H., and Reilly, P.J.A., eds. 1993. *Fish Fermentation Technology*. Tokyo, Japan: United Nations University Press.

Lee, E.J., Seo, J.E., Lee, J.K., Oh, S.W., and Kim, Y.J. 2008. Microbial and chemical properties of ready-to-eat skate in Korean market. *Journal of Food Hygiene and Safety* 23: 137–141.

Lee, G.B. 2001. *Nonya Favourites*. Hong Kong: Periplus Editions.

Lee, K.H. 1969. Microbiological and enzymological studies on the flavor components of seafood pickles. *Journal of the Korean Agriculture Chemical Society* 11: 1–27.

Lee, K.H., Chang, C.H., Mheen, T.I., Lee, S.R., Kwon, T.W., and Park, K.I. 1977. Korean Jeotkal. Proceedings, *Symposium on Indigenous Fermented Foods*. Bangkok, Thailand.

Leong, Q.L., Ab Karim, S., Selamat, J., Mohd Adzahan, N., Karim, R., and Rosita, J. 2009. Perceptions and acceptance of 'belacan' in Malaysian dishes. *International Food Research Journal* 16: 539–546.

Li, C.W., and Lai, C.Y. 1978. Studies on the proteolytic properties of the alkaline duck egg white. *Journal of the Chinese Agricultural Chemical Society* 16: 24.

Li, J.R., and Hsieh, Y.H. 2004. Traditional Chinese food technology and cuisine. *Asia Pacific Journal of Clinical Nutrition* 13: 147–155.

Liardon, R., and Hurrel, R.F. 1983. Amino acid racemization in heated and alkali-treated proteins. *Journal of Agricultural and Food Chemistry* 31: 432–437.

Lim, H.S. 2003. ACE inhibitory materials from *Raja kenojei*. *Korean Journal of Life Science* 13: 668–674.

Lin, C.W. 1983. The storage of eggs. In *The Chemistry and Utility of Eggs*, 2nd ed., pp. 98–121. Taipei, Taiwan: Hua Shiang Yuan Publishing Co.

Liu, L.Y., Yang, M.H., Lin, J.H., and Lee, M.H. 2005. Lipid profile and oxidative stability of commercial egg products. *Journal of Food and Drug Analysis* 13: 78–83.

Liu, Y.C., and Zhang, S.A. 1989. *Egg Processing and Technology*. Beijing, China: Agriculture Press.

Lopetcharat, K., Choi, Y.J., Park, J.W., and Daeschel, M.A. 2001. Fish sauce products and manufacturing: A review. *Food Reviews International* 17: 65–88.

Mah, J.H., Chang, Y.H., and Hwang, H.J. 2008. *Paenibacillus tyraminigenes* sp nov isolated from Myeolchi-jeotgal, a traditional Korean salted and fermented anchovy. *International Journal of Food Microbiology* 127: 209–214.

Majchzrak, D., Mitter, S., and Elmadfa, I. 2004. The effect of ascorbic acid on total antioxidant activity of black and green teas. *Food Chemistry* 88: 447–451.

Martin, C., and Sulit, J.I. 1955. Studies on the preparation of salted fish paste (bagoong), from dried dilis (*Stolephorus indicus*). *Philippine Journal of Fisheries* 3: 39.

Merican, Z., Lau, Y.Q., and Idrus, A.Z. 1980. *Malaysian Fermented Foods*: ASEAN Protein Project, Occasional paper no. 10.

Mheen, T.-I. 1993. Microbiology salt-fermented fishery products in Korea. In *Fish Fermentation Technology*, eds. C.-H. Lee, K.H. Steinkraus, and P.J.A. Reilly, pp. 231–247. Tokyo, Japan: United Nations University Press.

Michihata, T., Sado, Y., Yano, T., and Enomoto, T. 2000. Preparation of Ishiru (fish sauce) by a quick ripening process and changes in the composition of amino acids, oligopeptides and organic acids during processing. *Nippon Shokuhin Kagaku Kogaku Kaishi* 47: 369–377.

Miller, C.D., Louis, L., and Yonazawa, K. 1946. *Foods used by Filipinos in Hawaii* Honolulu: Agricultural experiment station bulletin 98, University of Hawaii.

Mizooku, Y., Yoshikawa, M., Tsuneyoshi, T., and Arakawa, R. 2003. Analysis of oxidized epigallocatechin gallate by liquid chromatography/mass spectrometry. *Rapid Communications in Mass Spectrometry* 17: 1915–1918.

Mizutani, T., Kimizuka, A., Ruddle, K., and Ishige, N. 1987. A chemical analysis of fermented fish products and discussion of fermented flavors in Asian cuisines. *Bulletin of the National Museum of Ethnology* 12: 801–864.

Mohammed, H.M.H. 2010. Nutritive value of fresh and salted fermented fish (*Alestes dentex*) Terkin. Proceedings, *Food Science and Technology*, p. 70. Khartoum, Sudan.

Mok, C., Lee, J., Song, K., Kim, S., Lim, S., and Woo, G. 2000. Changes in physicochemical properties of salted and fermented shrimp at different salt levels. *Korean Journal of Food Science and Technology* 32: 187–191.

Montano, N., Gavino, G., and Gavino, V.C. 2001. Polyunsaturated fatty acid contents of some traditional fish or shrimp paste condiments of the Philippines. *Food Chemistry* 75: 155–158.

Mori, K., Shinano, H., and Akiba, M. 1979. The aerobic bacteria in the ripening process of "ikashiokara." *Bulletin of the Japanese Society for the Science of Fish* 45: 771–779.

Myint, A., Aung, O., and Sein, M.M. 1997. Physiochemical and microbiological changes during the fermentation period in producing Myanmar shrimp paste. *Journal of Myanmar Chemical Society* 4: 25–35.

Nam, H.K., and Lee, M.K. 1999. Studies on the fatty acids and cholesterol level of Raja skate. *Journal of the Korean Oil Chemists' Society* 12: 55–58.

Namwong, S., Hiraga, K., Takada, K., Tsunemi, M., Tanasupawat, S., and Oda, K. 2006. A halophilic serine proteinase from *Halobacillus* sp. SR5–3 isolated from fish sauce: Purification and characterization. *Bioscience, Biotechnology, and Biochemistry* 70: 1395–1401.

Namwong, S., Tanasupawat, S., Kudo, T., and Itoh, T. 2011. *Haloarcula salaria* sp. nov. and *Haloarcula tradensis* sp. nov. from salt in Thai fish sauce. *International Journal of Systematic and Evolutionary Microbiology* 61: 231–236.

Namwong, S., Tanasupawat, S., Lee, K.C., and Lee, J.S. 2009. *Oceanobacillus kapialis* sp. nov., from fermented shrimp paste in Thailand. *International Journal of Systematic and Evolutionary Microbiology* 59: 2254–2259.

Namwong, S., Tanasupawat, S., Smitinont, T., Visessanguan, W., Kudo, T., and Itoh, T. 2005. Isolation of *Lentibacillus salicampi* strains and *Lentibacillus juripiscarius* sp. nov. from fish sauce in Thailand. *International Journal of Systematic and Evolutionary Microbiology* 55: 315–320.

Namwong, S., Tanasupawat, S., Visessanguan, W., Kudo, T., and Itoh, T. 2007. *Halococcus thailandensis* sp. nov., from fish sauce in Thailand. *International Journal of Systematic and Evolutionary Microbiology* 57: 2199–2203.

Okuzumi, M., and Fujii, T. 2000. Nutritional and functional properties of squid and cuttlefish. In *35th Anniversary Commemorative Publication*, pp. 205– 218. Tokyo, Japan: National Cooperative Association of Squid Processors.

Olympia, M.S.D. 1992. Fermented fish products in the Philippines. In *Applications of Biotechnology to Traditional Fermented Foods*, eds. E.L. Gaden, M. Bokanga, S. Harlander, C.W. Hesseltine, and K.H. Steinkraus, pp. 131–139. Washington, DC: National Academy Press.

Pakdeeto, A., Tanasupawat, S., Thawai, C., Moonmangmee, M., Kudo, T., and Itoh, T. 2007a. *Salinicoccus siamensis* sp. nov., isolated from fermented shrimp paste in Thailand. *International Journal of Systematic and Evolutionary Microbiology* 57: 2004–2008.

Pakdeeto, A., Tanasupawat, S., Thawai, C., Moonmangmee, S., Kudo, T., and Itoh, T. 2007b. *Lentibacillus kapialis* sp. nov., from fermented shrimp paste in Thailand. *International Journal of Systematic and Evolutionary Microbiology* 57: 364–369.

Park, J.N., Fukumoto, Y., Fujita, E., Tanaka, T., Washio, T., Otsuka, S., Shimizu, T., Watanabe, K., and Abe, H. 2001. Chemical composition of fish sauces produced in southeast and east Asian countries. *Journal of Food Composition and Analysis* 14: 113–125.

Petrus 2009. Improvement of Wadi Ikan Betok (*Anabas testudineus* Bloch) by incorporating palm sugar (*Arenga pinnata*) as food additive and acetic acid (CH2COOH) [Perbaikan mutu fermentasi Wadi Ikan Betok (*Anabas testudineus* Bloch) dengan penambahan food additive gula merah aren (*Arenga pinnata*) dan asam cuka (CH2COOH)]. M.Sc. thesis. Banjar, Indonesia: University of Lambung Mangkurat.

Petrus, Purnomo, H., Suprayitno, E., and Hardoko 2013. Physicochemical characteristics, sensory acceptability and microbial quality of Wadi Betok, a traditional fermented fish from South Kalimantan, Indonesia. *International Food Research Journal* 20: 933–939.

Phithakpol, B. 1993. Fish fermentation in Thailand. In *Fish Fermentation Technology*, eds. C.H. Lee, K.H. Steinkraus, and P.J.A. Reilly, pp. 155–168. Tokyo, Japan: United Nations University Press.

Phithakpol, B., Varanyanond, W., Reungmaneepaitoon, S., and Wood, H. 1995. *The Traditional Fermented Foods of Thailand*. Bangkok, Thailand: Institute of Food Research and Product Development, Kasetsart University.

Phrommao, E., Rodtong, S., and Yongsawatdigul, J. 2010. Identification of novel halotolerant bacillopeptidase F-like proteinases from a moderately halophilic bacterium, *Virgibacillus* sp. SK37. *Journal of Applied Microbiology* 110: 191–201.

Poernomo, A., Suryaningrum, T.D., Ariyani, F., and Putro, S. 1984. Studies on the nutritive value and microbiology of traditional fishery products. *Laporan Penelitian Teknologi Perikanan* 30: 9–19.

Putro, S. 1993. Fish fermentation technology in Indonesia. In *Fish Fermentation Technology*, eds. C.H. Lee, K.H. Steinkraus, and P.J.A. Reilly, pp. 107–128. Tokyo, Japan: United Nations University Press.

Rahayu, W.P., Ma'oen, S., Suliantari, and Fardiaz, D. 1992. *Teknologi Fermentasi Hasil Perikanan*. Bogor, Indonesia: PAU Pangan dan Gizi Institut Pertanian Bogor.

Rapsang, G.F., and Joshi, S.R. 2012. Bacterial diversity associated with tungtap, an ethnic traditionally fermented fish product of Meghalaya. *Indian Journal of Traditional Knowledge* 11: 134–138.

Riha, W.E.I., Izzo, H.V., Zhang, J., and Chi-Tang, H. 1996. Nonenzymatic deamidation of food proteins. *Critical Reviews in Food Science and Nutrition* 36: 225–255.

Roh, S.W., Nam, Y.D., Chang, H.W., Kim, K.H., Sung, Y., Kim, M.S., Oh, H.M., and Bae, J.W. 2009. *Haloterrigena jeotgali* sp. nov., an extremely halophilic archaeon from salt-fermented food. *International Journal of Systematic and Evolutionary Microbiology* 59: 2359–2363.

Roh, S.W., Nam, Y.D., Chang, H.W., Sung, Y., Kim, K.H., Oh, H.M., and Bae, J.W. 2007. *Halalkalicoccus jeotgali* sp. nov., a halophilic archaeon from shrimp jeotgal, a traditional Korean fermented seafood. *International Journal of Systematic and Evolutionary Microbiology* 57: 2296–2298.

Ruddle, K. 1990. *Fish Processing Technologies for Entrepreneurial Development in Bangladesh*. Bangkok: U.N. Economic and Social Commission for Asia and the Pacific.

Ruddle, K., and Ishige, N. 2010. On the origins, diffusion and cultural context of fermented fish products in Southeast Asia. In *Globalization, Food and Social Identities in the Asia Pacific region*, ed. J. Farrer, pp. 1–17. Tokyo, Japan: Sophia University Institute of Comparative Culture.

Saisithi, P. 1994. Traditional fermented fish: Fish sauce production. In *Fisheries Processing: Biotechnological Applications*, ed. A.M. Martin, pp. 111–129. London: Chapman & Hall.

Salampessy, J., Kailasapathy, K., and Thapa, N. 2010. Fermented fish products. In *Fermented Foods and Beverages of the World*, eds. J.P. Tamang, and K. Kailasapathy, pp. 289–307. Boca Raton, FL: CRC Press.

Sanchez, A.H. 1977. Old and new ways of producing patis. *Green Fields* 7: 4–8.

Sanchez, P.C. 2008. *Philippine Fermented Foods: Principles and Technology*. Quezon City, Philippines: The University of the Philippines Press.

Sanda, A. 1973. Analysis of free amino acids in HmyinNgan-pya-ye. M.Sc. thesis. Rangoon, Myanmar: Art and Science University.

Sands, A., and Crisan, E.V. 1974. Microflora of fermented Korean seafoods. *Journal of Food Science* 39: 1002–1005.

Sarnianto, P., Irianto, H.E., and Putro, S. 1984. Studies on the histamine contents of fermented fishery products. *Laporan Penelitian Teknologie Perikanan* 32: 35–39.

Sarojnalini, C., and Suchitra, T. 2009. Microbial profile of starter culture fermented fish product 'Ngari' of Manipur. *Indian Journal of Fisheries* 56: 123–127.
Sarojnalini, C., and Vishwanath, W. 1988. Composition and digestibility of fermented fish foods of Manipur. *Journal of Food Science and Technology* 25: 349–351.
Sein, M.M. 1999. The mode of fermentation of Myanmar shrimp paste. Ph.D. thesis. Yangon, Myanmar: Yangon Arts and Science University.
Sharif, R., Ghazali, A.R., Rajab, N.F., Haron, H., and Osman, F. 2008. Toxicological evaluation of some Malaysian locally processed raw food products. *Food and Chemical Toxicology* 46: 368–374.
Shi, L., Zhou, J., and Gunasekaran, S. 2008. Low temperature fabrication of ZnO-whey protein isolate nanocomposite. *Materials Letters* 62: 4383–4385.
Sinsuwan, S., Rodtong, S., and Yongsawatdigul, J. 2007. NaCl-activated extracellular proteinase from *Virgibacillus* sp. SK37 isolated from fish sauce fermentation. *Journal of Food Science* 72: C264–C269.
Sinsuwan, S., Rodtong, S., and Yongsawatdigul, J. 2010. A NaCl-stable serine proteinase from *Virgibacillus* sp. SK33 isolated from Thai fish sauce. *Food Chemistry* 119: 573–579.
Smriga, M., Mizukoshi, T., Iwahata, D., Eto, S., Miyano, H., Kimura, T., and Curtis, R.I. 2010. Amino acids and minerals in ancient remnants of fish sauce (garum) sampled in the 'Garum Shop' of Pompeii, Italy. *Journal of Food Composition and Analysis* 23: 442–446.
Soedarmo, P., and Sediaoetama, A.D. 1977. *Nutrition Science*. Jakarta, Indonesia: Penerbit Dian Rakyat (in Indonesian).
Steinkraus, K.H., ed. 1996. *Handbook of Indigenous Fermented Foods*, 2nd ed. New York: Marcel Dekker.
Su, H.P., and Lin, C.W. 1993. A new process for preparing transparent alkalized duck egg and its quality. *Journal of the Science of Food and Agriculture* 61: 117–120.
Subasinghe, S. 1993. Fish fermentation technology in Japan. In *Fish Fermentation Technology*, eds. C.H. Lee, K.H. Steinkraus, and P.J.A. Reilly, pp. 167–175. Tokyo, Japan: United Nations University Press.
Subba Rao, G.N. 1967. *Fish procession in the Indo-Pacific area*. Regional Studies No. 4, 231, Bangkok, Thailand: FAO Regional Office for Asia and the Far East.
Surano, I.S., and Hosono, A. 1994. Chemical and aerobic bacterial composition of Terasi, a traditional fermented product from Indonesia. *Journal of the Food Hygienic Society of Japan* 35: 299–304.
Takai, M., Kawai, Y., Inoue, N., and Shinano, H. 1992. Comparative studies on microbiological and chemical characteristics of Ika-shiokara akazukuri and Ika-shiokara kurozukuri. *Bulletin of the Japanese Society for the Science of Fish* 58: 2373–2378.
Tanasupawat, S. 1999. *Halophilic Bacteria in Salted Fermented Foods in Thailand*. Bangkok, Thailand: Department of Microbiology, Faculty of Pharmaceutical Sciences, Chulalongkorn University.
Tanasupawat, S., and Komagata, K. 1995. Lactic acid bacteria in fermented foods in Thailand. *World Journal of Microbiology and Biotechnology* 11: 253–256.

Tanasupawat, S., Namwong, S., Kudo, T., and Itoh, T. 2009. Identification of halophilic bacteria from fish sauce (nam-pla) in Thailand. *Journal of Culture Collections* 6: 69–75.

Tanasupawat, S., Pakdeeto, A., Namwong, S., Thawai, C., Kudo, T., and Itoh, T. 2006. *Lentibacillus halophilus* sp. nov., from fish sauce in Thailand. *International Journal of Systematic and Evolutionary Microbiology* 56: 1859–1863.

Tanasupawat, S., Taprig, T., Akaracharanya, A., and Visessanguan, W. 2011a. Characterization of *Virgibacillus* strain TKNR13-3 from fermented shrimp paste (ka-pi) and its protease production. *African Journal of Microbiological Research* 5: 4714–4721.

Tanasupawat, S., Thongsanit, J., Thawai, C., Lee, K.C., and Lee, J.S. 2011b. *Pisciglobus halotolerans* gen. nov., sp. nov., isolated from fish sauce in Thailand. *International Journal of Systematic and Evolutionary Microbiology* 61: 1688–1692.

Tapingkae, W., Tanasupawat, S., and Itoh, T. 2008. *Natrinema gari* sp. nov., a halophilic archaeon isolated from fish sauce in Thailand. *International Journal of Systematic and Evolutionary Microbiology* 58: 2378–2383.

Thongsanit, J., Tanasupawat, S., Keeratipibul, S., and Jatikavanich, S. 2002. Characterization and identification of *Tetragenococcus halophilus* and *Tetragenococcus muriaticus* strains from fish sauce (nam-pla). *Japanese Journal of Lactic Acid Bacteria* 13: 46–52.

Thongthai, C., and Gildberg, A. 2005. Asian fish sauce as a source of nutrition. In *Asian Functional Foods*, eds. J.S. Chi-Tang Ho, and F. Shahidi, pp. 215–265. New York: Marcel Dekker.

Thongthai, C., McGenity, T., Suntinanalert, J.P., and Grant, W.D. 1992. Isolation and characterization of an extremely halophilic archaeobacterium from traditionally fermented Thai fish sauce (nam pla). *Letters in Applied Microbiology* 14: 111–114.

Tyn, M.T. 1983. Recent development in nutrition and food technology in Burma. In *Burmese Encyclopedia*, supplement edition, pp. 65–73. Rangoon, Myanmar: Sarpaybeikhman Publishers.

Tyn, M.T. 1986. Traditional foods of Burma, Ngapi. In *Traditional Foods: Some Products and Technologies*, eds. V.H. Potty, J.V. Shankar, K.A. Ranganath, N. Chandrasekhara, V. Prakash, B.R. Sri Hari, and S.V.H. Dastur, pp. 97–113. Mysore, India: Central Food Technological Research Institute.

Tyn, M.T. 1993. Trends of fermented fish technology in Burma. In *Fish Fermentation Technology*, eds. C.H. Lee, K.H. Steinkraus, and P.J.A. Reilly, pp. 129–153. Tokyo, Japan: United Nations University Press.

Tyn, M.T. 2004. Industrialization of Myanmar fish paste and sauce fermentation. In *Industrialization of Indigenous Fermented Foods*, 2nd ed., ed. K.H. Steinkraus, pp. 737–762. New York: Marcel Dekker.

Udomsil, N., Rodtong, S., Tanasupawat, S., and Yongsawatdigul, J. 2010. Proteinase-producing halophilic lactic acid bacteria isolated from fish sauce fermentation and their ability to produce volatile compounds. *International Journal of Food Microbiology* 141: 186–194.

Van Veen, A.G. 1965. Seafood products in Southeast Asia. In *Fish as Food*, vol. 3, ed. G. Borgstrom, pp. 227–247. Boca Raton, FL: Taylor & Francis.

Wang, J., and Fung, D.Y. 1996. Alkaline-fermented foods: A review with emphasis on pidan fermentation. *Critical Reviews in Microbiology* 22: 101–138.

Weerasinghe, T.J., and Samaradiwakara, R. 1985. *Effects of Different Salt Concentrations on Microbiological and Chemical Changes in Jaadi Fermentation*. Colombo, Sri Lanka: Institute of Post Harvest Technology.

Wittanalai, S., Chandet, N., and Rakariyatham, N. 2010. Selection of starter cultures for the production of vegetarian kapi, a Thai fermented condiment. *Chiang Mai Journal of Science* 37: 314–325.

Wu, Q.L. 1992. New technology used in protection of pidan-2,3,4-oxy-Cabenzoate acid. *Journal of Food Science* 7: 32.

Yang, S.C., and Chen, K.H. 2001. The oxidation of cholesterol in the yolk of selective traditional Chinese egg products. *Poultry Science* 80: 370–375.

Yongsawatdigul, J., Choi, Y.J., and Udomporn, S. 2004. Biogenic amines formation in fish sauce prepared from fresh and temperature-abused Indian anchovy (*Stolephorus indicus*). *Journal of Food Science* 69: 312–319.

Yoon, J.H., Jung, S.Y., Kim, W., Nam, S.W., and Oh, T.K. 2006. *Nesterenkonia jeotgali* sp. nov., isolated from jeotgal, a traditional Korean fermented seafood. *International Journal of Systematic and Evolutionary Microbiology* 56: 2587–2592.

Yoon, J.H., Kang, K.H., and Park, Y.H. 2003a. *Psychrobacter jeotgali* sp. nov., isolated from jeotgal, a traditional Korean fermented seafood. *International Journal of Systematic and Evolutionary Microbiology* 53: 449–454.

Yoon, J.H., Kang, S.S., Lee, K.C., Kho, Y.H., Choi, S.H., Kang, K.H., and Park, Y.H. 2001. *Bacillus jeotgali* sp. nov., isolated from jeotgal, Korean traditional fermented seafood. *International Journal of Systematic and Evolutionary Microbiology* 51: 1087–1092.

Yoon, J.H., Lee, K.C., Weiss, N., Kang, K.H., and Park, Y.H. 2003b. *Jeotgalicoccus halotolerans* gen. nov., sp. nov. and *Jeotgalicoccus psychrophilus* sp. nov., isolated from the traditional Korean fermented seafood jeotgal. *International Journal of Systematic and Evolutionary Microbiology* 53: 595–602.

Yoshida, Y. 1998. Umami taste and traditional seasonings. *Food Reviews International* 14: 213–246.

Yuksel, Z., Avci, E., and Erdem, Y.K. 2010. Characterization of binding interactions between green tea flavonoids and milk proteins. *Food Chemistry* 121: 450–456.

Yunizal 1998. Processing of shrimp terasi. *Warta Penelitian dan Pengembangan Pertanian* 20: 4–6.

Zarei, M., Najafzadeh, H., Eskandari, M.H., Pashmforoush, M., Enayati, A., Gharibi, D., and Fazlara, A. 2012. Chemical and microbial properties of mahyaveh, a traditional Iranian fish sauce. *Food Control* 23: 511–514.

Zenitani, B. 1955. Studies on fermented fish products I. On the aerobic bacteria in "Shiokara." *Bulletin of the Japanese Society of Scientific Fisheries* 21: 280–283.

Zhang, B.F. 1988. *Chinese Traditional Egg Products*. Beijing, China: Chinese Commercial Press.

4

Microorganisms Predominating in Alkaline-Fermented Foods

Indigenous fermentation processes depend on microorganisms naturally existing in the raw materials, utensils, water, surrounding atmosphere, and wrapping material. Among these microorganisms, the most frequently isolated from alkaline-fermented foods (AFFs) are *Bacillus* and related genera, lactic acid bacteria, and yeasts. As industrial-scale production of AFFs usually requires pure starter to achieve reproducibility, consistent quality, and a high standard of public safety. The phenotypical and genotypical characterization of the fermenting microbes is essential for the selection of high-performing, and safe, starter cultures.

4.1 *Bacillus* and Related Genera

Charles Parkouda, Bréhima Diawara, and Kwaku Tano-Debrah

Alkaline-fermented foods (AFFs) are food products, processed from raw material (generally from seeds, leaves, or eggs) through spontaneous fermentation. These products, including natto, kinema, thua nao, douchi, bikalga, maari, afitin, dawadawa (iru or soumbala), ntoba mbodi, kawal, and pidan, are consumed in a number of African and Asian countries (Wang and Fung, 1996; Parkouda et al., 2009). The increase in pH during the production of AFFs is due to the ability of dominant microorganisms to hydrolyze proteins into amino acids and ammonia or due to alkali treatment during production. Investigations on the microbial diversity associated with alkaline fermentation of these products suggest that different species of bacteria are involved in the fermentation (Sarkar et al., 2002; Parkouda et al., 2010; Chadare et al., 2011). The fermentation is generally carried out by the aerobic mesophilic bacteria with the predominance of *Bacillus* spp. (Chukeatirote et al., 2010; Parkouda et al., 2010), accompanied in some cases by lactic acid bacteria (Wang and Fung, 1996; Parkouda et al., 2009).

4.1.1 Biodiversity

Various species of *Bacillus* and related genera are associated with the AFFs. These include *Bacillus subtilis*, *Bacillus cereus*, *Bacillus megaterium*, *Bacillus endophyticus*, *Bacillus licheniformis*, *Bacillus borstelensis*, *Bacillus pumilus*, *Bacillus coagulans*, *Bacillus circulans*, *Paenibacillus polymyxa*, *Lysinibacillus sphaericus*, and *Lysinibacillus fusiformis* (Odunfa, 1981; Sarkar et al., 2002; Ouoba et al., 2004, 2010; Azokpota et al., 2007; Parkouda et al., 2009).

The predominance of *Bacillus* spp. in such foods is assumed to be linked to different factors, such as long cooking time in their processing, which could select for heat resistant microorganisms, especially spore-forming bacteria (Ouoba et al., 2008), and recontamination after the cooking process and in handling prior to the fermentation, since these microorganisms are ubiquitous in the production environment. Indeed, these microorganisms are believed to originate from raw material and utensils and can be isolated from a variety of animal and plant products (Odunfa, 1981). *Bacillus* spp. are considered to be the main organisms responsible for AFF due to their properties, as reviewed by Wang and Fung (1996) and Parkouda et al. (2009).

The genus *Bacillus* includes a variety of phenotypically heterogeneous species exhibiting a wide range of nutritional requirements, physiological and metabolic diversity, and genotypic characteristics. Rössler et al. (1991) grouped nine species of *Bacillus* into four clusters. Ash et al. (1991) separated 51 species of *Bacillus* into five phylogenetically distinct clusters. The phylogenetic approach to *Bacillus* taxonomy has been accomplished largely by oligonucleotide sequencing analysis of the 16S rRNA gene (Figure 4.1).

4.1.2 Taxonomic Characteristics

Bacillus and related genera have been traditionally distinguished on the basis of phenotypic properties, structure, and physiology (Odunfa,

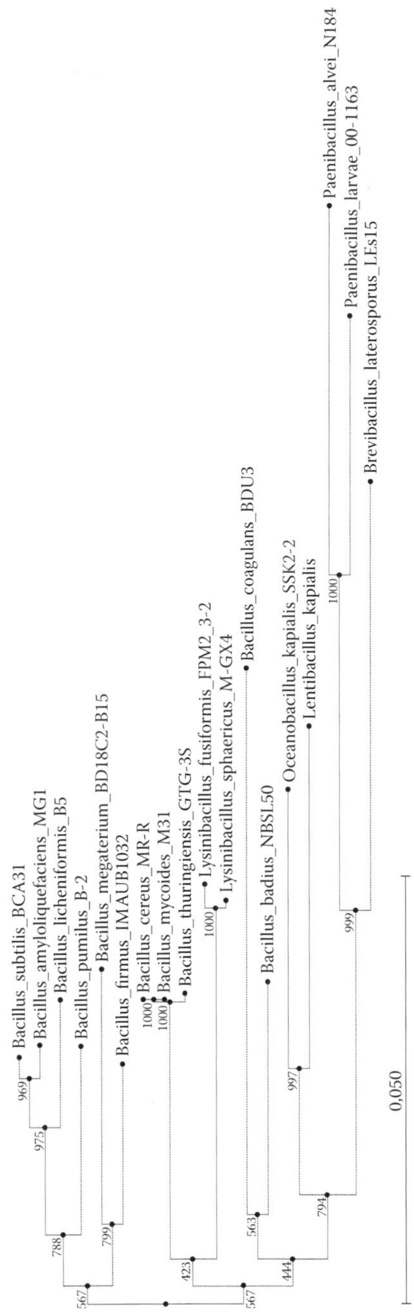

Figure 4.1 Phylogenetic relationship among *Bacillus* spp., based on partial (800 bp) 16S rRNA sequences. (Courtesy of Dennis S. Nielsen.)

1981). The use of only these characteristics resulted often in uncertain identification (Ouoba et al., 2004). Recent studies showed that to unambiguously identify the microorganisms, the methods used should include classical microbiological and phenotypic analysis along with molecular tools (Daffonchio et al., 1998; Parkouda et al., 2010; Thorsen et al., 2011). The current use of a polyphasic approach (molecular biological methods in microbiological research in combination with phenotypic methods) has allowed accurate and more reliable identification of microorganisms involved in these fermentations. In recent years, a number of new, valuable biochemical and molecular biological techniques for taxonomic purpose including *rpoB* and *gyrA* sequencing, 16S rDNA targeted probe analysis, DNA–DNA hybridizations, gelatinase tests, antibiotic susceptibility tests, and *gyrA* gene sequences have been developed, which could also be effectively utilized (Chun and Sook Bae, 2000; De Clerck and De Vos, 2004; Thorsen et al., 2011). The general scheme for identification includes isolation, phenotypic characterization, genotypic grouping, selection of representative of each genotypic group, and genotypic identification. The current classification of species within the genus *Bacillus* and related genera is well established and is based on a combination of numerous experimental approaches. Characterization at the genotypic and phenotypic levels of selected *Bacillus* spp. has led to the creation of several genera, including *Alicyclobacillus* (Wisotzkey et al., 1992), *Amphibacillus* (Niimura et al., 1990), *Brevibacillus* (Shida et al., 1996), *Paenibacillus* (Heyndrickx et al., 1996), *Oceanobacillus* (Namwong et al., 2009), *Geobacillus* (Nazina et al., 2001), and *Lentibacillus* (Pakdeeto et al., 2007).

4.1.2.1 Phenotypic Characteristics Preliminary analysis in microbial identification often involves one or more phenotypic methods. These methods are suitable for microorganisms that are cultivable and have well-established growth parameters, and physiological and biochemical profiles. Phenotypic characteristics were formerly considered suitable for the typing of individual strains within a species. It has been recognized for a long time that the genus *Bacillus* is phenotypically extremely diverse (Claus and Berkeley, 1986). Numerical classification based on a series of phenotypic characters has been used for the grouping of 368 strains of *Bacillus* into 79 clusters (Priest et al., 1988).

However, misidentification of some bacteria reported in some studies indicates that the identification of bacteria based on the phenotypic properties alone may not always be suitable (Towner and Cockayne, 1993; Ouoba et al., 2004).

4.1.2.1.1 Morphological Characteristics The morphological characteristics include colony and cell morphology obtained through simple isolation and culturing of the microorganisms and subsequent visual observation using microscopy. They also include shape, size, surface characteristics, pigmentation, Gram-staining, sporulation characteristics, and motility. *Bacillus* and related genera are generally Gram-positive, spore-forming, catalase-positive microorganisms, cultivable in several media (plate count agar, nutrient agar, tryptone soya agar, brain heart infusion agar, etc.), with the exception of insect-related *Bacillus* spp. The insect pathogens, *Paenibacillus larvae*, *Paenibacillus popilliae*, and *Paenibacillus lentimorbus*, are more fastidious and must be isolated on specific media (for example, J-agar). Furthermore, they are typically catalase-negative, and they require special media or inoculation into insect hosts for sporulation. Endospore formation, found in the group of *Bacillus*, is thought to be a strategy for survival in the soil environment, wherein these bacteria predominate (Moir, 2006). Aerial distribution of the dormant spores probably explains the occurrence of aerobic spore-formers in most habitats examined.

4.1.2.1.2 Biochemical, Physiological, and Metabolic Characteristics Phenotypic characteristics also include the biochemical profile and metabolic properties of microorganisms by testing their growth requirements and enzymatic activities. The biochemical tests use specific growth media, nutrients, chemicals, or growth conditions to elicit observable or measurable biochemical responses from the *Bacillus* and related genera, thereby enabling their characterization and further identification. These tests include utilization of carbon and nitrogen sources, growth requirements (anaerobic or aerobic, temperature optimum and range, and pH optimum and range), preferred osmotic conditions, generation of fermentation products, production of enzymes, etc. Examples of recognized tests include phenol red carbohydrate fermentation, catalase production, oxidase test, oxidation-fermentation, methyl red test, Voges-Proskauer reaction, nitrate reduction, starch hydrolysis, tryptophan hydrolysis, hydrogen sulfide production, citrate utilization, and litmus milk reactions. Today, several miniaturized and automated commercial systems are available with well-defined quality control procedures that allow rapid characterization of microorganisms, e.g., API 50 CHB system, *Microgen*™ *Bacillus ID* (Logan and Berkeley, 1984; Carpana et al., 1995).

There is a great diversity of physiology among *Bacillus* spp. Their collective features include degradation of most or all of the substrates derived from plant and animal sources, including cellulose, starch, pectin, proteins, agar, hydrocarbons, and others such as antibiotic production, alkaliphily, psychrophily, and thermophily. Most of the species of *Bacillus* require no organic growth factors and can be grown in defined or relatively simple complex media. For a few bacilli, e.g., *B. subtilis* and *B. megaterium*, minimal media have been established. Primary isolation can be performed on, for example, simple media such as nutrient agar or plate count agar. Others may require amino acids, B vitamins, or both. The majority of *Bacillus* species are mesophiles (e.g., *B. subtilis*), with temperature optima between 30°C and 45°C, but some are thermophiles (e.g., *Bacillus stearothermophilus*) with optimum as high as 65°C. Others are psychrotolerant (e.g., *Bacillus weihenstephanensis*) and are able to grow from below 7°C to 38°C (von Stetten et al., 1999). *Bacillus* spp. are found growing over a range of pH from 5 to 10 (Olajyigbe and Ajele, 2005).

4.1.2.2 Molecular Characteristics The conventional methods for identification of bacteria using only phenotypes are not always sufficiently discriminatory. Molecular biology has provided tools to study the genotypic relationships of microorganisms, reconstruct the evolution of these bacteria, and improve their taxonomy. Indeed, the development of a large number of DNA fingerprinting techniques has played an important role in increasing our understanding of the microorganisms' characteristics. For the genus *Bacillus*, several different genotyping techniques may be applied as tools for either species identification or differentiation of strains. The major advantages of these genotyping methods are based on their discriminatory power and their universal applicability. With molecular typing methods, closely related strains with similar phenotypic features may consistently be distinguished. Modern molecular techniques, including polymerase chain reaction

(PCR)-based and other genotyping methods, have become increasingly important for species identification or for the differentiation of probiotic strains. Molecular typing methods applicable to microorganisms include grouping and identification techniques. Methods that have been used in the identification and genotypic characterization of *Bacillus* and related genera include internal transcribed spacer PCR (ITS-PCR), random amplification polymorphic DNA-PCR (RAPD-PCR), repetitive sequence-based PCR (rep-PCR), pulsed field gel electrophoresis (PFGE), as well as sequencing of 16S rRNA, *gyrA*, *gyrB*, and *rpoB* genes (Liu et al., 1997; Daffonchio et al., 1998; De Silva et al., 1998; Yamada et al., 1999; Chun and Sook Bae, 2000; Herman and Heyndrickx, 2000; Mendo et al., 2000; Parkouda et al., 2010; Thorsen et al., 2011).

4.1.2.2.1 Genotypic Grouping Methods Several grouping methods have been used for the characterization and identification of *Bacillus* and related genera. The RAPD-PCR fingerprint method showed a high level of diversity among the species of *Bacillus*. The method has been successfully applied to *Bacillus* and is considered as a discriminatory tool for measuring diversity among strains belonging to the same species (Brousseau et al., 1993; Stephan et al., 1994; Stephan, 1996; Sarkar et al., 2002; Inatsu et al., 2006). The ITS-PCR has been found suitable for typing only at species level of *B. subtilis* of other origin as well as other *Bacillus* spp. (Gürtler and Stanisich, 1996; Johnson et al., 2000; Marten et al., 2000; Shangkuan et al., 2000). Ouoba et al. (2004) showed that ITS-PCR and ITS-PCR-RFLP are suitable for the typing of *Bacillus* isolates from soumbala at species level. The PFGE is a discriminative method reported to be suitable for typing of *B. subtilis*, *B. licheniformis*, and *B. cereus* at strain level (Johansen et al., 1996; Liu et al., 1997; Helgason et al., 2000; Marten et al., 2000; Mendo et al., 2000) using *Sfi*I and *Sma*I as restriction enzymes.

Comparing the uses of ITS-PCR, ITS-PCR-RFLP, and PFGE for typing *Bacillus* isolates from soumbala, Ouoba et al. (2004) reported that ITS-PCR and ITS-PCR-RFLP (using one restriction enzyme) are equally powerful and simple methods for typing the *Bacillus* spp. PFGE is more discriminative and allows identification at a level grouping isolates according to important technological properties and at strain level.

4.1.2.2.2 Gene Sequencing For the analysis of natural microbial populations in which an unknown diversity must be anticipated, the rRNA is being focused because they are functionally and evolutionarily homologous in all organisms. Indeed, the rRNAs are ancient molecules and extremely conserved in overall structure. The rRNA sequences were being established as a most useful molecular chronometer to infer phylogenetic relationship because they are present in all organisms and changes in the nucleotide sequences were deemed to occur in a clocklike manner (Woese, 1987). Several species of *Bacillus* were reclassified based on 16S rRNA sequence alignment. The sequencing of 16S rRNA has successfully been used for the identification of *B. subtilis* and *B. pumilus* in several investigations (Ash et al., 1991; Ash and Collins, 1992; Green et al., 1999; Siefert et al., 2000; El-Helow, 2001). Partial 16S rDNA sequence (Goto et al., 2000) and rRNA gene restriction patterns (Joung and Côté, 2002) have also been used for the rapid identification or classification of *Bacillus* species and related genera, respectively. However, sequencing of the 16S rRNA gene for identification of *Bacillus* shows limitations in differentiating closely related species and is used for basic classification for species delineation. It was not possible to distinguish *B. subtilis* from *B. licheniformis* or *B. cereus* from *Bacillus thuringiensis* by 16S rRNA sequencing (Azokpota et al., 2007; Parkouda et al., 2010). Both the groups have been reported to be so closely related that they could not easily be distinguished by sequencing the 16S rRNA (White et al., 1993; Rainey et al., 1994; Daffonchio et al., 1998). Chun and Sook Bae (2000) demonstrated that the different strains of *Bacillus* can show almost identical 16S rRNA sequences, but significantly low *gyrA* NT similarities. They concluded that the method for the amplification and sequencing of partial *gyrA* genes may be useful for the rapid identification of *B. subtilis* and allied taxa, especially organisms in these taxa, which cannot be differentiated by using conventional phenotypic tests and 16S rRNA analysis (Chun and Sook Bae, 2000). Additionally, sequencing of other genes, such as *rpoB* and *gyrB* has been reported to be more discriminatory for improved identification and differentiation of closely related *Bacillus* species (De Clerck and De Vos, 2004).

4.1.2.2.3 Culture-Independent Method The culture-independent-based method, PCR-DGGE (denaturing gradient gel electrophoresis) has been successfully used for bacterial identification in food microbiology (Ercolini, 2004). Differences in the *Bacillus*-related community structures between treatments were clearly detected using PCR-DGGE (Garbeva et al., 2003; Yi et al., 2012). Sequences affiliated to those of *Bacillus mycoides*, *B. pumilus*, *B. megaterium*, *B. thuringiensis*, and *Bacillus firmus*, as well as to related taxa, such as *Paenibacillus*, were obtained (Garbeva et al., 2003).

4.1.3 Ecology

Bacillus subtilis, the type species of the genus, encompasses aerobic, endospore-forming, rod-shaped bacteria, which are commonly found in soil, water bodies, and in association with plants (Claus and Berkeley, 1986). *Bacillus* is a diverse bacterial genus characterized by cells growing aerobically and forming dormant endospores (Maughan and Van der Auwera, 2011). *Bacillus* as well as most related aerobic spore-forming bacteria are ubiquitous and can be isolated from a large variety of sources, such as plant material, air, soil, water, sewage, and utensils, and can ferment or spoil food. The ubiquity of these bacteria can also be explained by an unusual resistance of their endospores to environmental stress, the developmental cycle of endospore formation, and the production of antibiotics (Ouoba et al., 2007b; Kaboré et al., 2012b). *Bacillus subtilis* has been reported to be capable of forming highly resistant dormant endospores in response to nutrient deprivation and other environmental stresses and these spores are easily made airborne and dispersed by wind (Earl et al., 2008). Hence, the presence of these bacteria in a certain environment is not necessarily an indication of habitat. However, it is generally accepted that the primary habitat of the aerobic endospore-forming bacilli is the soil.

Their resistance to environmental stress, as well as their long-term survival under adverse conditions, make *Bacillus* and related genera ubiquitous in the AFF production areas. *Bacillus* and related genera are reported to be able to degrade a series of biopolymers, such as proteins, lipids, and carbohydrates (Ouoba et al., 2003a, 2003b, 2007a). Owing to such abilities, these bacteria become metabolically active when suitable substrates for their growth are available in the production area, and most likely they form spores when their nutrients become exhausted.

4.1.4 Metabolism

Bacillus subtilis and its close relatives are among the most important industrial microorganisms. They are used for the production of antibiotics, enzymes such as proteases, amylases, lipases, or cellulases, as well as vitamins and proteins (Harwood, 1992). Some bacteria are characterized by their ability to adapt efficiently to a wide range of nutritional environments reflecting their use of overlapping regulatory systems that link gene expression to intracellular pools of a small number of key metabolites (Sonenshein, 2007). For example, the metabolism of *B. subtilis* evolves to adapt to a broad range of conditions of nutrient supply. This adaptation often involves changes in the transcription of the relevant genes (Schilling et al., 2007). *Bacillus subtilis* can use sugars or organic acids as a source of carbon and energy. These nutrients are metabolized by glycolysis, the pentose phosphate pathway, and the Krebs citric acid cycle (Schilling et al., 2007). In *B. subtilis*, the expression of the genes encoding the initial enzymes of the Krebs cycle, citrate synthase and aconitase (*citZ* and *citB*, respectively), is synergistically repressed by glucose and glutamate/ammonium (Jourlin-Castelli et al., 2000; Blencke et al., 2006).

Several regulators have been reported to play an important role in the control of metabolic pathways, but Goelzer et al. (2008) identified CcpA, TnrA, and CodY as the three most important regulators involved in the control of metabolic pathways. By integrating the activities of global regulators such as CcpA, CodY, and TnrA, *B. subtilis* directs traffic through two metabolic intersections that determine the flow of carbon and nitrogen to and from crucial metabolites such as pyruvate, 2-oxoglutarate, and glutamate (Sonenshein, 2007). Genes encoding enzymes of the Krebs cycle, such as *citZ* and *citB*, coding for citrate synthase and aconitase, respectively, are subject to a synergistic repression by glucose and a preferred nitrogen source such as glutamine (Blencke et al., 2006). On the other hand, ammonium assimilation can be initiated by its uptake by diffusion and by the ammonium transporter (AmtB) at, respectively, high and low ammonium concentrations (Detsch and Stülke, 2003).

Schilling et al. (2007) reported that *B. subtilis* prefers sugars and ammonia as carbon and nitrogen sources, respectively, and the preferred source of carbon and energy is glucose. This sugar is taken up and concomitantly phosphorylated by the bacterial phosphoenolpyruvate: sugar phosphotransferase system. The further metabolism involves glycolysis, the pentose phosphate pathway, and the Krebs cycle, resulting in the oxidation of glucose to carbon dioxide and the generation of ATP (Schilling et al., 2007). While glucose is present in excess, *B. subtilis* produces acetate to benefit from the additional substrate-level phosphorylation. However, the accumulation of acetate results in the acidification of the medium (Tobisch et al., 1999).

If glucose is provided as the only carbon source, the growth of *B. subtilis* depends on the activity of pyruvate carboxylase to replenish the Krebs cycle. A significant portion of pyruvate is used for oxaloacetate production. In the presence of glutamate and succinate, the anaplerotic reactions are less important (Schilling et al., 2007). In the same way, other carbohydrates, including simple sugar and polyols can be phosphorylated and catabolized in a similar manner (Stülke and Hillen, 2000).

Their utilization of organic acids requires the conversion of these acids to intermediates of the Krebs cycle, gluconeogenesis, and the pentose phosphate pathway to obtain both energy and metabolic precursors needed by the cell (Doan et al., 2003).

The central intermediate of nitrogen metabolism is glutamate, the universal amino-group donor for the biosynthesis of amino acids, nucleotides, and all other nitrogen-containing compounds in the cells (Fisher, 1999).

It has been reported that the glycolytic enzymes that are needed for reversible reactions are constitutively transcribed, whereas those that catalyze irreversible reactions (*pfkA*, *gapA*, and *pyk*, encoding phosphofructokinase, glyceraldehyde-3-phosphate dehydrogenase, and pyruvate kinase, respectively) are induced in the presence of glucose (Ludwig et al., 2001). Schilling et al. (2007) reported that if *B. subtilis* is supplied with succinate and glutamate in addition to glucose, the cells readjust their metabolism as determined by transcriptome and metabolic flux analyses. The portion of glucose-6-phosphate that feeds into the pentose phosphate pathway is significantly increased in the presence of organic acids. In the same way, important changes were detected at the level of pyruvate and acetyl coenzyme A. In the presence of organic acids, oxaloacetate formation is strongly reduced whereas the formation of lactate is significantly increased.

Microorganisms can interact by exchanging a number of metabolites. Many microbial consortia have been established upon metabolic interactions. Du et al. (2012), after studying the community dynamics of the *Ketogulonicigenium vulgare* and *B. megaterium* consortium, reported that *B. megaterium* was responsible for initiating the reproduction of *K. vulgare*, and meanwhile, *K. vulgare* could promote the growth of *B. megaterium*. They showed that the interaction between *B. megaterium* and *K. vulgare* is a synergistic combination of mutualism and antagonism. Zhou et al. (2011) identified a symbiotic interaction mechanism via metabolites in the ecosystem established by two isolated colonies of *B. megaterium* and *K. vulgare*.

Antimicrobial substances produced by *B. subtilis* strains were revealed (Kaboré et al., 2012b). Guo et al. (2006) showed that *B. subtilis* from fermented foods produces metabolites that inhibit *Escherichia coli*, *Salmonella typhimurium*, and *Staphylococcus aureus* and suggested *B. subtilis* as a promising alternative to antibiotics, which can also be used as a probiotic. N'dir et al. (1994) reported the ability of *Bacillus* strains isolated from nététou to produce antifungal substances, including iturin A, surfactin, and another compound with the same retention time as mycosubtilin and bacillomycin L. Nout et al. (1998) showed the capacity of *B. subtilis* to reduce growth of *B. cereus* and prevent its production of enterotoxin during soybean fermentation to kinema.

4.2 Lactic Acid Bacteria

Donatien Kaboré and Charles Parkouda

Various microorganisms are involved in common food fermentation processes. In particular, lactic acid bacteria (LAB) in food are a kind of biopreservation system. They not only contribute to the flavor of the food but are also able to control pathogenic and spoilage microorganisms through various ways including production of peroxidases, organic acids, and bacteriocins (Kaboré et al., 2012a). The present section reviews the LAB predominating in AFFs. The majority of the discussion will focus on their biodiversity, taxonomic characteristics, ecology, metabolism, genomics, and proteomics.

4.2.1 Biodiversity

Several LAB have been isolated from AFFs, as reported elsewhere (Parkouda et al., 2009; Ouoba et al., 2010). Table 4.1 summarizes the biodiversity of LAB in AFFs. Figure 4.2 shows some of these LAB with their phylogenetic relationships.

4.2.2 Taxonomic Characteristics

Enterococcus faecium and *Enterococcus faecalis* are Gram-positive, catalase-negative, non-spore-forming, facultative anaerobic, α-hemolytic or nonhemolytic bacteria in the Enterococcaceae family (Fisher and Phillips, 2009). Whereas the species *Ent. faecium* belongs to a

Table 4.1 Biodiversity of LAB in Alkaline-Fermented Foods

MICROORGANISMS	OCCURRENCE IN FOODS	REFERENCES
Enterococcus faecium	Maari	Parkouda et al. (2010)
	Ntoba mbodi, Soumbala, Bikalga	Ouoba et al. (2010)
	Kinema	Sarkar et al. (1994); Parkouda et al. (2009)
Enterococcus faecalis; Enterococcus avium	Ntoba mbodi	Ouoba et al. (2010)
Enterococcus casseliflavus	Maari	Parkouda et al. (2010)
	Ntoba mbodi	Ouoba et al. (2010)
Enterococcus hirae	Ntoba mbodi, Soumbala, Bikalga	Ouoba et al. (2010)
Pediococcus acidilactici	Douchi	Liu et al. (2012)
	Maari	Parkouda et al. (2010)
	Soumbala, Bikalga	Ouoba et al. (2010)
Tetragenococcus halophilus	Salted anchovies, Kecap	Sidhipol et al. (2013)
	Dawadawa	Uaboi-Egbenni et al. (2009)
Pediococcus pentosaceus	Douchi	Liu et al. (2012)
	Dawadawa	Uaboi-Egbenni et al. (2009)
	Ntoba mbodi	Ouoba et al. (2010)
Lactobacillus brevis	Douchi	Liu et al. (2012)
	Soumbala	Uaboi-Egbenni et al. (2009)
Lactobacillus plantarum	Douchi	Liu et al. (2012)
	Dawadawa	Uaboi-Egbenni et al. (2009)
	Ntoba mbodi	Ouoba et al. (2010)
	Soft white cheese	Ouadghiri et al. (2005)

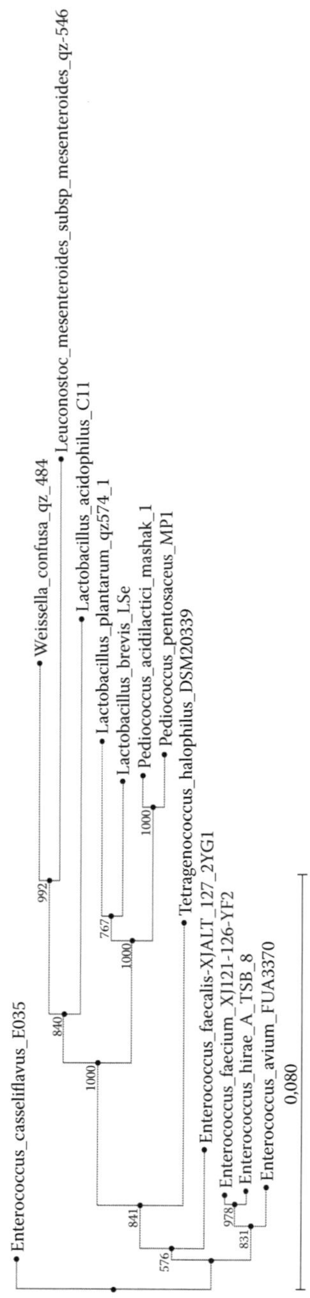

Figure 4.2 Phylogenetic relationship among LAB, based on partial (920 bp) 16S rRNA sequences. (Courtesy of Dennis S. Nielsen.)

group including *Ent. faecium*, *Enterococcus durans*, *Enterococcus hirae*, *Enterococcus mundtii*, *Enterococcus porcinus*, and *Enterococcus villorum*, *Ent. faecalis* species group includes *Ent. faecalis*, *Enterococcus haemoperoxidus*, and *Enterococcus moraviensis* (Klein, 2003; Fisher and Phillips, 2009). *Enterococcus hirae* (*hirae* of the intestine or gut) cells are ovoid and occur mostly in pairs or short chains (Farrow and Collins, 1985). *Enterococcus casseliflavus* is a Gram-positive, motile, catalase-negative bacterium (originally termed as *Streptococcus faecium* var. *casseliflavus*), appearing as coccoid cells grouped usually in pairs or short chains, are also part of the genus *Enterococcus*. *Enterococcus avium* is a Gram-positive, catalase-negative bacterium.

Pediococcus acidilactici is a Gram-positive, catalase-negative bacterium belonging to the Enterococcaceae family, Lactobacillales order, Bacilli class, and Firmicutes phylum (Bosley et al., 1990; Collins et al., 1990). *Pediococcus pentosaceus* are Gram-positive, non-motile, non-spore-forming microbes (Bosley et al., 1990; Collins et al., 1990; Department of Energy Joint Genome Project [DEJGP], 2007). *Pediococcus* is unique in that it form tetrads. These tetrads are formed via cell division in two perpendicular directions in a single plane (DEJGP, 2007).

The genus *Tetragenococcus* was created by Collins et al. (1990) after reclassification of the halophilic LAB *Pediococcus halophilus* as *Tetragenococcus halophilus*. Physiologically tetragenococci are distinguished from pediococci (and other LAB) mainly by their high salt tolerance (growth at >18% NaCl, w/v) and ability to grow at high pH values, i.e., up to pH 9.0, but not at pH 5.0 (Holzapfel et al., 2006). The type species *T. halophilus*, and others, namely *Tetragenococcus muriaticus* and *Tetragenococcus koreensis* are typically isolated from salt-rich environments such as fermented foods (Villar et al., 1985; Röling et al., 1994; Kobayashi et al., 2003; Chen et al., 2006).

Lactobacillus acidophilus (Latin, meaning acid-loving milk-bacterium) is a catalase-negative bacterium belonging to the Lactobacillaceae family. *Lactobacillus brevis* is a Gram-positive, catalase-negative, anaerobic, and rod-shaped bacterium (Sümengen et al., 2012). *Lactobacillus plantarum* is determined as Gram-positive, facultative anaerobic bacterium. The subspecies are *Lb. plantarum* subsp. *argentoratensis*, and *Lb. plantarum* subsp. *plantarum*.

Leuconostoc mesenteroides is a Gram-positive, asporogenous, non-motile, facultative anaerobic bacterium in the family Leuconostocaceae that belongs to the order Lactobacillales. Most strains in liquid culture appear as cocci, occurring singly or in pairs and short chains, but their morphology can vary with growth conditions. Cells grown in glucose or on solid media may have an elongated or rod-shaped morphology. *Leuconostoc mesenteroides* is a species which is divided further into subspecies including *Leuc. mesenteroides* subsp. *cremoris*, *Leuc. mesenteroides* subsp. *dextranicum*, and *Leuc. mesenteroides* subsp. *mesenteroides* (Reiter and Oram, 1982; Garvie, 1986).

Weissella confusa is a Gram-positive, catalase-negative, facultative anaerobic bacterium and is classified within the family Leuconostocaceae (Tajabadi et al., 2012).

4.2.3 Ecology

The origins of *Ent. faecium* and *Ent. faecalis* vary from environmental to animal and human sources. *Enterococcus faecium* is most common in the human gastrointestinal (GI) tract (Klein, 2003). The numbers of *Ent. faecium* and *Ent. faecalis* in human feces range from 10^4 to 10^5 per gram and 10^5 to 10^7 per gram, respectively. Studies on the ecology and epidemiology of *Enterococcus* have reported *Ent. faecium* being regularly isolated from cheese, fish, sausages, minced beef, and pork (Klein, 2003; Foulquie Moreno et al., 2006), while *Ent. faecalis* is regularly isolated from cheese, fish, sausages, minced beef, douchi, and pork (Klein, 2003; Foulquie Moreno et al., 2006; Liu et al., 2012). Foods, such as sausages and cheese, that are of animal origin are often associated with the contamination by *Ent. faecium*, as they are able to survive the heating process (Fisher and Phillips, 2009). *Enterococcus casseliflavus* is normally found in the GI tract and is capable of causing infection by colonization, especially after hospitalization (http://www.primerdesign.co.uk).

Tetragenococcus halophilus is a halophilic LAB active in the fermentation processes of soy sauce, miso, fish sauce, and salted anchovies (Sidhipol et al., 2013).

Lactobacillus acidophilus occurs naturally in the human and animal GI tract, mouth, and vagina (Ljungh and Wadström, 2006). *Lactobacillus plantarum* is a nonpathogenic bacterium naturally existing in human saliva and GI tract.

Weissella confusa was characterized from animal and human clinical samples. It has been isolated by Walter et al. (2001) from human feces in New Zealand. It was isolated from intestines of adult wild-caught and farmed seabass (*Lates calcarifer*) (Rengpipat et al., 2008) and traditional smoked and fish products of the Eastern Himalayas (Thapa et al., 2006).

4.2.4 Metabolism

Enterococcus faecium is a facultative anaerobe, which means it can make ATP by aerobic respiration if oxygen is present but will utilize fermentation if no oxygen is present. *Enterococcus* spp. are able to grow at a range of temperatures from 5°C to 50°C. The optimum, minimum, and maximum temperatures, are 42.7°C, 6.5°C, and 47.8°C, respectively, on brain heart infusion agar in aerobic conditions (Van den Berghe et al., 2006), although growth also occurs in anaerobic atmospheres (Domig et al., 2003). *Enterococcus faecium* and *Ent. faecalis* can survive heating at 60°C for 30 min (Foulquie Moreno et al., 2006) and are able to grow in a wide range of pH (4.6–9.9), with the optimum being 7.5 (Van den Berghe et al., 2006). *Enterococcus faecium* is known to produce antimicrobial activity. It was observed that *Ent. faecium*, isolated from maari, survived in 0.3% bile and exhibited antimicrobial activity against a broad spectrum of bacteria, including food spoilage and pathogenic organisms such as *Bacillus cereus*, *Salmonella* spp., *Listeria monocytogenes*, *Escherichia coli*, and *Micrococcus luteus* (Kaboré et al., 2012a).

Enterococcus faecalis is able to grow in 6.5% NaCl and has a cation homeostasis that is thought to contribute to its resistance to pH, salt, metals, and desiccation. The resistance of *Ent. faecalis* to a range of pH values is thought to be due to the durability of the membrane bound (Nakajo et al., 2005).

The production of amines is also closely related to the growth temperature and pH. The production of decarboxylases was optimum at acid pH, whereas biogenic amine production by *Ent. faecalis* EF37 decreased at low pH (Gardin et al., 2001). *Enterococcus* spp. are known to produce a range of enterocins. Enterocin EJ97 from *Ent. faecalis* S-47 has also shown antilisterial properties (Garcia et al., 2004).

The enterocin AS-48 (produced by *Ent. faecalis* A-48-32) inhibits growth of *Bacillus coagulans* vegetative cells at refrigeration and high temperatures (80°C–95°C) in canned fruits and vegetables (Lucas et al., 2006). Growth occurs at 10°C and 45°C, in 6.5% NaCl and at pH 9.6. L-Lactic acid is the major end product of glucose fermentation. All the strains hydrolyze esculin, are Voges-Proskauer, arginine dehydrolase, β-galactosidase, leucine arylamidase, and pyrrolidonyl arylamidase positive, and produce acid from amygdalin, arbutin, *N*-acetylglucosamine, cellobiose, fructose, galactose, glucose, β-gentiobiose, lactose, maltose, mannose, melibiose, ribose, salicin, sucrose, and trehalose. All the strains fail to produce acid from adonitol, D-arabinose, L-arabinose, D-arabitol, L-arabitol, dulcitol, erythritol, D-fucose, L-fucose, glycogen, gluconate, inositol, inulin, 2-keto-gluconate, 5-keto-gluconate, D-lyxose, mannitol, α-methyl-D-glucoside, α-methyl-D-mannoside, α-methylxyloside, rhamnose, sorbose, sorbitol, xylital, D-xylose, and L-xylose, but do not hydrolyze hippurate, and are alkaline phosphatase and β-glucuronidase negative. The GC content of the DNA ranges from 37 to 38 mol%, as determined by melting temperature. The strains form a distinct homology group, as determined by DNA–DNA hybridization (Farrow and Collins, 1985).

The study conducted by Barros et al. (2001) showed that *Ped. acidilactici* are resistant to vancomycin, positive for L-leucine aminopeptidase activity and hydrolysis of esculin in the presence of bile, and negative for pyrrolidonyl arylamidase activity. It was observed that *Ped. acidilactici*, isolated from maari, survived in 0.3% bile and exhibited antimicrobial activity against a broad spectrum of bacteria, including food spoilage and pathogenic organisms, such as *B. cereus*, *Salmonella* spp., *L. monocytogenes*, *E. coli*, and *M. luteus* (Kaboré et al., 2012a). *Ped. acidilactici* F isolated from fermented sausage produced an antimicrobial peptide "pediocin F," which was proteinaceous in nature and inhibitory to a variety of spoilage and pathogenic microorganisms as well as LAB often encountered in foods (Osmanagaoglu et al., 1998).

Tetragenococcus halophilus does not liquefy gelatin, nor reduce nitrate to nitrite. It is microaerophilic and does not grow at pH 4.4. Optimum growth occurs at NaCl concentrations between 6.5% and 10%, poor growth occurs in the absence of NaCl, and NaCl concentrations of

15% are tolerated. The temperature optimum is between 25°C and 30°C, with maximum at 40°C (Villar et al., 1985).

Pediococcus pentosaceus are anaerobic like most lactic acid producing bacteria (Pritchard and Coolbear, 1993) and produce energy via fermentation (Pritchard and Coolbear, 1993). It transforms hexose sugars such as glucose via the Embden-Meyerhof pathway, and lactic acid is the by-product of anaerobic fermentation (DEJGP, 2007). *Pediococcus pentosaceus* is industrially important due to its ability as a starter culture to ferment foods, such as various meats, vegetables, and cheeses (Hu et al., 2006). *Pediococcus pentosaceus* are cultured and researched for their ability to produce antimicrobial bacteriocins as well as their use in food preservation (Hu et al., 2006). They can produce bacteriocins (Makarova et al., 2006) against several species of *Lactobacillus*, *Lactococcus*, *Leuconostoc*, *Pediococcus*, *Staphylococcus*, *Enterococcus*, *Bacillus*, and *Listeria* (Osmanagaoglu et al., 2001). The bacteriocin obtained from *Ped. pentosaceus* was labeled "pediocin P" (Osmanagaoglu et al., 2001).

Lactobacillus acidophilus is a homofermentative species, fermenting sugars into lactic acid and grows readily at rather low pH values (<5.0) and has an optimum growth temperature of around 37°C (Bâati et al., 2000). They can be cultured at 35°C–40°C but are unable to grow at 50°C. These strains are commercially used in many dairy products, sometimes together with *Streptococcus thermophilus* and *Lactobacillus delbrueckii* subsp. *bulgaricus* in the production of acidophilus-type yogurt (Ljungh and Wadström, 2006).

Lactobacillus brevis produced high levels of extracellular and intracellular phytase in fermented cereal and soybean-based foods (Sümengen et al., 2012). A simple and effective fed-batch fermentation method was developed for *Lb. brevis* NCL912 to produce γ-aminobutyric acid. The results reveal that *Lb. brevis* NCL912 exhibits a great application potential in large-scale fermentation for the production of γ-aminobutyric acid (Li et al., 2010).

Lactobacillus plantarum is an aerotolerant bacterium that grows at 15°C but not at 45°C and produces both isomers (D and L) of lactic acid. This species and its related lactobacilli are unusual in that they can respire using oxygen but have no respiratory chain or cytochromes. In place of the protective enzyme superoxide dismutase present in almost all other oxygen-tolerant cells, this organism accumulates millimolar quantities of manganese polyphosphate. Manganese is also used by *Lb. plantarum* in a pseudo-catalase to lower reactive oxygen levels. Because the chemistry by which manganese complexes protect the cells from oxygen damage is subverted by iron, these cells contain virtually no iron atoms (http://en.wikipedia.org/wiki/Lactobacillus_plantarum).

Under microaerophilic conditions, heterolactic fermentation is carried out. Glucose and other hexose sugars are converted to equimolar amount of D-lactate, ethanol, and CO_2 via a combination of the hexose monophosphate and pentose phosphate pathways (Demoss et al., 1951; Garvie, 1986; Gottschalk, 1986). Other metabolic pathways include conversion of citrate to diacetyl and acetoin (Cogan et al., 1981) and production of dextran and levan from sucrose (Alsop, 1983). Viscous polysaccharides produced by *Leuc. mesenteroides* are widely recognized as causing product losses and processing problems in the production of sucrose from sugar cane and sugar beets (Tallgren et al., 1999). Commercial production of dextran and levan by *Leuc. mesenteroides*, for use in the biochemical and pharmaceutical industry, has been carried out for more than 50 years (Alsop, 1983; Sutherland, 2008).

4.2.5 Genomics

The genome of *Ent. faecium* has an estimated size of 2.8 Mbp (http://microbewiki.kenyon.edu/index.php/Enterococcus_faecium). For instance, the genome of *Ent. faecium* strain Aus0004, isolated from the bloodstream comprises a single circular chromosome of 2,955,294 bp (GC content, 38%) and three circular plasmids, Aus0004_p1, Aus0004_p2, and Aus0004_p3, with lengths of 56,520, 4119, and 3847 bp, and GC content of 35%, 37%, and 39%, respectively (Lam et al., 2012).

Enterococcus faecalis is a close relative of *Ent. faecium*, and its genome has been sequenced and analyzed. The sequencing of a vancomycin resistant *Ent. faecalis V583* revealed one circular chromosome and three plasmids. The chromosome consists of 3,218,031 bp, and the plasmids (pTEF1, pTEF2, and pTEF3) consist of 66,320, 57,660, and 17,963 bp, respectively. Two of the plasmids are pheromone-sensing conjugative plasmids. Also, found was a mobile conjugative transposon that encodes vancomycin resistance (http://microbewiki.kenyon.edu/index.php/Enterococcus_faecium).

The genome of *Ent. casseliflavus* is a linear DNA chromosome about 3.4 Mnt long, consisting of 3283 protein-coding genes (http://www.primerdesign.co.uk).

The genus *Tetragenococcus* was created by Collins et al. (1990) after reclassification of *Ped. halophilus* as *T. halophilus*.

The genome sequenced is made up of 1,832,387 nucleotides organized in a circular manner (DEJGP, 2007). The genome has 1755 protein-encoding genes and 72 RNA genes (DEJGP, 2007). It has 37.4% GC content. There are three to five resident plasmids (DEJGP, 2007) that give *Ped. pentosaceus* the ability to metabolize different compounds. *Pediococcus pentosaceus* are related to other LAB. In a comparison of ribosomal proteins, *Ped. pentosaceus* are related to *Lb. brevis* and *Lb. plantarum* (Makarova et al., 2006).

4.3 Yeasts

Line Thorsen, Elmer Nayra Kpikpi, and Lene Jespersen

4.3.1 Taxonomy, Ecology, and Functionality of Yeasts

Yeasts are eukaryotic organisms and can be distinguished from molds due to their predominantly unicellular morphology. Yeasts are divided into two taxa: the Ascomycota and the Basidiomycota. In both taxa, teleomorphic (perfect) and anamorphic (imperfect) yeasts are included, and yeasts may be referred to by either their teleomorphic or anamorphic name. Rules for the taxonomy of yeasts fall under the authority of the International Code of Nomenclature for algae, fungi, and plants (ICN; Melbourne Code, 2011). The current taxonomy of yeasts is described by Kurtzman et al. (2011). Attention should be paid to the fact that yeast taxonomy has significantly changed over recent years due to the fast development of molecular techniques and so have many yeast species names (Nielsen et al., 2010; Kurtzman et al., 2011). In the following, current species names will be used.

Traditional classification of yeasts relies on morphological, physiological, biochemical, and sometimes ecological characteristics. For sporogenous species the feature of the sexual cycle is included. However, within the last decade data for classification of yeasts are more and more provided by sequencing of the D1/D2 domain of the 26S rRNA gene followed by comparison of the sequence to sequences of known yeast species (Kurtzman et al., 2011). For typing at the strain level molecular techniques as pulse field gel electrophoresis (PFGE) and PCR-based techniques are often used (Hayford and Jespersen, 1999; Jespersen et al., 2000; Naumova et al., 2003; Petersen and Jespersen, 2004). For yeasts used as starter cultures in soybean-fermented foods, wine, beer, bread, and dairy products separation based on technological criteria might be used to characterize a particular yeast culture of industrial interest. Criteria often considered are flavor formation, flocculation, adhesion capacity, color formation, reproduction, attenuation, and several others depending on the food or beverage the yeast culture is used for.

Yeasts are widely distributed in nature and are generally found on leaves, flowers, and fruits. They also occur on the skin, hide, feathers, and alimentary canal of herbivores. Soil only serves as a temporary reservoir and is together with other natural habitats an important vehicle for carrying yeasts into food processing facilities and thus disseminating into foods (Botha, 2006; Deak, 2008). Most yeast species related to the production and spoilage of foods will be ascomycetous or anamorphic basidiomycetous yeasts. For yeast genera relevant to the production and spoilage of food and beverages, Romano et al. (2006) should be consulted. The yeast species being able to grow in a given food will be dependent on the ecology of the food product. Generally, yeasts prefer slightly acidic pH conditions; they grow best in the pH 4.5–7.0; however, some yeast species have been reported to grow at pH values as low as pH 2.0–2.5. Yeasts are generally tolerant to organic acids and do therefore generally grow very well together with LAB (Arneborg et al., 2000). However, alkali-tolerant strains of the species *Yarrowia lipolytica* and *Debaryomyces hansenii* have been reported with the ability to grow at pH 10 (Zvyagilskaya et al., 2001; Kurita and Yamazaki, 2002). Most yeasts prefer aerobic conditions even though yeasts, e.g., *Saccharomyces cerevisiae*, are facultative fermentative yeasts. Optimum temperature for most yeast species is 20°C–30°C; however, especially pathogenic yeasts and yeasts isolated from tropical environments are able to grow at temperatures above 37°C, even though very few yeasts grow at 40°C–50°C. Generally, yeasts accept a decreased water activity (a_w) and most yeasts grow in the a_w range 0.90–0.95. Several osmotolerant yeasts have been reported, e.g., the halotolerant yeast *D. hansenii* being of significant importance in the production of cheeses and sausages (Gori et al., 2007).

Yeasts take part in several fermented products where they are introduced either as starter cultures by back-slopping or take part in a spontaneous fermentation (Jespersen, 2003). Generally, yeasts are considered as beneficial microorganisms. However, they may also appear as spoilage organisms especially in foods where bacterial growth is restricted due to low pH, high acid content, low a_w, or high concentration of weak organic acids. This might cause severe problems such as discoloration, gas production, and off-flavors in foods (Deak, 2008). Spoilage of AFFs by yeasts has been suggested (Sarkar and Tamang, 1994; Chadare et al., 2011). Food-borne yeasts, except from very few and rare cases, are nonpathogenic (Klingberg et al., 2008; Perez-Torrado et al., 2012). A wide range of yeast species are involved in the fermentation of indigenous foods and beverages. As these products to a great extent are made by spontaneous fermentation, several different yeast species will be present (Amoa-Awua et al., 1997; Schwan et al., 2007). At the strain level a pronounced biodiversity is often found. The yeast population found will primarily depend on the raw materials and processing conditions but also the occurrence of other microorganisms may influence the composition, growth, and metabolic activities of the yeast population (Wah et al., 2013). *Saccharomyces cerevisiae* and *Candida krusei* are some of the most common yeasts in indigenous acidic fermented foods and beverages where they have been shown to be very important especially in the fermentation of cereals and alcoholic beverages (Jespersen, 2003; Halm et al., 2004; Glover et al., 2005, 2009; Schwan et al., 2007; Vieira-Dalode et al., 2007, 2008). Yeasts do also play a role in cash crops such as coffee and cocoa fermentation (Masoud et al., 2004; Jespersen et al., 2005; Nielsen et al., 2005, 2010; Masoud and Jespersen, 2006).

The functions of yeasts in fermented food and beverages are many. The mostly well-recognized functions of yeasts are the conversion of carbohydrates to ethanol and higher alcohols as well as their contribution to flavor development. Yeasts produce flavor compounds such as esters, organic acids, and carbonyls (Sørensen et al., 2011). Dependent on the interplay between the yeasts and the food matrix, the yeasts may have other functional properties such as production of tissue degrading enzymes, binding/degradation of mycotoxins, and cyanogenic glucosides (Masoud and Jespersen, 2006; Shetty et al., 2007). In indigenous fermentations yeasts often coexist with other microorganisms (Vieira-Dalode et al., 2007; Pedersen et al., 2012). Depending on the type of product a microbial succession involving both yeasts and other microorganisms will normally take place. Yeasts have been reported to stimulate the growth of other microorganisms including LAB by providing essential metabolites and may in some cases have an inhibitory effect on, e.g., mycotoxin producing molds. Yeasts have also been reported to improve the nutritional value of fermented products by symbiosis with LAB to increase the level of riboflavin, thiamine, niacin, and ascorbic acid (Onilude et al., 1999; Narvhus and Gadaga, 2003). Also, degradation of anti-nutritional factors, such as phytic acid, has been reported for several yeast species present in fermented food products (Nuobariene et al., 2011). Some yeasts may additionally have probiotic properties (van der Aa-Kühle et al., 2005; Moslehi-Jenabian et al., 2010; Pedersen et al., 2012).

Despite the fact that yeasts normally are associated with acidic foods, their presence has been noted in many AFFs. Whereas extensive knowledge on the functionality and application of yeasts in acid fermented foods and beverages as bread, cheese, wine, beer, etc. exists (Jakobsen et al., 2001), hardly nothing is known about the functions of yeasts in AFFs even though these foods play an important role in the diet of many people in especially Asia and Africa.

4.3.2 Yeasts in Legume-Based Alkaline-Fermented Foods

4.3.2.1 Soybean Products Yeasts have been reported in different soybean-based fermented food products as a part of a highly complex microbial consortium, which changes over time according to the changing environmental conditions. Depending on the product, the yeast metabolic activities during specific stages of fermentation and maturation may improve the nutritional and sensory quality or may oppositely contribute to spoilage.

Tungrymbai is produced by the Khasi tribe in Meghalaya, India. Dried soybeans are cooked for 45 min to 2 h until soft, drained for excess water, and left to ferment in a basket lined with fresh leaves of *Phrynium pubinerve* Blume (Marantaceae). The beans ferment at ambient temperature (25°C–30°C) for 3–4 days near or over a fire place. The use of tungrymbai differs from natto and kinema (described below) in that the fermented soybeans are ground in a mortar, before cooked together with ginger, garlic, and black beefsteak seeds (*Perilla frutescens* (L.) Britton) fried in mustard oil to produce the ready-to-eat tungrymbai (Thokchom and Joshi, 2012). The pH of tungrymbai has been reported to be slightly alkaline (6.95–7.05) (Thokchom and Joshi, 2012) to alkaline (7.6) (Tamang et al., 2012). Yeasts, *Bacillus subtilis* group spp., LAB, and enterobacteria have been isolated from tungrymbai (Thokchom and Joshi, 2012). Yeasts have been isolated from the final product (5.6×10^6 cfu/g) and also from the soaked soybeans (6.7×10^5 cfu/g) and from the mortar and pestle used for grinding the fermented soybeans (Sohliya et al., 2009). The detected yeast species include *S. cerevisiae*, *D. hansenii*, *Candida variabilis* (formerly *Pichia burtonii*) (Tamang et al., 2012), *Geotrichum candidum*, *Candida parapsilosis*, *Saccharomyces bayanus*, and *Saccharomycopsis fibuligera* (Sohliya et al., 2009). The role of the yeasts in tungrymbai has not been investigated. However, the yeast species occurring in tungrymbai are known from other foods, where they are known to influence the quality of these products. As an example, *C. variabilis* and *S. fibuligera* are considered important amylolytic yeasts in different indigenous fermentations, e.g., in Himalayan murcha (an amylolytic starter for alcoholic beverages) (Tsuyoshi et al., 2005; Takeuchi et al., 2006), and *S. bayanus* is associated with the ethanolic fermentation for different wines (Rojas et al., 2012).

Thua nao is a traditional Thai food condiment made from the spontaneous alkaline fermentation of soybeans. This low-cost source of protein can be eaten directly or is mixed with different spices before use (Chukeatirote et al., 2010). The technology used for the production of thua nao varies according to the region. The beans are either soaked overnight and then cooked for 3–4 h or are cooked directly for 7 h. Cooked beans are transferred to a bamboo basket (or similar) lined with fern or banana leaves and left at ambient temperature for 2–4 days to ferment spontaneously (Chukeatirote et al., 2010). The pH may increase from 6.3 to 8.0 during fermentation (Chukeatirote et al., 2006). Although *Bacillus* spp. are reported as the main fermenting microorganisms (Inatsu et al., 2006), high numbers of fungi (not further identified) have been reported (Chukeatirote et al., 2006). The ability of a *B. subtilis* starter culture to inhibit yeasts was studied, indicating that the occurrence of yeasts is undesirable. Low numbers of yeasts and molds (not further identified) were detected during thua nao formation using *B. subtilis* strain TN51 as starter culture (Dajanta et al., 2012).

Meitauza is a traditional fermented food product produced at household level in the Chinese provinces, Hubei, Henan, and Guangdong (Xu et al., 2012). Okara, a (soybean) waste product from tofu and soymilk productions, is used as substrate for meitauza productions (Kronenberg and Hang, 1984; O'Toole, 1999; Xu et al., 2012). The unfermented okara is rich in dietary fiber and also contains anti-nutritional factors such as phytic acid and indigestible carbohydrates such as stachyose and raffinose (O'Toole, 1999). Fermented okara contains lower levels of stachyose and raffinose (O'Toole, 1999). Traditional meitauza is spontaneously produced by indigenous microorganisms originating from rice straw used for covering the okara (Xu et al., 2012). The fermentation is for 8–14 days at temperatures below 20°C (Xu et al., 2012). Meitauza is also produced by *Actinomucor elegans* and *Zymomonas mobilis* starter cultures, resulting in an alkaline fermentation with pH values of 7.5–8.1 (Kronenberg and Hang, 1984; Xu et al., 2012). In other Asian countries, okara is fermented with different fungi, e.g., *Rhizopus oligosporus* (to produce tempe), as well as with *B. subtilis* (O'Toole, 1999; Zhu et al., 2008). The role of yeast spp. including *Candida guilliermondii* (formerly *Pichia guilliermondii*), *Kluyveromyces marxianus*, *Ogataea methanolica* (formerly *Pichia pinus*, *Pichia methanolica*), and *S. cerevisiae* has recently been investigated in okara fermentation trials (Rashad et al., 2011). The yeasts increased protein contents, reduced levels of crude fiber, and produced antioxidant activities in a species- and strain-dependent matter (Rashad et al., 2011).

Kinema is a soybean-based fermented food used as a meat substitute in the eastern Himalayan regions (Nepal, Darjeeling hills, Sikkim, and South Buthan). Soybeans are soaked overnight (12–20 h) at ambient temperature (10°C–25°C), cooked for 1.5–3 h until soft, drained, cracked in a mortar with a pestle, and mixed with 1% w/w ash (optional) (Sarkar and Tamang, 1994, 1995; Sarkar et al., 1994). The beans are wrapped in locally available leaves, covered with sack cloth, and are left for 1–3 days in bamboo baskets over an earthen oven to ferment spontaneously (Sarkar et al., 1994, 2002; Sarkar and Tamang, 1995; Tamang et al., 2012). Kinema has a final pH of 7.71–7.94 (Sarkar et al., 1994). Although *B. subtilis* is considered the main fermenting microorganism (Sarkar and Tamang, 1994), *C. parapsilosis* and *G. candidum* have been isolated from different kinema productions (Sarkar et al., 1994; Sarkar and Tamang, 1995; Tamang, 2003). Levels of 4.1–8.7 × 10^4 cfu/g of *C. parapsilosis* and 0.8–4 × 10^4 cfu/g of *G. candidum* were detected in 60%–80% and 40%–50%, respectively, of market samples obtained from different locations (Darjeeling, Kalimpong, Gangtok, and Rongli) (Sarkar et al., 1994). The yeast species have been found to originate from the mortar and pestle used for cracking the cooked soybeans before fermentation (Tamang, 2003). The role of the yeasts was investigated in different studies, and overall, they seem to have a negative effect on the nutritional and organoleptic quality of kinema. For example, the presence of *C. parapsilosis* and *G. candidum* resulted in significantly lower levels of free fatty acids, namely C18:1 (oleic), C18:2 (linoleic), and C18:3 (linolenic) (Sarkar et al., 1996) as well as significantly lower levels of free amino acids (e.g., lysine) as compared with fermentations performed with *B. subtilis* as monoculture (Sarkar et al., 1997). Kinema productions with *C. parapsilosis* and *G. candidum* were further associated with a rancid and objectionable smell (Sarkar and Tamang, 1994). The proteolytic and especially lipolytic activities of *G. candidum* are well known from other foods. However, contrary to kinema, the enzymatic activities of this yeast species are important in the production of different dairy products such as Spanish goat milk cheese, contributing to the ripening process and formation of desirable flavor (Sacristan et al., 2012). *Geotrichum candidum* and *C. parapsilosis* have additionally been identified in several acid-fermented cassava-based drinks and foods including caium (*C. parapsilosis*; Schwan et al., 2007) and agbelima (*G. candidum*; Amoa-Awua et al., 1997).

Soy dawadawa is an alkaline-fermented soybean condiment (final pH 8.25–8.41) produced in Ghana (Dakwa et al., 2005). Yeasts (not further identified) were found to increase in numbers during the first 24 to 48 h of fermentation along with high numbers of LAB and *Bacillus* spp. The yeasts were not detected at 72 h or in the final product and were therefore not considered to play a significant role in the fermentation (Dakwa et al., 2005).

Douchi is a Chinese fermented salty soybean food consumed as a soup or side dish (Tamang, 2012). Soybeans (preferably black) are soaked in water, cooked for 30–60 min, depending on the type of douchi (fungal or bacterial type) (Tamang, 2012). The fermentation is in two steps (primary and secondary fermentation). The primary fermentation (qu making) is initiated by inoculating beans with the desired fungi (*Mucor*, *Aspergillus*, or *Rhizopus* spp.) or is spontaneously fermented for 3–4 days at temperatures above 25°C (Tamang, 2012). Douchi, prepared using *Aspergillus*-type qu, is the most highly valued. In some qu productions, the soybeans may be mixed with wheat before fermentation (Chen et al., 2011a). Before initiation of the second fermentation step (ripening period) to produce douchi, the qu starter is mixed with >16% (w/w) salt and different spices (Zhang et al., 2007; Tamang, 2012). The second fermentation (ripening period) is in closed containers for different lengths of time, e.g., 7–40 days at 30°C–35°C (*Aspergillus*-type) or 10–12 months at 20°C (*Mucor*-type) (Tamang, 2012). Qu fermentation with *Aspergillus* spp. has a reported increase in pH from 6.88 to 8.43 (Wang et al., 2010), whereas during the second fermentation to produce douchi the pH decreases to 5.5–6.7 (Wang et al., 2010; Hu et al., 2012).

The microbiology of douchi production is highly complex, and the microbial composition varies according to the methodology, and the fermentation time and temperature (Chen et al., 2011a, 2011b, 2012; Hu et al., 2012). Variable levels of yeasts, molds, *Bacillus* spp., *Staphylococcus* spp., etc. are present during the fermentation (Chen et al., 2011a, 2011b, 2012). *Saccharomyces cerevisiae*, *Millerozyma farinosa* (formerly *Pichia farinosa*), and *C. guilliermondii* are commonly detected both during fermentation and in the final product (Chen et al., 2011b, 2012). The microbial succession patterns of *Aspergillus*-type douche qu (koji) and douchi (post-fermentation process) was determined using both a culture-dependent and a culture-independent approach (DGGE and sequencing of excised PCR products) (Chen et al., 2011a, 2011b). The fungi *A. oryzae* and *Absidia corymbifera* were detected together with different bacteria including *Bacillus* spp. during early stages of douchi qu (koji) fermentation (days 1–5, 30°C, 45%–65% RH), while the yeasts *S. cerevisiae*, *D. hansenii*, and *C. guilliermondii* occurred together with *A. oryzae* and different bacteria, including *Bacillus* spp. at a later stage (day 6, 20°C–27°C, 45%–65% RH) (Chen et al., 2011a). During the second fermentation step (ripening period) *C. guilliermondii*, *S. cerevisiae*, and *M. farinosa* occurred together with the mold *Absidia corymbifera* and different bacteria (Chen et al., 2011b). Other yeast species such as *C. variabilis* and *M. farinosa* have also been reported to be present in douchi (Yangjiang lobster sauce) (Qu et al., 2012). *Candida variabilis* was found to produce a high protease activity (up to 479 U/ml) in the pH range of 5–8 (optimum activity at pH 5.0), indicating that this species could be used as a starter culture for douchi production (Qu et al., 2012). Salt-tolerant yeasts and LAB are suggested to improve the aroma of douchi during the ripening process (Zhang et al., 2007). During the ripening process sugars and proteins are converted into lactic acid, ethanol, amino acids, and different aroma components (Zhang et al., 2007). Recently, there has been an interest in lowering the sodium chloride content of douchi. Fermented rice culture containing both yeasts and LAB (not further identified) were added together with 4% w/w NaCl to the douchi qu. The resulting low-salt douchi scored almost as high as traditional douchi in sensory tests (Hu et al., 2012).

Doenjang, a traditional Korean fermented food condiment made from soybeans, shows some similarities to Chinese douchi as it is produced in two fermentation steps (primary and secondary fermentation) and it is a salty product. Meju (the primary starter) is prepared by crushing cooked soybeans in a mortar and shaping into balls or bricks. While wrapped in rice straw, the beans are left to spontaneously ferment until covered with a white mold (1–3 months) (Ebine, 1986; Steinkraus, 1996; Choi et al., 2009). Meju may also be produced using fungal or bacterial starter cultures and using substrates other than soybeans (Kwon and Kim, 2005; Kim et al., 2010; Lee et al., 2012). The initial pH of meju (prepared with *A. oryzae* and *B. subtilis* starter cultures) has been reported to be between 4.97 and 5.07 and increases during the fermentation due to proteolytic activities of the fermenting microorganisms (Kim et al., 2010). The pH of 121 different traditional meju samples from different regions of Korea was on average 7.0 ± 0.8 (interior) and 6.9 ± 0.5 (exterior) (Yoo and Kim, 1998). To produce doenjang, the meju is dried and then mixed with brine (20% w/v NaCl) (Ebine, 1986). The reported length of the second fermentation (ripening period) varies greatly (Steinkraus, 1996; Nam et al., 2012). A ripening period of 6–12 months has been indicated (Ebine, 1986). At the end of fermentation the liquid part is separated and used as soy sauce (kanyang), while the solid part is doenjang (Lee, 1995).

The microbiology of various types of meju has been investigated in several studies using different methods. Recently, the microbiology was examined using DGGE and sequencing methods, and meju was found to harbor different LAB species, *Bacillus* spp. (mainly *B. subtilis*), *A. corymbifera*, *Aspergillus* spp., and *Candida rugosa*, with other yeast species as *M. farinosa* occurring less frequently (Lee et al., 2010). Lee et al. (1997b) isolated several different yeast spp. from meju using culturing techniques and phenotypic tests and characterized the yeasts enzymatic properties (Lee et al., 1997a). The following genera were reported: *Saccharomyces*, *Debaryomyces*, *Klyveromyces*, *Hansenula*, *Rhodoturola*, *Candida*, *Pichia*, and *Zygosaccharomyces* (Lee et al., 1997b). Almost all of the yeast isolates produced α-galactosidase activity, acidic and neutral protease activity, and invertase and inulinase in broth, suggesting that they could play a role in the fermentation of meju (Lee et al., 1997a). The bacterial composition was investigated for 11 doenjang samples (10.8%–14.9% NaCl, final pH 4.81–6.67) that had been collected in different regions of Korea using a next generation sequencing approach (Nam et al., 2012). It was found that *Bacillus* spp. dominated in samples from the central region of Korea, while LAB species dominated in other doenjang samples (Nam et al., 2012). Recently, the microbiota including fungi and yeasts of homemade as well as commercially made doenjang was determined using nested PCR-DGGE analysis (Kim et al., 2009). Apart from LAB species and *Bacillus* spp., several mold and yeast species were identified. The most frequently occurring yeast species was *D. hansenii*, but also *Zygosaccharomyces rouxii*, *Galactomyces geotrichum*, *Candida pelliculosa* (formerly *Pichia anomala/Hansenula anomala*), *Sterigmatomyces halophilus*, and different *Candida* spp. were identified (Kim et al., 2009). The yeast *D. hansenii* is also known from other (salty) food products and plays an important role in the surface ripening of cheeses (Petersen et al., 2002; Bockelmann et al., 2003); *D. hansenii* together with *C. pelliculosa* has been isolated from brines for olives fermentations (Nisiotou et al., 2010). A strain of the salt-tolerant *D. hansenii* isolated from soy sauce has shown the ability to grow in a wide range of pH (3.6–10.3) (Kurita and Yamazaki, 2002).

Yeasts have been proved to influence the flavor of fermented soybean products. Douchi, doenjang, and other soybean fermentations used to produce soy sauce all harbor different salt-tolerant yeast species (*D. hansenii*, *Z. rouxii*, etc.). These yeasts are fit for the environment of douchi, doenjang, and related fermentations, especially during later stages when the pH starts to decrease (Wah et al., 2013). Different yeast strains isolated from meju and doenjang were investigated for their ability to affect doenjang flavor (Kwon and Kim, 2005). One of the strains, *Zygosaccharomyces* Y-2-5 showed ability for growth in 10% w/v NaCl, and was found to produce ethanol and several higher alcohols important for doenjang flavor (Kwon and Kim, 2005). The best time for inoculation of the yeast was within the first 2 weeks of doenjang production (Kwon and Kim, 2005). In Japan (and also Korea, China, etc.), the production of soy sauce is industrialized and different starter cultures are added at appropriate times to fully utilize their potentials (Hayashida et al., 1997). For example, when producing soy sauce, koji (made from cooked defatted soybeans inoculated with *A. oryzae*) is added to a 22% w/v brine to make the shoyu mash, a specified halotolerant lactic acid bacterium is then added to produce acid to lower the pH, and after 1 week different yeasts including *Z. rouxii* are added (Hayashida et al., 1997). Acidification by halotolerant LAB during initial stages supports the growth and metabolic activities of halotolerant yeasts (Wah et al., 2013). *Zygosaccharomyces rouxii* is one of the most important yeasts in these types of fermentations. It is able to produce 2-hydroxy-2(or 5)-ethyl-5(or 2)-methyl-3(2*H*)-furanone, which tastes sweet and has a caramel-like flavor typical of soy sauce (Hayashida et al., 1997). It has together with *C. guilliermondii* been isolated from moromi fermentations to produce Thai soy sauce. Especially *Z. rouxii*, and *Z. rouxii* in co-culture with *C. guilliermondii* were found to play an important role in the formation of characteristic flavor components (3-methyl-1-butanol, 2-methyl-1-propanol, ethanol, and others) of soy sauce (Wah et al., 2013). *Candida variabilis* has been isolated from meju, and was suggested to contribute to flavor formation in the final product bealmijang (a short-term aged paste) (Kim et al., 2011).

4.3.2.2 Non-Soybean Legume Products Information on the occurrence of yeasts in alkaline-fermented non-soybean legume products is limited. If present, the role of the yeasts is largely unknown or the yeasts are suggested to be contaminants. Below are some examples of some food products where yeasts have been reported.

Kawal is a protein-rich food prepared by fermenting sicklepod (*Senna obtusifolia* (L.) H.S.Irwin & Barneby) leaves from the wild African legume in Sudan. Kawal is usually cooked in stews and soups and is used as a meat replacer or a meat extender (Dirar, 1993). The fermentation takes about 14 days, and the main microorganisms include *Bacillus* spp. and LAB. Yeasts including *C. krusei* and *Saccharomyces* spp. and molds including *Rhizopus* spp. are also involved (Dirar et al., 1985). Kawal is considered as an AFF, although the pH is reported to be between 6 and 7 (Wang and Fung, 1996).

Dawadawa, nététu, soumbala, afitin, iru, and sonru are traditional condiments produced by spontaneous and alkaline fermentation of African locust bean seeds (*Parkia biglobosa* (Jacq.) G.Don) in the West African countries like Nigeria, Senegal, Burkina Faso, and Benin. The product names vary according to the technology used for the production and the country (Antai and Ibrahim, 1986; Ouoba et al., 2004; Azokpota et al., 2006; Parkouda et al., 2009; Uaboi-Egbenni et al., 2009). Typically the beans are cooked for 12–24 h, dehulled, cooked again with an alkaline softening agent, drained, and fermented. Spices may be added during the process. For a more detailed explanation of the different production technologies, we refer to the review by Parkouda et al. (2009). Different *Bacillus* spp. are the main fermenting microorganisms, associated with *Staphylococcus* spp. and sometimes LAB (Azokpota et al., 2006; Parkouda et al., 2009). Yeasts seem to play a little or no role in these fermentations. For example, in afitin, iru, and sonru fermentations yeasts were below the detection limit (Azokpota et al., 2006). Yeasts (not further identified) were reported during the early stages of dawadawa production (log 5.3–6.6 cfu/g) (Uaboi-Egbenni et al., 2009), whereas another study did not detect any yeast in this product (Antai and Ibrahim, 1986).

Okpehe is a popular Nigerian spontaneously alkaline-fermented mesquite (*Prosopis africana* (Guill. & Perr.) Taub.) seeds condiment, frequently used as a seasoning agent in soups or as a meat substitute (Oguntoyinbo and Sanni, 2007). Seeds are cooked in water for 14–24 h until soft, manually dehulled, washed, and cooked for another 3–5 h. Cooked seeds are transferred to calabash or clay pots lined with leaves. After covering with the leaves and packing in jute bags, the seeds are left to ferment for 3–5 days at 30°C–35°C (Oguntoyinbo et al., 2007). *Saccharomyces cerevisiae* has been isolated from okpehe on different occasions (Balogun and Oyeyiola, 2011; Musa et al., 2011), although its role in the fermentation is unknown. The alkaline fermentation of the mesquite seeds with final pH of up to 9.0 (Ogunshe and Olasugba, 2008) is a result of the proteolytic activities of the main fermenting microorganisms, including various *Bacillus* spp. (Oguntoyinbo and Sanni, 2007; Oguntoyinbo et al., 2010).

Otiru is a Nigerian food condiment produced by the spontaneous fermentation of African yam bean (*Sphenostylis sternocarpa* (A.Rich.) Harms). *Saccharomyces cerevisiae* and *Candida vini* (formerly *Candida mycoderma*) have been isolated from otiru (Jeff-Agboola, 2007). The role of the yeasts in otiru is unknown, but they occurred at similar levels as those found for the bacteria (Jeff-Agboola, 2007). Yeasts were not detected in other AFFs, including owoh-type product (made from African yam bean; Ogbonna et al., 2001), oso (a traditional fermented *Cathormion altissimum* Hatch. seeds condiment produced in Southwest Nigeria; Popoola et al., 2004, 2006), and ugba (a salty, 6.5%–7% w/w NaCl fermented oil bean tree seeds; *Pentaclethra macrophylla* Benth.) food product (Oyeyiola, 1989).

4.3.3 Yeasts in Non-Legume Plant-Based Food Products

The reports on yeast in alkaline-fermented non-legume plant-based foods are relatively scarce, and for many of them, the yeasts are reported to be absent or they are present but the role of the yeasts are mostly unknown.

Tayohounta and dikouanyouri are traditional alkaline-fermented baobab (*Adansonia digitata* L.) seeds used as a flavoring agent in soups and sauces in Benin, West Africa. To produce tayohounta, baobab kernels are roasted, cooked, drained, and fermented in calabash or plastic containers covered with leaves for 2 days at 24°C–31°C (Chadare et al., 2010, 2011). Yeasts (not further identified) occurred (between 3.1 and 7.2 log cfu/g) together with high levels of different bacteria including *Bacillus* spp. and LAB (Chadare et al., 2010, 2011). The microbial composition of dikouanyouri, which is produced from whole seeds, is similar to that of tayohounta (Chadare et al., 2010). Trial fermentations performed with yeast isolates (not further identified) from tayohounta suggested that they might have a negative effect on the organoleptic properties of the product (Chadare et al., 2011). Maari is also an alkaline-fermented baobab seeds condiment produced in Burkina Faso. The microbiology has been investigated; however, yeasts were not reported (Parkouda et al., 2010). Chadare et al. (2010) suggested that depletion of nutrients by the rapidly growing bacteria, and formation of ammonia could restrain the growth of yeasts and molds.

Ogiri is a traditional Nigerian AFF condiment made from different types of seeds. Ogiri produced by the fermentation of castor bean (*Ricinus communis* L.) seeds and melon (*Citrullus lanatus* (Thunb.) Matsum. & Nakai) seeds did not harbor any yeasts (Barber et al., 1988; Nwosu and Ojimelukwe, 1993). The occurrence of yeasts (not further identified, numbers not reported) was, however, reported in ogiri made from groundnut (*Arachis hypogaea* L.) seeds (Nwosu and Ojimelukwe, 1993).

Cabuk is an alkaline-fermented sesame seeds product widely consumed in the southern and central parts of the Indonesian island Java. In this product, yeasts (not further identified) were detected at levels of 10 to 1000/g. The yeasts were considered as contaminants (Steinkraus, 1996).

The absence of yeast in Congolese ntoba mbodi (a fermented cassava, *Manihot esculenta* Crantz, leaf product) was suggested to be due to the limiting effect of the alkaline pH of 8.9 (Louembé et al., 2003). Other AFFs reported not to include yeasts, or where yeasts are thought to play no role include owoh (made from fermented cotton seeds, *Gossypium hirsutum* L.) in Nigeria (Sanni and Ogbonna, 1991) and semayi (fermented coconut residue) produced in Indonesia (Steinkraus, 1996).

4.3.4 Yeasts in Animal-Based Alkaline Food Products

African traditional fermented fish products reported with an alkaline pH include momone (pH 7.5–7.8) and lanhouin (pH 7.3–7.6) (Anihouvi et al., 2012). Momone is produced in Ghana using different fish species (catfish, barracuda, seabream, threadfin, and others) (Anihouvi et al., 2012). The fish is either used whole or cut into pieces, washed carefully, and salted the following day (Anihouvi et al., 2012). Fermentation is for 3–8 days, with salt being added again on the third day of fermentation. The fish is left to dry for 1–3 days when the fermentation period ends (Anihouvi et al., 2012). Momomi is used as a flavor enhancer for sauces (Sanni et al., 2002). The yeast species *D. hansenii* and *Hansenula anomala* were identified in momomi (pH 6.5–6.6), together with *Bacillus* spp., LAB, and *Aspergillus flavus* (Sanni et al., 2002). The levels of yeasts in momoni were, however, low (1.7×10^2 to 1.5×10^3 cfu/g) (Sanni et al., 2002). Yeasts were not detected in the following alkaline salt-fermented fish products; Korean salt-fermented shrimp products/sauces (pH 7.4–8.5) (Kim et al., 2003), Indonesian salt-fermented shrimp (or fish) paste terasi (pH 7.5) (Surono and Hosono, 1994a, 1994b), Beninese lanhouin (Anihouvi et al., 2012), and Swedish salt-fermented herring, surströmning (pH 7.5) (Kobayashi et al., 2000). Yeasts tend to occur in fermented fish products where the pH is below 7, including in adjuevan (pH 5.6–6.7) (a fermented fish product produced in the Ivory Coast), which was reported to harbor several different yeast species (*Kluyveromyces marxianus*, *C. pelliculosa*, *S. cerevisiae*, *Candida tropicalis*, *Candida zeylanoides*, *Pichia fermentans*, *D. hansenii*, *Hanseniaspora osmophila*, and *Rhodotorula glutinis*) (Clementine et al., 2012). Yeasts have also been isolated from traditional fermented meat products with pH near neutral. Examples are the Himalayan yak meat product kargyong (a fermented sausage, pH 6.9) (Rai et al., 2009, 2010), and Chinese jinhua ham (fermented dry cured ham, final pH 6.8) (He et al., 2008). *Schwanniomyces pseudopolymorphus* (formerly *Debaryomyces pseudopolymorphus*) was detected in yak kargyong (Rai et al., 2010), and *C. zeylanoides*, *D. hansenii*, *Wickerhamomyces sydoviorum* (formerly *Hansenula sydowiorum*), and *Rhodoturola glutinis* were detected in jinhua ham (He et al., 2008). Some of these yeast species (*D. hansenii* and *C. zeylanoides*) were also isolated from other fermented meat products (not necessarily alkaline), including, e.g., Italian dry cured Parma ham, showing a strain-dependent potential for proteolytic (*D. hansenii*) and lipolytic (*D. hansenii* and *C. zeylanoides*) activities (Simoncini et al., 2007). *Debaromyces hansenii* and especially *C. zeylanoides* may also be important as bio-protective cultures as they are able to produce antagonistic activity against ochratoxin A-producing *Penicillium nordicum*, a pathogenic fungus that is associated with cured pork meat (Virgili et al., 2012). The roles of *D. hansenii* and *C. zeylanoides* in meat fermentations are various. While the former is important for desirable flavor formation during ripening of cured meat products, e.g., Spanish cured

Iberian ham (Andrade et al., 2008), the latter may contrarily produce sulfur compounds associated to an unpleasant flavor of Iberian ham (Andrade et al., 2008). Iberian cured ham typically has a pH of 5.4–5.7 (Parra et al., 2010).

Pidan is made by alkali treatment of duck eggs, obtaining a final pH of 8.98. Yeasts were not reported in this product (Wang and Fung, 1996).

4.3.5 Current Use and Perspectives for Application of Yeasts in Alkaline-Fermented Foods

Although yeasts do not play a dominant role in the fermentation of many of the alkaline foods, their role should not be neglected. The diversity and functionality of yeasts, and their interactions with other microorganisms are best described for soybean-fermented foods, whereas the role and diversity of the yeasts occurring in AFFs based on legume (other than soybean), seeds, plant, and fish are hardly investigated. The studies performed, however, do indicate that diversity is seen, that several different yeast species may be present and perhaps contribute to the fermentations. Limited knowledge apparently exists at the subspecies level or strain level for yeast occurring in AFFs. Further, the possibility for identification of new valuable yeast species should not be neglected. In AFFs, yeasts might influence texture and aroma as many yeasts are known to produce important flavor compounds (Sørensen et al., 2011). By interacting with other prevailing microorganisms yeast might also provide vitamins and other nutrients as is known from other fermented products (Narvhus and Gadaga, 2003) or have other health beneficial effects (Pedersen et al., 2012).

In the future, investigations are needed to clarify the occurrence of yeasts in AFFs, their role during the fermentation as well as their functionalities and beneficial as well as spoilage effects. These should be investigated at both the species and the strain levels. Use of up-to-date molecular methods, e.g., sequencing of the D1/D2 domain of the 26S rRNA gene will ensure correct identifications of the yeasts species in these foods. For many of the AFFs their up-scaling is dependent on the use of appropriate starter cultures, e.g., in the form of combined multifunctional starter cultures. Among such starter cultures, yeasts may be introduced for relevant AFFs.

References

Alsop, R.M. 1983. Industrial production of dextrans. In *Progress in Industrial Microbiology*, ed. M.E. Bushell, pp. 1–42. New York: Elsevier.

Amoa-Awua, W.K., Frisvad, J.C., Sefa-Dedeh, S., and Jakobsen, M. 1997. The contribution of moulds and yeasts to the fermentation of "agbelima" cassava dough. *Journal of Applied Microbiology* 83: 288–296.

Andrade, M.J., Cordoba, J.J., Sanchez, B., Casado, E.M., and Rodriguez, M. 2008. Evaluation and selection of yeasts isolated from dry-cured Iberian ham by their volatile compound production. *Food Chemistry* 113: 457–463.

Anihouvi, V.B., Kindossi, J.M., and Hounhouigan, J.D. 2012. Processing and quality characteristics of some major fermented fish products from Africa: A critical review. *International Research Journal of Biological Sciences* 1: 72–84.

Antai, S.P., and Ibrahim, M.H. 1986. Micro-organisms associated with African locust bean (*Parkia filicoidea* Welw) fermentation for "dawadawa" production. *Journal of Applied Bacteriology* 61: 145–148.

Arneborg, N., Jespersen, L., and Jakobsen, M. 2000. Individual cells of *Saccharomyces cerevisiae* and *Zygosaccharomyces bailii* exhibit different short-term intracellular pH responses to acetic acid. *Archives of Microbiology* 174: 125–128.

Ash, C., and Collins, M.D. 1992. Comparative analysis of 23S ribosomal RNA gene sequences of *Bacillus anthracis* and emetic *Bacillus cereus* determined by PCR-direct sequencing. *FEMS Microbiology Letters* 94: 75–80.

Ash, C., Farrow, J.A.E., Wallbanks, S., and Collins, M.D. 1991. Phylogenetic heterogeneity of the genus *Bacillus* revealed by comparative analysis of small-subunit-ribosomal RNA sequences. *Letters in Applied Microbiology* 13: 202–206.

Azokpota, P., Hounhouigan, D.J., and Nago, M.C. 2006. Microbiological and chemical changes during the fermentation of African locust bean (*Parkia biglobosa*) to produce afitin, iru and sonru, three traditional condiments produced in Benin. *International Journal of Food Microbiology* 107: 304–309.

Azokpota, P., Moller, P.L., Houghouigan, J.D., and Jakobsen, M. 2007. Biodiversity of predominant *Bacillus* isolated from afitin, iru and sonru at different fermentation time. *International Journal of Biological and Chemical Sciences* 1: 211–222.

Bâati, L.L., Fabre-Gea, C., Auriol, D., and Blanc, P.J. 2000. Study of the cryotolerance of *Lactobacillus acidophilus*: Effect of culture and freezing conditions on the viability and cellular protein levels. *International Journal of Food Microbiology* 59: 241–247.

Balogun, M.A., and Oyeyiola, G.P. 2011. Microbiological and chemical changes during the production of Okpehe from *Prosopis africana* seeds. *Journal of Asian Scientific Research* 1: 390–398.

Barber, L., Achinewhu, S.C., and Ibiama, E.A. 1988. The microbiology of ogiri production from castor seed (*Ricinus communis*). *Food Microbiology* 5: 177–183.

Barros, R.R., Carvalho, M.D.G.S., Peralta, J.M., Facklam, R.R., and Teixeira, L.C.M. 2001. Phenotypic and genotypic characterization of *Pediococcus* strains isolated from human clinical sources. *Journal of Clinical Microbiology* 39: 1241–1246.

Blencke, H.M., Reif, I., Commichau, F.M., Detsch, C., Wacker, I., Ludwig, H., and Stülke, J. 2006. Regulation of citB expression in *Bacillus subtilis*: Integration of multiple metabolic signals in the citrate pool and by the general nitrogen regulatory system. *Archives of Microbiology* 185: 136–146.

Bockelmann, V., Willems, P., Rademaker, J., Noordman, W., and Heller, K.J. 2003. Cultures for surface ripening of smeared soft cheese. *Kieler Milchwirtschaftliche Forschungsberichte* 55: 277–299.

Bosley, G.S., Wallace, P.L., Moss, C.W., Steigerwalt, A., Brenner, D., Swenson, J.M., Herbert, G.A., and Facklam, R.R. 1990. Phenotypic characterization, cellular fatty acid composition and DNA relatedness of aerococci and comparison to related genera. *Journal of Clinical Microbiology* 28: 416–421.

Botha, A. 2006. Yeasts in soil. In *Biodiversity and Ecophysiology of Yeasts*, eds. C.A. Rosa, and G. Peter, pp. 221–240. Berlin, Germany: Springer-Verlag.

Brousseau, R., Saint-Onge, A., Préfontaine, G., Masson, L., and Cabana, J. 1993. Arbitrary primer polymerase chain reaction, a powerful method to identify *Bacillus thuringiensis* serovars and strains. *Applied and Environmental Microbiology* 59: 114–119.

Carpana, E., Marocchi, L., and Gelmini, L. 1995. Evaluation of the API 50CHB system for the identification and biochemical characterization of *Bacillus larvae*. *Apidologie* 26: 11–16.

Chadare, F.J., Gayet, D.P., Azokpota, P., Nout, M.J.R., Linnemann, A.R., Hounhouigan, J.D., and van Boekel, M.A.J.S. 2010. Three traditional fermented baobab foods from Benin, *Mutchayan, Dikouanyouri* and *Tayohounta*: Preparation, properties, and consumption. *Ecology of Food and Nutrition* 49: 279–297.

Chadare, F.J., Jonkman, J., Wolkers-Rooijackers, J., Nout, M.J.R., Hounhouigan, J.D., and Zwietering, M.H. 2011. Microbiota of *Tayohounta*, a fermented baobab flavour food of Benin. *African Journal of Biotechnology* 10: 15607–15615.

Chen, T., Jiang, S., Xiong, S., Wang, M., Zhu, D., and Wei, H. 2012. Application of denaturing gradient gel electrophoresis to microbial diversity analysis in Chinese Douchi. *Journal of the Science of Food and Agriculture* 92: 2171–2176.

Chen, T., Wang, M., Jiang, S., Xiong, S., Zhu, D., and Wei, H. 2011a. Investigation of the microbial changes during koji-making process of Douchi by culture-dependent techniques and PCR-DGGE. *International Journal of Food Science and Technology* 46: 1878–1883.

Chen, T., Xiong, S., Jiang, S., Wang, M., Wu, Q., and Wei, H. 2011b. Molecular identification of microbial community in Chinese douchi during post-fermentation process. *Food Science and Biotechnology* 20: 1633–1638.

Chen, Y.-S., Yanagida, F., and Hsu, J.-S. 2006. Isolation and characterization of lactic acid bacteria from suan-tsai (fermented mustard), a traditional fermented food in Taiwan. *Journal of Applied Microbiology* 101: 125–130.

Choi, K.S., Lee, H.J., and Kwo, D.J. 2009. Physicochemical and microbiological properties of Korean traditional Meju. *Korean Journal of Food Preservation* 16: 217–222.

Chukeatirote, E., Chainun, C., Siengsubchart, A., Moukamnerd, C., Chantawannaku, L.P., Lumyong, S., and Thakang, P. 2006. Microbiological and biochemical changes in Thua nao fermentation. *Research Journal of Microbiology* 1: 38–44.

Chukeatirote, E., Dajanta, K., and Apichartsrangkoon, A. 2010. Thua nao, indigenous Thai fermented soybean: A review. *Journal of Biological Sciences* 10: 581–583.

Chun, J., and Sook Bae, K. 2000. Phylogenetic analysis of *Bacillus subtilis* and related taxa based on partial gyrA gene sequences. *Antonie van Leeuwenhoek* 78: 123–127.

Claus, D., and Berkeley, R.C.W. 1986. Genus *Bacillus* Cohn 1872. In *Bergey's Manual of Systematic Bacteriology*, vol. 2, eds. P.H.A. Sneath, N.S. Mair, M.E. Sharpe, and J.G. Holt, pp. 1105–1140. Baltimore, MD: Williams and Wilkins.

Clementine, K.A., Mohamed, C., Epiphane, K., David, B.K., Dje, M.K., and Montet, D. 2012. Identification of yeasts associated with the fermented fish, adjuevan, of Ivory Coast by using the molecular technique of PCR-denaturing gradient gel electrophoresis (DGGE). *African Journal of Microbiology Research* 6: 4138–4145.

Cogan, T.M., O'Dowd, M., and Mellerick, D. 1981. Effects of sugar on acetoin production from citrate by *Leuconostoc lactis*. *Applied and Environmental Microbiology* 41: 1–8.

Collins, M.D., Williams, A.M., and Wallbanks, S. 1990. The phylogeny of *Aerococcus* and *Pediococcus* as determined by 16S rRNA sequences analysis: Description of *Tetragenococcus* gen. nov. *FEMS Microbiology Letters* 70: 255–262.

Daffonchio, D., Borin, S., Consolandi, A., Mora, D., Manachini, P.L., and Sorlini, C. 1998. 16S–23S rRNA internal transcribed spacers as molecular markers for the species of the 16S rRNA group I of the genus *Bacillus*. *FEMS Microbiology Letters* 163: 229–236.

Dajanta, K., Chukeatirote, E., and Apichartsrangkoon, A. 2012. Improvement of thua nao production using protein-rich soybean and *Bacillus subtilis* TN51 starter culture. *Annals of Microbiology* 62: 785–795.

Dakwa, S., Sakyi Dawson, E., Diako, C., Annan, N.T., and Amoa Awua, W.K. 2005. Effect of boiling and roasting on the fermentation of soybeans into dawadawa (soy-dawadawa). *International Journal of Food Microbiology* 104: 69–82.

De Clerck, E., and De Vos, P. 2004. Genotypic diversity among *Bacillus licheniformis* strains from various sources. *FEMS Microbiology Letters* 231: 91–98.

De Silva, S., Petterson, B., De Muro, M.A., and Priest, F.G. 1998. A DNA probe for the detection and identification of *Bacillus sporothermodurans* using the 16S–23S rDNA spacer region and phylogenetic analysis of some field isolates of *Bacillus* which form highly heat resistant spores. *Systematic and Applied Microbiology* 21: 398–407.

Deak, T. 2008. *Handbook of Food Spoilage Yeasts*. Boca Raton, FL: Taylor and Francis.

Demoss, R.D., Bard, R.C., and Gunsalus, I.C. 1951. The mechanism of heterolactic fermentation: A new route of ethanol formation. *Journal of Bacteriology* 62: 499–511.

Department of Energy Joint Genome Project (DEJGP). 2007. The Regents of the University of California. http://genomejgi-psforg/finished_microbes/pedpe/pedpehomehtml.

Detsch, C., and Stülke, J. 2003. Ammonium utilization in *Bacillus subtilis*: Transport and regulatory functions of NrgA and NrgB. *Microbiology* 149: 3289–3297.

Dirar, H. 1993. *The Indigenous Fermented Foods and Beverages of Sudan*. Wallingford, Oxon: CAB International.

Dirar, H.A., Harper, D.B., and Collins, M.A. 1985. Biochemical and microbiological studies on Kawal, a meat substitute derived by fermentation of *Cassia obtusifolia* leaves. *Journal of the Science of Food and Agriculture* 36: 881–892.

Doan, T., Servant, P., Tojo, S., Yamaguchi, H., Lerondel, G., Yoshida, K., Fujita, Y., and Aymerich, S. 2003. The *Bacillus subtilis ywkA* gene encodes a malic enzyme and its transcription is activated by the YufL/YufM two-component system in response to malate. *Microbiology* 149: 2331–2343.

Domig, K.J., Mayer, H.K., and Kneifel, W. 2003. Methods used for the isolation, enumeration, characterisation and identification of *Enterococcus* spp. 2. Pheno- and genotypic criteria. *International Journal of Food Microbiology* 88: 165–188.

Du, J., Zhou, J., Xue, J., Song, H., and Yuan, Y. 2012. Metabolomic profiling elucidates community dynamics of the *Ketogulonicigenium vulgare–Bacillus megaterium* consortium. *Metabolomics* 8: 960–973.

Earl, A.M., Losick, R., and Kolter, R. 2008. Ecology and genomics of *Bacillus subtilis*. *Trends in Microbiology* 16: 269–275.

Ebine, H. 1986. Miso. In *Legume-Based Fermented Foods*, eds. N.R. Reddy, M.D. Pierson, and D.K. Salunkhe, pp. 61–65. Boca Raton, FL: CRC Press.

El-Helow, E.R. 2001. Identification and molecular characterization of a novel *Bacillus* strain capable of degrading Tween-80. *FEMS Microbiology Letters* 196: 119–122.

Ercolini, D. 2004. PCR-DGGE fingerprinting: Novel strategies for detection of microbes in food. *Journal of Microbiological Methods* 56: 297–314.

Farrow, J.A.E., and Collins, M.D. 1985. *Enterococcus hirae*, a new species that includes amino acid assay strain NCDO 1258 and strains causing growth depression in young chickens. *International Jourrnal of Systematic Bacteriology* 35: 73–75.

Fisher, K., and Phillips, C. 2009. The ecology, epidemiology and virulence of *Enterococcus*. *Microbiology* 155: 1749–1757.

Fisher, S.H. 1999. Regulation of nitrogen metabolism in *Bacillus subtilis*: Vive la différence! *Molecular Microbiology* 32: 223–232.

Foulquie Moreno, M.R., Sarantinopoulos, P., Tsakalidou, E., and De Vuyst, L. 2006. The role and application of enterococci in food and health. *International Journal of Food Microbiology* 106: 1–24.

Garbeva, P., van Veen, J.A., and van Elsas, J.D. 2003. Predominant *Bacillus* spp. in agricultural soil under different management regimes detected via PCR-DGGE. *Microbial Ecology* 45: 302–316.

Garcia, M.T., Martinez Canamero, M., Lucas, R., Ben Omar, N., Pérez Pulido, R., and Galvéz, A. 2004. Inhibition of *Listeria monocytogenes* by enterocin EJ97 produced by *Enterococcus faecalis* EJ97. *International Journal of Food Microbiology* 90: 161–170.

Gardin, F., Martuscelli, M., Caruso, M.C., Galgano, F., Crudele, M.A., Favati, F., Guerzoni, M.E., and Suzzi, G. 2001. Effects of pH, temperature and NaCl concentration on the growth kinetics, proteolytic activity and biogenic amine production of *Enterococcus faecalis*. *International Journal of Food Microbiology* 64: 105–117.

Garvie, E.I. 1986. Genus *Leuconostoc*. In *Bergey's Manual of Systematic Bacteriology*, eds. P.H.A. Sneath, N.S. Mair, M.E. Sharpe, and J.G. Holt, pp. 1071–1075. Baltimore, MD: Williams and Wilkins.

Glover, R.L., Abaidoo, R.C., Jakobsen, M., and Jespersen, L. 2005. Biodiversity of *Saccharomyces cerevisiae* isolated from a survey of pito production sites in various parts of Ghana. *Systematic and Applied Microbiology* 28: 755–761.

Glover, R.L.K., Sawadogo-Lingani, H., Diawara, B., Jespersen, L., and Jakobsen, M. 2009. Utilization of *Lactobacillus fermentum* and *Saccharomyces cerevisiae* as starter cultures in the production of "dolo". *Journal of Applied Biosciences* 22: 1312–1319.

Goelzer, A., Bekkal, B.F., Martin-Verstraete, I., Noirot, P., Bessières, P., Aymerich, S., and Fromion, V. 2008. Reconstruction and analysis of the genetic and metabolic regulatory networks of the central metabolism of *Bacillus subtilis*. *BMC Systematic Biology* 2: 20.

Gori, K., Hebraud, M., Chambon, C., Mortensen, H.D., Arneborg, N., and Jespersen, L. 2007. Proteomic changes in *Debaryomyces hansenii* upon exposure to NaCl stress. *FEMS Yeast Research* 7: 293–303.

Goto, K., Omura, T., Hara, Y., and Sadaie, Y. 2000. Application of the partial 16S rDNA sequence as an index for rapid identification of species in the genus *Bacillus*. *Journal of General and Applied Microbiology* 46: 1–8.

Gottschalk, G. 1986. *Bacterial Metabolism*, 2nd ed. New York: Springer-Verlag.

Green, D.H., Wakeley, P.R., Page, A., Barnes, A., Baccigalupi, L., Ricca, E., and Cutting, S.M. 1999. Characterization of two *Bacillus* probiotics. *Applied and Environmental Microbiology* 65: 4288–4291.

Guo, X.H., Li, D.F., Lu, W.Q., Piao, X.S., and Chen, X.L. 2006. Screening of *Bacillus* strains as potential probiotics and subsequent confirmation of the *in vivo* effectiveness of *Bacillus subtilis* MA139 in pigs. *Antonie van Leeuwenhoek* 90: 139–146.

Gürtler, V., and Stanisich, V.A. 1996. New approaches to typing and identification of bacteria using the 16S–23S rDNA spacer region. *Microbiology* 142: 3–16.

Halm, M., Hornbaek, T., Arneborg, N., Sefa Dedeh, S., and Jespersen, L. 2004. Lactic acid tolerance determined by measurement of intracellular pH of single cells of *Candida krusei* and *Saccharomyces cerevisiae* isolated from fermented maize dough. *International Journal of Food Microbiology* 94: 97–103.

Harwood, C.R. 1992. *Bacillus subtilis* and its relatives: Molecular biological and industrial workhorses. *Trends in Biotechnology* 10: 247–256.

Hayashida, Y., Nishimura, K., and Slaughter, J.C. 1997. The influence of mash pre-aging on the development of the flavour-active compound, 4-hydroxy-2(or5)-ethyl-5(or2)-methyl-3(2H)-furanone (HEMF), during soy sauce fermentation. *International Journal of Food Science and Technology* 32: 11–14.

Hayford, A.E., and Jespersen, L. 1999. Characterization of *Saccharomyces cerevisiae* strains from spontaneously fermented maize dough by profiles of assimilation, chromosome polymorphism, PCR and MAL genotyping. *Journal of Applied Microbiology* 86: 284–294.

He, Z.-F., Zhen, Z.Y., Li, H.-J., Zhou, G.H., and Zhang, J.-H. 2008. Microorganisms flora study on Jinhua ham fermentation. *Food Science (China)* 29: 190–196.

Helgason, E., Økstad, O.A., Caugant, D.A., Johansen, H.A., Fouet, A., Mock, M., Hegna, I., and Kolstø, A.B. 2000. *Bacillus anthracis*, *Bacillus cereus*, and *Bacillus thuringiensis*—One species on the basis of genetic evidence. *Applied and Environmental Microbiology* 66: 2627–2630.

Herman, L., and Heyndrickx, M. 2000. The presence of intragenically located REP-like elements in *Bacillus sporothermodurans* is sufficient for REP-PCR typing. *Research in Microbiology* 151: 255–261.

Heyndrickx, M., Vandemeulebroecke, K., Hoste, B., Janssen, P., Kersters, K., De Vos, P., Logan, N.A., Ali, N., and Berkeley, R.C.W. 1996. Reclassification of *Paenibacillus* (formerly *Bacillus*) *pulvifaciens* (Nakamura 1984) Ash et al. 1994, a later subjective synonym of *Paenibacillus* (formerly *Bacillus*) *larvae* (White 1906) Ash et al. 1994, as a subspecies of *P. larvae*, with emended descriptions of *P. larvae* as *P. larvae* subsp. *larvae and P. larvae* subsp. *pulvifaciens*. *International Journal of Systematic Bacteriology* 46: 270–279.

Holzapfel, W.H., Franz, C.M.A.P., Ludwig, W., Back, W., and Dicks, L.M.T. 2006. The genera *Pediococcus and Tetragenococcus*. In *The Prokaryotes*, eds. E. Stackebrandt, and W.H. Holzapfel, pp. 229–266. New York: Spinger.

Hu, H., Hao, J., Cheng, Y., Yin, L., Ma, Y., Yu, Z., and Li, L. 2012. Effect of fermented rice culture on the microbiological, biochemical and sensory characteristics of low-salt douchi, a traditional Chinese fermented soybean condiment. *International Journal of Food Science and Technology* 47: 689–695.

Hu, Y., Xia, W., and Ge, C. 2006. Effect of mixed starter cultures fermentation on the characteristics of silver carp sausages. *World Journal of Microbiology and Biotechnology* 23: 1021–1031.

Inatsu, Y., Nakamura, N., Yuriko, Y., Fushimi, T., Watanasiritum, L., and Kawamoto, S. 2006. Characterization of *Bacillus subtilis* strains in Thua nao, a traditional fermented soybean food in northern Thailand. *Letters in Applied Microbiology* 43: 237–242.

Jakobsen, M., Larsen, M.D., and Jespersen, L. 2001. Production of bread, cheese and meat. In *The Mycota: Industrial Applications*, ed. H.D. Osiewacz, pp. 3–22. New York: Springer-Verlag.

Jeff-Agboola, Y.A. 2007. Microorganisms associated with natural fermentation of African yam bean (*Sphenostylis sternocarpa* Harms) seeds for the production of Otiru. *Research Journal of Microbiology* 2: 816–823.

Jespersen, L. 2003. Occurrence and taxonomic characteristics of strains of *Saccharomyces cerevisiae* predominant in African indigenous fermented foods and beverages. *FEMS Yeast Research* 3: 191–200.

Jespersen, L., Nielsen, D.S., Hønholt, S., and Jakobsen, M. 2005. Occurrence and diversity of yeasts involved in fermentation of West African cocoa beans. *FEMS Yeast Research* 5: 441–453.

Jespersen, L., van der Kuhle, A., and Petersen, K.M. 2000. Phenotypic and genetic diversity of *Saccharomyces* contaminants isolated from lager breweries and their phylogenetic relationship with brewing yeasts. *International Journal of Food Microbiology* 60: 43–53.

Johansen, K.A., Gill, R.E., and Vasil, M.L. 1996. Biochemical and molecular analysis of phospholipase C and phospholipase D activity in mycobacteria. *Infection and Immunity* 64: 3259–3266.

Johnson, Y.A., Nagpal, M., Krahmer, M.T., Fox, K.F., and Fox, A. 2000. Precise molecular weight determination of PCR products of the rRNA intergenic spacer region using electrospray quadrupole mass spectrometry for differentiation of *B. subtilis* and *B. atrophaeus*, closely related species of bacilli. *Journal of Microbiological Methods* 40: 241–254.

Joung, K.-B., and Côté, J.-C. 2002. Evaluation of ribosomal RNA gene restriction patterns for the classification of *Bacillus* species and related genera. *Journal of Applied Microbiology* 92: 97–108.

Jourlin-Castelli, C., Mani, N., Nakano, M.M., and Sonenshein, A.L. 2000. CcpC, a novel regulator of the LysR family required for glucose repression of the *citB* gene in *Bacillus subtilis*. *Journal of Molecular Biology* 295: 865–878.

Kaboré, D., Sawadogo-Lingani, H., Dicko, M.H., Diawara, B., and Jakobsen, M. 2012a. Acid resistance, bile tolerance and antimicrobial properties of dominant lactic acid bacteria isolated from traditional "maari" baobab seeds fermented condiment. *African Journal of Biotechnology* 11: 1197–1205.

Kaboré, D., Thorsen, L., Nielsen, D.S., Sune, B.T., Sawadogo-Lingani, H., Diawara, B., Dicko, M.H., and Jakobsen, M. 2012b. Bacteriocin formation by dominant aerobic sporeformers isolated from traditional maari. *International Journal of Food Microbiology* 154: 10–18.

Kim, H.E., Han, S.Y., and Kim, Y.S. 2010. Quality characteristics of Gochujang Meju prepared with different fermentation tools and inoculation time of *Aspergillus oryzae*. *Food Science and Biotechnology* 19: 1579–1585.

Kim, J.S., Shahidi, F., and Heu, M.S. 2003. Characteristics of salt-fermented sauces from shrimp processing byproducts. *Journal of Agricultural and Food Chemistry* 51: 784–792.

Kim, J.Y., Yeo, S.H., Baek, S.Y., and Cho, H.S. 2011. Molecular and morphological identification of fungal species isolated from Bealmijang Meju. *Journal of Microbiology and Biotechnology* 21: 1270–1279.

Kim, T.W., Lee, J.H., Kim, S.E., Park, M.H., Chang, H.C., and Kim, H.Y. 2009. Analysis of microbial communities in *doenjang*, a Korean fermented soybean paste, using nested PCR-denaturing gradient gel electrophoresis. *International Journal of Food Microbiology* 131: 265–271.

Klein, G. 2003. Taxonomy, ecology and antibiotic resistance of enterococci from food and the gastro-intestinal tract. *International Journal of Food Microbiology* 88: 123–131.

Klingberg, T.D., Lesnik, U., Arneborg, N., Raspor, P., and Jespersen, L. 2008. Comparison of *Saccharomyces cerevisiae* strains of clinical and nonclinical origin by molecular typing and determination of putative virulence traits. *FEMS Yeast Research* 8: 631–640.

Kobayashi, T., Kajiwara, M., Wahyuni, M., Kitakado, T., Hamada Sato, N., Imada, C., and Watanabe, E. 2003. Isolation and characterization of halophilic lactic acid bacteria isolated from "terasi" shrimp paste: A traditional fermented seafood product in Indonesia. *Journal of General and Applied Microbiology* 49: 279–286.

Kobayashi, T., Kimura, B., and Fujii, T. 2000. Strictly anaerobic halophiles isolated from canned Swedish fermented herrings (Surstromming). *International Journal of Food Microbiology* 54: 81–89.

Kronenberg, H.J., and Hang, Y.D. 1984. Biochemical changes in Okara during Meitauza fermentation. *Nutrition Reports International* 30: 439–443.

Kurita, O., and Yamazaki, E. 2002. Growth under alkaline conditions of the salt-tolerant yeast *Debaryomyces hansenii* IFO10939. *Current Microbiology* 45: 277–280.

Kurtzman, C.P., Fell, J.W., and Boekhout, T. eds. 2011. *The Yeasts, A Taxonomic Study*, 5th ed. Amsterdam, The Netherlands: Elsevier Science.

Kwon, D.J., and Kim, W.J. 2005. Isolation of higher alcohol-producing yeast as the flavor components and determination of optimal culture conditions. *Food Science and Biotechnology* 14: 576–580.

Lam, M.M.C., Seemann, T., Bulach, D.M., Gladman, S.L., Chen, H., Haring, V., Moore, R.J., Ballard, S., Grayson, M.L., Johnson, P.D.R., Benjamin, P., Howden, B.P., and Stinear, T.P. 2012. Comparative analysis of the first complete *Enterococcus faecium* genome. *Journal of Bacteriology* 194: 2334–2341.

Lee, J.H., Kim, T.W., Lee, H., Chang, H.C., and Kim, H.Y. 2010. Determination of microbial diversity in *meju*, fermented cooked soya beans, using nested PCR-denaturing gradient gel electrophoresis. *Letters in Applied Microbiology* 51: 388–394.

Lee, J.S., Yi, S.H., Kwon, S.J., Ahn, C., and Yoo, J.Y. 1997a. Enzyme activities and physiological functionality of yeasts from traditional Meju. *Korean Journal of Applied Microbiology and Bioengineering* 25: 448–453.

Lee, J.S., Yi, S.H., Kwon, S.J., Ahn, C., and Yoo, J.Y. 1997b. Isolation, identification and cultural conditions of yeasts from traditional Meju. *Korean Journal of Applied Microbiology and Biotechnology* 25: 435–441.

Lee, S.-Y., Park, N.Y., Kim, J.-Y., and Choi, H.-S. 2012. Quality characteristics of rice-doenjang during fermentation by differently shaped *Meju* and adding starter. *Korean Journal of Food & Nutrition* 25: 505–512.

Lee, S.S. 1995. Meju fermentation for a raw material of Korean traditional soy products. *The Korean Journal of Mycology* 23: 161–175.

Li, H., Qiu, T., Huang, G., and Cao, Y. 2010. Production of gamma-aminobutyric acid by *Lactobacillus brevis* NCL912 using fed-batch fermentation. *Microbial Cell Factories* 9: 85.

Liu, C.J., Gong, F.M., Li, X.R., Li, H.Y., Zhang, Z.H., Feng, Y., and Nagano, H. 2012. Natural populations of lactic acid bacteria in douchi from Yunnan Province, China. *Journal of Zhejiang University: Science B* 13: 298–306.

Liu, Y., Garceau, N.Y., Loros, J.J., and Dunlap, J.C. 1997. Thermally regulated translational control of FRQ mediates aspects of temperature responses in the *Neurospora* circadian clock. *Cell* 89: 477–486.

Ljungh, A., and Wadström, T. 2006. Lactic acid bacteria as probiotics. *Current Issues in Intestinal Microbiology* 7: 73–89.

Logan, N.A., and Berkeley, R.C.W. 1984. Identification of *Bacillus* strains using the API system. *Journal of General Microbiology* 130: 1871–1882.

Louembé, D., Kobawila, S.C., Kalou, G.B., and Kéléké, S. 2003. Étude microbiologique des feuilles fermentées de manioc: "Ntoba Mbodi". *Tropicultura* 121: 106–111.

Lucas, R., Grande, M.J., Abriouel, H., Maqueda, M., Ben Omar, N., Valdivia, E., Martinez-Canamero, M., and Gálvez, A. 2006. Application of the broad-spectrum bacteriocin enterocin AS-48 to inhibit *Bacillus coagulans* in canned fruit and vegetable foods. *Food and Chemical Toxicology* 44: 1774–1781.

Ludwig, H., Homuth, G., Schmalisch, M., Dyka, F.M., Hecker, M., and Stülke, J. 2001. Transcription of glycolytic genes and operons in *Bacillus subtilis*: Evidence for the presence of multiple levels of control of the gapA operon. *Molecular Microbiology* 41: 409–422.

Makarova, K., Slesarev, A., Wolf, Y., Sorokin, A., Mirkin, B., Koonin, E., Pavlov, A., Pavlova, N., Karamychev, V., Polouchine, N., Shakhova, V., Grigoriev, I., Lou, Y., Rohksar, D., Lucas, S., Huang, K., Goldstein, D.M., Hawkins, T., Plengvidhya, V., Welker, D., Hughes, J., Goh, Y., Benson, A., Baldwin, K., Lee, J.H., Diaz-Muniz, I., Dosti, B., Smeianov, V., Wechter, W., Barabote, R., Lorca, G., Altermann, E., Barrangou, R., Ganesan, B., Xie, Y., Rawsthorne, H., Tamir, D., Parker, C., Breidt, F., Broadbent, J., Hutkins, R., O'Sullllivan, D., Steele, J., Unlu, G., Saier, M., Klaenhammer, T., Richardson, P., Kozyavkin, S., Weimer, B., and Mills, D. 2006. Comparative genomics of the lactic acid bacteria. *Proceedings of the National Academy of Sciences of the United States of America* 103: 15611–15616.

Marten, P., Smalla, K., and Berg, G. 2000. Genotypic and phenotypic differentiation of an antifungal biocontrol strain belonging to *Bacillus subtilis*. *Journal of Applied Microbiology* 89: 463–471.

Masoud, W., Cesar, L.B., Jespersen, L., and Jakobsen, M. 2004. Yeast involved in fermentation of *Coffea arabica* in East Africa determined by genotyping and by direct denaturing gradient gel electrophoresis (DGGE). *Yeast* 21: 549–556.

Masoud, W., and Jespersen, L. 2006. Pectin degrading enzymes in yeasts involved in fermentation of *Coffea arabica* in East Africa. *International Journal of Food Microbiology* 110: 291–296.

Maughan, H., and van der Auwera, G. 2011. *Bacillus* taxonomy in the genomic era finds phenotypes to be essential though often misleading. *Infections Genetics and Evolution* 11: 789–797.

Mendo, S.A.L.V., Henriques, S.I., Correia, A.C.M., and Duarte, J.M.C. 2000. Genetic characterization of new thermotolerant *Bacillus licheniformis* strain. *Current Microbiology* 40: 137–139.

Moir, A. 2006. How do spores germinate? *Journal of Applied Microbiology* 101: 526–530.

Moslehi-Jenabian, S., Pedersen, L.L., and Jespersen, L. 2010. Beneficial effects of probiotic and food borne yeasts on human health. *Nutrients* 2: 449–473.

Musa, D.A., Omale, J., and Daikwo, A.M. 2011. Fungal contamination of fermented *Prosopis africana* (Okpehe) and toxicity screening of the crude extracts in albino rats (*Rattus norvegicus*). *Advances in Applied Science Research* 2: 108–113.

Nakajo, K., Iwami, Y., Komori, R., Ishikawa, S., Ueno, T., Suzuki, Y., and Takahashi, N. 2005. The resistance to acidic and alkaline environments of endodontic pathogen *Enterococcus faecalis*. *International Congress Series* 1284: 191–192.

Nam, Y.D., Lee, S.Y., and Lim, S.I. 2012. Microbial community analysis of Korean soybean pastes by next-generation sequencing. *International Journal of Food Microbiology* 155: 36–42.

Namwong, S., Tanasupawat, S., Lee, K.C., and Lee, J.S. 2009. *Oceanobacillus kapialis* sp. nov., from fermented shrimp paste in Thailand. *International Journal of Systematic and Evolutionay Microbiology* 59: 2254–2259.

Narvhus, J.A., and Gadaga, T.H. 2003. The role of interaction between yeasts and lactic acid bacteria in African fermented milks: A review. *International Journal of Food Microbiology* 86: 51–60.

Naumova, E.S., Korshunova, I.V., Jespersen, L., and Naumov, G.I. 2003. Molecular genetic identification of *Saccharomyces* sensu stricto strains from African sorghum beer. *FEMS Yeast Research* 3: 177–184.

Nazina, T.N., Tourova, T.P., Poltaraus, A.B., Novikova, E.V., Grigoryan, A.A., Ivanova, A.B., Lysenko, A.M., Petrunyaka, V.V., Osipov, G.A., Belyaev, S.S., and Ivanov, M.V. 2001. Taxonomic study of aerobic thermophilic bacilli: Descriptions of *Geobacillus subterraneus* gen. nov., sp. nov. and *Geobacillus uzenensis* sp. nov. from petroleum reservoirs and transfer of *Bacillus stearothermophilus*, *Bacillus thermocatenulatus*, *Bacillus thermoleovorans*, *Bacillus kaustophilus*, *Bacillus thermoglucosidasius* and *Bacillus thermodenitrificans* to *Geobacillus* as as the new combinations *G. stearothermophilus*, *G. thermocatenulatus*, *G. thermoleovorans*, *G. kaustophilus*, *G. thermoglucosidasius* and *G. thermodenitrificans*. *International Journal of Systematic and Evolutionary Microbiology* 51: 433–446.

N'dir, B., Hbid, C., Cornelius, C., Roblain, D., Jacques, P., Vanhentenryck, F., Diop, M., and Thonart, P. 1994. Propriétés antifongiques de la microflore sporulée du nététou. *Cahiers Agricultures* 3: 23–30.

Nielsen, D.S., Honholt, S., Tano-Debrah, K., and Jespersen, L. 2005. Yeast populations associated with Ghanaian cocoa fermentations analysed using denaturing gradient gel electrophoresis (DGGE). *Yeast* 22: 271–284.

Nielsen, D.S., Jakobsen, M., and Jespersen, L. 2010. *Candida halmiae* sp nov., *Geotrichum ghanense* sp nov. and *Candida awuaii* sp nov., isolated from Ghanaian cocoa fermentations. *International Journal of Systematic and Evolutionary Microbiology* 60: 1460–1465.

Niimura, Y., Koh, E., Yanagida, F., Suzuki, K.-I., Komagata, K., and Kozaki, M. 1990. *Amphibacillus xylanus* gen. nov., sp. nov., a facultatively anaerobic spore-forming xylan-digesting bacterium which lacks cytochrome, quinone, and catalase. *International Journal of Systematic Bacteriology* 40: 297–301.

Nisiotou, A.A., Chorianopoulos, N., Nychas, G.J.E., and Panagou, E.Z. 2010. Yeast heterogeneity during spontaneous fermentation of black Conservolea olives in different brine solutions. *Journal of Applied Microbiology* 108: 396–405.

Nout, M.J.R., Bakshi, D., and Sarkar, P.K. 1998. Microbiological safety of kinema, a fermented soya bean food. *Food Control* 9: 357–362.

Nuobariene, L., Hansen, A.S., Jespersen, L., and Arneborg, N. 2011. Phytase-active yeasts from grain-based food and beer. *Journal of Applied Microbiology* 110: 1370–1380.

Nwosu, C.D., and Ojimelukwe, P.C. 1993. Improvement of the traditional method of ogiri production and identification of the micro-organisms associated with the fermentation process. *Plant Foods for Human Nutrition* 43: 267–272.

O'Toole, D.K. 1999. Characteristics and use of okara, the soybean residue from soy milk production—A review. *Journal of Agricultural and Food Chemistry* 47: 363–371.

Odunfa, S.A. 1981. Microorganisms associated with fermentation of African locust bean (*Parkia filicoidea*) during iru preparation. *Journal of Plant Foods* 3: 245–250.

Ogbonna, D.N., Sokari, T.G., and Achinewhu, S.C. 2001. Development of an owoh-type product from African yam beans (*Sphenostylis stenocarpa*) (Hoechst (ex. A. Rich.) Harms.) seeds by solid substrate fermentation. *Plant Foods for Human Nutrition* 56: 183–194.

Ogunshe, A.A.O., and Olasugba, K.O. 2008. Microbial loads and incidence of food-borne indicator bacteria in most popular indigenous fermented food condiments from middle-belt and southwestern Nigeria. *African Journal of Microbiology Research* 2: 332–339.

Oguntoyinbo, F.A., Huch, M., Cho, G.S., Schillinger, U., Holzapfel, W.H., Sanni, A.L., and Franz, C.M.A.P. 2010. Diversity of *Bacillus* species isolated from *Okpehe*, a traditional fermented soup condiment from Nigeria. *Journal of Food Protection* 73: 870–878.

Oguntoyinbo, F.A., and Sanni, A.I. 2007. Determination of toxigenic potentials of *Bacillus* strains isolated from *Okpehe*, a Nigerian fermented condiment. *World Journal of Microbiology and Biotechnology* 23: 65–70.

Oguntoyinbo, F.A., Sanni, A.I., Franz, C.M.A.P., and Holzapfel, W.H. 2007. *In vitro* fermentation studies for selection and evaluation of *Bacillus* strains as starter cultures for the production of *Okpehe*, a traditional African fermented condiment. *International Journal of Food Microbiology* 113: 208–218.

Olajyigbe, F.M., and Ajele, J.O. 2005. Production dynamics of extracellular protease from *Bacillus* species. *African Journal of Biotechnology* 4: 776–779.

Onilude, A.A., Sanni, A.I., and Ighalo, M.I. 1999. Effect of process improvement on the physico-chemical properties of infant weaning food from fermented composite blends of cereal and soybeans. *Plant Foods for Human Nutrition* 54: 239–250.

Osmanagaoglu, O., Beyatli, Y., and Gunduz, U. 2001. Isolation and characterization of Pediocin producing *Pediococcus pentosaceus* Pep1 from vacuum-packed sausages. *Turkish Journal of Biology* 25: 133–143.

Osmanagaoglu, O., Gündüz, U., Beyatli, Y., and Çökmüs, C. 1998. Purification and characterization of pediocin F, a bacteriocin produced by *Pediococcus acidilactici* F. *Turkish Journal of Biology* 22: 217–228.

Ouadghiri, M., Amar, M., Vancanneyt, M., and Swings, J. 2005. Biodiversity of lactic acid bacteria in Moroccan soft white cheese (Jben). *FEMS Microbiology Letters* 251: 267–271.

Ouoba, L.I.I., Cantor, M.D., Diawara, B., Traore, A.S., and Jakobsen, M. 2003a. Degradation of African locust bean oil by *Bacillus subtilis* and *Bacillus pumilus* isolated from soumbala, a fermented African locust bean condiment. *Journal of Applied Microbiology* 95: 868–873.

Ouoba, L.I.I., Diawara, B., Amoa-Awua, W.K., Traore, A.S., and Møller, P.L. 2004. Genotyping of starter cultures of *Bacillus subtilis* and *Bacillus pumilus* for fermentation of African locust bean (*Parkia biglobosa*) to produce Soumbala. *International Journal of Food Microbiology* 90: 197–205.

Ouoba, L.I.I., Diawara, B., Christensen, T., Mikkelsen, J.D., and Jakobsen, M. 2007a. Degradation of polysaccharides and non-digestible oligosaccharides by *Bacillus subtilis* and *Bacillus pumilus* isolated from Soumbala, a fermented African locust bean (*Parkia biglobosa*) food condiment. *European Food Research and Technology* 224: 689–694.

Ouoba, L.I.I., Diawara, B., Jespersen, L., and Jakobsen, M. 2007b. Antimicrobial activity of *Bacillus subtilis* and *Bacillus pumilus* during the fermentation of African locust bean (*Parkia biglobosa*) for soumbala production. *Journal of Applied Microbiology* 102: 963–970.

Ouoba, L.I.I., Nyanga-Koumou, C.A.G., Parkouda, C., Sawadogo, H., Kobawila, S.C., Keleke, S., Diawara, B., Louembe, D., and Sutherland, J.P. 2010. Genotypic diversity of lactic acid bacteria isolated from African traditional alkaline-fermented foods. *Journal of Applied Microbiology* 108: 2019–2029.

Ouoba, L.I.I., Parkouda, C., Diawara, B., Scotti, C., and Varnam, A.H. 2008. Identification of *Bacillus* spp from bikalga, fermented seeds of *Hibiscus sabdariffa*: Phenotypic and genotypic characterization. *Journal of Applied Microbiology* 104: 122–131.

Ouoba, L.I.I., Rechinger, K.B., Barkholt, V., Diawara, B., Traore, A.S., and Jakobsen, M. 2003b. Degradation of proteins during the fermentation of African locust bean *Parkia biglobosa*, by strains of *Bacillus subtilis* and *Bacillus pumilus* for production of soumbala. *Journal of Applied Microbiology* 94: 396–402.

Oyeyiola, G.P. 1989. Microbiology of the fermentation of locally produced Ugba. *Ecology of Food and Nutrition* 23: 287–291.

Pakdeeto, A., Tanasupawat, S., Thawai, C., Moonmangmee, S., Kudo, T., and Itoh, T. 2007. *Lentibacillus kapialis* sp. nov., from fermented shrimp paste in Thailand. *International Journal of Systematic and Evolutionary Microbiology* 57: 364–369.

Parkouda, C., Nielsen, D.S., Azokpota, P., Ouoba, L.I.I., Amoa-Awua, W.K., Thorsen, L., Hounhouigan, J.D., Jensen, J.S., Tano-Debrah, K., Diawara, B., and Jakobsen, M. 2009. The microbiology of alkaline-fermentation of indigenous seeds used as food condiments in Africa and Asia. *Critical Reviews in Microbiology* 35: 139–156.

Parkouda, C., Thorsen, L., Compaore, C.S., Nielsen, D.S., Tano-Debrah, K., Jensen, J.S., Diawara, B., and Jakobsen, M. 2010. Microorganisms associated with Maari, a baobab seed fermented product. *International Journal of Food Microbiology* 142: 292–301.

Parra, V., Viguera, J., Sanchez, J., Peinado, J., Esparrago, F., Gutierrez, J.I., and Andres, A.I. 2010. Modified atmosphere packaging and vacuum packaging for long period chilled storage of dry-cured Iberian ham. *Meat Science* 84: 760–768.

Pedersen, L.L., Owusu-Kwarteng, J., Thorsen, L., and Jespersen, L. 2012. Biodiversity and probiotic potential of yeasts isolated from *Fura*, a West African spontaneously fermented cereal. *International Journal of Food Microbiology* 159: 144–151.

Perez-Torrado, R., Llopis, S., Jespersen, L., Fernandez-Espinar, T., and Querol, A. 2012. Clinical *Saccharomyces cerevisiae* isolates cannot cross the epithelial barrier in vitro. *International Journal of Food Microbiology* 157: 59–64.

Petersen, K.M., and Jespersen, L. 2004. Genetic diversity of the species *Debaryomyces hansenii* and the use of chromosome polymorphism for typing of strains isolated from surface-ripened cheeses. *Journal of Applied Microbiology* 97: 205–213.

Petersen, K.M., Westall, S., and Jespersen, L. 2002. Microbial succession of *Debaryomyces hansenii* strains during the production of Danish surface-ripened cheeses. *Journal of Dairy Science* 85: 478–486.

Popoola, T.O.S., Jolaoso, A.A., and Afolabi, R.O. 2004. Microbiology of the production of Oso, a condiment made by fermenting seeds of *Cathormion altissimum*. *Tropical Science* 44: 187–189.

Popoola, T.O.S., Jolaoso, A.A., Afolabi, R.O., and Akintokun, A.K. 2006. Microbiological studies on the production of Oso: A condiment made from fermenting seeds of *Cathormion altissimum*. *Journal of Herbs, Spices and Medicinal Plants* 12: 129–138.

Priest, F.G., Goodfellow, M., and Todd, C. 1988. A numerical classification of the genus *Bacillus*. *Journal of General Microbiology* 134: 1847–1882.

Pritchard, G., and Coolbear, T. 1993. The physiology and biochemistry of the proteolytic system in lactic acid bacteria. *FEMS Microbiology Reviews* 179–206.

Qu, Z., Jiang, A.M., Chen, M.T., Ockerman, H.W., and Basu, L. 2012. Potential starter culture isolated from traditionally fermented YangJiang lobster sauce and protease characteristics research. *Food Science and Technology Research* 18: 585–591.

Rai, A.K., Palni, U., and Tamang, J.P. 2009. Traditional knowledge of the Himalayan people on production of indigenous meat products. *Indian Journal of Traditional Knowledge* 8: 104–109.

Rai, A.K., Tamang, J.P., and Palni, U. 2010. Microbiological studies of ethnic meat products of the Eastern Himalayas. *Meat Science* 85: 560–567.

Rainey, F.A., Fritze, D., and Stackebrandt, E. 1994. The phylogenetic diversity of thermophilic members of the genus *Bacillus* as revealed by 16S rDNA analysis. *FEMS Microbiology Letters* 115: 205–211.

Rashad, M.M., Mahmoud, E.A., Abdou, M.H., and Nooman, U.M. 2011. Improvement of nutritional quality and antioxidant activities of yeast fermented soybean curd residue. *African Journal of Biotechnology* 10: 5750–5759.

Reiter, B., and Oram, J.D. 1982. Nutritional studies on cheese starter. Vitamin and amino acid requirements of single strain starters. *Journal of Dairy Research* 29: 63–68.

Rengpipat, S., Rueangruklikhit, T., and Piyatiratitivorakul, S. 2008. Evaluations of lactic acid bacteria as probiotics for juvenile seabass *Lates calcarifer*. *Aquaculture Research* 39: 134–143.

Rojas, I.B., Smith, P.A., and Bartowsky, E.J. 2012. Influence of choice of yeasts on volatile fermentation-derived compounds, colour and phenolics composition in Cabernet Sauvignon wine. *World Journal of Microbiology and Biotechnology* 28: 3311–3321.

Romano, P., Capece, A., and Jespersen, L. 2006. Taxonomic and ecological diversity of food and beverage yeasts. In *Yeast in Food and Beverages*, eds. A. Querol and G. Fleet, pp. 13–53. Berlin, Germany: Springer-Verlag.

Röling, W.F.M., Timotius, K.H., Prasetyo, A.B., Stouthamer, A.H., and Vanverseveld, H.W. 1994. Changes in microflora and biochemical composition during the Baceman stage of traditional Indonesian kecap (soy sauce) production. *Journal of Fermentation and Bioengineering* 77: 62–70.

Rössler, D., Ludwig, W., Schleifer, K.H., Lin, C., McGill, T.J., Wisotzkey, J.D., Jurtshuk Jr., P., and Fox, G.E. 1991. Phylogenetic diversity in the genus *Bacillus* as seen by 16S rRNA sequencing studies. *Systematic and Applied Microbiology* 14: 266–269.

Sacristan, N., Gonzalez, L., Ma Castro, J., Ma Fresno, J., and Tornadijo, M.E. 2012. Technological characterization of *Geotrichum candidum* strains isolated from a traditional Spanish goats' milk cheese. *Food Microbiology* 30: 260–266.

Sanni, A.I., Asiedu, M., and Ayernor, G.S. 2002. Microflora and chemical composition of *Momoni*, a Ghanaian fermented fish condiment. *Journal of Food Composition and Analysis* 15: 577–583.

Sanni, A.I., and Ogbonna, D.N. 1991. The production of *owoh*—A Nigerian fermented seasoning agent from cotton seed (*Gossypium hirsutum* L). *Food Microbiology* 8: 223–229.

Sarkar, P.K., Hasenack, B., and Nout, M.J.R. 2002. Diversity and functionality of *Bacillus* and related genera isolated from spontaneously fermented soybeans (Indian kinema) and locust beans (African soumbala). *International Journal of Food Microbiology* 77: 175–186.

Sarkar, P.K., Jones, L.J., Craven, G.S., Somerset, S.M., and Palmer, C. 1997. Amino acid profiles of kinema, a soybean-fermented food. *Food Chemistry* 59: 69–75.

Sarkar, P.K., Jones, L.J., Gore, W., Craven, G.S., and Somerset, S.M. 1996. Changes in soya bean lipid profiles during kinema production. *Journal of the Science of Food and Agriculture* 71: 321–328.

Sarkar, P.K., and Tamang, J.P. 1994. The influence of process variables and inoculum composition on the sensory quality of kinema. *Food Microbiology* 11: 317–325.

Sarkar, P.K., and Tamang, J.P. 1995. Changes in the microbial profile and proximate composition during natural and controlled fermentations of soybeans to produce kinema. *Food Microbiology* 12: 317–325.

Sarkar, P.K., Tamang, J.P., Cook, P.E., and Owens, J.D. 1994. Kinema—A traditional soybean fermented food: Proximate composition and microflora. *Food Microbiology* 11: 47–55.

Schilling, O., Frick, O., Herzberg, C., Ehrenreich, A., Heinzle, E., Wittmann, C., and Stülke, J. 2007. Transcriptional and metabolic responses of *Bacillus subtilis* to the availability of organic acids: Transcription regulation is important but not sufficient to account for metabolic adaptation. *Applied and Environmental Microbiology* 73: 499–507.

Schwan, R.F., Almeida, E.G., Souza-Dias, M.A.G., and Jespersen, L. 2007. Yeast diversity in rice-cassava fermentations produced by the indigenous Tapirape people of Brazil. *FEMS Yeast Research* 7: 966–972.

Shangkuan, Y.-H., Yang, J.-F., Lin, H.-C., and Shaio, M.-F. 2000. Comparison of PCR-RFLP, ribotyping and ERIC-PCR for typing *Bacillus anthracis* and *Bacillus cereus* strains. *Journal of Applied Microbiology* 89: 452–462.

Shetty, P.H., Hald, B., and Jespersen, L. 2007. Surface binding of aflatoxin B-1 by *Saccharomyces cerevisiae* strains with potential decontaminating abilities in indigenous fermented foods. *International Journal of Food Microbiology* 113: 41–46.

Shida, O., Takagi, H., Kadowaki, K., and Komagata, K. 1996. Proposal for two new genera, *Brevibacillus* gen. nov. and *Aneurinibacillus* gen. nov. *International Journal of Systematic Bacteriology* 46: 939–946.

Sidhipol, J., Tanasupawat, S., Tepkasikul, P., Yukphan, P., Tosukhowong, A., Itoh, T., Benjakul, S., and Visessanguan, W. 2013. Identification and histamine formation of *Tetragenococcus* isolated from Thai fermented food products. *Annals of Microbiology* 63: 745–753.

Siefert, J.L., Larios-Sanz, M., Nakamura, L.K., Slepcky, R.A., Paul, J.H., Moore, E.R.B., Fox, G.E., and Jurtshuk Jr., P. 2000. Phylogeny of marine *Bacillus* isolates from the gulf of Mexico. *Current Microbiology* 41: 84–88.

Simoncini, N., Rotelli, D., Virgili, R., and Quintavalla, S. 2007. Dynamics and characterization of yeasts during ripening of typical Italian dry-cured ham. *Food Microbiology* 24: 577–584.

Sohliya, I., Joshi, S.R., Bhagobaty, R.K., and Kumar, R. 2009. Tungrymbai—A traditional fermented soybean food of the ethnic tribes of Meghalaya. *Indian Journal of Traditional Knowledge* 8: 559–561.

Sonenshein, A.L. 2007. Control of key metabolic intersections in *Bacillus subtilis*. *Nature Reviews Microbiology* 5: 917–927.
Sørensen, L.M., Gori, K., Petersen, M.A., Jespersen, L., and Arneborg, N. 2011. Flavour compound production by *Yarrowia lipolytica*, *Saccharomyces cerevisiae* and *Debaryomyces hansenii* in a cheese-surface model. *International Dairy Journal* 21: 970–978.
Steinkraus, K.H., ed. 1996. *Handbook of Indigenous Fermented Foods*, 2nd ed. New York: Marcel Dekker.
Stephan, R. 1996. Randomly amplified polymorphic DNA (RAPD) assay for genomic fingerprinting of *Bacillus cereus* isolates. *International Journal of Food Microbiology* 31: 311–316.
Stephan, R., Schraft, H., and Untermann, F. 1994. Characterization of *Bacillus licheniformis* with the RAPD technique (randomly amplified polymorphic DNA). *Letters in Applied Microbiology* 18: 260–263.
Stülke, J., and Hillen, W. 2000. Regulation of carbon catabolism in *Bacillus* species. *Annual Reviews of Microbiology* 54: 849–880.
Sümengen, M., Dinçer, S., and Kaya, A. 2012. Phytase production from *Lactobacillus brevis*. *Turkish Journal of Biology* 36: 533–541.
Surono, I.S., and Hosono, A. 1994a. Chemical and aerobic bacterial composition of "Terasi", a traditional fermented product from Indonesia. *Journal of the Food Hygienic Society of Japan* 35: 299–304.
Surono, I.S., and Hosono, A. 1994b. Microflora and their enzyme profile in Terasi starter. *Bioscience, Biotechnology, and Biochemistry* 58: 1167–1169.
Sutherland, I.W. 2008. Extracellular polysaccharides. In *Biotechnology Vol. 6: Products of Primary Metabolism*, 2nd ed., eds. H.-J. Rehm, G. Reed, A. Puhler, and P. Stadler, pp. 613–657. New York: VCH.
Tajabadi, N., Mardan, M., Mustafa, S., Feizabadi, F., Nateghi, L., Rasti, B., and Manap, M.Y.A. 2012. *Weissella* sp. *Taj-Apis*, a novel lactic acid bacterium isolated from honey. *Journal of Food, Agriculture and Environment* 10: 263–267.
Takeuchi, A., Shimizu-Ibuka, A., Nishiyama, Y., Mura, K., Okada, S., Tokue, C., and Arai, S. 2006. Purification and characterization of an α-amylase of *Pichia burtonii*, isolated from traditional starter, "Murcha" in Nepal. *Bioscience, Biotechnology, and Biochemistry* 70: 3019–3024.
Tallgren, A.H., Airaksinen, U., von Weissenberg, R., Ojamo, H., Kuusisto, J., and Leisola, M. 1999. Exopolysaccharide-producing bacteria from sugar beets. *Applied and Environmental Microbiology* 65: 862–864.
Tamang, J.P. 2003. Native microorganisms in the fermentation of kinema. *Indian Journal of Microbiology* 43: 127–130.
Tamang, J.P. 2012. Plant-based fermented foods and beverages of Asia. In *Handbook of Plant-Based Fermented Food and Beverage Technology*, 2nd ed., eds. Y.H. Hui, and E.Ö. Evranuz, pp. 49–90. Boca Raton, FL: CRC Press.
Tamang, J.P., Tamang, N., Thapa, S., Dewan, S., Tamang, B., Yonzan, H., Rai, A.K., Chettri, R., Chakrabarty, J., and Kharel, N. 2012. Microorganisms and nutritional value of ethnic fermented foods and alcoholic beverages of North East India. *Indian Journal of Traditional Knowledge* 11: 7–25.

Thapa, N., Pal, J., and Tamang, J.P. 2006. Phenotypic identification and technological properties of lactic acid bacteria isolated from traditionally processed fish products of the Eastern Himalayas. *International Journal of Food Microbiology* 107: 33–38.

Thokchom, S., and Joshi, S.R. 2012. Microbial and chemical changes during preparation in the traditionally fermented soybean product *Tungrymbai* of ethnic tribes of Meghalaya. *Indian Journal of Traditional Knowledge* 11: 139–142.

Thorsen, L., Abdelgadir, W.S., Ronsbo, M.H., Abban, S., Hamad, S.H., Nielsen, D.S., and Jakobsen, M. 2011. Identification and safety evaluation of *Bacillus* species occurring in high numbers during spontaneous fermentations to produce gergoush, a traditional Sudanese bread snack. *International Journal of Food Microbiology* 146: 244–252.

Tobisch, S., Zühlke, D., Bernhardt, J., Stülke, J., and Hecker, M. 1999. Role of CcpA in regulation of the central pathways of carbon catabolism in *Bacillus subtilis*. *Journal of Bacteriology* 181: 6996–7004.

Towner, K.J., and Cockayne, A. 1993. *Molecular Methods for Microbial Identification and Typing*. London: Chapman & Hall.

Tsuyoshi, N., Fudou, R., Yamanaka, S., Kozaki, M., Tamang, N., Thapa, S., and Tamang, J.P. 2005. Identification of yeast strains isolated from marcha in Sikkim, a microbial starter for amylolytic fermentation. *International Journal of Food Microbiology* 99: 135–146.

Uaboi-Egbenni, P.O., Okolie, P.N., Sobande, A.O., Alao, O., Teniola, O., and Bessong, P.O. 2009. Identification of subdominant lactic acid bacteria in dawadawa (a soup condiment) and their evolution during laboratory-scale fermentation of *Parkia biglobosa* (African locust beans). *African Journal of Biotechnology* 8: 7241–7248.

Van den Berghe, E., De Winter, T., and De Vuyst, L. 2006. Enterocin A production by *Enterococcus faecium* FAIR-E 406 is characterised by a temperature- and pH-dependent switch-off mechanism when growth is limited due to nutrient depletion. *International Journal of Food Microbiology* 107: 159–170.

Van der Aa-Kühle, A., Skovgaard, K., and Jespersen, L. 2005. *In vitro* screening of probiotic properties of *Saccharomyces cerevisiae* var. *boulardii* and food borne *Saccharomyces cerevisiae* strains. *International Journal of Food Microbiology* 101: 29–39.

Vieira-Dalode, G., Jespersen, L., Hounhouligan, J., Moller, P.L., Nago, C.M., and Jakobsen, M. 2007. Lactic acid bacteria and yeasts associated with *gowe* production from sorghum in Benin. *Journal of Applied Microbiology* 103: 342–349.

Vieira-Dalode, G., Madode, Y.E., Hounhouigan, J., Jespersen, L., and Jakobsen, M. 2008. Use of starter cultures of lactic acid bacteria and yeasts as inoculum enrichment for the production of *gowe*, a sour beverage from Benin. *African Journal of Microbiology Research* 2: 179–186.

Villar, M., De Ruiz Holgado, A., Jorge, J., Sanchez, J.J., Trucco, R.E., and Oliver, G. 1985. Isolation and characterization of *Pediococcus halophilus* from salted anchovies (*Engraulis anchoita*). *Applied and Environmental Microbiology* 49: 664–666.

Virgili, R., Simoncini, N., Toscani, T., Leggieri, M.C., Formenti, S., and Battilani, P. 2012. Biocontrol of *Penicillium nordicum* growth and ochratoxin A production by native yeasts of dry cured ham. *Toxins* 4: 68–82.

von Stetten, F., Mayr, R., and Scherer, S. 1999. Climatic influence on mesophilic *Bacillus cereus* and psychrotolerant *Bacillus weihenstephanensis* populations in tropical, temperate and alpine soil. *Environmental Microbiology* 1: 503–515.

Wah, T.T., Walaisri, S., Assavanig, A., Niamsiri, N., and Lertsiri, S. 2013. Co-culturing of *Pichia guilliermondii* enhanced volatile flavor compound formation by *Zygosaccharomyces rouxii* in the model system of Thai soy sauce fermentation. *International Journal of Food Microbiology* 160: 282–289.

Walter, J., Hertel, C., Tannock, G.W., Lis, C.M., Munro, K., and Hammes, W.P. 2001. Detection of *Lactobacillus*, *Pediococcus*, *Leuconostoc*, and *Weissella* species in human feces by using group-specific PCR primers and denaturing gradient gel electrophoresis. *Applied and Environmental Microbiology* 67: 2578–2585.

Wang, J., and Fung, D.Y. 1996. Alkaline-fermented foods: A review with emphasis on pidan fermentation. *Critical Reviews in Microbiology* 22: 101–138.

Wang, L.J., Cheng, Y.Q., Yin, L.J., Bhandari, B., Saito, M., and Li, L.T. 2010. Changes during processing and sodium chloride supplementation on the physical and chemical properties of Douchi. *International Journal of Food Properties* 13: 131–141.

White, D., Sharp, R.J., and Priest, F.G. 1993. A polyphasic taxonomic study of thermophilic bacilli from a wide geographical area. *Antonie van Leeuwenhoek* 64: 357–386.

Wisotzkey, J.D., Jurtshuk Jr., P., Fox, G.E., Deinhard, G., and Poralla, K. 1992. Comparative sequence analyses on the 16S rRNA (rDNA) of *Bacillus acidocaldarius*, *Bacillus acidoterrestris*, and *Bacillus cycloheptanicus* and proposal for creation of a new genus, *Alicyclobacillus* gen. nov. *International Journal of Systematic Bacteriology* 42: 263–269.

Woese, C. 1987. Bacterial evolution. *Microbiological Reviews* 51: 221–271.

Xu, X., Liu, H., and Zhou, Y. 2012. Study on the Meitauza production from Okara by *Actinomucor elegans* and *Zymomonas mobilis*. In *Information Technology and Agricultural Engineering*, eds. E. Zhu, and S. Sambath, pp. 329–336. Berlin, Germany: Springer.

Yamada, S., Ohashi, E., Agata, N., and Venkateswaran, K. 1999. Cloning and nucleotide sequence analysis of gyrB of *Bacillus cereus*, *B. thuringiensis*, *B. mycoides*, and *B. anthracis* and their application to the detection of *B. cereus* in rice. *Applied and Environmental Microbiology* 65: 1483–1490.

Yi, J., Wu, H.Y., Wu, J., Deng, C.Y., Zheng, R., and Chao, Z. 2012. Molecular phylogenetic diversity of *Bacillus* community and its temporal-spatial distribution during the swine manure of composting. *Applied and Environmental Microbiology* 93: 411–421.

Yoo, J.Y., and Kim, H.G. 1998. Characteristics of traditional mejus of nationwide collection. *Journal of the Korean Society of Food Science and Nutrition* 27: 259–267.

Zhang, J.H., Tatsumi, E., Fan, J.F., and Li, L.T. 2007. Chemical components of *Aspergillus*-type Douchi, a Chinese traditional fermented soybean product, change during the fermentation process. *International Journal of Food Science and Technology* 42: 263–268.

Zhou, J., Ma, Q., Yi, H., Wang, L., Song, H., and Yuan, Y.J. 2011. Metabolome profiling reveals metabolic cooperation between *Bacillus megaterium* and *Ketogulonicigenium vulgare* during induced swarm motility. *Applied and Environmental Microbiology* 77: 7023–7030.

Zhu, Y.P., Fan, J.F., Cheng, Y.Q., and Li, L.T. 2008. Improvement of the antioxidant activity of Chinese traditional fermented okara (Meitauza) using *Bacillus subtilis* B2. *Food Control* 19: 654–661.

Zvyagilskaya, R., Andreishcheva, E., Soares, M.I.M., Khozin, I., Berhe, A., and Persson, B.L. 2001. Isolation and characterization of a novel leaf-inhabiting osmo-, salt-, and alkali-tolerant *Yarrowia lipolytica* yeast strain. *Journal of Basic Microbiology* 41: 289–303.

5
Quality Aspects of Alkaline-Fermented Foods

As with any food preparation, there are certain aspects of the fermentation process that can change the substrates to products with better sensory characteristics, nutritional value, and health-promoting effects. The quality of an alkaline-fermented food is of utmost importance for consumer acceptance as well as from the marketing point of view.

5.1 Sensory Characteristics

Paulin Azokpota

In general, aroma compounds are the major components of sensorial attributes of alkaline-fermented foods (AFFs). During alkaline fermentation, free amino and fatty acids resulting from microbial metabolism are the main precursors of the development of many volatile aroma compounds that contribute to odor, flavor, and texture of the fermented products (Owens et al., 1997; Azokpota et al., 2008, 2010).

Many works reported that the formation of odor, flavor, and texture in some traditional fermented African and Asian food condiments was related to the presence of some microorganisms involved in the condiments during the fermentation process. Several fermented products rely on the participation of various *Bacillus* spp., including *Bacillus subtilis* (Owens et al., 1997; Leejeerajumnean et al., 2001). Often, the finished products are of a very local character and exhibit sensory properties resulting from unique flora and processing technologies applied in small-scale, home-based fermentations. It has been reported that patented processes demonstrated the ability to produce a fermented flavoring composition with the use of *Bacillus* spp. (Leejeerajumnean et al., 2001; Beaumont, 2002). Hydrolyzed protein obtained after fermentation with *Bacillus* spp. is mixed with reactive flavor precursors, which are subsequently heated to induce flavor formation and can be dried to a powder format. The product of this patented process imparts a basic meaty flavor, with a reduced yet characteristic dawadawa-like aroma (Beaumont, 2002).

Fermentation with *B. subtilis* can produce very characteristic aromas in fermented products such as natto and dawadawa. Moreover, the hydrolytic capabilities of these microorganisms can result in a precursor-rich environment, which is useful for subsequent reactions leading to flavor production (Leejeerajumnean et al., 2001).

The controlled fermentation of African locust bean (*Parkia biglobosa* (Jacq.) G.Don) or soybean led particularly to an enormous increase in the concentration of volatile compounds due to the metabolic activities of the bacteria involved, as for the fermentations pure cultures of microorganisms are used (Owens et al., 1997; Leejeerajumnean et al., 2001; Ouoba et al., 2005). The *B. subtilis* strains were found to produce an array of hydrolases, including several proteases and lipases (Aderibigbe and Odunfa, 1990; Azokpota et al., 2006b). These hydrolases are particularly more active during controlled fermentation and contribute significantly to the flavor- and aroma-generating reactions (Tamang, 1998; Lasekan et al., 2001).

The widespread use of indigenous alkaline condiments in many African and Asian countries is due to their pleasant taste, characteristic odor, and important nutritional qualities (Dirar, 1994; Owens et al., 1997; Ouoba et al., 2005; Azokpota et al., 2008). Flavor of condiments resulting from alkaline fermentation has mainly been attributed to various volatile compounds produced through the microbial metabolic activities during the fermentation process (Leejeerajumnean et al., 2001; Beaumont, 2002; Ouoba et al., 2005; Azokpota et al., 2008).

Soluble nitrogenous constituents have been identified as one of the most important factors in the quality and taste of shoyu, a Japanese vegetable protein sauce from soybean (Steinkraus et al., 1965; Sanni and Ogbonna, 1991). The presence of glutamic acid is particularly important; the salt of this amino acid, monosodium glutamate, is used universally as an additive to enhance the flavor of soups. The sodium chloride traditionally added to iru or dawadawa may therefore, in addition to its preservative action, combine with this liberated glutamic acid to enhance the flavoring potential (Sundhagul, 1972; Odunfa and Oyeyiola, 1985; Antai and Ibrahim, 1986; Ohta, 1986; Sarkar et al., 1997a). The increase in glutamic acid in the AFFs may be explained in part by the ability of the fermenting microorganisms, notably *B. subtilis* and *Bacillus pumilus*, to accumulate glutamic acid extracellularly (Dulaney, 1967).

In Japan, the production of natto is done with selected pure cultures of *B. subtilis* (Ohta, 1986). The *Bacillus* spp. are strongly proteolytic, and the main biochemical changes in the beans are the hydrolysis of proteins and the metabolism of the resultant amino acids. This metabolism leads to the increase in pH values from 6.5–7.0 to 7.5–8.5 and often the development of a strong ammoniacal odor (Allagheny et al., 1996). Some hydrolyze fats and increase free fatty acid contents (Allagheny et al., 1996; Sarkar et al., 1996).

A variety of volatile aroma compounds have been identified in AFFs in which alkylpyrazines are the major compounds (Piggott and Paterson, 1994). The formation of these compounds is generally associated with heating. However, their presence in natto has attracted particular attention because they are formed in the absence of heating. Kosuge et al. (1961) were the first to identify tetramethylpyrazine in natto and cultures of *B. subtilis* var. *natto* and suggested that it had an aroma very similar to natto. Maga and Katz (1982) identified nine pyrazine compounds in laboratory-made and commercial samples of natto. The major volatile compounds in natto and thua nao have been reported to be ketones, acids, and pyrazine constituents. Free fatty acids are known to contribute positively to the production of characteristic flavor in these condiments; however, high levels of free fatty acids may easily cause rancidity (Maga and Katz, 1982; Leejeerajumnean et al., 2001). The compounds present at the highest concentrations were tetramethylpyrazine, trimethylpyrazine, and 2,5-dimethylpyrazine. Owens et al. (1997) identified a wide range of pyrazines in *B. subtilis*-fermented soybeans, including the methyl, trimethyl, tetramethyl, 2-ethyl-6-methyl, 3,5-dimethyl-2-ethyl, and 3,6-dimethyl-2-ethyl derivates as well as the parent compound.

A high production of acetoin, 2,5-dimethylpyrazine, trimethylpyrazine, 3-hydroxybutane, 2-methylbutanoic acid, pyrazines, dimethyldisulfide, and 2-pentylfuran by *B. subtilis* during the fermentation of soybeans to produce dawadawa has been reported (Owens et al., 1997; Leejeerajumnean et al., 2001; Dakwa et al., 2005). Some of these compounds have been identified in fermented soybeans after a pretreatment by either boiling or roasting before fermentation, particularly 3-methyl butan-1-ol, 2-methyl-1-propanol, benzaldehyde, 5-methyl-2-phenyl-2-hexenal, 3-methylbutyl pentanoate, hexadecanoic acid, trimethylpyrazine, and tetracosane, which have been found in soy-dawadawa (Owens et al., 1997; Leejeerajumnean et al., 2001).

During the fermentation of locust bean the amino acid content of dawadawa increases, in particular glutamic acid, which contributes to flavor enhancement of the final product as well as peptides (Beaumont, 2002). In afitin, iru or soumbala, and sonru, a lot of chemical groups of volatile compounds were found from which six major groups were distinguished: pyrazines, aldehydes, ketones, alkanes, esters, and alcohols (Ouoba et al., 2005; Azokpota et al., 2008, 2010). From these groups, compounds such as tetramethylpyrazine, trimethylpyrazine, 2,5-dimethylpyrazine, 3-methyl-butanol, 2-decanone, 3-hydroxy-2-butanone, heptadecane, linoleic acid ethyl ester, and 3,5-dimethyl benzene methanol were particularly found in higher concentrations. Afitin was characterized by high concentrations of 2-decanone and 3,5-dimethyl benzene methanol, whereas 3-methyl-butanol, tetramethylpyrazine, and 2,5-dimethylpyrazine were found specifically in higher concentration in both iru and sonru. The concentrations of heptadecane and linoleic acid ethyl esters were higher in sonru (Azokpota et al., 2008, 2010), whereas amines and pyridines dominated in soumbala (Ouoba et al., 2005).

In the condiments obtained from the controlled fermentation using pure cultures of *B. subtilis*, about 100 volatile compounds have been found, and more than 50 compounds were identified in relatively high concentrations and subdivided into seven main groups with the predominance of four: pyrazines, aldehydes, ketones, and alcohols (Azokpota et al., 2010). Compared with the spontaneously fermented condiments, volatile compounds identified in controlled fermented condiments have been found in high number and in concentrations that varied according to the strains of *B. subtilis* used. The condiments produced with starter cultures scored significantly higher for odor than the spontaneously fermented condiments. However, the overall acceptability of condiments was similar (Azokpota et al., 2010). It has been reported that aroma volatiles of fermented locust bean vary qualitatively and quantitatively according to the species and strain of *Bacillus* involved in the fermentation (Ouoba et al., 2005). The heat treatment during the production contributes to the flavor by formation of pyrazines originating from reactions between sugars and amino acids (Owens et al., 1997; Ouoba et al., 2005).

Important features of alkaline fermentations are changes that occur in the texture of the products during fermentation as the seeds soften. Njoku and Okemadu (1989) demonstrated the ability of *B. subtilis* to soften the tissue of African oil bean (*Pentaclethra macrophylla* Benth.) leading to the desired texture of ugba and suggested that the strain may possess pectic and proteolytic enzymes that readily hydrolyzed the pectin and protein components of the beans. They further suggested that the color of ugba developed during cooking due to non-enzymatic browning reactions involving the amino groups of amino acids and the non-reducing sugar constituents of the oil bean and microbial activity. Various alkaline fermentations involving *B. subtilis* as the main microorganism for the production of ugba, natto, kinema, soumbala, afitin, iru, and sonru are characterized by the extensive production of whitish mucilaginous polymers. In the case of ugba the polymers are assumed to be polysaccharides, which lead to sticking together of the slices (Njoku and Okemadu, 1989). For natto, the mucilage observed is composed of 44% of carbohydrates and 54% of proteins. A major part of the protein content is constituted of glutamic acid, as poly-γ-glutamic acid (Ohta, 1986).

Bacillus subtilis-fermented products exhibit a range of aromas. The *B. subtilis*-fermented product of roasted and autoclaved soybean cotyledons had an aroma somewhat similar to a sample of frozen commercial natto, but commercial dawadawa had a distinctly different aroma having some characteristics of southeast Asian fish sauces. Another fermented soybean product, thua nao, is also described as a fish sauce substitute (Sundhagul, 1972; Beaumont, 2002) and has an aroma with fish sauce-like characteristics. Hence, considerable interest attaches to how this range of different aromas is obtained from similar substrates using bacteria that so far have been regarded as similar.

Lipids are also considered as flavor or aroma generator during the alkaline fermentations. The beneficial effects of lipase in the development of characteristic flavor and aroma during the fermentation of African beans or soybeans to produce AAFs have been reported (Whitaker, 1978; Ouoba et al., 2005). During the fermentation extensive hydrolysis of lipids has been reported. Odunfa (1983b, 1985b) reported low levels of lipase activity in locust bean during dawadawa production and in melon seed fermentation. Similarly, Njoku and Okemadu (1989) observed minimal participation of lipase in African oil bean during ugba production. The significant lipolysis of legumes yields predominantly oleic, linoleic, and linolenic acids (Achinewhu and Ryley, 1986; Nout and Rombouts, 1990; Sarkar et al., 1996; Ouoba et al., 2003a). These unsaturated fatty acids are associated with nonspecific antitryptic activity (Winarno and Reddy, 1986). However, extensive hydrolysis could diminish nutritional quality. Limited lipase activity in some fermented foods has been considered desirable because of problems of objectionable taste and development of rancidity (Yong and Wood, 1977; Odunfa, 1983b, 1985b). An interesting area would involve the effects of change of lipid quantity and quality on the organoleptic characteristics of fermented condiments. Darweesh et al. (1991) have reported that aldehydes resulting from oxidation of lipids cause off-flavor and odor problems. However, organic solvent treatment of fermented condiments from locust bean, melon seed, and soybean on the objectionable odor, did not show any appreciation of the odor (Arogba et al., 1995). Not much can be said about the flavoring components of fermented vegetable proteins used as condiments as no research has been carried out on this aspect.

The consumers and producers have different perception and criteria for the judgment of the quality of AFFs: for example, kawal is a food product made from solid-state fermentation of fresh leaves of the legume called *Senna obtusifolia* (L.) H.S.Irwin & Barneby. It is a black product with a biting, peppery flavor. After 15 days of fermentation, a yellow liquid accumulates on the surface that is considered as nutritious as meat.

Pidan with a mud-coating can be examined by shaking and casting methods. In the shaking method, pidan is shaken near the ear. If any pidan has a watery sound, it is of bad quality (Wang and Fung, 1996). The casting method involves throwing a pidan into air by hand. When the pidan drops down with an elastic hand-feeling it is considered as good quality. Flash-candling is also good for examining the quality of pidan that is not mud-coated. Good-quality pidan, when candled, has a small air cell and penetrating area. The albumen is dark red in color and the yolk is intact. A standard for identifying good-quality pidan in China is available (Wang and Fung, 1996).

A unique odor, flavor, and notably stringy mucous material on the surface of soybeans are the features of ripe natto. According to Ohta (1986), whole soybeans with smooth seed coat and uniform size are preferred for natto preparation. String formation during natto production is characteristic of high-quality product. The string material is a polypeptide of glutamic acid and fructan, produced by *B. subtilis* var. *natto* (Ohta, 1986). The best-quality natto can be obtained by incubating the inoculated beans at 85% relative humidity for 12–16 h (Ohta, 1986). *Bacillus subtilis* var. *natto* was first isolated by Sawamura in 1905 (Kanno et al., 1984) and was further studied by Kato (1985) and Sulistyo et al. (1988) who showed that *B. subtilis* var. *natto* produced better-quality natto with respect to string formation, aroma, and flavor than *B. subtilis*.

Thua nao is a fermented food product made from whole soybeans. It is a viscous product of colorless material and is accompanied by a pungent ammonia odor (Wang and Fung, 1996). Traditionally, the spontaneously produced afitin, iru, and sonru are appreciated and evaluated by the consumers primarily on the basis of odor and secondly for their flavor and texture (Azokpota et al., 2008). Moreover, according to the producers the odor is the major commercial factor of acceptability of the condiments so that a condiment with a strong or inadequate odor is systematically rejected (Gutierrez et al., 2000; Azokpota, 2005). Due to their long fermentation time, iru and sonru are very soft products as they are preferred by some consumers, whereas afitin is preferred with cotyledons having less soft texture (Azokpota et al., 2006a).

5.2 Nutritional Value

Shawn Mark Somerset

AFFs comprise a diverse group of region-specific food products that, although widely consumed in parts of Asia and Africa, are not generally recognized outside of their regions of traditional production and consumption. This apparent obscurity should not, however, undermine the significant role these foods play in the nutritional intakes of communities in these regions. The fermentation process transforms somewhat mundane, unpalatable, potentially toxic, and difficult to digest plant foods into culinary diverse and desirable foods that deliver essential nutrients (particularly protein, amino acids, and minerals) within otherwise marginal diets. Further, foods such as African locust bean (*Parkia biglobosa* (Jacq.) G.Don) dawadawa are made during the dry season in parts of Africa (Wang and Fung, 1996), maintaining some level of protein intake during periods of potential food insecurity. Consequently, AFFs play an important role in achieving dietary sufficiency in a range of traditional settings. From a dietary intake perspective, this fermentation process has a range of distinct advantages common to all forms of fermentation: preservation against microbiological spoilage, enhanced depth and diversity of flavor, enhanced dietary versatility and palatability, altered nutritional value, and decreased toxicity (Wang and Fung, 1996), all of which have dietary significance (Table 5.1).

AFFs are characterized by the breakdown, during fermentation (either bacterial or enzymatic), of protein resulting in the release of ammonia (i.e., proteolytic alkaline fermentation), thus raising the pH of the food to alkaline levels (Parkouda et al., 2009). Two features of

Table 5.1 Key Nutritional and Dietary Features of Alkaline Fermentation

Major increase in free amino acid content
Major breakdown of complex carbohydrate
Minor lipolysis resulting in higher content of free essential fatty acids
Increase in riboflavin content
Increase in niacin content
Increase in vitamin B_{12} content
Improved vitamin K status
Enhanced Ca bioavailability through γ-polyglutaminic acid production
Reduction in anti-nutrient and toxin content
Reduced allergenicity
Improved preservation (shelf life)
Enhanced organoleptic qualities
Enhanced culinary and dietary diversity
Reduced reliance on meat as protein source
Strengthened regional food cultures
Insulation against seasonal variations in food security

this process are particularly significant in terms of nutrition and health impacts of these foods. First, the major bacterial enzymes produced during fermentation result in the hydrolysis of macronutrient molecules (proteins, complex carbohydrates, and triglycerides), enhancing digestibility and nutrient bioavailability of the fermented product compared with the raw food substrate (Wang and Fung, 1996). Early compositional studies on natto, kinema, dawadawa, and other AFFs highlighted their protein and mineral contents (Reddy et al., 1982) and thus the potential importance of these foods in traditional African and Asian diets. *Bacillus subtilis* is the most common microbial species used for AFF production. The effectiveness of *B. subtilis* fermentation on soybean proteins and polysaccharides is amply demonstrated by the production of low-molecular-weight, water-soluble cleavage products that then require only minimal further degradation by gastrointestinal enzymes (Kiers et al., 2000). Second, enzymatic breakdown also significantly reduces levels of naturally occurring toxins, allergens, and anti-nutrients present in the non-fermented food substrate, transforming inedible silage into a palatable, nutritious, and highly valued cultural delicacy. In the African cultural context, there are often several different indigenous fermented foods available within each region (Iwuoha and Eke, 1996), thus contributing to a complex traditional dietary diversity, which has important food security and health implications (Parkouda et al., 2009).

Soybean is by far the most common food substrate for AFFs, and is the basis for products such as dawadawa (Western Africa), natto (Japan), kinema (India, Nepal), and thua nao (Thailand). Raw soybean is a highly nutritious plant food with high levels of protein and fat (Bouchenak and Lamri-Senhadji, 2013). However, raw soybeans are difficult to digest due to high contents of complex oligosaccharides and anti-nutrients such as phytates and tannins (Reddy and Pierson, 1994), as well as potential allergens (Phromraksa et al., 2008), and thus require some form of processing to liberate nutrients for human digestion and absorption.

Of the AFFs produced using soybean, natto and kinema have the most extensive range of high-quality compositional data available to date. Natto is the only AFF produced on a significant industrial scale, although a range of studies to improve the consistency, shelf life, and nutritional quality of dawadawa have been published toward the development of commercial scale production. Research work toward a standardized dawadawa product conducive to industrial production has focused on bacterial strain selection (Terlabie et al., 2006) and identification of optimal condition for fermentation (Dakwa et al., 2005).

Parallels can be seen between AFFs and foods such as European cheeses, which started as highly region-specific artisanal products with highly variable (and highly sought after) organoleptic and nutrition attributes, but through industrialization developed into a highly consistent mass-market product with substantial international market reach (Tregeara et al., 2007).

5.2.1 Asian Foods

Alkaline-fermented plant foods of Asia almost exclusively use soybean as a substrate, reflecting the importance and agricultural suitability of this crop across the Asian region. Fish and other seafood are also widespread as fermentation substrates for the production of sauce and paste condiments across a range of traditional Asian cultures (Van Veen and Steinkraus, 1970). Pidan is somewhat unique among AFFs, using duck eggs as a substrate. Pidan is produced in some parts of China and also stands apart from other AFFs, since it is produced through chemical rather than bacterial fermentation (Wang and Fung, 1996). Of the AFFs that rely on bacterial fermentation, the soy-based natto and kinema have been most extensively studied for their composition, reflecting the importance of these foods in terms of total consumption.

Attention to compositional features of natto has intensified in response to epidemiological studies in the Japanese population. Higher consumption of this Japanese traditional food is positively associated with bone density in Japanese females (Katsuyama et al., 2002). The proposed mechanism involves increases in serum vitamin K and γ-carboxylated osteocalcin, both of which facilitate bone mineralization (Tsukamoto et al., 2000). However, it is unclear whether this increase in vitamin K status is due to direct intake or enhancement of production by colonic microbiota. The effects of natto consumption on gut microbiota have also been investigated in a bovine model (Sun et al., 2012), with measurable effects on milk production and the bacterial species profiles of the rumen. Notwithstanding the substantial potential for probiotic and prebiotic effects in humans, no studies on effects by natto consumption on the human microbiota have been reported to date. In relation to water-soluble vitamins, fermentation can result in a 3-fold increase in thiamine and riboflavin contents and a 5-fold increase in vitamin B_{12} content, compared with the raw substrate (Reddy et al., 1982).

Kinema is a traditional soybean-based food produced in Nepal and parts of India with significant dietary importance as a meat substitute. In terms of nutrient composition, kinema has been extensively studied (Sarkar and Tamang, 1995; Sarkar et al., 1997a, 1997b, 1998). Free amino acid content increases 60-fold compared with the original substrate. In the context of its dietary role as a meat substitute, tryptophan, cyst(e)ine, and methionine are the major limiting amino acids, although protein quality is comparable to animal sources such as milk and egg protein (Sarkar et al., 1997b). Although vitamin B_{12} levels have not been reported, changes in the content of other B vitamins in kinema vary from the original soybean substrate, with increases in riboflavin and niacin, but a decrease in thiamin levels (Sarkar et al., 1998).

The Thai soybean-based thua nao is used as a substitute for fermented fish (Wang and Fung, 1996). As well as the fresh fermented product, which lasts several days, thua nao can be dried to extend its shelf-life substantially (Parkouda et al., 2009), thus enhancing protein food security. Dajanta et al. (2011b) found that thua nao production increases the content of free amino acids, reflecting a strong hydrolytic enzyme activity. The content of essential amino acids (EAAs) varies according to the microbiological starter culture, with pure strains of *Bacillus* spp. producing higher levels of EAAs than found in domestically fermented products.

Doenjang, a fermented soybean paste, is an important protein source and flavoring agent in traditional Korean food culture. Although fermentation for 80 days leads to a relative increase in threonine and methionine compared with raw substrate, these amino acids remain, limiting and complementation with cereal foods is a viable dietary strategy to assure adequate EAA intake (Namgung et al., 2010). Fermentation has also been shown to enhance mineral and fatty acid content, with the major fatty acids being linoleic acid, oleic acid, and α-linolenic acid (Jung et al., 2006).

Douchi is a Chinese fermented soy product with a long history of use as both a condiment and a traditional medicine. Li et al. (2007) found that lipid and amino acid contents of douchi remain similar to the raw substrate, although protein was significantly solubilized. As is the case for soybean in general, threonine is the limiting amino acid. The predominant fatty acids in douchi are palmitic, oleic, linoleic, and α-linolenic acids.

There is a range of fermented fish and other seafood products consumed throughout traditional food cultures in Asia. These foods generally have high levels of protein with high biological value, but are often consumed more as condiments than staples because of their high salt contents (15%–25%), although variations with lower (10%–12%) contents occur and may be consumed in higher amounts (Van Veen and Steinkraus, 1970). Specific compositional studies have been conducted only for a small number of these foods, and center on amino acid profiles. For example, a study by Peralta et al. (2008) found a fermented fish paste from the Philippines had higher levels of the free amino acids taurine, glycine, alanine, leucine, lysine, and arginine, and lower levels of glutamic acid than the original substrate.

5.2.2 African Foods

African alkaline-fermented plant foods are used either as meat substitutes or as condiments to enhance the organoleptic qualities of cereal staple dishes. The specific use seems to reflect the production quantities of the various substrates. Dawadawa is a Nigerian condiment that is used to flavor soups and stews. The nutritional composition of dawadawa varies since it can be produced from a range of substrates including soybean, locust bean, and bambara groundnut (*Vigna subterranea* (L.) Verdc.) (Achi, 2005a; Ijarotimi and Esho, 2009). Soy dawadawa has significantly higher content of moisture and protein compared with the original substrate (Terlabie et al., 2006), with a proximate distribution for protein, fat, and carbohydrate of 40%, 35%, and 15%, respectively (Reddy et al., 1982). Dakwa et al. (2005) found that fermentation led to an increase in the contents of protein and the amino acids isoleucine, leucine, lysine, phenylalanine, arginine, and proline, and a decrease in fat and ash. Protein, fat, and carbohydrate hydrolyses are the main biological changes (Terlabie et al., 2006).

The locust bean is inedible, and in fact toxic, to humans in its raw state (Wang and Fung, 1996). However, alkaline fermentation results in a significant reduction in anti-nutrient content and enhanced amino acid profile of the bean flour (Ijarotimi and Keshinro, 2012), and this has generated interest in transitioning dawadawa to an industrialized food (Steinkraus, 1996b; Odibo et al., 2008). Sulfur-containing amino acids are limiting in this type of dawadawa (Campbell-Platt, 1980), as they are in ogiri, which is an AFF made from melon (*Citrullus lanatus* (Thunb.) Matsum. & Nakai) seeds (Wang and Fung, 1996). However, these amino acids are complemented by consumption with the staples sorghum and millet (Wang and Fung, 1996). In addition, fermentation results in a 3-fold increase in riboflavin content, a 6-fold increase in thiamine content, but a 30% reduction in niacin content, compared with the raw substrate (Reddy et al., 1982). The organoleptic enhancement of locust bean during fermentation no doubt is an incentive to add this condiment to these grain staples, although excessive fermentation may lead to overproduction of ammonia-like odor, which is a potential barrier to consumption (Ouoba et al., 2005).

Ugba (ukana), a Nigerian AFF, uses African oil bean (*Pentaclethra macrophylla* Benth.) seeds as a substrate and is added to soup, tapioca, garden egg leaves, yam, and bitter cola (Ogueke et al., 2010). Although nutrition quality is highly variable (Isu and Ofuya, 2000), ugba is recognized as an inexpensive protein source produced largely through an unregulated cottage industry. In general, vitamin and mineral (except Zn) contents decrease during fermentation although, as is characteristic of most AFFs, the digestibility and biological value of protein are enhanced through fermentation. Lipase production in AFFs is comparatively low. However, ugba production has been shown to enhance the content of free essential fatty acids (linoleic and linoleic acids) (Ouoba et al., 2003a). In ugba, lipase activity results in an increase in the ω-6-linoleic acid and a decrease in the monounsaturated oleic acid. Niacin and riboflavin contents decrease, corresponding to a progressive reduction in cyanide, oxalate, phytate, and tannin (Ogueke et al., 2010). Ugba is consumed either as a snack or sidedish, or soup condiment depending on duration of fermentation. This versatility has also generated studies to develop more controlled production processes to increase market availability (Isu and Ofuya, 2000).

Ogiri-okpei is produced in West Africa by fermentation of seeds from the tree *Prosopis africana* (Guill. & Perr.) Taub. This AFF is comparable to dawadawa and is used as a condiment. Fermentation leads to an increase in soluble sugars and free amino acids and significant increases in calcium, phosphorus, potassium, and zinc (Odibo et al., 2008). The fatty acid profile is also modified during fermentation, with significant decreases in saturated fat and increases in both ω-3 and ω-6 essential fatty acids and the monounsaturated oleic acid. Although there is a slight decrease in most amino acids, the lysine content of the fermented product increases significantly after 72 h of fermentation.

Cotton (*Gossypium hirsutum* L.) seed meal is used extensively in livestock production by virtue of its high protein and fat content. However, its use is limited by the presence of gossypol and other toxic components (Randel et al., 1992). Owoh is a fermented cotton seed food produced for human consumption in Nigeria. Its production highlights the important role of alkaline fermentation in removing various anti-nutrients and the toxic phenol gossypol (Wang and Fung, 1996) rendering a potentially toxic, inedible plant product into an organoleptically desirable and culturally important condiment. This toxin reduction is coupled with an increase in mineral content and extensive proteinase activity (Sanni and Ogbonna, 1992).

Another product tonou, which is prepared from locust bean, peanut, and soybean has been shown to be protein-rich (≥40%), and contain high levels of iron, magnesium, and zinc. Fermentation by *Bacillus* spp. also increases free amino acid content (Konlani et al., 1999). In an extensive review of the composition of traditional AFFs, Ferguson et al. (1993) indicated that the addition of foods such as dawadawa into the diet can enhance mineral intake in otherwise marginal diets.

In Benin, the baobab (*Adansonia digitata* L.) is an important forest food source, which is processed into a range of traditional foods (Chadare et al., 2008; Buchmann et al., 2010). These include the AFFs dikouanyouri and tayohounta. Dikouanyouri is prepared from the whole baobab fruit and is used as a condiment for cereal porridges. Fermentation leads to increases in crude lipid and ash and free amino acids from protein degradation. Tayohounta is prepared from baobab kernels and thus has slightly different flavor and nutrient characteristics. It is used as a condiment for sauces and stews, and similar to dikouanyouri, fermentation leads to increased crude lipid and free nitrogen content. Although both of these foods are consumed in relatively small amounts because of their enhanced flavor characteristics they facilitate the consumption of a range of staple foods and thus have an important role in traditional diets (Chadare et al., 2010). Chadare et al. (2009) also published a comprehensive review comparing the nutrient contents of various parts of the baobab plant (seeds, leaves, and pulp) in their raw state and have noted high levels of vitamin C, calcium, and lipids, and proteins with a chemical score of 0.81. However, they did not analyze the effects of fermentation on these nutrient contents.

The leaves of various plants are also fermented to produce food products in parts of Africa. Ntoba mbodi is produced in Congo and central Africa from cassava (*Manihot esculenta* Crantz) leaves. This food is relatively rich in protein, despite a slight reduction in free nitrogen during fermentation, most likely due to liberation of hydrogen cyanide. Methionine and phenylalanine are the limiting amino acids. Extensive cellulase activity enhances the bioavailability of carbohydrates. Fermentation also increases the contents of potassium, calcium, and magnesium compared with the raw substrate (Mokemiabeka et al., 2011). Fermentation of the leaves of sickle pod (*Senna obtusifolia* (L.) H.S.Irwin & Barneby) results in the production of kawal, a wild African plant found during the rainy season in Chad and Sudan. Kawal is used as either a meat substitute or flavoring agent. Fermentation results in a 50% protein loss; however, this is outweighed by the substantial increase in palatability and digestibility. Histidine, threonine, and cysteine are the limiting amino acids, but kawal is often combined with the complementary proteins in cereal foods (Mbaiguinam et al., 2005).

5.2.3 Other Putative Health Benefits

Various epidemiological studies have shown an association between natto consumption and bone health, implying some enhancement of nutrition quality due to fermentation. The enhanced bone health has been attributed not so much to increased calcium content per se, but to a combination of bacterial production of γ-polyglutaminic acid (GPGA), which enhances calcium bioavailability, and the production of vitamin K, which plays a role in osteoblast bone mineralization (Tsukamoto et al., 2000). Similarly, thua nao consumption has also been implicated in a potential preventive role in osteoporosis through the production of GPGA, which is known to enhance calcium absorption (Inatsu et al., 2006). During fermentation, a profibrinolytic enzyme (subtilisin NAT) is produced. This enzyme is thought to be involved in reducing thrombogenesis (Inatsu et al., 2006), potentially lowering coronary risk in humans. The potential for proteolytic alkaline fermentation in thua nao production to reduce the allergenicity of soy protein has also shown promise (Phromraksa et al., 2008). Bacterial enzymes produced during thua nao production can also reduce the allergenicity of protein foods such as milk and wheat (Phromraksa et al., 2008), indicating a potential nutritional advantage in combining thua nao consumption with these other foods.

Jung et al. (2006) found that Korean doenjang was protective in in vivo murine cancer and metastatic models. This effect was attributed to the increased contents of genestein (and possibly linoleic acid) that occur with fermentation.

The potential health benefits presented by traditional AFFs have stimulated the development of functional food analogues. For example, in Brazil a functional food based on alkaline fermentation of soymilk has been shown to inhibit tumor growth in a murine adenocarcinoma model (Kinouchi et al., 2012). In that study, nitrous oxide and tumor necrosis factor (TNF)-α production was also shown to be impaired. These are interesting preliminary findings that may ultimately be tested in humans in conjunction with traditional versions of AFFs.

To date, no human studies have been conducted on the effects of AFF consumption on variations in the gut microbiome in humans. A single study of the effects of natto on the rumen microbiota in cattle (Sun et al., 2012) indicates a potential role of AFFs as a prebiotic, although this line of research is in its very early stages of development.

5.2.4 Reliability of Alkaline-Fermented Food Compositional Data

The major nutritional shifts associated with alkaline fermentation compared with the composition of raw food substrate are summarized in Table 5.2. With the exception of natto, all AFFs are produced at the household or artisanal level rather than through widespread industrial production and distribution. As a result, the organoleptic

Table 5.2 Summary of Published Nutritional Changes in Alkaline Fermented Plant Foods Compared to Raw Substrates[a]

		AMINO ACID		FATTY ACID			MONO-	VITAMIN					MINERAL							
FOOD	SUBSTRATE	FREE	ESSENTIAL	C18:1	C18:2	C18:3	SACCHARIDE	B_1	B_2	B_3	B_{12}	K	Ca	Cu	P	Fe	Zn	Mn	Mg	Na
Natto	Soybean	+							+		+	+								
		18							18		18	27								
Kinema	Soybean	+	+				+	–	+	+			–	–	–	–	–	–	–	
		24	24				23	25	25	25			25	25	25	25	25	25	25	
Doenjang	Soybean	+	+	+	+	+														
		14	14	14	14	14														
Douchi	Soybean	+																		
		11																		
Thua nao	Soybean	+																		
		20																		
Dawadawa	Soybean	+	+					+	+	–										
		4, 26	21					21	21	21										
Dawadawa,	Locust bean	+					+		+				+				+			
iru		5					16		5				6				6			
Dawadawa	Bambara		+										+		–	+			+	–
	groundnut		8										9		9	9			9	9
Ogiri-okpei	Mesquie seed	+		+	+	+	+						+		+		+	+		
		15		15	15	15	15						15		15		15	15		
Owoh	Cotton seed	+											+	+		+	+	+	+	
		22											22	22		22	22	22	22	
Ogiri	Melon seed	+						+	+	+										
		16						2	2	2										

(continued)

Table 5.2 (Continued) Summary of Published Nutritional Changes in Alkaline Fermented Plant Foods Compared to Raw Substrates[a]

				FATTY ACID			MONO-	VITAMIN					MINERAL							
FOOD	SUBSTRATE	FREE	ESSENTIAL	C18:1	C18:2	C18:3	SACCHARIDE	B_1	B_2	B_3	B_{12}	K	Ca	Cu	P	Fe	Zn	Mn	Mg	Na
Ogiri-ugu	Fluted pumpkin	+						+	+	+			–	–	–	+	+			
	seed	7						2	2	2			1	1	1	1	1			
Ugba	African oil	+			+	+		+	–	–										
	bean	10			19	19		2	17	17										
Ntoba mbodi	Cassava leaf												+						+	
													13						13	
Kawal	Sickle pod leaf	–																		
		12																		
Fururndu	Roselle seed	+											–	–		–		–		
		28											28	28		28		28		
Dikouanyouri,	Baobab seed	+																		
tayohounta		3																		

[a] +, increase; –, decrease; 1, Achinewhu, *S.C., Food Chemistry*, 19, 105–116, 1986a; 2, Achinewhu, S.C., and Ryley, J., *Food Chemistry*, 20, 243–252, 1986; 3, Chadare, F.J. et al., *Ecology of Food and Nutrition*, 49, 279–297, 2010; 4, Dakwa, S. et al., *International Journal of Food Microbiology*, 104, 69–82, 2005; 5, Eka, O.W., *Food Chemistry*, 5, 303–308, 1980; 6, Ferguson, E.L. et al., *Journal of Food Composition and Analysis*, 6, 87–99, 1993; 7, Giami, Y.S., and Isichei, I., *Plant Foods for Human Nutrition*, 54, 67–77, 1999; 8, Ijarotimi, O.S., and Esho, T.R., *British Food Journal*, 111, 376–386, 2009; 9, Ijarotimi, O.S., and Keshinro, O.O., *Acta Scientiarum Polonorum, Technologia Alimentaria*, 11, 151–165, 2012; 10, Isu, N.R., and Ofuya, C.O., *International Journal of Food Microbiology*, 59, 235–239, 2000; 11, Li, H. et al., *Asia Pacific Journal of Clinical Nutrition*, 16, 1, 215–221, 2007; 12, Mbaiguinam, M. et al., *African Journal of Biotechnology*, 4, 1080–1083, 2005; 13, Mokemiabeka, S. et al., *Advance Journal of Food Science and Technology*, 3, 418–423, 2011; 14, Namgung, H.J. et al., *Journal of the Science of Food and Agriculture*, 90, 1926–1935, 2010; 15, Odibo, F.J.C. et al., *Journal of Industrial Microbiology and Biotechnology*, 35, 947–952, 2008; 16, Odunfa, S.A., *Plant Foods for Human Nutrition*, 32, 11–18, 1983a; 17, Ogueke, C.C. et al., *Pakistan Journal of Biological Sciences*, 13, 489–496, 2010; 18, Ohta, T., Natto, in *Legume-Based Fermented Foods*, eds. N.R. Reddy et al., pp., 85–93. Boca Raton, FL: CRC Press, 1986; 19, Ouoba, L.I.I. et al., *Journal of Applied Microbiology*, 95, 868–873, 2003a; 20, Phromraksa, P. et al., *Journal of Food Science*, 73: M189–M195, 2008; 21, Reddy, N.R. et al., *Critical Reviews of Food Science and Nutrition*, 17, 335–370, 1982; 22, Sanni, A.I., and Ogbonna, D.N., *Food Microbiology*, 9, 177–183, 1992; 23, Sarkar, P.K. et al., *Letters in Applied Microbiology*, 24, 337–339, 1997a; 24, Sarkar, P.K. et al., *Food Chemistry*, 59, 69–75, 1997b; 25, Sarkar, P.K. et al., *Journal of the Science of Food and Agriculture*, 78, 498–502, 1998; 26, Terlabie, N.N. et al., *International Journal of Food Microbiology*, 106, 145–152, 2006; 27, Tsukamoto, Y. et al., *Journal of Bone and Mineral Metabolism*, 18, 216–222, 2000; 28, Yagoub, A.E.G.A., and Mohammed, M.A., *Pakistan Journal of Nutrition*, 7, 352–358, 2008.

and compositional quality and consistency of such products are highly variable within and between the various food products.

The high prevalence of local variation in AFFs in terms of diversity of substrates and methods of production impacts upon specific knowledge about nutrient composition. First, available funding for compositional studies often is limited compared with that available for foods with broader consumption bases. Thus, compositional studies have often relied on small sample sizes analyzed in laboratories with limited analytical resources. Second, the large variations in substrate composition (Chadare et al., 2009) and fermentation procedures (Dakwa et al., 2005; Li et al., 2007; Odibo et al., 2008; Namgung et al., 2010) yield extensive compositional variation. Many compositional studies have noted changes in ash content in proximate analysis, implying change in mineral content. However, many of these studies were conducted in laboratories with limited capacity to determine variations of specific minerals within the changes in ash content. One study has speculated that the changes in ash content were due to *B. subtilis* being able to produce divalent cations (Odibo et al., 2008). However, no clear conclusions have been made. Ferguson et al. (1993) have noted that nutrient (particularly minerals) content estimates may be overestimated in some published African food composition tables. Rather than focusing on empirical values for nutrient content of AFFs, it may be more relevant to understand the transition in nutrient content as the various plant substrates are fermented, as shown in Table 5.2.

On a cautionary note, some studies indicate prolonged fermentation can actually destroy some essential amino acids (e.g., lysine) (Winarno and Reddy, 1986), so preparation techniques are integral to nutrient content. Iwuoha and Eke (1996) have identified that household-based production of indigenous fermented foods may compromise food safety through lack of hygiene, lack of cultural knowledge and know-how, and inadequate production conditions.

5.2.5 Conclusion

Dietary diversity has been identified as an important determinant of health, not only through assuring adequate intake of a wide range of nutrients and minimizing the intake of specific toxins by spreading risk across a range of foods, but also by adding culinary diversity into an otherwise mundane and unappetizing culinary landscape. The nutritional contribution of AFFs to dietary intake in Asian and African cultures centers on their role within the total diet, rather than being considered as standalone foods. In this context, AFFs add significantly to dietary diversity, nutrient bioavailability, and food security in many regions of the world with marginal nutrition intakes and fragile food security.

5.3 Health-Promoting Effects

Alkaline-fermented foods (AFFs) have increasingly gained interest in food applications because of their possible nutritional and health benefits and the potential use as food-flavoring agents. Soybeans are regarded as a healthy food in several Asian countries and are widely consumed as soymilk, tofu, and a variety of fermented products. In soybeans and soy products, several phytochemicals are found and they appear to be the active compounds causing many beneficial health effects. The present chapter summarizes various findings on the biological activities of AFFs and/or the microorganisms involved therein that can impact favorably on human health.

5.3.1 Reduction of Anti-Nutritional Factors

Paulin Azokpota

Fermented foodstuffs and condiments remain key constituents of diets throughout many parts of Asia and Africa. In cases where the process of fermentation evolved for the development of taste or aroma, it often resulted in enhanced nutrition, stabilization of the original raw materials, and detoxification of anti-nutritional factors (Okafor, 1977). Many fruits and vegetables contain naturally occurring toxins and anti-nutritional compounds. These can be removed or detoxified by the action of microorganisms during processing, including fermentation. For instance, the fermentation process that produces Sudanese kawal removes the toxins from the leaves of sickle pod (*Senna obtusifolia* (L.) H.S.Irwin & Barneby), ensuring safety of the product (Okafor, 1977; Odunfa, 1985a).

Alkaline fermentation decreases and/or eliminates the toxic components of some seeds including African locust bean (*Parkia biglobosa* (Jacq.) G.Don) seeds and soybeans (Odunfa, 1985b). For example, the toxicity of African oil bean (*Pentaclethra macrophylla* Benth.) seeds decreased in the fermented product, ugba (Odunfa, 1985b). Toxicological evaluation of ogiri (fermented castor oil bean (*Ricinus communis* L.) seeds) using chicken embryo showed that the toxicity of the beans, possibly due to toxic proteins, decreased significantly in the fermented product (Odunfa, 1985b).

In central Africa, particularly in Congo, the leaves of cassava (*Manihot esculenta* Crantz) are traditionally fermented and consumed as a delicious product known as ntoba mbodi. Although rich in proteins, minerals, and vitamins, cassava leaves contain high amount of cyanogenic glucosides, linamarin and lotaustralin (Montogomery, 1980). The hydrolysis of these glucosides by the endogenous or microbial linamarase releases toxic cyanhydric acid that causes goiter, dwarfism, and tropical ataxic neuropathy (Balagopalan et al., 1988). During the traditional fermentation of cassava leaves to kawal, the cyanogenic glucoside content is reduced significantly by 70%–75% (Kobawila et al., 2003, 2005). Fermentation of cassava leaves is mainly carried out by *Bacillus*, causing the alkaline pH. Certain strains of *Bacillus*, notably *B. pumilus*, have the ability to use cyanhydric acid for their nutrition (Knowles, 1976), thereby causing reduction of the cyanide content in leaves.

Raffinose family oligosaccharides (RFOs) are present in significant amounts in mature legume seeds (31%–76% of the total soluble sugar content) (Reddy et al., 1984). The production of flatus in monogastric animals is due to colonic fermentation of sugars that escape digestion and absorption in the stomach and small intestine. The RFOs are soluble, low-molecular weight, and consist of linear chains of galactopyranosyl residues attached to the glucose moiety of sucrose via an α-1,6-galactopyranosidic linkage. Since mucus in the small intestine is deprived of α-galactosidase capable of hydrolyzing this linkage, RFOs are not digested but accumulate in large intestine where they undergo anaerobic fermentation by bacteria, leading to the production of gases, such as carbon dioxide, hydrogen, and methane (Jood et al., 1985). Therefore, it is desirable to remove these flatulence factors before consumption. A significant decrease in the concentration of RFOs in soybeans was demonstrated at different processing steps of kinema-making. When beans were soaked, drained, cooked, and again drained, there was a 4.8-times reduction of RFOs, which were completely removed at the subsequent fermentation step carried out by *Bacillus subtilis* (Sarkar et al., 1997a).

Trypsin inhibitors (TIs), also known as serine protease inhibitors (serpins), are the largest and most diverse family of protease inhibitors. Serpins control the activation and catabolism of proteins by the inhibition of serine proteases in vivo. Soybean is one of the four natural and commercial sources of TIs, which act as competitive substrate analogues and bind with serine protease to form an inactive complex, thereby rendering the protease inactive. Soy TIs have also been found to contribute to pancreatic hypertrophy in rats. Miyagi et al. (1997) measured average TI activity in whole soybean as 4819 U/100 g, and the remaining activity in Japanese natto produced by *B. subtilis* against that of whole soybean as 0.7%. The level of TI activity reduced significantly from 2.9 mg/g unfermented locust bean to 0.32 mg/g dawadawa after 72 h of fermentation (Esenwah and Ikenebomeh, 2008).

Phytate is the principal storage form of phosphorus and is particularly abundant in legumes (Reddy et al., 1989). It chelates divalent cations such as calcium, magnesium, zinc, and iron, thereby reducing their bioavailability, because humans have a limited ability to hydrolyze phytate (Sandberg, 2002). During the fermentation of locust bean, the phytate content reduced by 62.5% from 2.4 to 0.9 mg/g dawadawa (Esenwah and Ikenebomeh, 2008). During the fermentation of African oil bean seeds, the decrease was 75.8% from 11.2 to 2.7 g/kg ugba (Kingsley, 1995).

Tannins are water-soluble phenolic compounds with a molecular weight of greater than 500 Da. They cause precipitation of proteins by binding with them, thus interfering with digestion by displaying anti-trypsin and anti-amylase activities. The tannin content reduced significantly by 58.8% (from 40.0 mg/g to 16.1 mg/g) during the production of dawadawa (Esenwah and Ikenebomeh, 2008). During the production of ugba, the tannin content in African oil bean reduced by 51.5% after cooking and further by 56.8% after fermentation (Kingsley, 1995).

In Nigeria, seeds of cotton plant (*Gossypium hirsutum* L.) are traditionally fermented to produce owoh. Cotton seeds are rich in gossypol, which is a phenolic aldehyde that permeates cells and acts as an inhibitor for several dehydrogenases. Sanni and Ogbonna (1992) found a reduction in the gossypol content from 55 μg/g unfermented cotton seeds to 15 μg/g owoh.

Hence, the traditional technologies in producing alkaline-fermented foods reduce and/or eliminate the anti-nutritional factors in legumes and non-leguminous plants as well. Strategies should be devised to reduce their levels further to enhance nutritional status.

5.3.2 Production of Vitamins

Paulin Azokpota

While the western world can afford to enrich its foods with synthetic vitamins, the developing world must rely upon biological enrichment for its vitamins (Steinkraus, 1994). Alkaline-fermented foods have been reported to be a good source of vitamin B_2 (riboflavin) and vitamin B_3 (niacin) (Sarkar et al., 1998). There is a general agreement that riboflavin content in natto is higher than that of unfermented soybeans. A significant increase in riboflavin content during kinema production is consistent with previous reports on other fermentations (Van Veen and Steinkraus, 1970). During kinema production, *Bacillus* fermentation enhanced thiamine, riboflavin, and niacin levels by 45%, 71%, and 23%, respectively. This shows that *Bacillus subtilis* has a great synthetic capacity for all these vitamins (Sarkar et al., 1998).

Achinewhu and Ryley (1986) have shown that fermentation significantly increased the content of thiamine, riboflavin, and niacin in the African oil bean (*Pentaclethra macrophylla* Benth.) during the production of ugba. Similar changes were observed during the fermentation of melon (*Citrullus lanatus* (Thunb.) Matsum. & Nakai) seeds and fluted pumpkin (*Telfairia occidentalis* Hook.f.) seeds at the time of ogiri production (Achinewhu, 1986a, 1986b). The levels of B-group vitamin, especially riboflavin, in ugba and dawadawa (iru) are much higher than in the unfermented cooked seeds (Eka, 1980; Odunfa and Oyewole, 1998).

The biosynthesis of riboflavin was studied in *B. subtilis* using classical genetics and r-DNA technology. Cloning and DNA nucleotide sequencing indicated that the enzymes required for biosynthesis of riboflavin are encoded by gene organization as a single 4.3-kb operon (Survase et al., 2006).

5.3.3 Increase in Digestibility

Paulin Azokpota

Apart from increasing shelf life, nutritive value, and flavor, and reducing the level of anti-nutritional factors, fermentation markedly improves digestibility of the raw materials (Odunfa, 1985b; Barimalaa et al., 1994; Reddy and Pierson, 1994). For example, in the recapitulative scientific studies on the fermentation of the carbohydrate-rich substrates, the following findings were noted by Steinkraus (1996a): (1) the lactic acid fermentation process has been reported to improve the in vitro protein digestibility of non-tannin cereal grains; (2) phytate was shown to be completely hydrolyzed after fermentation of germinated white sorghum, and as a result, the amount of soluble iron was found to be strongly increased; (3) protein digestibility was reported to increase from 47% to 73% after lactic acid fermentation of whole-grain sorghum; (4) over a 9-month period, consumption of acid-fermented gruels reduced the incidence of diarrheal episodes in a group of school children, because these foods can be easily digested; and (5) the growth of rats fed with fermented wheat product improved significantly over those fed with unfermented wheat, there is an increase in availability of lysine during fermentation. A feature of many of these fermentations is that the fermented foods are capable of improving the digestibility of raw materials and at the same time destroying factors that are toxic, or at least those that might inhibit digestion (Steinkraus, 1996a).

Similar findings have been reported concerning the fermentation of protein-rich substrates to produce food condiments. The extent of protein hydrolysis is one of the most important factors in the changes in texture and flavor in most fermented high-protein products (Achi, 2005b). Soluble low-molecular-weight peptides and amino acids that contribute to flavor are produced through the enzymatic breakdown of proteins (Njoku and Okemadu, 1989; Odunfa and Oyewole, 1998; Ogbonna et al., 2001; Ouoba et al., 2003a). A good-quality condiment has a characteristic strong smell of ammonia, a dark color, and is semi hard. Amino acids produced, because of protein metabolism, are responsible for the gradual pH increase toward alkaline side (Achi, 1992, 2005a; Sarkar et al., 1997b).

Ogiri, dawadawa, and kinema are deficient in sulfur-containing amino acids but rich in alanine, lysine, and glutamic acid (Odunfa, 1981, 1985a; Sarkar et al., 1997b). Soluble products increased during fermentation of melon seeds resulting in high digestibility of the fermented product. Alanine, lysine, and glutamic acid were the predominant amino acids, with arginine and proline occurring in small amounts (Odunfa, 1983a; Aidoo, 1986; Achi, 2005a, 2005b). The improved nutritive values were attributed to the increase in amino acid profiles due to fermentation. Increased levels of amino acids, except for arginine, leucine, and phenylalanine, were also reported. Similar results were reported for other seed legumes (Odunfa, 1985b; Achinewhu and Ryley, 1986; Achi, 2005a). Because of increased soluble products during fermentation, the products become highly digestible. The fermented foods also have improved digestibility, protein efficiency ratio (PER), net protein utilization, and the biological value of fermented beans is much higher than those of the cooked seeds (Fetuga et al., 1973; Umoh and Oke, 1974). The increased digestibility is in part due to extensive hydrolysis of proteins to amino acids and the hydrolysis of galacto-oligosaccharides to simpler sugars.

Classical techniques for assessment of bioavailability of nutrients generally involve in vitro tests with animals. Protein utilization measured as PER and the digestibility of soybean and some legumes are hardly improved by fermentation (Nout and Rombouts, 1990; Achi, 2005a). However, Achinewhu (1983a, 1983b) and Achi (2005b) reported that fermentation and heat treatment improved apparent digestibility, feed conversion efficiency, and PER when rats were fed on diets of African oil bean (*Pentaclethra macrophylla* Benth.) seed rations. Eka (1980) and Achi (2005a, 2005b) studied the effect of fermentation on the nutrient content of African locust bean (*Parkia biglobosa* (Jacq.) G.Don) seeds and reported that protein and fat increased when fermented, whereas the quantity of carbohydrates decreased.

It was reported that some enzymes such as amylase, galactanase, galactosidase, glucosidase, and fructofuranosidase are produced by the fermented microorganisms, which are involved in the degradation of carbohydrates (Achi, 2005a, 2005b). Microbial amylases hydrolyze carbohydrates into sugars, which are then readily digestible by humans; galactanases soften the texture of the seeds and liberate sugars for digestion (Achi, 2005a, 2005b).

Most legumes contain large amounts of nondigestible carbohydrates, which may include arabinogalactan, raffinose, stachyose, and verbascose, which are associated with abdominal distention and flatulence in humans (Irvine, 1961; Odunfa, 1985b; Achi, 2005a, 2005b). Fermentation reduced total flatus factors from 16.5 mg/g to 2.0 mg/g in soybeans and from 0.16% to less than 0.1% in groundnut (Sarkar et al., 1997a; Achi, 2005a). Trypsin inhibitor activity of bambara groundnut (*Vigna subterranea* (L.) Verdc.) was also reduced during fermentation (Barimalaa et al., 1994; Naczk et al., 1997; Achi, 2005a, 2005b).

It was shown that *Bacillus* spp. were able to degrade soybean macromolecules to a large extent resulting in water-soluble low-molecular-weight compounds (Kiers et al., 2000; Achi, 2005a, 2005b). During *Bacillus* fermentation of autoclaved soybeans for 48 h, the quantity of soluble and dialyzable matters increased from 22% and 6% up to 65% and 40%, respectively (Kiers et al., 2000). Protein and carbohydrate degradation during fermentation of soybean with several *Bacillus* spp. was investigated and appeared to be substantial during the first 18 h of fermentation, resulting in the release of high levels of peptides and oligosaccharides (Kiers et al., 2000).

In vitro digestibility was increased from 29% up to 33%–43% after *Bacillus* fermentation for 48 h. In vitro digestion of *Bacillus*-fermented soybeans using gastrointestinal enzymes only slightly increased the amount of dialyzable matter, which clearly demonstrated the beneficial effect of *Bacillus* fermentation on food nutrient availability (Kolawole and Okonkwo, 1985).

Microorganisms contain some enzymes, particularly cellulases, which are incapable of being synthesized by humans. Microbial cellulases hydrolyze cellulose into sugars, which are then readily digestible by humans. Similarly pectinases soften the texture of foods and liberate sugars for digestion. Fermented foods are often more easily digestible than unfermented ones (Kuboye, 1985).

It was known that lactic acid-fermented weaning foods have been reported to traditionally produce in developing countries, both to improve the safety of the food and to improve its digestibility. Starchy porridges are commonly fed to weaning infants in developing countries. The consistency of these gruels, combined with the small capacity of the infants' stomach, means that it is physically impossible for the child to consume adequate energy to meet its high demands. By acidifying the porridge through lactic acid fermentation, starch is hydrolyzed into shorter chains of glucose and dextrose, which reduce the viscosity of the porridge and increase its energy density. Thus, the child is more able to meet its energy requirements (Kuo et al., 2013).

In addition, protein digestibility of the uncooked glutinous rice-like sorghum was 65.7% and was reduced to 42% after steaming and was increased to 52%–81%, respectively, after the fermentation–distillation processes (Kuo et al., 2013).

Before cooking, red glumes sorghum contained 6.5 mg/g polyphenols, which were reduced to 0.7 mg/g after cooking. Repeated fermentation and distillation increased polyphenol contents from 1.8 to 4.5 mg/g. Thus, koji's solid-state fermentation method might substantially improve nutrient values of sorghum as judged from its in vitro protein digestibility, especially important in underdeveloped Africa as protein source (Kuo et al., 2013).

The most significant activity during alkaline fermentation is the rapid and steady increase in the quantity of free amino acids throughout the fermentation. This is due to a consistently active proteinase activity by the fermentative microorganisms (Azokpota et al., 2006b). The number and quantities of each amino acid usually increase in the fermented beans. Glutamic acid, valine, aspartic acid, and alanine were liberated during alkaline fermentation (Odunfa, 1985a). High proteinase activity has frequently been reported in the fermentation of protein-rich seeds as in the production of ogiri (Odunfa and Oyewole, 1998), afitin, iru, and sonru (Azokpota et al., 2006b), and kinema (Sarkar et al., 1993). The strong ammonia smell, characteristic of the latter stages of iru production, may also be an end product of proteinase activity as reported by Whitaker (1978).

Globally, the enzymatic activities in the fermenting locust beans are somewhat related to the enzymatic properties of the microbial isolates from the iru or dawadawa. The fermenting microorganisms isolated from alkaline-fermented foods (*Bacillus subtilis*, *Bacillus licheniformis*, and *Bacillus pumilus*) were proteolytic, lipolytic, or amylolytic (Odunfa and Oyewole, 1998). The most significant change during the fermentation is the increase in soluble products, particularly amino acids.

This increase improves the digestibility of locust bean or soybean, which in its unfermented state is not normally used as food. Lipase activity constitutes 30%–31% of the locust bean and fairly low especially when compared with tempeh production (Nout and Rombouts, 1990; Steinkraus, 1996a) and natto (Kiuchi et al., 1976). This is desirable since a high amount of fatty acids in foods generally imparts an objectionable taste. Free fatty acids in foods have also been found to inhibit trypsin activity as well as several glycolytic enzymes (Shibasaki and Hesseltine, 1962; Wang et al., 1975).

Pronounced biochemical changes have been observed during the fermentation of locust bean to produce soumbala or dawadawa (Odunfa and Oyewole, 1998; Ouoba et al., 2003a). The total as well as essential free amino acids, namely valine, leucine, isoleucine, tryptophan, phenylalanine, lysine, methionine, threonine, and histidine, increased during 72-h fermentation of locust bean to produce soumbala. The increase was more pronounced between 24 and 48 h. The quantity of essential free amino acids depends on the strains of *B. subtilis* used for the fermentation. Interestingly, aspartic acid, glutamic acid, threonine, serine, proline, glycine, alanine, valine, histidine, tryptophan, ornithine, lysine, and arginine were common to both unfermented and fermented locust bean proteins, whereas cysteine, methionine, leucine, isoleucine, tyrosine, and phenylalanine were not detected in the unfermented locust bean protein (Ouoba et al., 2003b).

5.3.4 Medicinal Benefits

Many of the AFFs possess remarkable prophylactic as well as therapeutic properties. The ethnic groups in different countries of the world believe that the consumption of AFFs protects them from diseases. There is a sound scientific basis to these assertions, as will be discussed below.

5.3.4.1 Antioxidative Activity

Toshirou Nagai

Soybean itself is well known to contain a wide variety of antioxidative components like saponins, isoflavones, tocopherols (E vitamins), and peptides. In soybean the main forms of isoflavones include glucosidic conjugates genistin and malonyl genistin, daidzein and malonyl daidzein, glycitin, and malonyl glycitin (Wang and Murphy, 1994). The aglycone forms, daidzein, glycitein, and genistein are developed by the hydrolytic action of the enzyme β-glucosidase on the glucosidic isolfavone precursors (Matsuura et al., 1989).

The composition of isoflavones in soybeans is greatly changed during the process of natto production, especially in both boiling and fermentation steps (Shimakage et al., 2006). Almost all isoflavone malonyl glucosides lose their malonyl groups to become glucosides during the steaming process, and during the next process 42% of the glucosides are succinylated by *Bacillus subtilis* (*natto*). The content and composition of isoflavones in commercial natto products are predominantly (>76%) isoflavone glucosides (Wei et al., 2008). Isoflavone aglycones may be absorbed faster and in higher amounts in the human intestine than their glucosides. Loss of isoflavones from soybeans is negligible, but during the fermentation of another type of natto, hikiwari-natto (natto from split soybeans; see Figure 2.1b), 20% of isoflavones are lost when soybeans are split.

The first report of antioxidative evaluation during the production of natto came from the experiments dealing with okara (soybean residue after extraction of soymilk) for utilization of food-processing waste (Kawabata et al., 1986). In this experiment, natto was produced from okara supplemented with sugar for sufficient growth of *B. subtilis* (*natto*). Before fermentation, okara showed an antioxidative activity (suppression of increase of a peroxide value and a thiobarbituric acid value) to some extent. Fermentation by *B. subtilis* (*natto*) increased the antioxidative activity in okara. Total tocopherol content in okara rose from 5.67 up to 6.36 mg/100 g dry weight, and the authors speculated that the tocopherols along with coexisting isoflavones and peptides expressed the antioxidative activity in okara and okara-natto. *Bacillus subtilis*, isolated from meitauza, produces water-soluble antioxidative activities in okara and soybeans, accompanied by the degradation of soy proteins and liberation of peptides (Zhu et al., 2008b). The peptides were thought to have a part of the antioxidative activity. Tamura and Takenaka (1999) also detected an additional antioxidative activity in okara-natto, other than isoflavones and tocopherols. The activity was not identified, but they supposed that peptides produced during fermentation could have the activity, based on their studies on antioxidative peptides in okara.

Through a detailed analysis of antioxidative compounds in boiled soybeans and natto, the contents of isoflavones and tocopherols could not explain the great difference between antioxidative activities in natto and boiled soybeans (Esaki et al., 1990). The antioxidative activities in natto were attributed to three unknown compounds in an 80% methanol extract from natto after separation through a size exclusion chromatography.

Hattori et al. (1995a) also found that an ethanol-soluble fraction from natto (NTX) had an antioxidative activity against oxidation of unsaturated fatty acids and that NTX prevented the cells from the death caused by oxidation. NTX contained no tocopherols but had an antioxidative acidity equal to or stronger than that of tocopherols. NTX depressed rat foot-edema induced by free radicals, indicating that NTX works as radical scavenger in vivo (Hattori et al., 1995b), and also it depressed atherosclerosis development in cholesterol-fed rabbits via inhibition of oxidation of low-density lipoprotein (Yokota et al., 1996). However, the antioxidative identity in NTX remains to be unclear, but it was found to be a compound with a molecular size of 1500 kDa and absorbed into the liver. Iwai et al. (2002a, 2002b) found that the water-soluble fractions from natto (a viscous fraction with molecular weight of <100 kDa and an extract from beans of natto) had antioxidative activities against oxidation of low-density lipoprotein (LDL). All of the investigations on antioxidative activities in natto indicated that the activities in natto are composed of those from soybean itself and those produced by *B. subtilis* (*natto*) during fermentation; the latter activities remain to be identified.

In doenjang (a Korean soybean paste prepared after fermentation of soybeans by *B. subtilis*), the contents of daidzein and genistein were increased to maximum level of 764.5 and 561.4 mg/kg, respectively, after 30 days of fermentation (Choi et al., 2007). The change of daidzein and genistein contents was similar to the pattern observed in β-glucosidase activity. Therefore, the contents of daidzein and genistein in doenjang were increased due to the hydrolytic action of β-glucosidase. For Koreans, doenjang appears to be the second most important source of isoflavone intake, next to tofu. The intake of doenjang with a high level of isoflavone might be responsible for the low incidence of hormone-related diseases such as breast and prostate cancer.

In an extract of douchi (a Chinese fermented soybeans) with organic solvent, 14 antioxidative compounds were isolated and identified: 4 phenols, 1 flavanone, 8 isoflavones, and 1 novel pyrone (Chen et al., 2005). Of them, five compounds were produced during fermentation. A water extract of douchi also contained antioxidative and chelating activities, which could work in vitro and in vivo, resulting in mitigation of distortion of hepatocytes (Wang et al., 2008). This substance(s) was not identified, but peptides or amino acids might be ascribed to the activities.

In thua nao (non-salted fermented soybean, produced by mainly *B. subtilis* in Thailand), antioxidative activities higher than those in boiled soybeans were detected, but the activities did not correlate with the free radical-scavenging activities in them (Dajanta et al., 2011a). Thua nao contained the highest antioxidative activity, but at the same time, the lowest free radical-scavenging activity. The components of the activities remain to be identified.

During kinema (fermented soybean product in India) production using *B. subtilis*, lipid peroxidation inhibitory activity, radical-scavenging activity, metal-chelating activity, and reducing power in methanolic extract are shown to be positively correlated with total phenol content, which increases during fermentation, but the mechanisms and the chemical identities remain unclear (Moktan et al., 2008).

In oso (a fermented food in Nigeria from the seeds of *Cathormion altissimum* Hutch.), *Bacillus*, *Leuconostoc*, and *Staphylococcus* species are concerned with the fermentation (Popoola et al., 2006). Oso has a low peroxide value, indicating that some components in oso resist oxidation (Jolaoso et al., 2012). However, the antioxidative activities were limited in power and the peroxide value of oso increases during preservation.

Antioxidative activities associated with enzymatic hydrolysis or fermentation of marine animals have been reported. Fermented blue mussel, tuna, carp, shrimp, and krill products were found to produce peptides showing enhanced antioxidant activities such as superoxide, hydroxyl, and radical-scavenging, metal-chelating, Fe^{3+}-reducing, and lipid peroxidation (Rajapakse et al., 2005; He et al., 2006; Je et al., 2007; Faithong et al., 2010).

In conclusion, it is well known that soybean itself contains various kinds of antioxidants (saponins, isoflavones, tocopherols, and peptides). Besides, fermentation by microorganisms enhances the antioxidative activities of soybean due to production of antioxidative compounds or conversion of non-antioxidative components in soybean to antioxidative ones, which might work synergistically with the antioxidative activities of soybean itself. Although some of the antioxidative compounds are discerned, many remain unidentified. Fermented soybean is an attractive food in terms of achievement of human health with no adverse effect, which is proven by native people eating that for a long time.

5.3.4.2 Probiotic Activity

Toshirou Nagai

Folklores or proverbs on the health benefits of natto that have been passed down through history include intestinal healing effects, laxative effects, smooth bowel movements, and so on. Since ancient times, the Japanese people empirically have recognized natto as a regulator on the function of the intestines. Subjects of the studies on probiotic activities and on antagonistic activities against pathogens are separated into three types: antibacterial activities against pathogens, probiotic effects on domestic animals instead of chemical antibiotics, and probiotic effects on human health. At that time, natto as a probiotic was studied for its effect on the quality of life of human beings. However, the mechanisms were not fully clarified because of the complicated results of the experiments, which varied greatly depending on the conditions.

5.3.4.2.1 Antibacterial Activity of Bacillus subtilis (natto) The first report of *Bacillus subtilis* (*natto*) as an in vivo antibacterial agent against a human pathogen came from the study of medical treatment to a patient infected with *Salmonella* paratyphi B, who could not be cured by chemotherapy (Eguchi and Kumashiro, 1931). Oral administration of both natto and *B. subtilis* (*natto*) resulted in no detection of *Salmonella* in his feces after 4 days. Interestingly, administration of only *B. subtilis* (*natto*) was not sufficient to drive out *Salmonella* from his intestines. *B. subtilis* (*natto*) had an antibacterial activity against *Salmonella* paratyphi A or paratyphi B on agar plates or in broth, but not against *Escherichia coli* (Matsumura,

1934b). The culture filtrate also exhibited the activity that was lost by heating. The author presumed that the identity of the activity might be an enzyme because of its heat lability. *Bacillus subtilis* (*natto*) injected intraperitoneally with targets (*Salmonella* paratyphi A, paratyphi B, or *Salmonella enterica* subsp. *enterica* serovar. typhi) to mice showed no antagonistic activities against the targets, but inhibited the growth of the targets if the targets were injected 24 to 48 h after intraperitoneal injection of *B. subtilis* (*natto*) (Matsumura, 1934c). However, in vivo antagonism of *B. subtilis* (*natto*) was observed against neither *E. coli* nor *Shigella dysenteriae*. The dead cells and the culture filtrate of *B. subtilis* (*natto*) had no inhibitory activity. In vitro and in vivo inhibitory activities against *Sal. enterica* subsp. *enterica* serovar. typhi were also observed by Saito (1938).

Arima (1937) reported that *B. subtilis* (*natto*) protected mice from death caused by *Sh. dysenteriae* in intestines via oral administration. Intraperitoneal injection of dead or living *B. subtilis* (*natto*) cells also exhibited the effect on *Sh. dysenteriae* to some extent. However, *B. subtilis* (*natto*) did not completely drive out *Sh. dysenteriae* from the intestines. The author concluded that *B. subtilis* (*natto*) was not effective to curing patients of dysentery.

It was reported that *B. subtilis* (*natto*) cured ringworm, which was caused by *Trichophyton* fungus (Senbon, 1940). *Bacillus subtilis* (*natto*) also inhibited the growth of *Candida albicans* (Ozawa et al., 1979). It might have antagonistic activities against not only bacteria, but also against fungi including yeasts, although their effects vary depending on species, or strains of targets. Dipicolinic acid, which is associated to spores of *B. subtilis,* was found to have growth inhibitory activity against bacteria (Udo, 1936; Sumi and Ohosugi, 1999). One of the mechanisms of probiotics of *B. subtilis* (*natto*) could be ascribed to dipicolinic acid, although dipicolinic acid has a wide spectrum of growth inhibition, including *B. subtilis* itself.

Bacillus subtilis (*natto*) itself had no toxicity on the health of living bodies of mice, rabbits, and guinea pigs after injection intraperitoneally or intravenously (Matsumura, 1934a).

5.3.4.2.2 Probiotic Activities of Bacillus subtilis (natto) For body gain, health gain, and a decrease in morbidity, antibiotics were fed to domestic animals. After residual antibiotics in the body of the animals became a problem, researchers, or livestock farmers desired the replacement of these agents. In these movements, *B. subtilis* (*natto*) has attracted their attention because of the effects of natto on human health, antagonism of *B. subtilis* (*natto*) against pathogens described above, and the safety of natto. Recently, *B. subtilis* (*natto*) is thought to be a probiotic, beneficially effective on human intestinal microflora.

Bacillus subtilis (*natto*) favors body weight gain and resistance against diseases of chickens (Ebata et al., 1960; Kimura, 1979) and swine (Kimura, 1979; Saka, 2008), indicating that natto or *B. subtilis* (*natto*) could replace antibiotics and give hygienic benefit to animals. Natto decreased the number of *Bifidobacterium* cells in feces of rats, increased that of Enterobacteriaceae, however, did not affect that of *Lactobacillus* (Isshiki and Onozaki, 1992). Continuous feeding with *B. subtilis* C-3102 excluded *Campylobacter, Salmonella,* Enterobacteriaceae, and *Clostridium* from chicken intestinal tracts in the experiments carried out in the laboratory or fields, but increased the number of *Lactobacillus* (Maruta et al., 1996a). In the studies, *Bifidobacterium* was not detected in experimental and control chickens at all. Feeding pigs with *B. subtilis* C-3102 resulted in an increase in the number of *Lactobacillus* and *Bifidobacterium* in their feces, increase in water content of the feces to a beneficial level, and decrease in the number of *Clostridium* and *Streptococcus* (Maruta et al., 1996b). When *B. subtilis* C-3102 was administered to sows, the bacteria reduced the occurrence of diarrhea and mortality rate of their piglets (Maruta et al., 1996b). Saka (2008) also showed that administration of natto paste solution to piglets for 1 week increased the number of *Lactobacillus* in feces and the weight of piglets and decreased the number of coliform bacteria in feces and occurrence of diarrhea. Capsules containing *B. subtilis* C-3102 are sold as supplement for human health by a yogurt company in Japan.

Daily administration of natto (50 g/day) to male adults resulted in an increase in cell counts of *B. subtilis* (*natto*) and *Bifidobacterium* in feces as expected, with a decrease in counts of lecithinase-positive clostridia (Terada et al., 1999). Also, *B. subtilis* (*natto*) decreased fecal concentrations of malodorant substances (phenol, skatol, ammonia, and cresol). Ingestion of natto (50 g/day) produced by *B. subtilis* (*natto*) K-2, which has been mutated to grow more slowly than the parent strain under 20°C, has beneficial effects on bowel movement and the number of *Bifidobacterium* cells in feces, and excess intake of natto

(150 g/day) showed no adverse effect on the body in blood tests (Mitsui et al., 2006). Tablets containing *B. subtilis* C-3102 spores decreased the number of Enterobacteriaceae cells and contents of ammonia and putrefactive products in feces, but had no effect on the cell number of lactic acid bacteria (LAB) (Suzuki et al., 2004). In the experiments, 3 of 21 human volunteers showed no changes in the number of bacteria in each bacterial group and of chemical composition of putrefactive products.

Boiled soybeans, which contain oligosaccharides beneficial to LAB, had no significant effect on microbiota in human feces but decreased the pH of feces (Mitsui et al., 2006). Administration of boiled soybeans to rats also did not affect the number of LAB cells in feces but decreased that of Enterobacteriaceae cells (Isshiki and Onozaki, 1992). Takemura et al. (2009) also observed no significant effect on microbiota using boiled soybeans after enzymatic digestion of oligosaccharides as control. Moreover, oligosaccharides in boiled soybeans were digested during fermentation (Kanno et al., 1982). These findings indicate that *B. subtilis* (*natto*) or the substances that *B. subtilis* (*natto*) produced on natto exerts probiotic effect on the intestinal microbiota.

The effects of *B. subtilis* (*natto*) on microbiota in intestines are summarized in Table 5.3. The levels of the effects vary depending on tested individuals (sex, age, and dietary habit), target species, regions in intestines, feed, microbiota before experiments, sampling time, and other experimental conditions; the effects are not so drastic in many experiments. Since under experimental conditions the human subject is hard to control, this situation makes the results complicated to interpret. However, it is possible to conclude that *B. subtilis* (*natto*) functions to some degree as a probiotic, beneficial for the growth of *Lactobacillus* and *Bifidobacterium* in intestines by unknown mechanisms.

5.3.4.2.3 Behavior of Bacillus subtilis (natto) *in Digestive System* How do *B. subtilis* (*natto*) spores behave in the intestine, where the environment is microaerobic to anaerobic? The answer to the question is important to elucidate the mechanism of probiotic effects of *B. subtilis* (*natto*). If *B. subtilis* (*natto*) exerts its effect in the intestine, some substances might need to be produced by the bacteria or leak out from the cells during germination or cell lysis.

Spores of *B. subtilis* (*natto*) strain BN germinated in the ileum of dogs and then died, but they could not germinate in their colon

Table 5.3 Probiotic Effect of *Bacillus subtilis* (*natto*) on Intestinal Microbiota

					MICROORGANISM								
STRAIN	FORM	CONTROL	TARGET	SPECIMEN	*Cam*	*Sal*	Ent	*Cp*	*Str*	*Lb*	*Bf*	Bac	REFERENCE
B. subtilis C-3102	Cell	No cell	Chicken	Cecal content, feces	–	–	–~±	–	±	+	ND	±	Maruta et al. (1996a)
B. subtilis C-3102	Cell	Before administration	Sow	Feces	NT	NT	±	–	–	±	+	NT	Maruta et al. (1996b)
B. subtilis C-3102	Cell	Before administration	Gilt	Feces	NT	NT	±	–	–	+	+	NT	Maruta et al. (1996b)
B. subtilis C-3102	Cell	Before administration	Sow	Feces	NT	NT	–	–	–	+	±	–	Maruta et al. (1996b)
B. subtilis (*natto*) BN	Spore	No spore	Piglet	Intestinal content	NT	NT	NT	NT	+	±	+	NT	Ozawa et al. (1981)
B. subtilis (*natto*)	Natto	Before administration	Human	Feces	NT	NT	±	–	±	±	+	±`	Terada et al. (1999)
B. subtilis (*natto*) BN	Cell	No cell	Piglet	Feces, intestine	NT	NT	± (*Ec*)	NT	NT	+	NT	NT	Murata et al. (1977)
B. subtilis (*natto*) OUV23481	Natto	Before administration, boiled beans without oligosaccaharide	Human, female	Feces	NT	NT	±	±	±	±	±~+	±	Takemura et al. (2009)
B. subtilis	Cell	No cell	Rat	Cecal content	NT	NT	±	NT	+	±	+	NT	Tsuchihashi et al. (1986)
B. subtilis (*natto*)	Natto paste suspension	Water	Piglet	Feces	NT	NT	– (Cb)	±	NT	+	NT	NT	Saka (2008)
B. subtilis (*natto*) K-2	Natto	Before administration, boiled beans	Human	Feces	NT	NT	±	±	NT	±	+	NT	Mitsui et al. (2006)
B. subtilis C-3102	Spore	Before administration	Human	Feces	NT	NT	–	NT	NT	±	±	±	Suzuki et al. (2004)
B. subtilis (*natto*)	Natto	Boiled beans, no natto	Rat	Feces	NT	NT	+	NT	–	±	–	±	Isshiki and Onozaki (1992)

Note: +, a positive effect on the count of bacteria; –, a negative effect; ±, no effect; NT, not tested; ND, not detected; *Cam, Campylobacter*; *Sal, Salmonella*; Ent, Enterobacteriaceae; *Ec, Escherichia coli*; Cb, Coliform bacteria; *Cp, Clostridium perfringens* or *Clostridium* spp.; *Str, Streptococcus*; *Lb, Lactobacillus, Bf, Bifobacterium*; Bac, Bacteroidaceae.

(Hisanaga et al., 1978). Some substances in the contents of the ileum also seemed to inhibit the growth of *B. subtilis* (*natto*) after germination (Hisanaga et al., 1978). The strains might germinate and multiply in piglet intestines because of increased recovery of *B. subtilis* (*natto*) in feces, compared with ingested dosage amount (Ozawa et al., 1981).

Hosoi et al. (1999) supposed that the spores of *B. subtilis* (*natto*) germinated and the cells grew in the small intestine of humans. In vitro growth of *Lactobacillus murinus* was enhanced by intact spores of *B. subtilis* (*natto*), but not by autoclaved ones (Hosoi et al., 1999). This result might indicate that germination and growth of *B. subtilis* (*natto*) are important factors in functioning as probiotic.

Bacillus subtilis (*natto*) does not merely run through digestive organs after ingestion, but it can germinate in intestines and exert its probiotic activity there by producing bioactive substances.

5.3.4.2.4 Conclusion The probiotic effects of natto or *B. subtilis* (*natto*) have been studied to improve the health of human beings and farm animals. In early studies, it was found that *B. subtilis* (*natto*) inhibited the growth of human pathogens antagonistically. Recently, the bacterium is used for intestinal regulation. Although boiled soybeans have oligosaccharides, which are known to promote the growth of LAB in intestines, the oligosaccharides seem not to influence intestinal microbiota and moreover are degraded during fermentation. Spores of *B. subtilis* (*natto*) do germinate in digestive organs and might produce some bioactive substances there. One candidate substance responsible for probiotic effects of *B. subtilis* (*natto*) is an antibiotic, dipicolinic acid. However, the mechanism of probiotic effects of *B. subtilis* (*natto*) or natto has not been elucidated because of the complicated results, which vary depending on experimental conditions. At least, it can be said that *B. subtilis* (*natto*) can exert probiotic effects on intestinal microbiota to some extent, but not drastically.

5.3.4.3 Immunomodulating Activity

Toshirou Nagai

5.3.4.3.1 Nonspecific Immunity Conferred by Bacillus subtilis (natto) Initially, immunomodulating activity of natto (including *Bacillus subtilis* (*natto*)) was investigated mainly using mice. First, Ohkuro et al.

(1974b) injected cells of *B. subtilis* (*natto*) intraperitoneally to mice to give them nonspecific immunity. In their experiments, *Staphylococcus aureus* cells were injected 3, 8, or 14 days after injection of *B. subtilis* (*natto*) live or dead cells, or *S. aureus* dead cells. As expected, the dead cells of *S. aureus* conferred mice resistance to infectious morbidity caused by *S. aureus*. The living cells of *B. subtilis* (*natto*) also conferred mice the resistance to *S. aureus*, but the resistance was reduced 14 days after injection of *B. subtilis* (*natto*). Dead cells of *B. subtilis* (*natto*) were also effective against attack by *S. aureus* 3 days after the injection. However, the resistance of *B. subtilis* (*natto*) dead cells vanished 8 or 14 days after injection. The phenomenon of nonspecific immunity of *B. subtilis* (*natto*) depended on strains, but not on forms of cells (vegetative cells or spores) (Komatsuzaki et al., 1976, 1979; Ohkuro et al., 1980).

Enhancement of the nonspecific immunity in mice had a peak at 1 to 3 days after pretreatment with *B. subtilis* (*natto*) (Komatsuzaki et al., 1978). At the peak, *B. subtilis* (*natto*)-induced resistance to infectious morbidity by *S. aureus* live cells was stronger than that induced by *S. aureus* dead cells. The resistance was not observed at all 30 min after injection with *B. subtilis* (*natto*). Although *B. subtilis* (*natto*) cells remained in organs of mice for 1 to 3 days after injection, the bacteria seemed not to inhibit directly the growth of *S. aureus* in vivo because of no influence on growth of *S. aureus* in the culture with *B. subtilis* (*natto*).

The mechanism of the nonspecific immunity induced by *B. subtilis* (*natto*) was unclear, although the authors thought that some enzymes could be involved in the nonspecific immunity (Komatsuzaki et al., 1976, 1979).

Although the intraperitoneal injection of *B. subtilis* (*natto*) had no adverse effects on mice even though the bacteria were detected in blood and organs of mice (Ohkuro et al., 1974a), a few *B. subtilis* (*natto*) strains were found to be harmful to mice (Komatsuzaki et al., 1976).

5.3.4.3.2 Mechanism of Immunomodulation Natto, *B. subtilis* (*natto*), or substances produced by the bacteria has immunomodulating activities as described below with some examples.

Bacillus subtilis (*natto*) strains induced the production of interferon type II in mice after intravenous injection (Sato, 1980). The active fraction was on the cell surface and/or in the cells, but was not secreted from *B. subtilis* (*natto*) cells. A fraction with molecular weight of more than 10 kDa, corresponding to an endotoxin fraction prepared from Gram-negative bacteria, showed a high level of induction but the other fraction also showed induction, indicating that there was a variety of substances that could induce the production of interferon in *B. subtilis* (*natto*). The induction of interferon was enhanced by injection of cycloheximide, and the enhancement proved that *B. subtilis* (*natto*)-induced interferon was a type II interferon. The active fraction in *B. subtilis* (*natto*) also enhanced induction of interferon type I like as an adjuvant when mixed with measles virus.

Bacillus subtilis (*natto*) cells, which had been heated at 100°C and treated with formalin, raised a natural killer activity of mice spleen cells against target BW5147 cells (mouse lymphoma) and a productivity of antibody to sheep red blood cells in a time-dependent manner (Hamajima et al., 1983).

Chicks, fed a diet with *B. subtilis* (*natto*) cells, showed an increase in antibody production against sheep red blood cells, depending on the concentration of the bacteria and the length of feeding before intravenous injection of the antigen (Inooka and Kimura, 1983). This effect might be due to stimulation of lymphoid organs in the intestinal tract by the bacteria or substances produced by the bacteria (Inooka and Kimura, 1983). In their experiment, the most important point is that natto administered orally, neither injected nor implanted, is effective on enhancement of antibody production.

Bacillus subtilis (*natto*) cells induced cytokine, interleukin (IL)-6 and IL-8, secretion by epithelial-like human colon carcinoma caco-2 cells (Hosoi et al., 2003). The IL-6 induces B-cell proliferation and secretion of immunoglobulins, and IL-8 is a leukocyte chemotactic factor and activates chemotaxis of T cells and inhibits IL-4-induced IgE production in B cells.

Autoclaved vegetative cells of *B. subtilis* (*natto*), autoclaved spores of them, and mucilage produced by *B. subtilis* (*natto*) induced murine macrophages, J744.1 and RAW264.7, to produce IL-12 p40 and tumor necrosis factor (TNF)-α, respectively (Xu et al., 2006). The active fraction in the natto mucilage to stimulate the macrophages via TLR (Toll-like receptor) 4 was found to be levan, which is a polysaccharide with a backbone of β-(2,6)-linked fructoses. Oral administration of levan to mice resulted in decrease of both IgE and IL-4 levels, indicating that levan or natto may suppress allergic diseases.

Bacillus subtilis (*natto*) TTCC865 cells (vegetative cells, spores, or disrupted cells) induced IL-12 p40 in the experiments using a murine J774.1 macrophage-like cell line (Kawane, 2007). From 623 strains of *B. subtilis* (*natto*), the authors screened out 4 strains inducing IL-12 p40 at a high level and producing natto with good quality. *Bacillus subtilis* (*natto*) reduced inflammation and IgE, and increased IgG2a and interferon (INF)-γ in model mice with atopic dermatitis (Kawane, 2007).

The extract of metabolic products of *B. subtilis* AK, which was a mutant of *B. subtilis* (*natto*), enhanced natural killer cell activity in peripheral blood mononuclear cells of 3 rats of 6 via oral administration (Takeda et al., 2005). Oral administration of the encapsulated extract for a week also enhanced natural killer cell activity of healthy adults in a dose-dependent matter. The effect varied among individuals tested. The activation of natural killer cells was supposed to be not due simply to an increase in the number of natural killer cells in blood but to enhancement of the activity of natural killer cells themselves.

5.3.4.3.3 Anti-Tumor Activity The anti-tumor activity of food was the center of the scientists' attention for a long time. For the first time, Kameda et al. (1968) showed an anti-tumor activity of *B. subtilis* (*natto*) KMD 1126 experimentally. They injected *B. subtilis* (*natto*) cells to groins of mice, where Ehrlich carcinoma cells had been implanted 2 days before. The carcinoma tumors treated with *B. subtilis* (*natto*) cells decreased their weight remarkably more than the non-treated tumors. The anti-tumor activity of *B. subtilis* (*natto*) might be ascribed to its cytolytic substances, which are composed of surfactin, protease, and an unknown acidic substance (Kameda et al., 1971, 1972, 1978).

Sixty-five percent of mice implanted intraperitoneally with both MM102 tumor cells and *B. subtilis* (*natto*) cells survived for 35 days, whereas all of the tumor-implanted mice without *B. subtilis* (*natto*) cells treatment died by 16 days (Hamajima et al., 1983). This anti-tumor activity was effective only at an early stage in carcinogenesis.

Hexane extract of natto contained anti-tumor-promoting activity, which was detected by a dye-transfer method, in which the activity was determined as the recovery of reduced dye transfer through gap junctions between cultured cells caused by a tumor promoter (Takahashi et al., 1995). The activity was ascribed to straight-chain saturated hydrocarbons with C_{30} to C_{32}, especially C_{31} (hentriacontane). Interestingly, the hydrocarbons were not detected in hexane extracts from soybeans or *B. subtilis* (*natto*) cells cultured in nutrient broth, indicating that fermentation of soybean by *B. subtilis* (*natto*) or some growth conditions of the bacteria may be essential to production of the hydrocarbons. One package of natto gives enough concentration of the hydrocarbons in human blood to reduce the risk of tumor promotion, if the hydrocarbons can be absorbed into blood.

The anti-tumor effects of natto may be associated with the enhancement effects on the human immune system, but detailed investigation remains to be carried out to elucidate their mechanisms in vivo.

5.3.4.3.4 Conclusion *Bacillus subtilis* (*natto*) and therefore natto have immunomodulating activities, of which anti-tumor activities are attractive ones to the public. Although the mechanisms to exert the activities remain unclear, levan (fructose polymer), surfactin, straight-chain saturated hydrocarbons, and enzymes are found to be concerned with the immunomodulating activites. It is worth noting that the activities can be developed after oral administration in some cases. Natto manufacturers have bred *B. subtilis* (*natto*) to produce natto with stronger immunomodulating activities. Natto as an immunopotentiator is a fascinating foodstuff for medical and food scientists and for the food industry because its safety is ensured by the fact that the Japanese people have been taking it for a long time without any adverse effects.

5.3.4.4 Fibrinolytic Activity

Toshirou Nagai

Thrombolytic agents or enzymes (streptokinase, urokinase, tissue plasminogen activator, etc.) are used in clinical practice on thrombotic diseases, but the agents have disadvantages in terms of price, half-life period in blood, or side effect (Sumi, 1991). If thrombolytic agents that are safe, cheap, and easy to take become available instead of expensive ones, they will be hugely beneficial to humans.

Sumi et al. (1987) observed that one bean of natto developed a clear zone on a fibrin plate, indicating that insoluble fibrin around the bean was digested by an unknown enzyme produced by *Bacillus subtilis* (*natto*). From the extract of natto with saline, the active enzyme was purified and named nattokinase after the combination of "natto" and "urokinase." Nattokinase could digest a synthetic plasmin substrate directly, but was not inhibited by anti-plasminic agents. Extract from 1 g of natto had fibrinolytic activites of 40 CU (caseinolytic unit) plasmin or 1600 IU (international unit) urokinase. Oral administration of 200 g natto to healthy adults at breakfast resulted in shortening of euglobrin lysis time (ELT) and enhancement of euglobrin fibrinolytic activity (EFA) in plasma after 2–4 h, all indicating activation of fibrinolytic system (Sumi et al., 1990). Encapsulated purified nattokinase (1.3 g each after 3 meals a day) also increased the amount of fibrin/fibrinogen degradation products (FDP) in serum, tissue plasminogen activator (TPA), and EFA (Sumi et al., 1990). Thrombi produced in a blood vessel of a dog were also dissolved after oral administration of nattokinase. Nattokinase or its degraded products seems to activate the production of TPA but not to react directly with thrombi.

When nattokinase was injected intravenously to chemically induced thrombosis rats, thrombolysis in the rats was observed with detection of degraded products from the thrombi by nattokinase (Fujita et al., 1995). The thrombolytic activity of nattokinase was stronger than that of plasmin or elastase on a molar basis, when the enzymes were administered intravenously.

Nattokinase was purified and its basic characteristics were investigated in detail: molecular weight, 28 kDa; pI 8.7; 257 amino acid residues having no cysteine; highly homologous with subtilisin (a serine protease produced by *B. subtilis*) (Fujita et al., 1993). Nattokinase is stable at pH ranging 7 to 12 and below 60°C. Elastase, one from porcine pancreas is a medical drug for hyperlipidemia, but is produced in natto (Muramatsu et al., 1994). The elastase was found to be identical to nattokinase, based on amino acid sequence at their N-teminals, molecular weight, and pI (Sumi et al., 1999). Subtilisin, *B. subtilis* elastase, and nattokinase may be the same enzyme, differing by the researchers' point of view of the enzyme.

Three or more pro-urokinase activators were found in natto, which could be separated by gel permeation chromatography. These pro-urokinase activators were different from nattokinase in molecular weight and strength of activity (Sumi et al., 1996). Oral administration of dried natto containing a high level of the pro-urokinase activators enhanced plasma fibrinolysis of adults. The factors, also found to be serine proteases, were not studied in detail. Nattokinase also catalyzed activation of pro-urokinase (Sumi et al., 1992).

The mechanism of fibrinolysis activated by natto is depicted in Figure 5.1. Nattokinase and pro-urokinase activators activate pro-urokinase or tissue plasminogen activator indirectly by reactions, which were unclear in details. Nattokinase can digest fibrin directly in vitro or in vivo.

In Japan, mutants of *B. subtilis* (*natto*) or fermentation process for production of nattokinase in large quantity have been developed, and

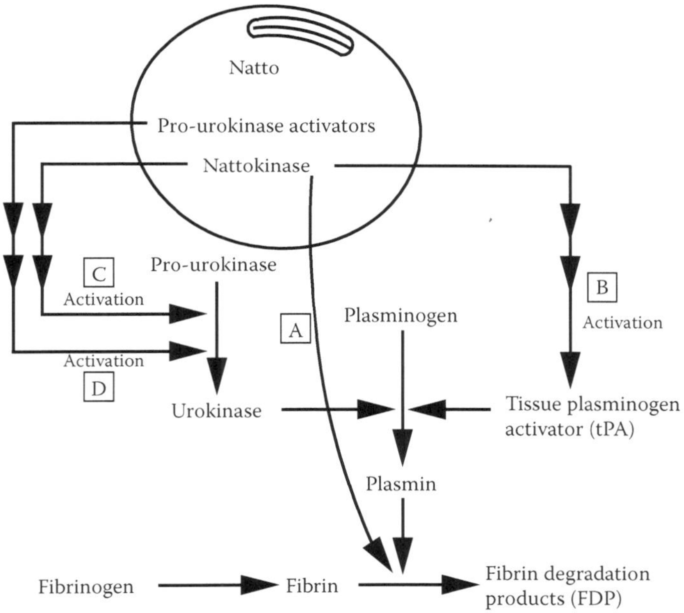

Figure 5.1 Fibrinolytic activities in natto. Line A: in vitro digestion of fibrin (from Sumi, H. et al., *Experientia* 43: 1110–1111, 1987) or in vivo direct hydrolysis of fibrin in a vessel after intravenous injection (from Fujita, M. et al., *Biological and Pharmaceutical Bulletin* 18: 1387–1391, 1995); Line B: activation of tissue plasminogen activator by nattokinase (from Sumi, H., *Bio Industry* 7: 724–730, 1990); Line C: activation of pro-urokinase by nattokinase (from Sumi, H. et al., *Fibrinolysis* 6 (Supplement 2): 86, 1992); Line D: activation of pro-urokinase by pro-urokinase activators (from Sumi, H. et al., *Nippon Shokuhin Kagaku Kogaku Kaishi* 43: 1124–1127, 1996). Double arrowheads indicate unknown pathways.

natto containing more nattokinase is offered commercially (Kiuchi and Suzuki, 1991). Kumada et al. (1993) isolated a *B. subtilis* (*natto*) To 9 strain having a high fibrinolytic activity from rice straw in Japan. Natto produced by To 9 strain increased a fibrinolytic activity in plasma of human adults 2 h after oral administration (Kumada et al., 1994). Interestingly, the authors pointed out that colonies of strains with high fibrinolytic activities were distinguishable from those with low activities in morphology: the former formed a mound in the center of colonies due to production of poly-γ-glutamic acid (PGA), and the latter formed a flat rough colony. *Bacillus subtilis* (*natto*) NN-1 strain with a high productivity of PGA and protease was found to produce more nattokinase than other commercial starter strains (Kiuchi and Suzuki, 1991; Nagai et al., 1994). On the other hand, Sumi (1990) mutated *B. subtilis* (*natto*) to produce less vitamin K, which is an antagonist of anticoagulant warfarin. The mutant produced more nattokinase but lost the productivity of PGA during mutation, indicating that there is no direct correlation between nattokinase and PGA productivity.

Culture broth containing a high activity of nattokinase is also available as a supplement, a food additive, and a medical drug (Takaoka, 2003). Nattokinase will continue to attract attention from food and medical business.

After surveying fibrinolytic activities in fermented foodstuff sold in Japanese markets, the activities were found in hama-natto (salted fermented soybeans using *Aspergillus oryzae*), miso (salted fermented soybean paste using *A. oryzae*), shiokara (salted fermented fish guts), tofuyo (tofu fermented by *Monascus purpureus*), kimchi (Korean pickled Chinese cabbages with hot peppers, fermented by lactic acid bacteria), and gorgonzola cheese, but not in tempeh (non-salted fermented soybeans using *Rhizopus oligosporus*), and other types of cheese (Noda, 2005). A wide variety of fermented foodstuff was found to have fibrinolytic activities presumably produced by fermentation microorganisms. The fibrinolytic activities, however, were not confirmed by in vivo experiments. It was also reported that tempeh purchased in Indonesia had a fibrinolytic acitivity (Sumi and Okamoto, 2003). This discrepancy may be ascribed to contamination of Indonesian tempeh with some bacteria (Kuswanto, 2004) because tempeh produced in Japan is fermented by a pure starter culture of *R. oligosporus*.

The discovery of nattokinase was presented at the annual meeting of the Agricultural Chemical Society of Japan in 1988, and the presentation was broadcast on television. The impact was so high that Japanese people rushed to purchase natto, and the sale of natto increased significantly. Natto has been taken by the Japanese over 1000 years, proving that natto is a very safe foodstuff. Natto is also a safe and cheap thrombolytic agent for oral administration. These days, some health-food companies sell encapsulated nattokinase as supplements.

In conclusion, one grain of natto can digest fibrin on a turbid fibrin plate to make a clear zone. From natto, the enzyme-digesting fibrin was purified and characterized to be the same enzyme subtilisin as a serine protease. The enzyme, nattokinase, dissolves thrombi by activating the production of tissue plasminogen activator, not by digesting them directly, although it was observed that intact nattokinase injected in a blood vessel decomposed thrombi. Natto has also pro-urokinase activators, which also contribute to fibrinolysis. Nattokinse is now commercially available as a supplement. Interestingly, types of fermentation products other than natto are also found to have natto-kinase-like enzymes. Nattokinase might be a cheap and safe alternative to medical thrombolytic agent.

5.3.4.5 Anti-α-Glucosidase (Antidiabetic) Activity

Li-Te Li and Yan-Li Ma

5.3.4.5.1 Diabetes Mellitus Diabetes mellitus is a metabolic disorder syndrome characterized by hyperglycemia and insufficient secretion or action of endogenous insulin. It is one of the most serious chronic diseases occurring with the increase in obesity and aging in the general population. Diabetes is mainly divided into four types by the World Health Organization: type I diabetes (insulin-dependent diabetes mellitus), type II diabetes (non-insulin-dependent diabetes mellitus), gestational diabetes, and other types of diabetes.

Type II diabetes is a heterogeneous metabolic disorder characterized by both insulin deficiency and peripheral insulin resistance. In addition to intrinsic metabolic abnormalities such as hyperglycemia, patients with type II diabetes frequently develop multisystem complications such as retinopathy, nephropathy, and neuropathy. The primary defect in type II diabetes is increased β-cell apoptosis, which causes a decrease in β-cell mass. Therefore, the progressive nature of type II diabetes is due to β-cell failure. Attenuating insulin resistance, improving β-cell function, and increasing β-cell mass are important for preventing and delaying the progression of type II diabetes.

Diabetes mellitus has become a common disease not only in developed countries but also in developing countries due to the changes in people's lifestyle and dietary habits. The occurrences of type II diabetes mellitus are increasing at an alarming rate worldwide; type II diabetes accounts for almost 1 in 10 of all deaths globally each year. Indeed, epidemiological data suggest that 9 of 10 cases of type II diabetes mellitus could be attributed to habits and forms of modifiable behavior (Hu et al., 2001). One of the therapeutic approaches to decrease post-prandial hyperglycemia is to retard the absorption of glucose by inhibition of carbohydrate hydrolyzing enzymes, for example, α-amylase and α-glucosidase in the digestive organs.

There are reports of established α-glucosidase inhibitors such as acarbose and voglibose from microorganisms, nojirimycin, and 1-deoxynojirimycin from plants and their effects on blood glucose levels after food intake (Truscheit et al., 2003). Some drugs have been developed for diabetes, and the best way to control post-prandial plasma glucose level is with a medication in combination with dietary restriction. Because of the incurability, drug resistance, and side effects of drugs for diabetes, food with hypoglycemic function is very important for diabetic patients.

5.3.4.5.2 α-Glucosidase Inhibitors and Diabetes Mellitus It is well known that α-glucosidases are exo-acting carbohydrases, which catalyze the release of α-D-glucopyranose from the non-reducing ends of various carbohydrate substrates. These enzymes play an important role in the biochemical processes of glycoproteins and glycolipids. Presence of α-glucosidase inhibitor in diets can inhibit the activity of α-glucosidase and reduce absorption of dietary carbohydrates (Zhu et al., 2008c). The α-glucosidase inhibitors are widely used in the treatment of patients with type II diabetes. The action mechanism of α-glucosidase inhibitor is shown in Figure 5.2. α-Glucosidase inhibitors delay the absorption of carbohydrates from the small intestine

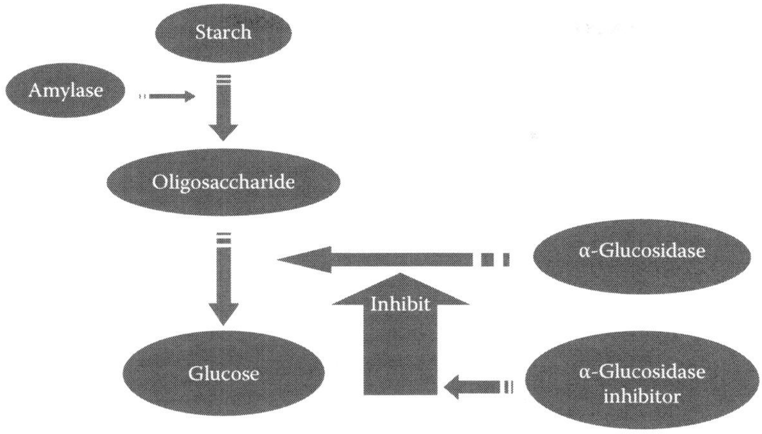

Figure 5.2 Mechanism of controlling blood glucose by α-glucosidase inhibitor.

and thus have a lowering effect on post-prandial blood glucose and insulin levels.

Generally, the α-glucosidase inhibitors can be isolated naturally from plants or food products. However, they can also be synthesized chemically or produced by microorganisms. Several chemical synthetic compounds such as sulfonamide, xanthone derivatives, and deoxy salacious, have been reported to exhibit inhibitory effects against α-glucosidase (Zhu et al., 2008c). The water extract of douchi exhibited α-glucosidase inhibitory activity. Agents with α-glucosidase inhibitory activity delay carbohydrate digestion in the small intestinal tract and thereby reducing meal-induced increases in blood glucose and plasma insulin levels.

Acarbose and voglibose are typical therapeutic agents that display such a mechanism of action on post-prandial blood glucose levels and are useful oral antidiabetic additions to the drugs currently available for the treatment of patients with type II diabetes. α-Glucosidase inhibitor is usually used to prevent or medically treat type II diabetes. These inhibitors combine with intestine α-glucosidase and block the uptake of post-prandial blood glucose. Although powerful synthetic α-glucosidase inhibitors (i.e., voglibose) are available, they usually cause hepatic disorders and other negative gastrointestinal symptoms. Hence, natural α-glucosidase inhibitors from food sources have become an attractive therapeutic approach for treating post-prandial hyperglycemia.

Under normal circumstances, α-glucosidases are responsible for the hydrolysis of complex carbohydrates (starch and other oligosaccharides) to absorbable simple sugars (monosaccharides, most importantly glucose). α-Glucosidases are usually found along the length of the intestine, but for some people they are only active in the upper small intestine. α-Glucosidase inhibitors bind with high affinity and specificity to α-glucosidases found in the brush border of the small intestine. The effect of α-glucosidase inhibitors is to delay the digestion of oligosaccharides in the small intestine so that the release and absorption of glucose take place over a longer time across the length of the small intestine (Hanefeld, 2007). The result is that α-glucosidase inhibitor therapy directly reduces post-prandial hyperglycemia.

5.3.4.5.3 Main Source of α-Glucosidase Inhibitors There are three main ways to obtain α-glucosidase inhibitors: isolating from natural sources, chemical synthesis, and microbial synthesis. Diet therapy plays an important role in the treatment of diabetes. Therefore, the separation of α-glucosidase inhibitors from food has aroused the interest of many researchers (Asano, 2003). Some foods having α-glucosidase inhibitory activity have already been reported. Except for the isolation from natural sources, a variety of α-glucosidase inhibitors have been synthesized with the rapid development of the technology to produce synthetic drugs (Borges de Melo et al., 2006).

However, chemical synthesis usually has problems of complex synthetic process, high cost, technical requirements, and side effects (Murai et al., 2002). Separation of α-glucosidase inhibitors from natural sources has also been limited by many conditions, such as sources of natural products, extraction yield, efficacy of extract, and so on.

Compared with isolation from natural sources and chemical synthesis, microbial synthesis of α-glucosidase inhibitors has many advantages. For example, some hypoglycemic active ingredients have been found from microbial resources (Table 5.4). These ingredients include acarbose (Schmidt et al., 1979), nojirimycin (Niwa et al., 1970), 1-deoxynojirimycin (Ezure et al., 1985), valiolamine (Kameda et al., 1980), dibutyl phthalate (Lee, 2000), cyclophellitol (Atsumi et al., 1990), and CKD-711 (Kwon et al., 2002). Currently, three α-glucosidase inhibitors used in clinical (acarbose, miglitol, and voglibose) are related to microorganisms. Comparatively, synthesis by

Table 5.4 α-Glucosidase Inhibitors from Microorganisms

INHIBITOR INGREDIENT	MICROORGANISM	LITERATURE SOURCE
Acarbose	*Actinoplanes* sp. SE50	Schmidt et al. (1979)
Norjirimycin	*Streptomyces roseochromogenes* R-468, *S. lavendulae* SF-425	Niwa et al. (1970)
1-Deoxynojirimycin	*Streptomyces, Bacillus*	Schmidt et al. (1979), Murao and Miyata (1980), Ezure et al. (1985)
Valiolamine	*Streptomyces hygroscopicus*	Kameda et al. (1980)
Dibutyl phthalate	*Streptomyces* sp.	Lee et al. (2000)
Cyclophellitol	*Phellinus* sp.	Atsumi et al. (1990)
CKD-711	*Streptomyces* sp.	Kwon et al. (2002)

microorganisms is an effective strategy to produce cost-effective and productive α-glucosidase inhibitors. It has been reported that some microorganisms, including species of *Streptomyces* (Iwasa et al., 1970), *Actinoplanes* (Schmidt et al., 1977), and *Flavobacterium* (Kameda et al., 1980), were able to synthesize α-glucosidase inhibitors. For example, acarbose, isolated from the fermentation broth of *Actinoplanes* spp. SE-50 (Schmidt et al., 1977), is a very popular α-glucosidase inhibitor and has been utilized as a medicine for treatment of type II insulin-independent diabetes. It is also found that *Bacillus subtilis* B2 can produce α-glucosidase inhibitors.

5.3.4.5.4 α-Glucosidase Inhibitory Activity of Alkaline-Fermented Foods The presence of α-glucosidase inhibitor in diets can inhibit the activity of the α-glucosidase enzyme, thus reducing the absorption of dietary carbohydrates. This is beneficial for the treatment of carbohydrate-mediated diseases such as diabetes. Studies have shown that some AFFs exhibit α-glucosidase inhibitory activity. The anti-α-glucosidase activities of aqueous douchi extracts were investigated, and it was found that 31 samples of douchi, collected from different parts of China, exerted various degrees of inhibitory activity against rat intestinal α-glucosidase. The anti-α-glucosidase activities of douchi were affected by the production methods. Table 5.5 shows the effect of salt on the anti-α-glucosidase activities of douchi samples inoculated with *Aspergillus oryzae*, *Aspergillus elegans*, and *Rhizopus arrhizus* during the post-fermentation period. The anti-α-glucosidase activity of douchi qu fermented with *A. oryzae*

Table 5.5 Effect of Microorganisms, Salt Concentration, and Post-Fermentation Time on the Relative Anti-α-Glucosidase Activity of Douchi during the Maturation Period

MICROBIAL STRAIN	SALT (%, W/W)	POST-FERMENTATION PERIOD (WEEK) 1	2	3	4
Aspergillus oryzae 3.951	5.0	2.55 ± 0.44	2.98 ± 0.18	5.64 ± 0.15	6.85 ± 0.41
	7.5	1.78 ± 0.12	2.85 ± 0.35	3.68 ± 0.12	5.45 ± 0.49
	10.0	0.16 ± 0.05	1.80 ± 0.09	3.88 ± 0.30	3.97 ± 0.63
	12.5	0.35 ± 0.14	1.58 ± 0.20	2.32 ± 0.09	3.88 ± 0.46
Actinomucor elegans 3.118	5.0	0.76 ± 0.04	0.90 ± 0.05	1.03 ± 0.21	0.94 ± 0.05
	7.5	0.34 ± 0.02	1.35 ± 0.05	1.01 ± 0.08	1.40 ± 0.13
	10.0	0.34 ± 0.08	1.00 ± 0.03	1.22 ± 0.06	1.30 ± 0.15
	12.5	0.46 ± 0.07	0.77 ± 0.04	1.00 ± 0.20	0.99 ± 0.08
Rhizopus arrhizus 3.078	5.0	0.67 ± 0.06	0.73 ± 0.02	1.04 ± 0.09	1.32 ± 0.07
	7.5	0.96 ± 0.07	1.38 ± 0.02	1.01 ± 0.04	1.60 ± 0.13
	10.0	0.58 ± 0.03	1.00 ± 0.06	1.17 ± 0.23	1.31 ± 0.09
	12.5	0.46 ± 0.05	0.77 ± 0.07	1.20 ± 0.14	1.34 ± 0.11

Source: Chen, J. et al., *Food Chemistry*, 103, 1091–1096, 2007.

was higher than that of *A. elegans* and *R. arrhizus*, and the highest anti-α-glucosidase activity was observed in douchi qu fermented with *A. orzyzae* at 5.0% and 7.5% salt levels (Chen et al., 2007). Hiroyuki et al. (2001) found that the mean blood glucose levels of mild diabetes can be decreased from 9.31 to 8.61 mmol/l after taking 0.9 g doses of douchi extract for 6 months. Meanwhile, there is no symptom of side effect.

Meitauza is another type of AFF of which studies have shown that specific microbial fermentation can improve its α-glucosidase inhibitory activity. Medium containing okara fermented by *B. subtilis* B2 possessed a very strong α-glucosidase inhibitory activity, nearly two times than that of medium without okara supplement and contained little glucose and sucrose (Zhu et al., 2008c). α-Glucosidase inhibitor from okara fermented by *B. subtilis* B2 has been purified, and its chemical structure has been identified (Zhu et al., 2010). The mass spectrometry (MS) and nuclear magnetic resonance (NMR) spectrometry data showed that the purified R-glucosidase inhibitor was 1-deoxynojirimycin with a molecular weight of 163 Da (Figure 5.3). Therefore, *B. subtilis* B2 might be used as a starter strain for the production of food-derived 1-deoxynojirimycin products as functional foods for diabetic patients.

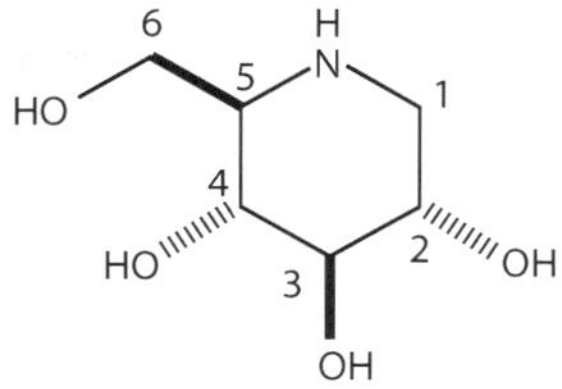

Figure 5.3 Structure of 1-deoxynojirimycin.

5.3.4.5.5 Prospects Isolation of suitable α-glucosidase inhibitors from natural sources and their future chemical synthesis will provide biochemical tools for the prevention and treatment of diabetes. Such knowledge is fundamental to the discovery of lead compounds because of their promising therapeutic potential. Additional studies in both humans and animals are required to identify which components and constituents have beneficial roles in glucose metabolism. The benefits and mechanism of action of AFFs for preventing or treating diabetes are an area of much needed research that would include experimental animal models and clinical human studies.

5.3.4.6 Antihypertensive Activity

Li-Te Li and Yan-Li Ma

5.3.4.6.1 Hypertension Cardiovascular disease is the leading cause of death worldwide, accounting for an estimated 30% and 10% of all deaths and disability, respectively. Despite recent substantial downward trends in the prevalence of cardiovascular disease within many industrialized countries, the global burden is rising because of its increasing prevalence in lower and middle income countries, where 80% of all global cardiovascular disease-related deaths occur. Hypertension is one of the leading causes of cardiovascular disease and premature mortality, and the lifetime risk of developing the condition exceeds 50% for most populations. It is a worldwide problem of epidemic proportions that affects 15%–20% of all adults. It is the most common serious chronic health problem because it carries a high risk factor for stroke, myocardial infarction, and renal failure.

Hypertension is an increasingly important medical and public health burden worldwide due to its high prevalence and associated morbidity and mortality. It has been suggested that the prevalence of hypertension is increasing rapidly and has been predicted to rise with economic development. The current prevalence in many developing countries, particularly in urban societies, is said to be already as high as those seen in developed countries. In addition, as the incidence of hypertension tends to increase with age, aging of the world's population is a key driver of the expected increase. Kearney et al. (2005) predicted that the burden of hypertension would increase by 60% to approximately 1.56 billion in the year 2025.

It is suggested that hypertension is closely related to food components, and hypertension control may be associated with the presence of an antihypertensive peptide motif. Awareness, prevention, treatment, and control of hypertension are significant public health measures. With nearly one quarter of the world's adult population suffering from hypertension, one can conclude that this modifiable, preventable, and controllable risk factor has reached epidemic proportions. Theoretically, the knowledge necessary to prevent and control hypertension exists, but control rates are dismal in every part of the world. From a public health point of view, the prevention and control of hypertension are cost-effective interventions, particularly in older patients. Risk factors for hypertension include dietary habits such as high sodium or low potassium intake, high alcohol consumption, low levels of physical activity, and overweight. Efforts to reverse these habits are the pivotal aspect of suggested lifestyle changes.

5.3.4.6.2 Angiotensin-Converting Enzyme Inhibitors and Hypertension
Angiotensin I-converting enzyme (ACE) is a zinc metallo peptidase that is activated by chloride and has a broad in vitro substrate specificity. ACE exists in two distinct isoforms, the somatic and smaller testicular types, which are transcribed from a single gene at alternative initiation sites. ACE belongs to the class of zinc proteases and is located in the vascular endothelial lining of lungs. ACE acts as an exopeptidase that cleaves dipeptides from the C-terminus of various oligopeptides. It can elevate blood pressure by converting the inactive angiotensin I to the potent vasoconstrictor, angiotensin II, and by catalyzing the degradation of bradykinin (Li et al., 2004). In the renin–angiotensin system (RAS), ACE plays a crucial role in the regulation of blood pressure as well as cardiovascular function (Figure 5.4). Within the enzyme cascade of the RAS, ACE converts

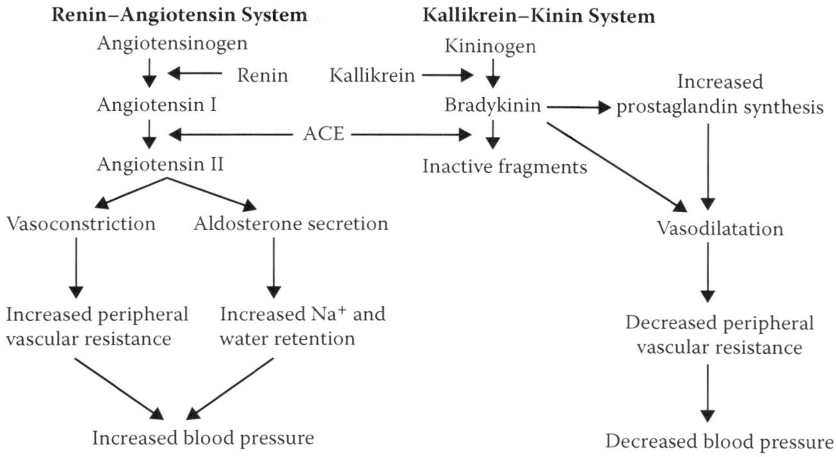

Figure 5.4 Role of angiotensin-converting enzyme (ACE) in blood pressure regulation. (From Li, G.H. et al., *Nutrition Research*, 24, 469–486, 2004.)

the inactive decapeptide angiotensin I by cleaving dipeptide from the C-terminus into the potent vasoconstricting octapeptide angiotensin II. This potent vasoconstritor is also involved in the release of a sodium-retaining steroid, aldosterone, from the adrenal cortex, which has a tendency to increase blood pressure. ACE is a multifunctional enzyme that also catalyzes the degradation of bradykinin, a blood pressure lowering nonapeptide, in the kallikrein–kinin system (Li et al., 2004).

Moreover, ACE has been shown to degrade neuropeptides, which may interact with the cardiovascular system. Inhibition of ACE is considered to be a useful therapeutic approach in the treatment of hypertension. Therefore, in the development of drugs to control high blood pressure, ACE inhibition has become an important target. A large number of highly potent and specific ACE inhibitors have been developed as orally active drugs that are used in the treatment of hypertension and congestive heart failure (Li et al., 2004). ACE catalyzes the conversion of angiotensin I to vasoconstrictor angiotensin II, and also inactivates the antihypertensive vasodilator bradykinin. Inhibition of ACE mainly results in an overall antihypertensive effect.

Although several synthetic ACE inhibitors are currently used as clinical antihypertensive drugs, they cause adverse effects, such as coughing, allergic reactions, taste disturbances, and skin rashes.

Therefore, in recent years, safer, innovative, and economical ACE inhibitors from food resources have been paid a lot of attention for the prevention and remedy of hypertension.

5.3.4.6.3 ACE Inhibitory Peptides Derived from Fermented Foods ACE plays a key physiological role in the control of blood pressure via the renin–angiotensin system. Since the discovery of ACE inhibitors in snake venom, many studies have been directed toward the attempted synthesis of ACE inhibitors, such as captopril, enalapril, alacepril, and lisinopril, which are currently used extensively in the treatment of inevitable hypertension and heart failure in humans. However, these synthetic drugs are believed to have certain side effects. Therefore, a search for ACE inhibitors from food has become a major area of research. Various ACE inhibitors, such as nicotianamine from soy sauce, phytic acid and peptide from miso paste, soybean paste, and tofuyo have been isolated. Among them, ACE inhibitory peptides were paid more attention. Many ACE inhibitory peptides have been isolated from various food proteins.

Besides supplying amino acids and energy that are essential for growth and maintenance, food proteins can act as an important source of biologically active peptides with antihypertensive activity. These peptides are inactive within the sequence of parent proteins, but they can be released by enzymatic proteolysis in vivo or in vitro, for example, during gastrointestinal digestion or during food processing. Once bioactive peptides are liberated, they may act as physiological modulators with hormone-like activity. Thus, these peptides represent potential health-enhancing nutraceuticals for food and pharmaceutical applications (Li et al., 2004). ACE inhibitory peptides derived from food proteins have attracted focused attention for its antihypersensitive activity and thus have been studied most comprehensively for their ability to prevent hypertension. Oshima et al. (1979) first separated six potent ACE inhibitory peptides by digesting gelatin with bacterial collagenase. Subsequently, a series of peptides with ACE inhibitory activity were isolated and identified. These peptides could be used as a potent functional food additive and represent a healthier and natural alternative to ACE inhibitory drugs.

Fermentation is an effective way to generate ACE inhibitory peptides, which are released from long-chain protein of the food matrix in the role of protease secreted by microorganisms. Studies showed that fermented foods, such as yogurt (Quirós et al., 2007), cheese (Pripp et al., 2006), fermented soybean food (Kuba et al., 2003; Zhang et al., 2006), fermented fish sauce (Ichimura et al., 2003), fermented oyster sauce (Je et al., 2005a), grape wine (Pozo-Bayón et al., 2007), and sake and sake lees (Saito et al., 1994) contained ACE inhibitory peptide. These peptides that originate from food components can be classified into three groups: (1) inhibitor type; IC_{50} values of peptides that are not affected after preincubation with ACE; (2) substrate type; peptides that are hydrolyzed by ACE to give peptides with weaker activity; and (3) prodrug-type inhibitor; these peptides are converted to true inhibitors by ACE or gastrointestinal proteases (Iroyukifujita et al., 2008). Most of the ACE inhibitory activity peptides belong to competitive inhibitor type. Table 5.6 lists some ACE inhibitory peptides derived from fermented food. Hata et al. (1996) confirmed that milk fermented with a starter containing *Lactobacillus helveticus* and *Saccharomyces cerevisiae* contains ACE inhibitory peptides Val-Pro-Pro and Ile-Pro-Pro. The systolic and diastolic blood pressure of hypertension patients significantly lowered after taking this yogurt for 8 weeks. Fermented soybean foods are good sources of ACE inhibitory peptides. It has been reported that soybean peptide fraction, His-His-Leu, isolated from Korean fermented soybean paste, exhibited ACE inhibitory activity in vitro and exerted antihypertensive activity in vivo (Shin et al., 2001). Two ACE inhibitory peptides, Ile-Phe-Leu (IC_{50}, 44.8 μM) and Trp-Leu (IC_{50}, 29.9 μM), were isolated and identified from tofuyo in Japan (Kuba et al., 2003). Both of the peptides are noncompetitive inhibitors. The inhibitory activity of Trp-Leu was completely preserved after a treatment with pepsin, chymotrypsin, or trypsin. Even after successive digestion by these gastrointestinal proteases, the activity remained at 29% of the

Table 5.6 Some Bioactive Peptides Identified in Fermented Foods

AMINO ACID SEQUENCES OF BIOACTIVE PEPTIDES	SOURCE	REFERENCE
Val-Pro-Pro, Ile-Pro-Pro	Fermented milk	Hata et al. (1996)
His-His-Leu	Korean soybean paste	Shin et al. (2001)
Ile-Phe-Leu, Trp-Leu	Tofuyo in Japan	Kuba et al. (2003)
Leu-Val-Gln-Gly-Ser	Soy sauce	Rho et al. (2009)
Ala-Phe, Phe-Ile, Ile-Phe	Soy sauce	Zhu et al. (2008a)

original value. Three dipeptides, Ala-Phe, Phe-Ile, and Ile-Phe, were successfully identified from the salt-free soy sauce as transportable inhibitory peptides (Zhu et al., 2008a). Rho et al. (2009) prepared a fermented soybean extract through a rapid fermentation at an elevated temperature to accelerate proteolytic hydrolysis and identified an ACE inhibitory peptide, Leu-Val-Gln-Gly-Ser, in which its IC_{50} value was 22 µg/ml (43.7 µM). Peptides could be used as a potent functional food additive and represent a healthier and natural alternative to ACE inhibitor drugs.

5.3.4.6.4 ACE Inhibitory Activity of Alkaline-Fermented Foods ACE inhibitors from some AFFs have been studied for their ability to prevent and alleviate hypertension. Douchi is a class of AFF that originated from China, and was reported to exhibit significant ACE inhibitory activities. ACE inhibitory activities of commercial douchi samples, expressed as IC_{50} values, are shown in Table 5.7. All the aqueous douchi extracts exhibited in vitro ACE inhibitory activities. There was no significant difference in the ACE inhibitory activities between *Aspergillus*- and *Mucor*-type douchi, but significant differences were observed between douchi produced by fungi and that produced by bacteria. Higher ACE inhibitory activities were found in two *Aspergillus*-type douchis, namely sample 15 and 17 with IC_{50} values of 0.31 and 0.40 mg/ml, respectively, and in three *Mucor*-type douchis, namely sample 21, 25, and 27 with IC_{50} values of 0.39, 0.25, and 0.38 mg/ml, respectively (Li et al., 2010). Fan et al. (2009) also detected ACE inhibitory activities of some douchi samples and a novel ACE inhibitory peptide, His-Leu-Pro, was then isolated from a *Mucor*-type douchi and exhibited the most potent ACE inhibitory activity (IC_{50}, 0.204 mg/ml). The peptide is a competitive inhibitor and maintained its inhibitory activity even after incubation with some gastrointestinal proteases.

In addition to douchi, some other AFFs have also been reported to exhibit ACE inhibitory activity. Je et al. (2005b) investigated ACE inhibitory activity of fermented blue mussel sauce. The IC_{50} value of fermented blue mussel sauce for ACE activity was 1.01 mg/ml. An ACE inhibitory peptide was purified and the sequence was identified as EVMAGNLYPG. The purified peptide can make blood pressure of spontaneously hypertensive rats decrease significantly after oral

Table 5.7 Brands, Origins in China, Types of Soybean Used, Classification, and ACE Inhibitory Activities of Commercial Salt-Fermented Douchi Samples

BRAND	ORIGIN[a]	TYPE OF SOYBEAN	CLASSIFICATION	IC_{50} (mg/ml)
LYPX	Hunan P (middle)	Black	*Aspergillus*-type	0.69 ± 0.02
TPQ	Hunan P (middle)	Black	*Aspergillus*-type	0.88 ± 0.01
QW	Hunan P (middle)	Black	*Aspergillus*-type	2.20 ± 0.11
KX	Hunan P (middle)	Black	*Aspergillus*-type	0.94 ± 0.04
TMS	Hunan P (middle)	Black	*Aspergillus*-type	1.83 ± 0.01
DXY	Jiangxi P (middle)	Black	*Aspergillus*-type	1.05 ± 0.05
YHIX	Jiangxi P (middle)	Black	*Aspergillus*-type	0.97 ± 0.02
DXIX	Jiangxi P (middle)	Black	*Aspergillus*-type	0.65 ± 0.01
NF	Jiangxi P (middle)	Black	*Aspergillus*-type	2.46 ± 0.14
JX	Jiangxi P (middle)	Black	*Aspergillus*-type	0.88 ± 0.07
KXED	Jiangxi P (middle)	Black	*Aspergillus*-type	0.71 ± 0
HP	Guangxi A (south)	Black	*Aspergillus*-type	0.83 ± 0.02
RK	Guangxi A (south)	Black	*Aspergillus*-type	2.03 ± 0.12
YZ	Guangxi A (south)	Black	*Aspergillus*-type	1.25 ± 0.04
YJQ	Guangdong P (south)	Black	*Aspergillus*-type	0.31 ± 0.01
HSQ	Guangdong P (south)	Black	*Aspergillus*-type	0.62 ± 0.03
XQ	Guangdong P (south)	Black	*Aspergillus*-type	0.40 ± 0
JF	Guangdong P (south)	Black	*Aspergillus*-type	0.58 ± 0.03
KWK	Shanghai C (east)	Black	*Aspergillus*-type	0.53 ± 0.01
ZLW	Guizhou P (middle)	Black	*Aspergillus*-type	0.50 ± 0
YC	Chongqing C (middle-west)	Yellow	*Mucor*-type	0.39 ± 0.02
RM	Chongqing C (middle-west)	Yellow	*Mucor*-type	1.17 ± 0.03
QM	Chongqing C (middle-west)	Yellow	*Mucor*-type	0.94 ± 0.03
YCJX	Chongqing C (middle-west)	Yellow	*Mucor*-type	1.20 ± 0.01
ZT	Chongqing C (middle-west)	Yellow	*Mucor*-type	0.25 ± 0
YX	Chongqing C (middle-west)	Yellow	*Mucor*-type	0.95 ± 0.04
TC	Sichuan P (middle-west)	Yellow	*Mucor*-type	0.38 ± 0
WNH	Sichuan P (middle-west)	Yellow	*Mucor*-type	2.10 ± 0.09
CH	Sichuan P (middle-west)	Yellow	*Mucor*-type	1.09 ± 0.34
DDS	Guizhou P (middle)	Yellow	Bacteria-type	7.47 ± 0.05
FW	Sichuan P (middle-west)	Yellow	Bacteria-type	11.80 ± 1.02
WX	Sichuan P (middle-west)	Yellow	Bacteria-type	11.51 ± 0.79

(*continued*)

Table 5.7 (Continued) Brands, Origins in China, Types of Soybean Used, Classification, and ACE Inhibitory Activities of Commercial Salt-Fermented Douchi Samples

BRAND	ORIGIN[a]	TYPE OF SOYBEAN	CLASSIFICATION	IC_{50} (mg/ml)
RC	Sichuan P (middle-west)	Yellow	Bacteria-type	7.83 ± 0.51
WYZ	Shandong P (east)	Black	Bacteria-type	1.60 ± 0.08
BS	Yunnan P (south-west)	Yellow	Bacteria-type	1.61 ± 0.07
LSF	Henan P (middle)	Yellow	Bacteria-type	4.22 ± 0.21

Source: Li, F.J. et al., *International Journal of Food Properties*, 13, 512–524, 2010.
Note: Results are the means ± SD of three determinations.
[a] P, province; A, autonomous region; C, city.

administration. Natto is a popular traditional Japanese food made by fermenting steamed soybeans with *Bacillus subtilis* (*natto*). It has long been thought and proven that natto is effective in preventing hypertension (Ibe et al., 2009).

The ACE inhibitory activity of fermented oyster sauce was also investigated, and the IC_{50} value was determined to be 2.45 mg/ml. An ACE inhibitor was isolated and purified from fermented oyster sauce. The purified inhibitor had an IC_{50} value of 0.09 mg/ml, and it exhibited competitive inhibition against ACE. The purified peptide was evaluated for its antihypertensive effect in spontaneously hypertensive rats following oral administration. Rat blood pressure significantly decreased after inhibitor injection (Je et al., 2005a).

5.3.4.6.5 Prospects Hypertension, which is also called the "silent killer," is a major risk factor for cardiovascular disease. Almost one third of individuals with hypertension do not know that they have it. Those who do know, usually cannot control it. There are some drugs that have been proved to be effective in lowering blood pressure, but they usually have side effects. Therefore, the search for diet-related preventive measures for hypertension has attracted much attention. Food proteins derived from ACE inhibitory peptides are just the right candidates for such products.

AFFs are a good source of ACE inhibitory peptides. Although the ACE inhibitory potencies of these peptides are not as great as those of drugs commonly used in the treatment of hypertension, they are naturally derived from food protein sources that are consumed daily, and thus are considered to be milder and safer without the side effects associated with the drugs. On the other hand, these peptides usually have multifunctional properties and are easily absorbed. Therefore, these AFF-derived ACE inhibitory peptides show a great promise in the development of a novel physiologically functional food for preventing hypertension as well as for therapeutic purposes. Furthermore, they perhaps can be considered as starting compounds for development of antihypertensive drugs against ACE (Li et al., 2004).

Despite the observed potential antihypertensive effects of some ACE inhibitory peptides derived from AFF, it should also be noted that the inhibitory activities of all the peptides are determined in vitro, and the blood pressure-lowering activities of most of these peptides have not yet been confirmed in vivo. Efforts are still needed to examine whether these peptides are efficacious in vivo in all cases and to elucidate the mechanism underlying the antihypertensive activities shown in these peptides. Extended clinical trials in human volunteers are also necessary to evaluate further their long-term efficacy and safety before considering the exploitation of these peptides in physiologically functional foods for preventing hypertension as well as for therapeutic purposes (Li et al., 2004).

5.3.4.7 Tooth Enamel Hardness

Chanya Chuenarrom

Tooth enamel is the strongest substance in the human body. The main inorganic material is calcium hydroxyapatite, which is composed of calcium phosphate. Enamel covers and protects softer dentin from acid attack. The role of the teeth in food digestion is to start tearing and crushing the food into small enough pieces for swallowing. The strength and hardness of teeth are very important to human beings.

Many dental problems are related to dietary consumption. One of the major common problems found in both children and adults is tooth erosion. Dental erosion was documented as long ago as 1892 (Darby, 1892), a few years after the introduction of various soft drinks. The erosive effect of dietary acids and the means of erosion prevention are still to be investigated.

Although tooth erosion differs from tooth caries, both are characterized by the progressive loss of tooth mineral by acid degradation. Caries is a result of acid produced by oral bacteria from the fermentation of sugars and other carbohydrates, which accumulate in plaque, and then attack the teeth. However, dental erosion is the loss of hard tissue by the chemical process of direct acidic attack. Undoubtedly, dietary acids are one of the main causes of dental erosion. The frequent consumption of acidic foods and drinks damages tooth enamel permanently, and the dentin loses some of its protection. The microscopic tubes inside the dentin allow hot, cold, sour, or sweet foods to stimulate nerves within the tooth resulting in painful sensitivity.

Tooth enamel is composed primarily of hydroxyapatite, $Ca_{10}(PO_4)_6(OH)_2$. In the mouth, teeth are bathed in a neutral pH of saliva, and the natural level of calcium phosphate in saliva is sufficient to supersaturate it with respect to hydroxyapatite. The reaction is expressed by the following equation:

$$Ca_{10}(PO_4)_6(OH)_2 \rightarrow 10Ca^{2+} + 6PO_4^{3-} + 2OH^-$$

However, consumption of dietary acids results in the decrease of calcium and phosphate ions in saliva, and changes the reaction equilibrium between enamel hydroxyapatite and calcium phosphate in saliva. Hydroxyapatite is forced to release calcium and phosphate into saliva to maintain equilibrium, which leads to tooth dissolution, otherwise known as demineralization. The critical pH of tooth dissolution is pH 5.5, which means that foods or drinks with a pH below 5.5 are able to erode tooth enamel.

Acidic foods and drinks, including citrus fruits and juices, and vinegary foods such as pickles, wine, energy drinks, and carbonated drinks, affect the tooth surface and hardness. Studies on the causes and prevention of tooth erosion from dietary acids indicate that most sport drinks and carbonated soft drinks have pH levels below the critical pH of teeth dissolution (Benjakul and Chuenarrom, 2011). The sufficient calcium concentration in drinks is one of the factors that assist in reducing the rate of hydroxyapatite dissolution (Hooper et al., 2004). Calcium supplementation in acidic drinks is one of the methods for reducing the erosive potential of acidic drinks (Parry et al., 2001); however, this also seems to reduce the pleasantness of the taste.

For dental health, fermented foods have always been linked to dental erosion because they are associated with acidity. Basically, sour-tasting fermented foods or beverages can damage tooth enamel if their pH value is below the critical pH of tooth dissolution. Interestingly, however, some fermented milk with a pH of 3.5–3.8 were found to promote only superficial mineral loss rather than dental enamel erosion (Lodi et al., 2010). Several types of yogurt were studied and found to have no erosive potential (Caglar et al., 2006; Kargul et al., 2007). Calcium and phosphate content in these products may be the important key to reducing erosive potential.

Although there are numerous studies on the effects of acidic agents on dental erosion and prevention, there are only a few on the effects of alkaline agents on dental health. It was found that the alkaline pH of swimming pool water causes the proteins in the mouth to break down and combine with minerals to form a hard, yellowish-brown mineral deposit on teeth. Frequent swimmers may be at risk for developing stains on their teeth. However, there are no clinical reports on the effects of alkaline food on teeth. It is generally assumed that in terms of dental health, AFFs are not harmful to the surface of teeth. Interestingly, a recent study proposes that AFFs with high calcium content are beneficial in enamel erosion prevention (Chuenarrom and Benjakul, 2010).

Calcium is one of the most important nutrients for overall health. In the form of calcium phosphate, calcium is continually deposited into bones and teeth, giving them their strength, structure, and density. After intake, calcium is ionized by stomach acid before it is absorbed into other parts of the body. It was found that dietary calcium may be able to maintain tooth hardness immediately upon eating. In situ studies revealed that cheese or milk consumption increased the hardness of the enamel after the enamel was softened by a cola-type drink. This increase in hardness was presumably due to deposited organic and mineral materials on the enamel surface from the dairy consumption (Gedalia et al., 1991).

Dairy products are considered good sources of calcium in the western hemisphere. However, around 95% of Asian adults are unable to consume dairy products because of their inability to digest the lactose in milk (Robbins, 2001), so they seek calcium from other sources. Thus, soybean and seafood are the essential source of calcium for Asian people. Seafood is important as a source of both protein and calcium, particularly in countries along the coast. To compare, the calcium content in fishbone is around 175 g/kg (Toppe et al., 2007), 30 g/kg in shrimp shell (Heu et al., 2003), and 8 g/kg in cheese (Gonzalez-Martin et al., 2011). Unfortunately, fishbone and shrimp shells are not directly edible, and they are also difficult to incorporate into a delicious meal. In Southeast Asia, whole fish or shrimp is fermented in a variety of sauces and pastes, with no removal of fishbone or shrimp shell. A number of fermented fish and shrimp sauces and pastes are prepared as various condiments to complement many dishes. For example, fermented fish products such as pla-dek (Laos), pla-ra (Thailand), and budu (Malaysia) are important ingredients of many traditional dishes. Moreover, shrimp paste is a common ingredient in various dishes in Thailand, Laos, Myanmar, Malaysia, Singapore, Indonesia, Vietnam, and the Philippines.

Shrimp paste has a high content of calcium because it is produced from whole, small, silver shrimp. The shrimp is salted, fermented, and dried until broken down into a dark colored, foul-smelling paste that is rich in protein, calcium, vitamins, and various extractable compounds (Heu et al., 2003). In our research, we analyzed the calcium content of three modern trade name packages and three traditional packages of shrimp paste in Thailand using ICP-OES (inductively coupled plasma-optical emission spectrometry). The calcium content was found to be 11.1–17.9 g/kg, indicating a remarkably high content of calcium in shrimp paste, higher than that found in most dairy products.

A recent study investigated the effects of temperature of acidic Thai foods on dental erosion (Benjakul et al., 2011). It was found that the tooth specimens, immersed at higher temperatures in the sour-tasting curry "kangsom," showed less dental erosion than those exposed to lower temperature kangsom – a finding that seems to contrast with other foods tested in the experiment. Kangsom is the most popular curry dish in Thailand. The main ingredients of kangsom sauce include chilli, shallots, garlic, turmeric, tamarind juice, kapi (Thai fermented shrimp paste), salt, and water. The main ingredients in kangsom sauce were studied, and it was determined that the only ingredients associated with tooth enamel hardness were tamarind juice and kapi. It was then determined that without kapi in kangsom sauce, tamarind juice eroded tooth enamel quicker than if there was kapi. At 25°C, 37°C, and 55°C, the pH of 1% kapi in water were 7.4, 7.3, and 7.0, respectively, whereas the pH of kangsom without kapi were 3.0, 2.8, and

2.4, respectively. Adding 1% kapi in kangsom increased the respective pH to 4.1, 4.0, and 3.7, and also reduced the erosive potential of tamarind juice, which resulted in a reduction of enamel surface softening. Figure 5.5 shows the changes in microhardness of tooth enamel after being immersed in kangsom with and without kapi for 105 min at different temperatures. At 55°C, the hardness of enamel immersed in kangsom without kapi decreases to 17.8% of the initial value, which is clearly different from the results when immersed in kangsom with kapi (Benjakul et al., 2011). The effect can also be explained that the high temperatures may accelerate the dissociation of calcium and phosphate compounds in kapi, providing saliva with sufficient calcium and phosphate to supersaturate and reduce the solubility of hydroxyapatites.

Calcium in food inhibits tooth enamel loss when calcium ions interact with the carboxyl group of dietary acid to create calcium salt. When the number of available carboxyl groups of dietary acid decreases, calcium is extracted from the enamel surface at a lower rate. For this reason, a diet with sufficient calcium to saturate saliva in oral cavity is important; the saturated solution prevents further dissolution or erosion (Featherstone and Lussi, 2006).

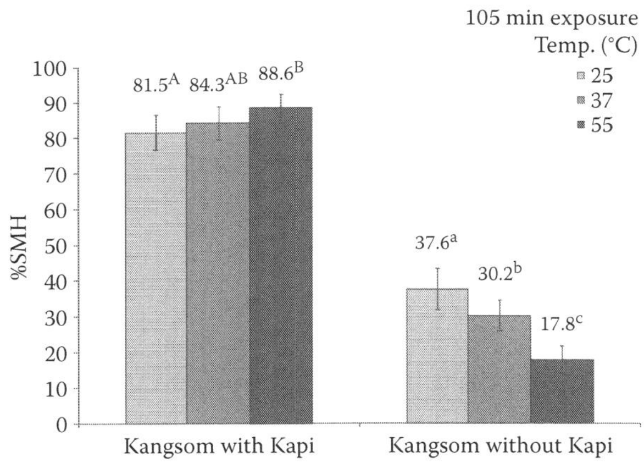

Figure 5.5 Percentage of surface microhardness (SMH) change of tooth enamel after exposure to assigned agents for 105 min at different temperatures. SMH is expressed as fraction (%) of the initial hardness before exposure. Means identified with the same letter, capital letters in kangsom with 1 g/100 g kapi groups or lowercase letters in kangsom without kapi are not significantly different ($p < 0.05$). (Adapted from Benjakul, P. et al., *Songklanakarin Journal of Science and Technology*, 33, 419–423, 2011.)

In our research, only 0.65 g of shrimp paste added to 100 ml of sour tamarind juice demonstrated that a tiny amount of shrimp paste added to food can significantly decrease the rate of enamel loss, and the more shrimp paste added in sour food, the more the protection against erosion. Not only does it reduce the erosive potential of acidic food, but shrimp paste in a non-acidic food also promotes enamel hardness (Chuenarrom and Benjakul, 2010). Other AFFs such as soybean condiments with high calcium content are eaten frequently. In the same way, as shrimp paste, these foods act to reduce the erosive potential from acidic foods and promote enamel hardness from non-acid food, even though they have less calcium content than fish or shrimp products.

In addition to the nutrients in AFFs, microorganisms in fermented foods may be another factor to prevent caries from forming on the enamel surface. Recently, the application of probiotic therapy in oral health has been investigated. An in vivo, long-term study of children consuming milk supplemented with a probiotic strain of *Lactobacillus rhamnosus* had fewer dental caries than the control group (Näse et al., 2001). Several bacterial strains in the dairy products, including *Streptococcus thermophilus*, *Lactobacillus lactis* subsp. *lactis*, *L. rhamnosus*, *Lactobacillus casei*, *Lactobacillus bulgaricus*, and *Lactobacillus reuteri* were investigated to find the capacity in which they inhibit development of the cariogenic species or the formation of *Streptococcus mutans* biofilm (Meurman et al., 1995; Näse et al., 2001; Comelli et al., 2002; Nikawa et al., 2004; Petti et al., 2008). *Lactobacillus paracasei*, *Lactobacillus plantarum*, and *L. rhamnosus*, isolated from saliva and plaque, demonstrated interference capacity against mutans streptococci (Simark-Mattsson et al., 2007). The mechanism of probiotic action is still unknown; however, it has been proposed that the probiotic effect protects against caries by first adhering to the tooth surface, where cariogenic bacteria reside, and second, becoming part of the biofilm that develops on teeth. Finally, these bacteria compete with cariogenic bacteria, thus reducing the level of their colonization (Bonifait et al., 2009).

Recently, probiotic supplement products, chewing gum, and probiotic drinks have been developed to reduce the cariogenic bacteria in the oral cavity. Probiotic supplements in food need the ascertainment of their particular effectiveness. Since acid production from sugar is harmful to teeth, care must be taken not to select strains with high fermentation capacity (Stamatova and Meurman, 2009). Cultured fermented foods provide the best alternative to unclear probiotic supplements. Recently, *Weissella cibaria* and *L. plantarum* strains, which have the capacity to inhibit *S. mutans* (Kang et al., 2006; Simark-Mattsson et al., 2007), were found in ntoba mbodi, an African AFF obtained from cassava leaves (Ouoba et al., 2010). Although the benefit for oral health of these bacterial strains in fermented foods has not been studied, these findings encourage further research to investigate and select suitable probiotic strains in other AFFs for oral health. The nutritional values of AFFs and prolonged ingestion of oral probiotics should further be studied to obtain better understanding of the benefits of indigenous fermented foods on oral health.

References

Achi, O.K. 1992. Microorganisms associated with natural fermentation of *Prosopis africana* seeds for the production of okpiye. *Plant Foods for Human Nutrition* 42: 297–304.

Achi, O.K. 2005a. Traditional fermented protein condiments in Nigeria. *African Journal of Biotechnology* 4: 1612–1621.

Achi, O.K. 2005b. The upgrading of traditional fermented foods through biotechnology. *African Journal of Biotechnology* 4: 375–380.

Achinewhu, S.C. 1983a. Chemical and nutritional composition of fermented products from plant foods. *Nigerian Food Journal* 1: 115–117.

Achinewhu, S.C. 1983b. Protein quality of African oil bean seed (*Pentaclethra macrophylla*). *Journal of Food Science* 48: 1374–1375.

Achinewhu, S.C. 1986a. The effect of fermentation on carbohydrate and fatty acid composition of African oil bean seeds (*Pentaclethra macrophylla*). *Food Chemistry* 19: 105–116.

Achinewhu, S.C. 1986b. Some biochemical and nutritional changes during the fermentation of fluted pumpkin (*Telferia occidentalis*). *Plant Foods for Human Nutrition* 36: 97–106.

Achinewhu, S.C., and Ryley, J. 1986. Effect of fermentation on the thiamin, riboflavin and niacin contents of melon seed (*Citrullus vulgaris*) and African oil bean seed (*Pentaclethra macrophylla*). *Food Chemistry* 20: 243–252.

Aderibigbe, E.Y., and Odunfa, S.A. 1990. Growth and extracellular enzyme production by strains of *Bacillus* species isolated from fermenting African locust bean, iru. *Journal of Applied Bacteriology* 69: 662–671.

Aidoo, K.E. 1986. Lesser-known fermented plant foods. *Tropical Science* 26: 249–258.

Allagheny, N., Obanu, Z.A., Campbell-Platt, G., and Owens, J.D. 1996. Control of ammonia formation during *Bacillus subtilis* fermentation of legumes. *International Journal of Food Microbiology* 29: 321–333.

Antai, S.P., and Ibrahim, M.H. 1986. Micro-organisms associated with African locust bean (*Parkia filicoidea* Welw) fermentation for "dawadawa" production. *Journal of Applied Bacteriology* 61: 145–148.

Arima, S. 1937. Experimental study of antagonistic action of *Bacillus* "natto" towards dysentery *Bacillus*. Part II. Antagonistic action of *Bacillus* "natto" towards Shiga dysentery *Bacillus* in vivo. *Bulletin of the Naval Medical Association* 26: 398–419.

Arogba, S.S., Ademola, A., and Elum, M. 1995. The effect of solvent treatment on the chemical composition and organoleptic acceptability of traditional condiments from Nigeria. *Plant Foods for Human Nutrition* 48: 31–38.

Asano, N. 2003. Glycosidase inhibitors: Update and perspectives on practical use. *Glycobiology* 13: 93R–104R.

Atsumi, S., Umezawa, K., Iinuma, H., Naganawa, H., Nakamura, H., Iitaka, Y., and Takeuchi, T. 1990. Production, isolation and structure determination of a novel β-glucosidase inhibitor, cyclophellitol, from *Phellinus* sp. *The Journal of Antibiotics* 43: 49–53.

Azokpota, P. 2005. Étude de la dynamique physico-chimique et microbiologique de la fermentation des graines de néré pour la production de l'afitin, de l'iru et du sonru. Ph.D. thesis. p. 172. Cotonou, Benin: Université d'Abomey-Calavi; Université Royale Agricole et Vétérinaire de Copenhague au Danemark (in French).

Azokpota, P., Hounhouigan, D.J., and Nago, M.C. 2006a. Microbiological and chemical changes during the fermentation of African locust bean (*Parkia biglobosa*) to produce afitin, iru and sonru, three traditional condiments produced in Benin. *International Journal of Food Microbiology* 107: 304–309.

Azokpota, P., Hounhouigan, D.J., Nago, M.C., and Jakobsen, M. 2006b. Esterase and protease activities of *Bacillus* spp. from afitin, iru and sonru; three African locust bean (*Parkia biglobosa*) condiments from Benin. *African Journal of Biotechnology* 5: 265–272.

Azokpota, P., Hounhouigan, J.D., Annan, N.T., Nago, M.C., and Jakobsen, M. 2008. Diversity of volatile compounds of afitin, iru and sonru, three fermented food condiments from Benin. *World Journal of Microbiology and Biotechnology* 24: 879–885.

Azokpota, P., Hounhouigan, J.D., Annan, N.T., Odjo, T., Nago, M.C., and Jakobsen, M. 2010. Volatile compounds profile and sensory evaluation of Beninese condiments produced by inocula of *Bacillus subtilis*. *Journal of the Science of Food and Agriculture* 90: 438–444.

Balagopalan, C., Padmaja, G., Nanda, S.K., and Moorthy, S.N. 1988. *Cassava in Food, Feed, and Industry*. Boca Raton, FL: CRC Press.

Barimalaa, I.S., Achinewhu, S.C., Yibatama, I., and Amadi, E.N. 1994. Studies on the solid substrate fermentation of bambara groundnut (*Vigna subterranea* (L.) Verdc). *Journal of the Science of Food and Agriculture* 66: 443–446.

Beaumont, M. 2002. Flavouring composition prepared by fermentation with *Bacillus* spp. *International Journal of Food Microbiology* 75: 189–196.

Benjakul, P., and Chuenarrom, C. 2011. Association of dental enamel loss with the pH and titratable acidity of beverages. *Journal of Dental Sciences* 6: 129–133.

Benjakul, P., Prommontri, P., Chuenarrom, C., and Leggat, U. 2011. Effects of sour curry temperature with fermented shrimp paste on surface hardness of tooth enamel. *Songklanakarin Journal of Science and Technology* 33: 419–423.

Bonifait, L., Chandad, F., and Grenier, D. 2009. Probiotics for oral health: Myth or reality? *Journal of the Canadian Dental Association* 75: 585–590.

Borges de Melo, E., da Silveira Gomes, A., and Carvalho, I. 2006. α- and β-Glucosidase inhibitors: Chemical structure and biological activity. *Tetrahedron* 62: 10277–10302.

Bouchenak, M., and Lamri-Senhadji, M. 2013. Nutritional quality of legumes, and their role in cardiometabolic risk prevention: A review. *Journal of Medicinal Food* 16: 185–198.

Buchmann, C., Prehsler, S., Hartl, A., and Vogl, C.R. 2010. The importance of baobab (*Adansonia digitata* L.) in rural West African subsistence – Suggestion of a cautionary approach to international market export of baobab fruits. *Ecology of Food and Nutrition* 49: 145–172.

Caglar, E., Lussi, A., Kargul, B., and Ugur, K. 2006. Fruit yogurt: Any erosive potential regarding teeth? *Quintessence International* 37: 647–651.

Campbell-Platt, G. 1980. African locust bean (parkia species) and its west african fermented food product, dawadawa. *Ecology of Food and Nutrition* 9: 123–132.

Chadare, F.J., Gayet, D.P., Azokpota, P., Nout, M.J.R., Linnemann, A.R., Hounhouigan, J.D., and van Boekel, M.A.J.S. 2010. Three traditional fermented baobab foods from Benin, *Mutchayan, Dikouanyouri* and *Tayohounta*: Preparation, properties, and consumption. *Ecology of Food and Nutrition* 49: 279–297.

Chadare, F.J., Hounhouigan, J.D., Linnemann, A.R., Nout, M.J.R., and van Boekel, M.A.J.S. 2008. Indigenous knowledge and processing of *Adansonia digitata* L. food products in Benin. *Ecology of Food and Nutrition* 47: 338–362.

Chadare, F.J., Linnemann, A.R., Hounhouigan, J.D., Nout, M.J.R., and van Boekel, M.A.J.S. 2009. Baobab food products: A review on their composition and nutritional value. *Critical Reviews in Food Science and Nutrition* 49: 254–274.

Chen, J., Cheng, Y.Q., Yamaki, K., and Li, L.T. 2007. Anti-α-glucosidase activity of Chinese traditionally fermented soybean (douchi). *Food Chemistry* 103: 1091–1096.

Chen, Y.-C., Sugiyama, Y., Abe, N., Kuruto-Niwa, R., Nozawa, R., and Hirota, A. 2005. DPPH radical-scavenging compounds from dou-chi, a soybean fermented food. *Bioscience, Biotechnology, and Biochemistry* 69: 999–1006.

Choi, Y.M., Kim, Y.S., Ra, K.S., and Suh, H.J. 2007. Characteristics of fermentation and bioavailability of isoflavones in Korean soybean paste (doenjang) with application of *Bacillus* sp KH-15. *International Journal of Food Science and Technology* 42: 1497–1503.

Chuenarrom, C., and Benjakul, P. 2010. Dental erosion protection by fermented shrimp paste in acidic food. *Caries Research* 44: 20–23.

Comelli, E.M., Guggenheim, B., Stingele, F., and Neeser, J.R. 2002. Selection of dairy bacterial strains as probiotics for oral health. *European Journal of Oral Sciences* 110: 218–224.

Dajanta, K., Apichartsrangkoon, A., and Chukeatirote, E. 2011a. Antioxidant properties and total phenolics of thua nao (a Thai fermented soybean) as affected by *Bacillus*-fermentation. *Journal of Microbial and Biochemical Technology* 3: 56–59.

Dajanta, K., Apichartsrangkoon, A., Chukeatirote, E., and Frazier, R.A. 2011b. Free-amino acid profiles of thua nao, a Thai fermented soybean. *Food Chemistry* 125: 342–347.

Dakwa, S., Sakyi Dawson, E., Diako, C., Annan, N.T., and Amoa Awua, W.K. 2005. Effect of boiling and roasting on the fermentation of soybeans into dawadawa (soy-dawadawa). *International Journal of Food Microbiology* 104: 69–82.

Darby, E.T. 1892. Dental erosion and the gouty diathesis: Are they usually associated? *Dental Cosmos* 34: 629–638.

Darweesh, L.L., Toma, R.B., Lee, H.C., and Weiss, T.J. 1991. Chemical and sensory evaluation of lipid blends used in pie crusts. *Food Chemistry* 39: 87–98.

Dirar, H.A. 1994. Commentary. The fermented foods of the Sudan. *Ecology of Food and Nutrition* 32: 207–218.

Dulaney, E.L. 1967. Microbial production of amino acids. In *Microbial Technology*, ed. H.J. Peppler, pp. 308–343. New York: Reinhold Publishing Corporation.

Ebata, N., Matui, K., and Uno, R. 1960. The efficaciousness of *Bacillus natto*. *Science Report of the Faculty of Liberal Arts and Education, Gifu University. Natural Science* 2: 384–387 (in Japanese).

Eguchi, Y., and Kumashiro, N. 1931. A cured case of paratyphoid B bacilli-career by eating of natto (first report). *Bulletin of the Naval Medical Association* 20: 245–247 (in Japanese).

Eka, O.W. 1980. Effect of fermentation on the nutrient status of locust beans. *Food Chemistry* 5: 303–308.

Esaki, H., Nohara, Y., Onozaki, H., and Osawa, T. 1990. Antioxidative activity of natto. *Nippon Shokuhin Kogyo Gakkaishi* 37: 474–477.

Esenwah, C.N., and Ikenebomeh, M.J. 2008. Processing effects on the nutritional and anti-nutritional contents of African locust bean (*Parkia biglobosa* Benth.) seed. *Pakistan Journal of Nutrition* 7: 214–217.

Ezure, Y., Maruo, S., Miyazaki, K., and Kawamata, M. 1985. Moranoline (1-deoxynojirimycin) fermentation and its improvement. *Agricultural and Biological Chemistry* 49: 1119–1125.

Faithong, N., Benjakul, S., Phatcharat, S., and Bindan, W. 2010. Chemical composition and antioxidative activity of Thai traditional fermented shrimp and krill products. *Food Chemistry* 119: 133–140.

Fan, J., Hu, X., Tan, S., Zhang, Y., Tatsumi, E., and Li, L.T. 2009. Isolation and characterisation of a novel angiotensin I-converting enzyme-inhibitory peptide derived from douchi, a traditional Chinese fermented soybean food. *Journal of the Science of Food and Agriculture* 89: 603–608.

Featherstone, J.D., and Lussi, A. 2006. Understanding the chemistry of dental erosion. *Monographs in Oral Science* 20: 66–76.

Ferguson, E.L., Gibson, R.S., Opare-Obisaw, C., Osei-Opare, F., Stephen, A.M., Lehrfeld, J., and Thompson, L.U. 1993. The zinc, calcium, copper, manganese, nonstarch polysaccharide and phytate content of seventy-eight locally grown and prepared African foods. *Journal of Food Composition and Analysis* 6: 87–99.

Fetuga, B.L., Babatunde, G.M., and Oyenuga, V.A. 1973. Proteins quality of some Nigerian feedstuffs. I. Chemical assay of nutrients and amino acid composition. *Journal of the Science of Food and Agriculture* 24: 1505–1514.

Fujita, M., Hong, K., Ito, Y., Fujii, R., Kariya, K., and Nishimuro, S. 1995. Thrombolytic effect of nattokinase on a chemically induced thrombosis model in rat. *Biological and Pharmaceutical Bulletin* 18: 1387–1391.

Fujita, M., Nomura, K., Hong, K., Ito, Y., Asada, A., and Nishimuro, S. 1993. Purification and characterization of a strong fibrinolytic enzyme (nattokinase) in the vegetable cheese natto, a popular fermented soybean fermented food in Japan. *Biochemical and Biophysical Research Communications* 197: 1340–1347.

Gedalia, I., Dakuar, A., Shapira, L., Lewinstein, I., Goultschin, J., and Rahamim, E. 1991. Enamel softening with Coca-Cola and rehardening with milk or saliva. *American Journal of Dentistry* 4: 120–122.

Giami, Y.S., and Isichei, I. 1999. Preparation and properties of flours and protein concentrates from raw, fermented and germinated fluted pumpkin (*Telfairia occidentalis* Hook) seeds. *Plant Foods for Human Nutrition* 54: 67–77.

Gonzalez-Martin, I., Hernandez-Hierro, J.M., Revilla, I., Vivar-Quintana, A., and Ortega, I.L. 2011. The mineral composition (Ca, P, Mg, K, Na) in cheeses (cow's, ewe's and goat's) with different ripening times using near infrared spectroscopy with a fibre-optic probe. *Food Chemistry* 127: 147–152.

Gutierrez, M.L., Maizi, P., Nago, C.M., and Hounhouigan, D.J. 2000. *Production et commercialisation de l'Afitin dans la région d'Abomey-Bohicon au Bénin*. Montpellier, France: CERNA, CNEARC, CIRAD: Librairie de CIRAD (in French).

Hamajima, K., Yamada, S., Okuda, K., and Tadokoro, I. 1983. Immunoregulatory effects of so-called *Bacillus natto*. *Yokohama Medical Journal* 34: 139–142 (in Japanese).

Hanefeld, M. 2007. Cardiovascular benefits and safety profile of acarbose therapy in prediabetes and established type 2 diabetes. *Cardiovascular Diabetology* 6: 20.

Hata, Y., Yamamoto, M., Ohni, M., Nakajima, K., Nakamura, Y., and Takano, T. 1996. A placebo-controlled study of the effect of sour milk on blood pressure in hypertensive subjects. *The American Journal of Clinical Nutrition* 64: 767–771.

Hattori, T., Ohishi, H., Yokota, T., Ohoami, H., and Watanabe, K. 1995a. Antioxidative effect of crude antioxidant preparation from soybean fermented by *Bacillus natto*. *LWT - Food Science and Technology* 28: 135–138.

Hattori, T., Ohishi, H., Yokota, T., Ohoami, H., and Watanabe, K. 1995b. Beneficial effect of crude antioxidant preparation from fermented soybean food on xanthine oxidase-hypoxanthine-induced foot-edema in rats. *LWT - Food Science and Technology* 28: 169–173.

He, H., Chen, X., Sun, C., Zhang, Y., and Gao, P. 2006. Preparation and functional evaluation of oligopeptide-enriched hydrolysate from shrimp (*Acetes chinensis*) treated with crude protease from *Bacillus* sp. SM98011. *Bioresource Technology* 97: 385–390.

Heu, M.S., Kim, J.S., and Shahidi, F. 2003. Components and nutritional quality of shrimp processing by-products. *Food Chemistry* 82: 235–242.

Hiroyuki, F., Tomohide, Y., and Kazunori, O. 2001. Efficacy and safety of Touchi extract, an α-glucosidase inhibitor derived from fermented soybeans, in non-insulin-dependent diabetic mellitus. *The Journal of Nutritional Biochemistry* 12: 351–356.

Hisanaga, S., Mase, M., Mizuno, A., Hayakawa, Y., Ohkubo, T., and Hachisuka, Y. 1978. Germination of *Bacillus natto* spores in the canine intestine. *Japanese Journal of Bacteriology (Nippon Saikingaku Zasshi)* 33: 689–696 (in Japanese).

Hooper, S., West, N.X., Sharif, N., Smith, S., North, M., De'Ath, J., Parker, D.M., Roedig-Penman, A., and Addy, M. 2004. A comparison of enamel erosion by a new sports drink compared to two proprietary products: A controlled, crossover study in situ. *Journal of Dentistry* 32: 541–545.

Hosoi, T., Ametani, A., Kiuchi, K., and Kaminogawa, S. 1999. Changes in fecal microflora induced by intubation of mice with *Bacillus subtilis* (*natto*) spores are dependent upon dietary components. *Canadian Journal of Microbiology* 45: 59–66.

Hosoi, T., Hirose, R., Saegusa, S., Ametani, A., Kiuchi, K., and Kaminogawa, S. 2003. Cytokine responses of human intestinal epithelial-like Caco-2 cells to the nonpathogenic bacterium *Bacillus subtilis* (*natto*). *International Journal of Food Microbiology* 82: 255–264.

Hu, F.B., Manson, J.A.E., Stampfer, M.J., Colditz, G., Liu, S., Solomon, C.G., and Willett, W.C. 2001. Diet, lifestyle, and the risk of type 2 diabetes mellitus in women. *New England Journal of Medicine* 345: 790–797.

Ibe, S., Yoshida, K., Kumada, K., Tsurushin, S., Furusho, T., and Otobe, K. 2009. Antihypertensive effects of natto, a traditional Japanese fermented food, in spontaneously hypertensive rats. *Food Science and Technology Research* 15: 199–202.

Ichimura, T., Hu, J.N., Aita, D.Q., and Maruyama, S. 2003. Angiotensin I-converting enzyme inhibitory activity and insulin secretion stimulative activity of fermented fish sauce. *Journal of Bioscience and Bioengineering* 96: 496–499.

Ijarotimi, O.S., and Esho, T.R. 2009. Comparison of nutritional composition and anti-nutrient status of fermented, germinated and roasted bambara groundnut seeds (*Vigna subterranea*). *British Food Journal* 111: 376–386.

Ijarotimi, O.S., and Keshinro, O.O. 2012. Comparison between the amino acid, fatty acid, mineral and nutritional quality of raw, germinated and fermented african locust bean (*Parkia biglobosa*) flour. *Acta Scientiarum Polonorum, Technologia Alimentaria* 11: 151–165.

Inatsu, Y., Nakamura, N., Yuriko, Y., Fushimi, T., Watanasiritum, L., and Kawamoto, S. 2006. Characterization of *Bacillus subtilis* strains in Thua nao, a traditional fermented soybean food in northern Thailand. *Letters in Applied Microbiology* 43: 237–242.

Inooka, S., and Kimura, M. 1983. The effect of *Bacillus natto* in feed on the sheep red blood cell antibody response in chickens. *Avian Diseases* 27: 1086–1089.

Iroyukifujita, H., Eiichiyokoyama, K., and Yoshikawa, M. 2008. Classification and antihypertensive activity of angiotensin I-converting enzyme inhibitory peptides derived from food proteins. *Journal of Food Science* 65: 564–569.

Irvine, F.R. 1961. *Woody Plants of Ghana, With Special Reference to Their Uses*. London: Oxford University Press.

Isshiki, S., and Onozaki, H. 1992. Effects of natto and steamed soybeans on intestinal microflora in rats. *Journal of Sugiyama Jogakuen University (Part 1)* 23: 273–283 (in Japanese).

Isu, N.R., and Ofuya, C.O. 2000. Improvement of the traditional processing and fermentation of African oil bean (*Pentaclethra macrophylla* Bentham) into a food snack – "ugba." *International Journal of Food Microbiology* 59: 235–239.

Iwai, K., Nakaya, N., Kawasaki, Y., and Matsue, H. 2002a. Antioxidative functions of natto, a kind of fermented soybeans: Effect on LDL oxidation and lipid metabolism in cholesterol-fed rats. *Journal of Agricultural and Food Chemistry* 50: 3597–3601.

Iwai, K., Nakaya, N., Kawasaki, Y., and Matsue, H. 2002b. Inhibitory effect of natto, a kind of fermented soybeans, on LDL oxidation in vitro. *Journal of Agricultural and Food Chemistry* 50: 3592–3596.

Iwasa, T., Yamamoto, H., and Shibata, M. 1970. Studies on validamycins, new antibiotics. I. *Streptomyces hygroscopicus* var. *limoneus* nov. var., validamycin-producing organism. *The Journal of Antibiotics* 23: 595–602.

Iwuoha, C.I., and Eke, O.S. 1996. Nigerian indigenous fermented foods: Their traditional process operation, inherent problems, improvements and current status. *Food Research International* 29: 527–540.

Je, J.Y., Park, J.Y., Jung, W.K., Park, P.J., and Kim, S.K. 2005a. Isolation of angiotensin I converting enzyme (ACE) inhibitor from fermented oyster sauce, *Crassostrea gigas*. *Food Chemistry* 90: 809–814.

Je, J.Y., Park, P.J., Byun, H.G., Jung, W.K., and Kim, S.K. 2005b. Angiotensin I converting enzyme (ACE) inhibitory peptide derived from the sauce of fermented blue mussel, *Mytilus edulis*. *Bioresource Technology* 96: 1624–1629.

Je, J.Y., Qian, Z.J., Byun, H.G., and Kim, S.K. 2007. Purification and characterization of an antioxidant peptide obtained from tuna backbone protein by enzymatic hydrolysis. *Process Biochemistry* 42: 840–846.

Jolaoso, A.A., Ajayi, J.O., Ogunmuyiwa, S.I.O., and Albert, O.M. 2012. Changes in functional properties as a measure of biochemical deterioration of oso (fermented seeds of *Cathormion altissimum*). *Journal of Emerging Trends in Engineering and Applied Sciences* 3: 608–613.

Jood, S., Mehta, U., Singh, R., and Bhat, C.M. 1985. Effect of processing on flatus-producing factors in legumes. *Journal of Agricultural and Food Chemistry* 33: 268–271.

Jung, K.O., Park, S.Y., and Park, K.Y. 2006. Longer aging time increases the anticancer and antimetastatic properties of doenjang. *Nutrition* 22: 539–545.

Kameda, Y., Asano, N., Teranishi, M., and Matsui, K. 1980. New cyclitols, degradation of validamycin A by *Flavobacterium saccharophilum*. *The Journal of Antibiotics* 33: 1573–1574.

Kameda, Y., Kanatomo, S., Kameda, Y., and Saito, Y. 1968. A contact antitumor activity of *Bacillus natto* on solid type Ehrlich carcinoma cells. *Chemical and Pharmaceutical Bulletin* 16: 186–187.

Kameda, Y., Kanatomo, S., Matsui, K., Nakabayashi, T., Ueno, K., Nagai, S., and Ohki, K. 1978. Antitumor activity of *Bacillus natto*. VI. Analysis of cytolytic activity on Ehrlich ascites carcinoma cells in the culture medium of *Bacillus natto* KMD 1126. *Yakugaku Zasshi* 98: 1432–1435 (in Japanese).

Kameda, Y., Matsui, K., Kato, H., Yamada, T., and Sagai, H. 1972. Antitumor activity of *Bacillus natto*. III. Isolation and characterization of a cytolytic substance on Ehrlich ascites carcinoma cells in the culture medium of *Bacillus natto* KMD 1126. *Chemical and Pharmaceutical Bulletin* 20: 1551–1557.

Kameda, Y., Sagai, H., Kamada, T., Kanatomo, S., and Matsui, K. 1971. Antitumor activity of *Bacillus natto*. II. Formation of cytolytic substances on Ehrlich ascites carcinoma in *Bacillus natto* KMD 1126. *Chemical and Pharmaceutical Bulletin* 19: 2572–2578.

Kang, M.S., Chung, J., Kim, S.M., Yang, K.H., and Oh, J.S. 2006. Effect of *Weissella cibaria* isolates on the formation of *Streptococcus mutans* biofilm. *Caries Research* 40: 418–425.

Kanno, A., Takamatsu, H., and Takano, N. 1984. Determination of several volatile components produced by *Bacillus natto* in commercial natto. *Nippon Shokuhin Kogyo Gakkaishi* 31: 587–595.

Kanno, A., Takamatsu, H., Takano, N., and Akimoto, T. 1982. Studies on natto. Part I. Change of saccharides in soybeans during manufacturing of natto. *Nippon Shokuhin Kogyo Gakkaishi* 29: 105–110 (in Japanese).

Kargul, B., Caglar, E., and Lussi, A. 2007. Erosive and buffering capacities of yogurt. *Quintessence International* 38: 381–385.

Kato, E. 1985. The manufacture of natto, traditional and modern. Proceedings, *Asian Symposium on Non-salted Soybean Fermentation*. Tsukuba Science City, Ibaraki, Japan.

Katsuyama, H., Ideguchi, S., Fukunaga, M., Saijoh, K., and Sunami, S. 2002. Usual dietary intake of fermented soybeans (Natto) is associated with bone mineral density in premenopausal women. *Journal of Nutritional Science and Vitaminology* 48: 207–215.

Kawabata, M., Taguchi, K., and Ohtsuki, K. 1986. Food chemical evaluation of fermented okara products, okara-natto and okara-tempeh. *Scientific Reports of Kyoto Prefectural University. Natural Science and Living Science* 37: 39–45 (in Japanese).

Kawane, M. 2007. Development of *Bacillus subtilis* (*natto*) controlling immune function. *Shokuhin to Gijutsu* 433: 11–17 (in Japanese, translated title).

Kearney, P.M., Whelton, M., Reynolds, K., Muntner, P., Whelton, P.K., and He, J. 2005. Global burden of hypertension: Analysis of worldwide data. *Lancet* 365: 217–223.

Kiers, J.L., van Laeken, A.E.A., Rombouts, F.M., and Nout, M.J.R. 2000. *In vitro* digestibility of *Bacillus* fermented soya bean. *International Journal of Food Microbiology* 60: 163–169.

Kimura, M. 1979. Effect of a feed-additive, *Bacillus natto* (BN strain) product, on economic performances of animals and poultry. *Animal-Husbandry* 33: 31–36 (in Japanese, translated title).

Kingsley, M.O. 1995. Effect of processing on some antinutritive and toxic components and on the nutritional composition of the African oil bean seed (*Pentaclethra macrophylla* Benth). *Journal of the Science of Food and Agriculture* 68: 153–158.

Kinouchi, F.L., Maia, D.C.G., de Abreu Ribeiro, L.C., Placeres, M.C.P., de Valdez, G.F., Colombo, L.L., Rossi, E.A., and Carlos, I.Z. 2012. A soy-based product fermented by *Enterococcus faecium* and *Lactobacillus helveticus* inhibits the development of murine breast adenocarcinoma. *Food and Chemical Toxicology* 50: 4144–4148.

Kiuchi, K., Ohta, T., Itoh, H., Takabayashi, T., and Ebine, H. 1976. Lipids of natto. *Journal of Agricultural and Food Chemistry* 24: 404–407.

Kiuchi, K., and Suzuki, H. 1991. The current state on development of natto containing a high activity of nattokinase. *Food Science* 33: 85–89 (in Japanese, translated title).

Knowles, C.J. 1976. Microorganisms and cyanide. *Bacteriological Reviews* 40: 652–680.

Kobawila, S.C., Louembé, D., Keleke, S., Hounhouigan, D.J., and Gamba, C. 2005. Reduction of the cyanide content during fermentation of cassava roots and leaves to produce bikedi and ntoba mbodi, two food products from Congo. *African Journal of Biotechnology* 4: 689–696.

Kobawila, S.C., Louembé, D., Keleke, S., and Traore, A.S. 2003. Aspects physico-chimiques et biochimiques de la fermentation des feuilles de manioc en ntoba mbodi. *Procédés Biologiques et Alimentaires* 1: 106–119 (in French).

Kolawole, D.O., and Okonkwo, B.A. 1985. Microbiology of traditional fermentation of oil bean (*Pentaclethra macrophylla* Benth) seeds for ukpaka, a Nigerian condiment. *Nigerian Food Journal* 3: 149–152.

Komatsuzaki, T., Ohkuro, I., Kawashima, M., and Ito, S. 1978. Non-specific immunity within a short time after pretreatment with natto bacilli. *Medicine and Biology* 97: 459–463 (in Japanese).

Komatsuzaki, T., Ohkuro, I., Kawashima, M., and Kuriyama, S. 1979. The difference in unspecifically immunizing effects among strains of natto bacilli. *Medicine and Biology* 99: 343–345 (in Japanese).

Komatsuzaki, T., Ohkuro, I., Kohno, T., and Ito, S. 1976. Virulence and lipase of *Bacillus natto*. *Medicinice and Biology* 93: 403–406 (in Japanese).

Konlani, S., Lamboni, C., Ameyapoh, Y., and De Souza, C. 1999. Microbiology and biochemical composition of "Tonou", a food condiment from fermented seeds of *Parkia biglobosa* (locust bean) and soybean. *Food Biotechnology* 13: 217–225.

Kosuge, T., Kamiya, H., and Adachi, T. 1961. Nattô no níoí seíbun ní tsuíte (An odorous component of natto, fermented soybeans). *Yakugaku Zasshi (Journal of the Pharmaceutical Society of Japan)* 82: 190 (in Japanese).

Kuba, M., Tanaka, K., Tawata, S., Takeda, Y., and Yasuda, M. 2003. Angiotensin I-converting enzyme inhibitory peptides isolated from tofuyo fermented soybean food. *Bioscience, Biotechnology, and Biochemistry* 67: 1278–1283.

Kuboye, A.O. 1985. Traditional fermented foods and beverages of Nigeria. In *Development of Indigenous Fermented Foods and Food Technology in Africa*, pp. 224–237. Stockholm, Sweden: International Foundation for Science.

Kumada, K., Onga, T., Hoshino, H., Tomoyose, R., and Koike, K. 1993. Isolation of *Bacillus subtilis natto* which shows high fibrinolytic activity. *Medicine and Biology* 126: 299–303 (in Japanese).

Kumada, K., Onga, T., Hoshino, H., Tomoyose, R., and Koike, K. 1994. The effect of natto possessing a high fibrinolytic activity in human plasma. *Medicine and Biology* 128: 117–119 (in Japanese).

Kuo, H.T., Wu, P.H., Chung, H.H., Jiang, C.M., and Wu, M.C. 2013. Koji fermentation improve the protein digestibility of sorghum. *Journal of Food Processing and Preservation* 37: 419–423.

Kuswanto, K.R. 2004. Industrialization of tempe fermentation. In *Industrialization of Indigenous Fermented Foods*, 2nd ed., ed. K.H. Steinkraus, pp. 587–635. New York: Marcel Dekker.

Kwon, Y.I., Son, H.J., Kyoung, S.I.K.M., Kim, J.K., Kim, J.G., Chun, H.S., Soon, K.I.L.A.H.N., and Chung, I.L.H. 2002. Novel α-glucosidase inhibitors, CKD-711 and CKD-711a produced by *Streptomyces* sp. CK-4416. II. Biological properties. *The Journal of Antibiotics* 55: 462–466.

Lasekan, O.O., Feijiao Teixeira, J.P., and Salva, T.J.G. 2001. Volatile flavour compounds of cooked acha (*Digitaria exilis* Stapf). *Food Chemistry* 75: 333–337.

Lee, D.S. 2000. Dibutyl phthalate, an α-glucosidase inhibitor from *Streptomyces melanosporofaciens*. *Journal of Bioscience and Bioengineering* 89: 271–273.

Leejeerajumnean, A., Duckham, S.C., Owens, J.D., and Ames, J.M. 2001. Volatile compounds in *Bacillus* fermented soybeans. *Journal of the Science of Food and Agriculture* 81: 525–529.

Li, F.J., Yin, L.J., Lu, X., and Li, L.T. 2010. Changes in angiotensin I-converting enzyme inhibitory activities during the ripening of douchi (a Chinese traditional soybean product) fermented by various starter cultures. *International Journal of Food Properties* 13: 512–524.

Li, G.H., Le, G.W., Shi, Y.H., and Shrestha, S. 2004. Angiotensin I-converting enzyme inhibitory peptides derived from food proteins and their physiological and pharmacological effects. *Nutrition Research* 24: 469–486.

Li, H., Feng, F.-Q., Shen, L.-R., Xie, Y., and Li, D. 2007. Nutritional evaluation of different bacterial douchi. *Asia Pacific Journal of Clinical Nutrition* 16: Suppl. 1, 215–221.

Lodi, C.S., Sassaki, K.T., Fraiz, F.C., Delbem, A.C., and Martinhon, C.C. 2010. Evaluation of some properties of fermented milk beverages that affect the demineralization of dental enamel. *Brazilian Oral Research* 24: 95–101.

Maga, J.A., and Katz, I. 1982. Pyrazines in foods: An update. *CRC Critical Reviews in Food Science and Nutrition* 16: 1–48.

Maruta, K., Miyazaki, H., Masuda, S., Takahashi, M., Marubashi, T., Tadano, Y., and Takahashi, H. 1996a. Exclusion of intestinal pathogens by continuous feeding with *Bacillus subtilis* C-3102 and its influence on the intestinal microflora in broilers. *Animal Science and Technology (Japan)* 67: 273–280.

Maruta, K., Miyazaki, H., Tadano, Y., Masuda, S., Suzuki, A., Takahashi, H., and Takahashi, M. 1996b. Effects of *Bacillus subtilis* C-3102 intake on fecal flora of sows and on diarrhea and mortality rate of their piglets. *Animal Science and Technology (Japan)* 67: 403–409.

Matsumura, T. 1934a. Experimentelle Studien zur Bekämpfung der Bakterienträger mittels der Antagonistischen Wirkung anderer Bakterien. Teil I. Untersuchung von *Bacillus natto*. Versuche in Reagenzgläsern. *Mitteilungen aus der Medizinischen Akademie zu Kioto* 12: 38–52 (in Japanese).

Matsumura, T. 1934b. Experimentelle Studien zur Bekämpfung der Bakterienträger mittels der Antagonistischen Wirkung anderer Bakterien. Teil II. Versuche über die Antagonistische Wirkung des Natto-Bazillus gegen die Typhusbazillengruppe. Versuche in Reagenzgläsern. *Mitteilungen aus der Medizinischen Akademie zu Kioto* 12: 53–89 (in Japanese).

Matsumura, T. 1934c. Experimentelle Studien zur Bekämpfung der Bakterienträger mittels der Antagonistischen Wirkung anderer Bakterien. Teil III. Versuche über die Antagonistische Wirkung des Natto-Bazillus gegen die typhusbazillengruppe im Tierkörper. Versuche in Reagenzgläsern. *Mitteilungen aus der Medizinischen Akademie zu Kioto* 12: 1185–1210 (in Japanese).

Matsuura, M., Obata, A., and Fukushima, D. 1989. Objectionable flavor of soymilk developed during the soaking of soybeans and its control. *Journal of Food Science* 54: 602–605.

Mbaiguinam, M., Mahmout, Y., Tarkodjie, M., Delobe, B., and Bessiere, J.M. 2005. Constituents of Kawal, fermented *Cassia obtusifolia* leaves, a traditional food from Chad. *African Journal of Biotechnology* 4: 1080–1083.

Meurman, J.H., Antila, H., Korhonen, A., and Salminen, S. 1995. Effect of *Lactobacillus rhamnosus* strain GG (ATCC 53103) on the growth of *Streptococcus sobrinus* in vitro. *European Journal of Oral Sciences* 103: 253–258.

Mitsui, N., Kajimoto, O., Tsukahara, M., Murasawa, H., Tamura, M., Nishimura, A., Kajimoto, Y., and Benno, Y. 2006. Effect of natto including *Bacillus subtilis* K-2 (spore) on defecation and fecal microbiota, and safety of excessive ingestion in healthy volunteers. *Japanese Pharmacology and Therapeutics* 34: 135–148 (in Japanese).

Miyagi, Y., Shinjo, S., Nishida, R., Miyagi, C., Takamatsu, K., Yamamoto, T., and Yamamoto, S. 1997. Trypsin inhibitor activity in commercial soybean products in Japan. *Journal of Nutritional Science and Vitaminology (Tokyo)* 43: 575–580.

Mokemiabeka, S., Dhellot, J., Kobawila, S.C., Diakabana, P., Ntietie Loukombo, R.N., Nyanga-Koumou, A.G., and Louembe, D. 2011. Softening and mineral content of cassava (*Manihot esculenta* Crantz) leaves during the fermentation to produce Ntoba Mbodi. *Advance Journal of Food Science and Technology* 3: 418–423.

Moktan, B., Saha, J., and Sarkar, P.K. 2008. Antioxidant activities of soybean as affected by *Bacillus*-fermentation to kinema. *Food Research International* 41: 586–593.

Montogomery, R.D. 1980. Cyanogens. In *Toxic Constituents of Plant Foodstuffs*, ed. E. Liener, pp. 143–160. New York: Academic Press.

Murai, A., Iwamura, K., Takada, M., Ogawa, K., Usui, T., and Okumura, J. 2002. Control of postprandial hyperglycaemia by galactosyl maltobionolactone and its novel anti-amylase effect in mice. *Life Sciences* 71: 1405–1415.

Muramatsu, K., Kanai, Y., Kimura, N., Miura, N., Yoshida, K., and Kiuchi, K. 1994. Production of natto with high elastase activity. *Nippon Shokuhin Kagaku Kogaku Kaishi* 41: 123–128 (in Japanese).

Murao, S., and Miyata, S. 1980. Isolation and characterization of a new trehalase inhibitor, S-GI. *Agricultural and Biological Chemistry* 44: 219–221.

Murata, H., Yaguchi, H., Sano, H., and Namioka, S. 1977. Effect of administration of *Bacillus natto* upon weanling piglets. *Journal of the Japan Veterinary Medical Association* 30: 645–649.

Naczk, M., Amarowicz, R., and Shahidi, F. 1997. α-Galactosides of sucrose in foods: Composition, flatulence-causing effects, and removal. Washington, DC: ACS. *Symposium Series* 662: 127–151.

Nagai, T., Nishimura, K., Suzuki, H., Banba, Y., Sasaki, H., and Kiuchi, K. 1994. Isolation and characterization of a *Bacillus subtilis* strain producing natto with strong umami-taste and high viscosity. *Nippon Shokuhin Kogyo Gakkaishi* 41: 123–128 (in Japanese).

Namgung, H.J., Park, H.J., Cho, I.H., Choi, H.K., Kwon, D.Y., Shim, S.M., and Kim, Y.S. 2010. Metabolite profiling of doenjang, fermented soybean paste, during fermentation. *Journal of the Science of Food and Agriculture* 90: 1926–1935.

Näse, L., Hatakka, K., Savilahti, E., Saxelin, M., Pönkä, A., Poussa, T., Korpela, R., and Meurman, J.H. 2001. Effect of long-term consumption of a probiotic bacterium, *Lactobacillus rhamnosus* GG, in milk on dental caries and caries risk in children. *Caries Research* 35: 412–420.

Nikawa, H., Makihira, S., Fukushima, H., Nishimura, H., Ozaki, Y., Ishida, K., Darmawan, S., Hamada, T., Hara, K., Matsumoto, A., Takemoto, T., and Aimi, R. 2004. *Lactobacillus reuteri* in bovine milk fermented decreases the oral carriage of mutans streptococci. *International Journal of Food Microbiology* 95: 219–223.

Niwa, T., Inouye, S., Tsuruoka, T., Koaze, Y., and Niida, T. 1970. "Nojirimycin" as a potent inhibitor of glucosidase. *Agricultural and Biological Chemistry* 34: 966–968.

Njoku, H.O., and Okemadu, C.P. 1989. Biochemical changes during the natural fermentation of the African oil bean seed (*Pentaclethra macrophylla*) for the production of ugba. *Journal of the Science of Food and Agriculture* 49: 457–465.

Noda, H. 2005. Study of physiologically active substance related to blood in fermented foods. *Bulletin of Ichinomiya Women's Junior College* 44: 11–17. (in Japanese).

Nout, M.J.R., and Rombouts, F.M. 1990. A review: Recent developments in tempe research. *Journal of Applied Bacteriology* 69: 609–633.

Odibo, F.J.C., Ezeaku, E.O., and Ogbo, F.C. 2008. Biochemical changes during the fermentation of *Prosopis africana* seeds for ogiri-okpei production. *Journal of Industrial Microbiology and Biotechnology* 35: 947–952.

Odunfa, S.A. 1981. Microbiology and amino acid composition of Ogiri–A food condiment from fermented melon seeds. *Food/Nahrung* 25: 811–816.

Odunfa, S.A. 1983a. Biochemical changes during production of ogiri, a fermented melon (*Citrullus vulgaris* Schrad) product. *Plant Foods for Human Nutrition* 32: 11–18.

Odunfa, S.A. 1983b. Carbohydrate changes in fermenting locust bean during iru preparation. *Plant Foods for Human Nutrition* 32: 3–10.

Odunfa, S.A. 1985a. Biochemical changes in fermenting African locust bean (*Parkia biglobosa*) during iru fermentation. *Journal of Food Technology* 20: 295–303.

Odunfa, S.A. 1985b. Microbiological and toxicological aspects of fermentation of castor oil seeds for ogiri production. *Journal of Food Science* 50: 1758–1759.

Odunfa, S.A., and Oyewole, O.B. 1998. African fermented foods. In *Microbiology of Fermented Foods*, 2nd ed., vol. 2, ed. B.J.B. Wood, pp. 713–752. London: Blackie Academic & Professional.

Odunfa, S.A., and Oyeyiola, G.F. 1985. Microbiological study of the fermentation of ugba, a Nigerian indigenous fermented food flavour. *Journal of Plant Foods* 6: 155–163.

Ogbonna, D.N., Sokari, T.G., and Achinewhu, S.C. 2001. Development of an owoh-type product from African yam beans (*Sphenostylis stenocarpa*) (Hoechst (ex. A. Rich.) Harms.) seeds by solid substrate fermentation. *Plant Foods for Human Nutrition* 56: 183–194.

Ogueke, C.C., Nwosu, J.N., Owuamanam, C.I., and Iwouno, J.N. 2010. Ugba, the fermented African oilbean seeds: Its production, hemical composition, preservation, safety and health benefits. *Pakistan Journal of Biological Sciences* 13: 489–496.

Ohkuro, I., Komatsuzaki, T., Kuriyama, S., and Kawashima, M. 1980. Nonspecific immunizing effects of vegetative cells and spores of natto bacilli. *Medicine and Biology* 100: 351–355 (in Japanese).

Ohkuro, I., Suzuki, K., Ito, S., and Komatsuzaki, T. 1974a. Influences of intraperitoneal injection of esculent bacteria ("*Bacillus natto*" and *Lactobacillus*) on the mouse. *Medicine and Biology* 88: 115–120 (in Japanese).

Ohkuro, I., Suzuki, K., Ito, S., and Komatsuzaki, T. 1974b. Resistance of mice pretreated with "*Bacillus natto*" to lethal challenge with *Staphylococcus aureus*. *Medicine and Biology* 89: 35–40 (in Japanese).

Ohta, T. 1986. Natto. In *Legume-Based Fermented Foods*, eds. N.R. Reddy, M.D. Pierson, and D.K. Salunkhe, pp. 85–93. Boca Raton, FL: CRC Press.

Okafor, N. 1977. Micro-organisms associated with cassava fermentation for garri production. *Journal of Applied Bacteriology* 42: 279–284.

Oshima, G., Shimabukuro, H., and Nagasawa, K. 1979. Peptide inhibitors of angiotensin I-converting enzyme in digests of gelatin by bacterial collagenase. *Biochimica et Biophysica Acta (BBA)-Enzymology* 566: 128–137.

Ouoba, L.I.I., Cantor, M.D., Diawara, B., Traore, A.S., and Jakobsen, M. 2003a. Degradation of African locust bean oil by *Bacillus subtilis* and *Bacillus pumilus* isolated from soumbala, a fermented African locust bean condiment. *Journal of Applied Microbiology* 95: 868–873.

Ouoba, L.I.I., Diawara, B., Annan, N.T., Poll, L., and Jakobsen, M. 2005. Volatile compounds of Soumbala, a fermented African locust bean (*Parkia biglobosa*) food condiment. *Journal of Applied Microbiology* 99: 1413–1421.

Ouoba, L.I.I., Nyanga-Koumou, C.A.G., Parkouda, C., Sawadogo, H., Kobawila, S.C., Keleke, S., Diawara, B., Louembe, D., and Sutherland, J.P. 2010. Genotypic diversity of lactic acid bacteria isolated from African traditional alkaline-fermented foods. *Journal of Applied Microbiology* 108: 2019–2029.

Ouoba, L.I.I., Rechinger, K.B., Barkholt, V., Diawara, B., Traore, A.S., and Jakobsen, M. 2003b. Degradation of proteins during the fermentation of African locust bean *Parkia biglobosa*, by strains of *Bacillus subtilis* and *Bacillus pumilus* for production of soumbala. *Journal of Applied Microbiology* 94: 396–402.

Owens, J.D., Allagheny, N., Kipping, G., and Ames, J.M. 1997. Formation of volatile compounds during *Bacillus subtilis* fermentation of soya beans. *Journal of the Science of Food and Agriculture* 74: 132–140.

Ozawa, K., Yabu-uchi, K., Yamanaka, K., Yamashita, Y., Ueba, K., and Miwatani, T. 1979. Antagonistic effects of *Bacillus natto* and *Streptococcus faecalis* on growth of *Candida albicans*. *Microbiology and Immunology* 23: 1147–1156.

Ozawa, K., Yokota, H., Kimura, M., and Mitsuoka, T. 1981. Effects of administration of *Bacillus subtilis* strain BN on intestinal flora of weanling piglets. *Japanese Journal of Veterinary Science* 43: 771–775.

Parkouda, C., Nielsen, D.S., Azokpota, P., Ouoba, L.I.I., Amoa-Awua, W.K., Thorsen, L., Hounhouigan, J.D., Jensen, J.S., Tano-Debrah, K., Diawara, B., and Jakobsen, M. 2009. The microbiology of alkaline-fermentation of indigenous seeds used as food condiments in Africa and Asia. *Critical Reviews in Microbiology* 35: 139–156.

Parry, J., Shaw, L., Arnaud, M.J., and Smith, A.J. 2001. Investigation of mineral waters and soft drinks in relation to dental erosion. *Journal of Oral Rehabilitation* 28: 766–772.

Peralta, E.M., Hatate, H., Kawabe, D., Kuwahara, R., Wakamatsu, S., Yuki, T., and Murata, H. 2008. Improving antioxidant activity and nutritional components of Philippine salt-fermented shrimp paste through prolonged fermentation. *Food Chemistry* 111: 72–77.

Petti, S., Tarsitani, G., and Simonetti D'Arca, A. 2008. Antibacterial activity of yoghurt against viridans streptococci in vitro. *Archives of Oral Biology* 53: 985–990.

Phromraksa, P., Nagano, H., Boonmars, T., and Kamboonruang, C. 2008. Identification of proteolytic bacteria from Thai traditional fermented foods and their allergenic reducing potentials. *Journal of Food Science* 73: M189–M195.

Piggott, J.R., and Paterson, A. eds. 1994. *Understanding Natural Flavours*. London: Blackie Academic & Professional.

Popoola, T.O.S., Jolaoso, A.A., Afolabi, R.O., and Akintokun, A.K. 2006. Microbiological studies on the production of Oso: A condiment made from fermenting seeds of *Cathormion altissimum*. *Journal of Herbs, Spices and Medicinal Plants* 12: 129–138.

Pozo-Bayón, M.A., Alcaíde, J.M., Polo, M.C., and Pueyo, E. 2007. Angiotensin I-converting enzyme inhibitory compounds in white and red wines. *Food Chemistry* 100: 43–47.

Pripp, A.H., Sørensen, R., Stepaniak, L., and Sørhaug, T. 2006. Relationship between proteolysis and angiotensin-I-converting enzyme inhibition in different cheeses. *LWT—Food Science and Technology* 39: 677–683.

Quirós, A., Ramos, M., Muguerza, B., Delgardo, M.A., Miguel, M., Aleixandre, A., and Recio, I. 2007. Identification of novel antihypertensive peptides in milk fermented with *Enterococcus faecalis*. *International Dairy Journal* 17: 33–41.

Rajapakse, N., Mendis, E., Jung, W.K., Je, J.Y., and Kim, S.K. 2005. Purification of a radical scavenging peptide from fermented mussel sauce and its antioxidant properties. *Food Research International* 38: 175–182.

Randel, R.D., Chase Jr., C.C., and Wyse, S.J. 1992. Effects of gossypol and cottonseed products on reproduction of mammals. *Journal of Animal Science* 70: 1628–1638.

Reddy, N.R., and Pierson, M.D. 1994. Reduction in antinutritional and toxic components in plant foods by fermentation. *Food Research International* 27: 281–290.

Reddy, N.R., Pierson, M.D., Sathe, S.K., and Salunkhe, D.K. 1982. Legume-based fermented foods: Their preparation and nutritional quality. *Critical Reviews of Food Science and Nutrition* 17: 335–370.

Reddy, N.R., Pierson, M.D., Sathe, S.K., and Salunkhe, D.K. 1984. Chemical, nutritional and physiological aspects of dry bean carbohydrates—A review. *Food Chemistry* 13: 25–68.

Reddy, N.R., Pierson, M.D., Sathe, S.K., and Salunkhe, D.K. 1989. Occurrence, distribution, content and dietary intake of phytate. In *Phytates in Cereals and Legumes*, eds. N.R. Reddy, M.D. Pierson, S.K. Sathe, and D.K. Salunkhe, pp. 39–56. Boca Raton, FL: CRC Press.

Rho, S.J., Lee, J.S., Chung, Y.I., Kim, Y.W., and Lee, H.G. 2009. Purification and identification of an angiotensin I-converting enzyme inhibitory peptide from fermented soybean extract. *Process Biochemistry* 44: 490–493.

Robbins, J. 2001. *The Food Revolution, How Your Diet Can Help Save Your Life and The World*. Berkeley, CA: Conari Press.

Saito, T. 1938. Antagonism action of *Bacillus subtilis* (*natto*) against *S. enterica* var *enterica* serovar Typhi. *Acta Medica Hokkaidonensia* 16: 82–92. (in Japanese, translated title).

Saito, Y., Wanezaki, K., Kawato, A., and Imayasu, S. 1994. Structure and activity of angiotensin I converting enzyme inhibitory peptides from sake and sake lees. *Bioscience, Biotechnology, and Biochemistry* 58: 1767–1771.

Saka, N. 2008. Effects of natto that gives to health of piglets. *All About Swine* 33: 15–19. (in Japanese).

Sandberg, A.S. 2002. Bioavailability of minerals in legumes. *British Journal of Nutrition* 88: S281–S285.

Sanni, A.I., and Ogbonna, D.N. 1991. The production of Owoh—A Nigerian fermented seasoning agent from cotton seed (*Gossypium hirsutum* L). *Food Microbiology* 8: 223–229.

Sanni, A.I., and Ogbonna, D.N. 1992. Biochemical studies on owoh—A Nigerian fermented soup condiment from cotton seed. *Food Microbiology* 9: 177–183.

Sarkar, P.K., Cook, P.E., and Owens, J.D. 1993. *Bacillus* fermentation of soybeans. *World Journal of Microbiology and Biotechnology* 9: 295–299.

Sarkar, P.K., Jones, L.J., Craven, G.S., and Somerset, S.M. 1997a. Oligosaccharide profiles of soybeans during kinema production. *Letters in Applied Microbiology* 24: 337–339.

Sarkar, P.K., Jones, L.J., Craven, G.S., Somerset, S.M., and Palmer, C. 1997b. Amino acid profiles of kinema, a soybean-fermented food. *Food Chemistry* 59: 69–75.

Sarkar, P.K., Jones, L.J., Gore, W., Craven, G.S., and Somerset, S.M. 1996. Changes in soya bean lipid profiles during kinema production. *Journal of the Science of Food and Agriculture* 71: 321–328.

Sarkar, P.K., Morrison, E., Tinggi, U., Somerset, S.M., and Craven, G.S. 1998. B-group vitamin and mineral contents of soybeans during kinema production. *Journal of the Science of Food and Agriculture* 78: 498–502.

Sarkar, P.K., and Tamang, J.P. 1995. Changes in the microbial profile and proximate composition during natural and controlled fermentations of soybeans to produce kinema. *Food Microbiology* 12: 317–325.

Sato, T. 1980. Studies on induction of interferon by "*Bacillus natto*." *Journal of Tokyo Medical College* 38: 815–826 (in Japanese).

Schmidt, D.D., Frommer, W., and Junge, B. 1977. α-Glucosidase inhibitors. New complex oligosaccharides of microbial origin. *Naturwissenschaften* 64: 535–536.

Schmidt, D.D., Frommer, W., Müller, L., and Truscheit, E. 1979. Glucosidase-inhibitoren aus Bazillen. *Naturwissenschaften* 66: 584–585.

Senbon, S. 1940. Anwendung des *Bazillus natto* zur Therapie gegen Trichophytia. *Journal of the Medical Association of Formosa* 39: 14–17 (in Japanese).

Shibasaki, K., and Hesseltine, C.W. 1962. Miso fermentation. *Economic Botany* 16: 180–195.

Shimakage, A., Shinbo, M., Yamada, S., and Ito, H. 2006. Changes in isoflavone content in soybeans during the manufacturing processes of tubu-natto and hikiwari-natto. *Nippon Shokuhin Kagaku Kogaku Kaishi* 53: 185–188 (in Japanese).

Shin, Z.I., Yu, R., Park, S.A., Chung, D.K., Ahn, C.W., and Nam, H.S. 2001. His-His-Leu, an angiotensin I converting enzyme inhibitory peptide derived from Korean soybean paste, exerts antihypertensive activity in vivo. *Journal of Agricultural and Food Chemistry* 49: 3004–3009.

Simark-Mattsson, C., Emilson, C.G., Hakansson, E.G., Jacobsson, C., Roos, K., and Holm, S. 2007. *Lactobacillus*-mediated interference of mutans streptococci in caries-free vs. caries-active subjects. *European Journal of Oral Sciences* 115: 308–314.

Stamatova, I., and Meurman, J.H. 2009. Probiotics: Health benefits in the mouth. *American Journal of Dentistry* 22: 329–338.

Steinkraus, K.H. 1994. Nutritional significance of fermented foods. *Food Research International* 27: 259–267.

Steinkraus, K.H., ed. 1996a. *Handbook of Indigenous Fermented Foods*, 2nd ed. New York: Marcel Dekker.

Steinkraus, K.H. 1996b. Indigenous fermented foods involving an alkaline fermentation. In *Handbook of Indigenous Fermented Foods*, 2nd ed., ed. K.H. Steinkraus, pp. 349–362. New York: Marcel Dekker.

Steinkraus, K.H., Lee, C.Y., and Buck, P.A. 1965. Soybean fermentation by the Ontjom mold *Neurospora*. *Food Technology* 19: 1301–1302.

Sulistyo, J., Taya, N., Funane, K., and Kiuchi, K. 1988. Production of natto starter. *Nippon Shokuhin Kogyo Gakkaishi* 35: 278–283.

Sumi, H. 1990. Nattokinase and development of healthy natto. *Bio Industry* 7: 724–730 (in Japanese, translated title).

Sumi, H. 1991. Nattokinase and fibrinolytic system. *Kagaku to Seibutsu* 29: 119–123 (in Japanese).

Sumi, H., Banba, T., and Kishimoto, N. 1996. Strong pro-urokinase activators proved in Japanese soybean cheese natto. *Nippon Shokuhin Kagaku Kogaku Kaishi* 43: 1124–1127 (in Japanese).

Sumi, H., Hamada, H., Nakanishi, K., and Hiratani, H. 1990. Enhancement of fibrinolytic activity in plasma by oral administration of nattokinase. *Acta Haematologica* 84: 139–143.

Sumi, H., Hamada, H., Tsushima, H., Mihara, H., and Muraki, H. 1987. A novel fibrinolytic enzyme (nattokinase) in the vegetable cheese natto; a typical and popular soybean food in the Japanese diet. *Experientia* 43: 1110–1111.

Sumi, H., and Ohosugi, T. 1999. Anti-bacterial component dipicolic acid measured in natto and natto bacilli. *Journal of the Agricultural Chemical Society of Japan* 73: 1289–1291 (in Japanese).

Sumi, H., and Okamoto, T. 2003. Thrombolytic activity of an aqueous extract of tempe. *Journal of Home Economics of Japan* 54: 337–342 (in Japanese).

Sumi, H., Taya, N., Nakajima, N., and Hiratani, H. 1992. Structure and fibrinolytic properties of nattokinase. *Fibrinolysis* 6 (Supplement 2): 86.

Sumi, H., Yoshikawa, M., Baba, T., Matsuda, K., and Kubota, H. 1999. Elastase activity in natto, and its relation to nattokinase. *Journal of the Agricultural Chemical Society of Japan* 73: 1187–1190 (in Japanese).

Sun, P., Wang, J.Q., and Deng, L.F. 2012. Effects of *Bacillus subtilis* natto on milk production, rumen fermentation and ruminal microbiome of dairy cows. *Animal* June: 1–7.

Sundhagul, M. 1972. Thua-nao: A fermented soybean food of northern Thailand. I. Traditional processing method. *Thai Journal of Agriculture Science* 5: 43–46.

Survase, S.A., Bajaj, I.B., and Singhal, R.S. 2006. Biotechnological production of vitamins. *Food Technology and Biotechnology* 44: 381–396.

Suzuki, H., Watabe, J., Takeuchi, H., Tadano, Y., Masuda, S., and Maruta, K. 2004. Effect of *Bacillus subtilis* C-3102 intakes on the composition and metabolic activity of fecal microflora of humans. *Journal of Intestinal Microbiology* 18: 93–99.

Takahashi, C., Kikuchi, N., Katou, N., Miki, T., Yanagida, F., and Umeda, M. 1995. Possible anti-tumor–promoting activity of components in Japanese soybean fermented food, natto: Effect on gap junctional intercellular communication. *Carcinogenesis* 16: 471–476.

Takaoka, S. 2003. Function and market trend of NSK (extract containing a high activity of nattokinase from culture broth of *Bacillus natto*). *New Food Industry* 45: 25–30. (in Japanese, translated title).

Takeda, K., Ohshiro, T., Ohshiro, T., and Okumura, K. 2005. Enhancement of natural killer cell activity by supplementation of extract of metabolic products of *Bacillus subtilis* AK (EMBSAK). *Japanese Journal of Complementary and Alternative Medicine* 2: 127–133 (in Japanese).

Takemura, H., Shioya, N., Komori, Y., and Tho, Y. 2009. Effect of fermented soybeans containing *Bacillus subtilis* MC1 on defecation, fecal properties and fecal microflora of healthy female volunteers. *Journal of Urban Living and Health Association* 53: 11–18.

Tamang, J.P. 1998. Role of microorganisms in traditional fermented foods. *Indian Food Industry* 17: 162–167.

Tamura, Y., and Takenaka, T. 1999. Antioxidative activity of water soluble extracts from okara fermented with *Bacillus natto* and *Rhizopus oligosporus*. *Nippon Shokuhin Kagaku Kogaku Kaishi* 46: 561–569. (in Japanese).

Terada, A., Yamamoto, M., and Yoshimura, E. 1999. Effect of the fermented soybean product "natto" on the composition and metabolic activity of the human fecal flora. *Japanese Journal of Food Microbiology* 16: 221–230.

Terlabie, N.N., Sakyi Dawson, E., and Amoa Awua, W.K. 2006. The comparative ability of four isolates of *Bacillus subtilis* to ferment soybeans into dawadawa. *International Journal of Food Microbiology* 106: 145–152.

Toppe, J., Albrektsen, S., Hope, B., and Aksnes, A. 2007. Chemical composition, mineral content and amino acid and lipid profiles in bones from various fish species. *Comparative Biochemistry and Physiology B-Biochemistry & Molecular Biology* 146: 395–401.

Tregeara, A., Arfinib, F., Bellettic, G., and Marescottic, A. 2007. Regional foods and rural development: The role of product qualification. *Journal of Rural Studies* 23: 12–22.

Truscheit, E., Frommer, W., Junge, B., Müller, L., Schmidt, D.D., and Wingender, W. 2003. Chemistry and biochemistry of microbial α-glucosidase inhibitors. *Angewandte Chemie (International Edition in English)* 20: 744–761.

Tsuchihashi, N., Watanabe, T., Shimizu, K., and Takai, Y. 1986. Effect of living cells of *Bacillus* on rats. *Bulletin of Chiba College of Health Science* 5: 3–8 (in Japanese).

Tsukamoto, Y., Ichise, H., Kakuda, H., and Yamaguchi, M. 2000. Intake of fermented soybean (natto) increases circulating vitamin K_2 (menaquinone-7) and γ-carboxylated osteocalcin concentration in normal individuals. *Journal of Bone and Mineral Metabolism* 18: 216–222.

Udo, S. 1936. Composition of natto (1) dipicolic acid in natto and its activity. *Journal of the Agricultural Chemistry Society of Japan* 12: 386–394 (in Japanese, translated title).

Umoh, I.B., and Oke, O.L. 1974. Nutritive value of some lesser known oilseeds in rats. *Nutrition Reports International* 9: 453–460.

Van Veen, A.G., and Steinkraus, K.H. 1970. Nutritive value and wholesomeness of fermented foods. *Journal of Agricultural and Food Chemistry* 18: 576–578.

Wang, D., Wang, L.J., Zhu, F.X., Zhu, J.Y., Chen, X.D., Zou, L., Saito, M., and Li, L.T. 2008. *In vitro* and *in vivo* studies on the antioxidant activities of the aqueous extracts of douchi (a traditional Chinese salt-fermented soybean food). *Food Chemistry* 107: 1421–1428.

Wang, H., and Murphy, P.A. 1994. Isoflavone content in commercial soybean foods. *Journal of Agricultural and Food Chemistry* 42: 1666–1673.

Wang, H.L., Swain, E.W., Wallen, L.L., and Hesseltine, C.W. 1975. Free fatty acids identified as antitryptic factor in soybeans fermented by *Rhizopus oligosporus*. *The Journal of Nutrition* 105: 1351–1355.

Wang, J., and Fung, D.Y. 1996. Alkaline-fermented foods: A review with emphasis on pidan fermentation. *Critical Reviews in Microbiology* 22: 101–138.

Wei, Q.K., Chen, T.R., and Chen, J.T. 2008. Use of *Bacillus subtilis* to enrich isoflavone aglycones in fermented natto. *Journal of the Science of Food and Agriculture* 88: 1007–1011.

Whitaker, J.R. 1978. Biochemical changes during the fermentation of high protein foods. *Food Technology* 32: 175–190.

Winarno, F.G., and Reddy, N.R. 1986. Tempe. In *Legume-Based Fermented Foods*, eds. N.R. Reddy, M.D. Pierson, and D.K. Salunkhe, pp. 95–117. Boca Raton, FL: CRC Press.

Xu, Q., Yajima, T., Li, W., Saito, K., Ohshima, Y., and Yoshikai, Y. 2006. Levan (β-2, 6-fructan), a major fraction of fermented soybean mucilage, displays immunostimulating properties via Toll-like receptor 4 signalling:

Induction of interleukin-12 production and suppression of T-helper type 2 response and immunoglobulin E production. *Clinical and Experimental Allergy* 36: 94–101.

Yagoub, A.E.G.A., and Mohammed, M.A. 2008. Furundu, a meat substitute from fermented roselle (*Hibiscus sabdariffa* L.) seed: Investigation on amino acids composition, protein fractions, minerals content and HCl-extractability and microbial growth. *Pakistan Journal of Nutrition* 7: 352–358.

Yokota, T., Hattori, T., Ohishi, H., Hasegawa, K., and Watanabe, K. 1996. The effect of antioxidant-containing fraction from fermented soybean food on atherosclerosis development in cholesterol-fed rabbits. *LWT – Food Science and Technology* 29: 751–755.

Yong, F.M., and Wood, B.J.B. 1977. Biochemical changes in experimental soy sauce koji. *International Journal of Food Science and Technology* 12: 163–175.

Zhang, J.H., Tatsumi, E., Ding, C.H., and Li, L.T. 2006. Angiotensin I-converting enzyme inhibitory peptides in douchi, a Chinese traditional fermented soybean product. *Food Chemistry* 98: 551–557.

Zhu, X.L., Watanabe, K., Shiraishi, K., Ueki, T., Noda, Y., Matsui, T., and Matsumoto, K. 2008a. Identification of ACE-inhibitory peptides in salt-free soy sauce that are transportable across caco-2 cell monolayers. *Peptides* 29: 338–344.

Zhu, Y.P., Fan, J.F., Cheng, Y.Q., and Li, L.T. 2008b. Improvement of the antioxidant activity of Chinese traditional fermented okara (Meitauza) using *Bacillus subtilis* B2. *Food Control* 19: 654–661.

Zhu, Y.P., Yamaki, K., Yoshihashi, T., Ohnishi Kameyama, M., Li, X.T., Cheng, Y.Q., Mori, Y., and Li, L.T. 2010. Purification and identification of 1-deoxynojirimycin (DNJ) in okara fermented by *Bacillus subtilis* B2 from Chinese traditional food (Meitaoza). *Journal of Agricultural and Food Chemistry* 58: 4097–4103.

Zhu, Y.P., Yin, L.J., Cheng, Y.Q., Yamaki, K., Mori, Y., Su, Y.C., and Li, L.T. 2008c. Effects of sources of carbon and nitrogen on production of α-glucosidase inhibitor by a newly isolated strain of *Bacillus subtilis* B2. *Food Chemistry* 109: 737–742.

6

Safety Aspects of Alkaline-Fermented Foods

Ggot-Im Lee and Cherl-Ho Lee

6.1 Alkaline-Fermented Foods Are Generally Safe

Alkaline-fermented foods (AFFs) are less known and studied than alcohol fermentation and acid fermentation even though they are widely consumed in various regions of the world. These foods have been good diet sources for many people who have difficulty in accessing to enough food and proper storage facilities.

AFF refers to the food containing alkaline substances that are formed during the fermentation process. For instance, fermented legume seeds and fish, which contain abundant amount of protein, are AFFs. Korean chongkukjang and doenjang, Japanese natto, Chinese douchi, Thai thua nao, and Nepalese and Indian kinema, which are all made from soybean, belong to this class of fermented foods. Nigerian dawadawa, Ivory Coast soumbara (alternatively soumbala), and Beninese iru, sonru, and afitin are made from locust bean (*Parkia biglobosa* (Jacq.) G.Don), and they are also AFFs. Other AFFs include Nigerian ogiri made by fermenting melon (*Citrullus lanatis* (Thunb.) Matsum. & Nakai) seeds, ogiri-ugbo from castor bean (*Ricinus communis* L.) seeds, ogiri-nwan from fluted pumpkin (*Telfairia occidentalis* Hook.f.) seeds, and ugba made from African oil bean (*Pentaclethra macrophylla* Benth.), Burkina Faso maari from baobab (*Adansonia digitata* L.) seeds, and Sierra Leone ogiri-saro from the sesame seed (*Sesamum indicum* L.). A typical alkaline fermentation using fish is Korean hongeohoe made from flat fish, the skate (Steinkraus, 1997; Lee et al., 2008).

The essential and main microorganisms responsible for making AFFs are *Bacillus subtilis* and related bacilli. The organisms are highly proteolytic, so that the proteins are hydrolyzed to peptides and amino acids. For example, the concentration of free amino acids of dawadawa indicates around 2.4 mg/g before fermentation; however, it surprisingly goes up to around 84.3 mg/g after 72 h of fermentation (Omafuvbe et al., 2000). Ammonia concentration of thua nao is 0.5 g/kg before fermentation, but it increases to 5.0 g/kg at 72 h of fermentation (Leejeerajumnean, 2000). The released ammonia triggers the rapid pH increase of as high as 8.0 or higher. This makes the products keep quite stable and well-preserved even though they are often manufactured in an unhygienic environment.

These foods contain large amounts of ammonia and amino acids, which are formed by the strong proteolytic activity of the indigenous enzymes and microorganisms. The released ammonia causes a strong smell and contributes to the typical taste of the AFF. Unlike other fermented foods, such as acid and alcoholic fermented foods that contain strong antimicrobial substances formed during the fermentation, AFFs do not have enough protective agents against harmful microorganisms. Furthermore, alkaline fermentation of protein-rich substrates can produce biogenic amines and other hazardous compounds. However, well-made traditional AFFs are safe and have been widely consumed, especially in Asia and Africa, as condiments and favorite traditional foods.

A number of researches support that AFFs are safe, although they are usually produced and consumed in developing countries where hygienic measures are not properly practiced. It indicates that AFFs use their own protective systems during the fermentation process. Several mechanisms of how they work will be explained below.

6.1.1 Initial Thermal Treatment

Most of the AFFs are made from beans and seeds, which have hard seed coats. Therefore, raw materials first need to be soaked in water and boiled for a long time so that the seed coat can be softened before fermentation. This process may take more than 12 h depending on the condition and the kind of the material. Soaking and boiling steps prior to the fermentation help removing the anti-nutritional factors of plant materials such as trypsin inhibitor, hemagglutinin, phytic acid, and also other harmful microorganisms (Khokhar and Chauhan,

1986; Egounlety and Aworh, 2003). For instance, locust beans contain highly toxic substances such as oxalic acid, phytic acid, and cyanhydric acid. However, they disappear during the fermentation process (Parkouda et al., 2009).

The boiling process is quite important for setting up a good environment for fermentation. It makes the fermentation environment favorable to useful microorganisms while keeping biological hazards away from the whole fermentation step. Most of non-spore-forming pathogenic bacteria are killed during the boiling process, although some spore-forming bacteria, mainly *Bacillus* spp., survive. Table 6.1 shows an example of bacterial occurrence through all steps of iru production. Most microorganisms, except *Staphylococcus epidermidis*, are present in the raw locust beans. However, pathogenic microorganisms, except *Fusarium* sp., are all gone after initial boiling step (Olayinka and Omobayonle, 2006). After dehulling with the boiled beans, potential pathogens, except *Clostridium* sp., disappear during the second boiling step. In the case of hongeohoe, which does not have a cooking process, raw fish is mixed with ash or buried in the ash pile, by which alkaline environment is made at the initial stage of fermentation process (Kim, 2006). The pH reaches the degree of up to 9.4 when the fermentation is completed (Lee et al., 2008).

6.1.2 *Growth of* Bacillus subtilis

The surviving microorganisms after cooking, mainly *Bacillus* spp., grow rapidly in the system because of their thermophilic nature, of which the optimal growth temperature is around 40°C, and become the dominant flora. Among the *Bacillus* spp., *B. subtilis* is known to be the most dominant organism for AFFs such as chongkukjang, natto, kinema, soumbara, dawadawa, iru, and sonru (Sarkar et al., 2002; Azokpota et al., 2006). *Bacillus subtilis* grows significantly during the fermentation process (Sarkar and Tamang, 1995; Parkouda et al., 2010).

Several reports demonstrated the ability of *B. subtilis* to produce metabolites that inhibit the growth of pathogenic bacteria (N'dir et al., 1994; Guo et al., 2006). During kinema production, *B. subtilis* appeared to have the ability to reduce the growth of *Bacillus cereus*

Table 6.1 Distribution of Isolated Microorganims in Samples Collected at Different Stages during Traditional Processing of Locust Beans to Iru

	WHOLE BEANS		COTYLEDONS				COTYLEDONS FERMENTED AT DIFFERENT TIME PERIOD			
ISOLATE	RAW	BOILED	FRESHLY DEHULLED	WASHED	BOILED	DRAINED	12 h	24 h	36 h	PRODUCT (IRU)
Bacillus subtilis	+	+	+	+	+	+	+	+	+	+
Bacillus pumilus	+	+	+	+	+	+	+	+	+	+
Bacillus licheniformis	+	–	–	+	+	+	+	+	+	+
Clostridium sp.	+	–	–	+	+	+	–	–	–	+
Micrococcus sp.	+	–	+	+	–	–	–	–	–	+
Staphylcoccus aureus	+	–	+	+	–	–	–	–	–	+
Staphylococcus epidermidis	–	–	+	+	–	–	–	–	–	+
Aspergillus niger	+	–	+	+	–	+	–	–	–	–
Saccharomyces saccharolyticus	+	–	+	–	–	+	+	–	–	–
Fusarium sp.	+	+	+	+	–	+	–	–	–	–
Neurospora sp.	+	–	–	+	–	–	–	–	–	–
Escherichia coli	+	–	+	+	–	+	–	–	–	–
Lactobacillus amylovorus	+	–	+	+	–	+	+	–	+	+
Leuconostoc mesenteroides	+	+	+	+	–	+	+	+	+	+
Rhizopus oligosporus	+	+	+	+	–	+	–	–	–	+

Source: Olayinka, E.M., and Omobayonle, F., *Research Journal of Microbiology*, 1, 503–511, 2006.

and to prevent it from producing enterotoxin (Nout et al., 1998). Therefore, the strains of *B. subtilis* have been used as a good starter for the controlled production of natto and kinema (Sarkar et al., 1993; Owens et al., 1997).

Table 6.2 shows that *B. subtilis* and *Bacillus pumilus*, isolated from soumbala, inhibit both Gram-positive and Gram-negative bacteria including *Micrococcus luteus*, *Staphylococcus aureus*, *B. cereus*, *Enterococcus faecium*, *Listeria monocytogenes*, *Escherichia coli*, *Salmonella typhimurium*, *Shigella dysenteriae*, *Yersinia enterocolitica*, as well as ochratoxigenic molds such as *Aspergillus ochraceus* (Ouoba et al., 2007).

The *Bacillus* species not only have an antibiotic effect, especially *B. subtilis*, but they also produce strong proteolyic enzymes that hydrolyze the protein in the beans and seeds into peptides and amino acids and finally into ammonia. The combination of high pH and high concentration of free ammonia, along with rapid growth of *B. subtilis* in the high temperature condition (above 40°C), makes it very difficult for other pathogenic microorganisms to grow (Steinkraus, 1997).

Table 6.2 Inhibition of Pathogenic Bacteria after 24 h by *Bacillus* spp. Isolated from Soumbala

	B. subtilis STRAIN			*B. pumilus* STRAIN	
TARGET BACTERIUM	B7	B9	B15	B6	B10
Micrococcus luteus A1	++	+	+	+	+
Bacillus cereus A4	+	–	+	–	–
Bacillus cereus A10	++	++	+	++	++
Staphylococcus aureus A11	–	–	+	+	+
Staphylococcus aureus A12	++(p)	–	+++	+	+
Enterococcus faecium A19	–	–	–	+	+
Listeria monocytogenes A18	++	+	+	+	+
Escherichia coli A13	++	–	+(p)	–	++(p)
Salmonella enteritidis A14	+	–	+	–	–
Salmonella typhimurium A15	++	–	+	–	+
Shigella flexneri A16	+	–	–	–	–
Shigella dysenteriae A17	++	–	+	–	++
Yersinia enterocolitica A21	++(p)	–	++	–	+

Source: Ouoba, L.I.I. et al., *Journal of Applied Microbiology*, 102, 963–970, 2007.

Note: –, No inhibition (<1 mm); +, inhibition 1–<3 mm; ++, inhibition 3–5 mm; +++, inhibition ≥5 mm; p, partial (inhibition zone not totally clear).

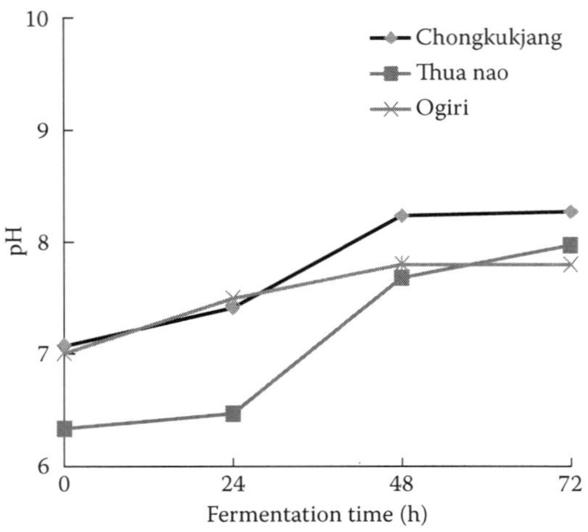

Figure 6.1 The pH change of several alkaline-fermented foods. (From Odunfa, S.A., *Journal of Food Science*, 50, 1758–1759, 1985; Chukeatirote, E. et al., *Research Journal of Microbiology*, 1, 38–44, 2006; Ann, Y.G., *Korean Journal of Food and Nutrition*, 24, 124–131, 2011.)

6.1.3 pH Change

The increase in pH during fermentation plays a key role in the safety aspects of AFFs. The pH before fermentation begins at 6.0–7.0, and it reaches 7.5–9.0 at the final stage of fermentation. Most AFFs such as chongkukjang, thua nao, kinema, and ogiri show rapid pH changes until 48 h of fermentation, and then they become stable around 8.0 (Figure 6.1). This is a quite fast and vigorous fermentation that it can

Table 6.3 Approximate pH Tolerance of Some Microorganisms

	pH TOLERANCE	
ORGANISMS	MINIMUM	MAXIMUM
Escherichia coli	4.4	9.0
Salmonella typhi	4.5	8.0
Campylobacter jejuni	2.3 and 5.8	–
Shigella sp.	4.5	8.0
Streptococcus lactis	4.3–4.8	–
Lactobacillus sp.	3.0	7.2
Yeasts	1.5	8.0–8.5
Molds	1.5–2.0	11.0

Source: Blaser, M.J. et al., *Journal of Clinical Microbiology*, 11, 309–313, 1980; Gill, C.O., and Harris, L.M., *Applied and Environmental Microbiology*, 44, 259–263, 1982; Jay, J.M., *Modern Food Microbiology*, 3rd ed., Van Nostrand Reinhold, New York, 1986.

be completed in 2–3 days (Odunfa, 1985; Chukeatirote et al., 2006; Ann, 2011).

Many of the pathogenic and spoilage microorganisms are inhibited by the higher pH over 8.0. Table 6.3 shows that *Salmonella typhi* and *Shigella* sp. cannot grow at the degree of pH over 8.0 and *E. coli* at over 9.0 (Mensah, 1997).

6.1.4 Cooking Process

Most of the AFFs are consumed as forms of soups or condiments through the cooking process. Boiling or stewing is normally applied for the preparation of these kinds of meal. Thanks to this heat treatment; the pathogenic microorganisms that possibly contaminated during and after fermentation or storage can be removed. This is how indigenous AFFs have been consumed relatively safely in Asia and Africa.

6.2 Hazards in Alkaline-Fermented Foods

AFFs have beneficial effects and can be considered safe as mentioned above. However, safety issues can be raised, although it appears to be not only the problem of AFFs but also that of all types of traditional fermented foods. The hazards occurring in AFFs can be divided into two groups, namely microbiological and chemical. The possible microbiological hazards include *B. cereus* contamination and mycotoxin in raw materials. The chemical hazards include the formation of biogenic amines during the fermentation and contamination of pesticides in raw materials. These safety issues are discussed in more detail below.

6.2.1 Microbiological Hazards

6.2.1.1 Bacillus cereus Although the initial step of alkaline fermentation includes a long cooking process, some hazardous spore-forming bacteria like *B. cereus* can survive if the raw material is heavily contaminated with these bacteria. It has been reported that *B. cereus* exists in various AFFs like iru, ogiri (10^{6-8} cfu/g; Oguntoyinbo and Oni, 2004), and chongkukjang (10^{5-6} cfu/g; Oguntoyinbo and Oni,

2004; Lee, 2010). Pathogens found in fermented foods mainly come from the raw materials, handlers, or poor kitchen utensils. This bacterium is known to produce emetic toxin (cereulide) and enterotoxins that cause vomiting and diarrhea (Granum, 2001; Ehling-Schulz et al., 2004; Ouoba et al., 2008). Therefore, it is important to avoid cereulide production in foods because cereulide, unlike enterotoxins, is heat-stable in a wide range of pH (2–11), and is not degraded by the digestive enzymes like pepsin and trypsin (Shinagawa et al., 1995, 1996). However, it was reported that cereulide production of an emetic *B. cereus* isolated from sonru appeared to be dependent on the fermentation temperature as being strongly inhibited at 40°C (Parkouda et al., 2009).

Bacillus subtilis has been regarded as a dominant microorganism for ugba production. However, *B. cereus*, rather than *B. subtilis*, has been recently reported to be the dominant strain in the production of this food (Sanni et al., 2002; Ahaotu et al., 2013). All the *B. cereus* strains isolated from ugba-expressed enterotoxin genes, but no genes encoding cereulide were detected (Ahaotu et al., 2013). Consumption of foods contaminated with *B. cereus* does not always cause diarrhea or vomiting. However, it should be well cooked for a suitable period of time to prevent the risk of enterotoxins from ugba. In addition, an investigation regarding toxin production by *Bacillus* spp., other than *B. cereus*, noticed that some strains of *B. subtilis*, *B. pumilus*, and *Bacillus fusiformis* can produce either cytotoxins or ring-formed emetic toxins. Therefore, special care is also required to select strains for starter culture even though *Bacillus* spp. related to AFFs does not show pathogenic activities (From et al., 2005).

6.2.1.2 Other Potential Pathogens Apart from *B. cereus*, the presence of other potential pathogenic microorganisms like *Staphylococcus* spp., *E. coli*, *Salmonella* spp., *Shigella* spp., *Clostridium* spp., and *Staphylococcus* spp. were found in iru and chongkukjang (Azokpota et al., 2006; Lee, 2010). *Bacillus cereus*, Enterobacteriaceae, and coliform bacteria were detected over 10^4 cfu/g in kinema. *Salmonella* was detected in papad, legume-based popular fermented foods in India (Roy et al., 2007). It was reported that *Staphylococcus sciuri* involves in the fermentation of baobab seeds to produce maari (Parkouda et al., 2010). *Staphylococcus sciuri* is well known as a highly pathogenic bacterial species responsible for endocarditis, peritonitis, septic shock, urinary tract infection, endophthalmitis, pelvic inflammatory disease, and most wound infections (Dakić et al., 2005). Recently, people are more interested in the capacity of this species to carry antimicrobial resistance determinants. Further investigation is required to illuminate the exact role of pathogenic microorganisms mentioned above, because the presence of microorganims itself does not mean expression of toxin substances. AFFs, especially, have several inhibitory systems to suppress gene-encoding toxins.

6.2.1.3 Mycotoxins Mycotoxins are toxic secondary metabolites produced by fungal contamination, which is a very hazardous factor causing diseases including cancer. Most of the AFFs are made in warm and humid areas such as Africa and Asia using seeds and beans, which are high-protein foods. These conditions may cause molds to grow quickly and amplify the chance of mold contamination, and the handling and storage of fermented foods accordingly should be processed with care. Contamination of mold on grains during storage and transportation is liable to be migrated rapidly to another country due to the increasing world trade. For this reason, many international agencies are trying to set up universal standard limits for mycotoxins.

There are two possible sources of mycotoxin contamination: (1) toxin from the contaminated cereals and beans themselves and (2) toxin produced by the mold contaminated during fermentation (Lee and Lee, 2002). Table 6.4 summarizes mycotoxins, origins of fungi, and related symptoms (Bullerman, 1986).

Aflatoxins are one of the representative mycotoxins produced by *Aspergillus flavus* and *Aspergillus parasiticus*, and they have four types called B_1, B_2, G_1, and G_2. *Aspergillus flavus* is the most common contaminant in agricultural products. *Fusarium*, *Alternaria*, and *Acremonium* species are typical parasites that exist on roots and stems of plants and produce zearalenone, alternaria, and tremorgenic toxins (Betina et al., 1984).

The traditional fermentation starters used in Asia and Africa have a high risk of mycotoxin contamination because natural fermentation environment of cereals and beans often leaves foods exposed to numerous kinds of molds. *Penicillium* and *Aspergillus* are the major fungi related to the toxin production on various foods, which often

Table 6.4 Mycotoxins Produced by Fungi

TOXIN	PRODUCER FUNGI	SYMPTOMS
Aflatoxin	*Aspergillus flavus*, *Aspergillus parasiticus*, *Aspergillus nomius*	Hepatotoxic, hepatocarcinogenic, mutagenic, teratogenic
Zearalenones	*Fusarium roseum*, Other *Fusarium* spp.	Estrogenic, infertility in cattle/swine/poultry
Vomitoxin	*Fusarium* spp.	Weight loss in most livestock, feed refusal, vomiting, immunosuppression
T-2 toxin	Pink *Fusarium* spp.	Feed refusal, vomiting, nervous system dysfunction, immunosuppression
Trichothecenes	*Fusarium* spp.	Gastrointestinal tract hemorrhaging, alimentary toxic aleukia, feed rejection, vomiting, immunotoxicity
Fumonisins	*Fusarium moniliforme*	Esophageal carcinogenic in human (ELEM) in equines, hepatocarcinogenic in rats
Orchatoxins	*Aspergillus* spp., *Penicillium* spp.	Nephrotoxic, teratogenic, immunotoxic, suspected carcinogen
Sterigmatocystin	*Aspergillus versicolor*, Other *Aspergillus* spp.	Hepatotoxic, hepatocarcinogenic
Citrinin	*Aspergilllus* spp., *Penicillium* spp.	Nephrotoxic
Rubratoxin	*Penicillium* spp.	Hepatotoxic

Source: Bullerman, L.B., *Food Technology*, 40, 59–66, 1986.

grow during storage and natural fermentation. *Penicillum glandicola* produces ochratoxin A on cereals containing over 19% moisture, *Penicillium expansum* produces patulin on apples and pear, and *Penicillium roqueforti* produces citrinin on blue cheese (Lee and Lee, 2002). *Aspergillus flavus* and *A. parasiticus* produce aflatoxin B and G on corn, peanut, and rice, and *A. ochraceus* produces ochratoxin. *Aspergillus flavus* can also produce mycotoxin cyclopiazonic acid.

It is an unavoidable problem that AFFs are contaminated with mycotoxin. The aflatoxin was detected at the level of 0.20 μg/g in melon seeds sold in the markets of Nigeria (Ekundayo and Idzi, 1990). Three mold species, *Rhizopus stolonifer*, *Aspergillus fumigatus*, and *Triscelophorus monosporus*, were found in ogiri sold in the local markets (Ijabadeniyi, 2007). A report showed that *R. stolonifer* and other *Rhizopus* isolates have powerful (>95%) detoxification activity of ochratoxin A in 16 days, assuming that *Rhizopus* sp. have a role of suppressing mycotoxins during fermentation (Varga et al., 2005). However, *A. fumigatus* is well known to produce cytotoxic mycotoxin and gliotoxin in marine conditions and to cause more than 80% of aspergillosis (Gareis and Wernery, 1994). *Triscelophorus monosporus* is known to be related to iru production (Evans et al., 2013).

The incidence of aflatoxin B_1 in ogiri was examined using TLC (thin layer chromatography) in Nigeria in 2005 and 2006. The results showed that 25% of the total samples were contaminated with aflatoxin B_1. The mean values of aflatoxin B_1 indicated 8.9 and 9.0 μg/kg in 2005 and 2006, respectively (Bankole et al., 2010). The range of contamination varied from 2 to 13 μg/kg in 2005 and from 4.1 to 14.3 μg/kg in 2006. Another result reported in 2004 showed a small increase in aflatoxin levels in melon seeds: 14.8 μg/kg in the forest and 11.3 μg/kg in the savanna regions (Bankole et al., 2004). This difference may be due to different processing manuals. None of the samples had more than 20 μg/kg, which is within the Nigerian foods standard, but daily total dose should be considered.

Numerous studies have been carried out to control the mycotoxin contamination in fermented soybean products in Korea. It has been proven that aflatoxins (B_1 and G_1) decreased to 80%–90% after 2 months of fermentation and to 100% after 3 months of fermentation when the target mold was introduced into the meju, a fermented soybean loaf, which is used for doenjang preparation (Park et al., 2003). Indeed, the doenjang samples were collected from various markets to detect aflatoxin levels by ELISA (enzyme-linked immunosorbent assay)/HPLC (high-performance liquid chromatography). The results showed that aflatoxin levels were not detected in any of the samples. Another analysis was conducted using 10 commercial samples of chongkukjang (Lee, 2010). There also aflatoxin could not be detected in any of the samples. Previous report performed in 2004–2005 showed that the aflatoxin B_1 was detected (0.05–0.17 μg/kg) in 13 of 56 samples of doenjang (Chun et al., 2006). They were under the permitted level (10 μg/kg) of Korea Foods Standards (Table 6.5). Although those are within the safe levels, well-controlled manufacturing process and proper management practices (including cultivating and storage technology) are important to minimize contamination of mycotoxin, including aflatoxin and ochratoxin, in every stage of production (Lee et al., 2012). Table 6.5 shows the different standards limits for mycotoxin. According to foods consumed frequently in each country, the food standards for mycotoxins are determined.

Table 6.5 Regulation Levels of Mycotoxin Applied in Different Countries (μg/kg)

MYCOTOXINS	CODEX	USA	CANADA	JAPAN	AUSTRALIA AND NEW ZEALAND	EU	KOREA
Aflatoxin B1	–	–	–	10	–	0.1–8	10
Aflatoxins	10–15	20	15	–	15	4–15	15
Aflatoxin M1	0.5	0.5	–	–	–	0.025–0.05	0.5
Ochratoxin A	5	–	–	–	–	0.5–10	5–20
Deoxynivalenol	–	1000	2000	1100	–	200–1750	1000–2000
Zearalenone	–	–	–	–	–	20–400	200–1000
Patulin	50	–	–	50	50	10–50	50
Fumonisins	–	2000–3000	–	–	–	200–4000	1000–4000
Ergot	–	–	–	–	500,000	–	–

Source: Oh, S.S., Strategy for the advancement of risk evaluation on food contaminants. *National Institute of Food and Drug Safety Evaluation (Korea NIFDS) research report*, 2011.

Several factors can be considered in degrading aflatoxins during fermentation. It was reported that NH_3 produced during the fermentation, light, microbial competitions with *Bacillus* spp., and addition of charcoal or vitamin C contribute to reducing aflatoxins from soybean products. Furthermore, it was found that increasing concentrations of CO_2 or N_2 suppress the formation and growth of aflatoxin producers, and it is because aflatoxin-producing molds are aerobic organisms, requiring O_2. The most important is to suppress aflatoxin producers during the fermentation process because it is hard to remove aflatoxin even by heating. Therefore, it was suggested that storage temperature should be kept at 0°C–7.5°C, as optimum temperature for aflatoxin production is 25°C–30°C. For this reason, food industries in Korea are launching the refrigerated soybean produce system to enhance safety. Current refrigerated soybean paste takes about 16% of the entire brand. It was also reported that this method removes off-flavor and improves taste.

6.2.2 Chemical Hazards: Biogenic Amines

Biogenic amines are biologically active amines that are formed by microbial decarboxylation of amino acids. They are frequently found in fish and fish products, alcohols, dairy products, and fermented foods (BIOHAZ, 2011). Biogenic amines are naturally produced in the human body and are related to various biological activities such as synaptic transmission, blood pressure control, allergy response, and cellular growth control. Nonetheless, biogenic amines are regarded as toxic substances that cause diseases and food poisoning when their levels in the body reach a critical threshold (Ladero et al., 2010). Common symptoms of intoxication of biogenic amines are headache, sweating, burning nasal secretion, facial flushing, red rashes, nausea, dizziness, itching, diarrhea, blood pressure disorders, neurological disorders, vomiting, and carcinogenic effects (Rice et al., 1976).

Amino acid decarboxylases convert amino acids to amines, and microorganisms that have amino acid decarboxylases are responsible for the formation of biogenic amines in foods (Beutling, 1996). It has been reported that decarboxylase activity is found in some microbial groups: *Bacillus*, *Citrobacter*, *Clostridium*, *Klebsiella*, *Escherichia*, *Proteus*, *Pseudomonas*, *Salmonella*, *Shigella*, *Photobacterium*, *Lactobacillus*, *Pediococcus*, and *Streptococcus* (Rice et al., 1976; Ten Brink et al., 1990). Amino acid decarboxylation takes place by removing a carboxyl group from amino acid, forming the corresponding amine. The important biogenic amines found frequently in food are histamine (2-(1*H*-imidazol-4-yl)ethanamine), tyramine (4-(2-aminoethyl) phenol), putrescine (butane-1,4-diamine), cadaverine (1,5-diaminopentane), and phenylethylamine (2-phenyl-2-ethanamine), which are products of the decarboxylation of histidine, tyrosine, ornithine, lysine, and phenylalanine, respectively. Putrescine can also be formed by the deimination of agmatine (1-(4-aminobutyl)guanidine – decarboxylation product of arginine). The precursors of the main biogenic amines are described in Table 6.6.

Occurrences of biogenic amines have been reported in AFFs such as natto, chongkukjang, and doenjang (Cho et al., 2006; Tsai et al., 2007a, 2007b; Kim et al., 2012). Contents of biogenic amines in different fermented soybean products are listed in Table 6.7.

Shalaby (1996) reported that miso contained high levels of histamine (4620 mg/kg), putrescine (12,340 mg/kg), cadverine (6340 mg/kg), and tyramine (35,680 mg/kg). Cho et al. (2006) reported that the levels of histamine, tyramine, and putrescine in traditional Korean doenjang were as high as 596.4, 669.5, and 462.6 mg/kg, respectively. Another study examined using four traditional doenjang and six commercial doenjang showed similar results that the content of putrescine

Table 6.6 Main Biogenic Amines and Their Precursors

COMPOUND NAME	PRECURSOR	STRUCTURE	MOLECULAR WEIGHT
Agmatine	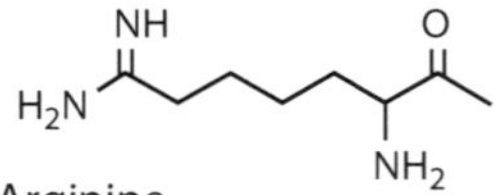 Arginine	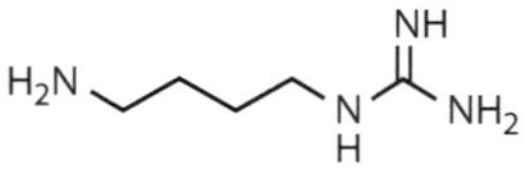	130.2
Tryptamine	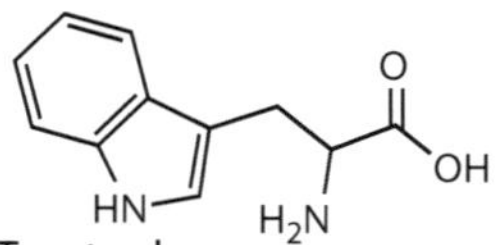 Tryptophan	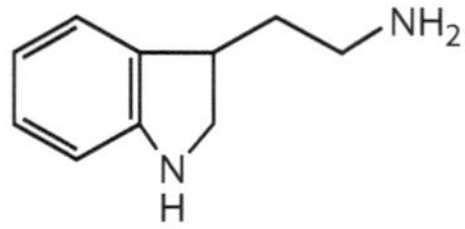	160.2
2-Phenylethyl amine	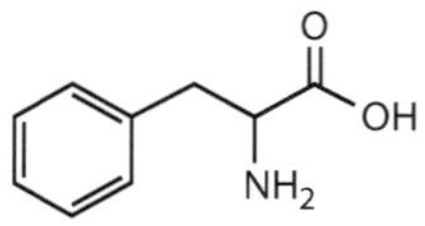 Phenylalanine	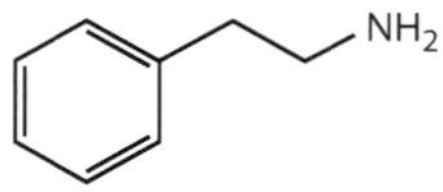	121.2
Putrescine	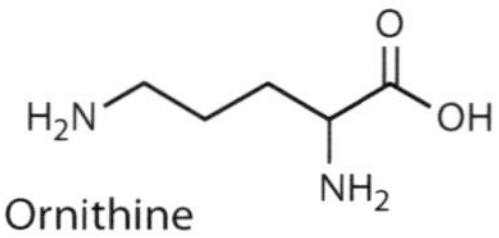 Ornithine	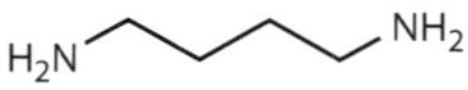	88.2
Cadaverine	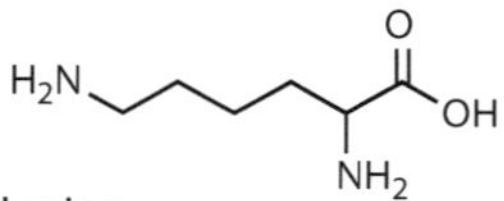 Lysine		202.2
Histamine	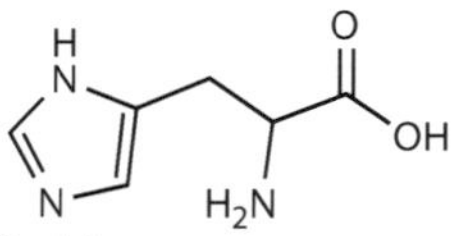 Histidine	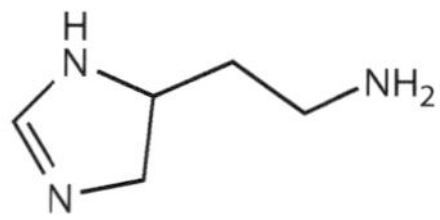	111.0
Tyramine	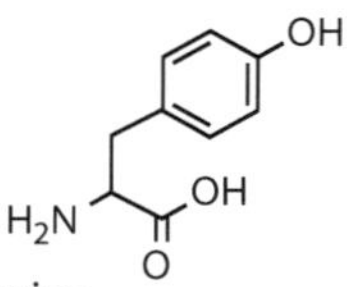 Tyrosine	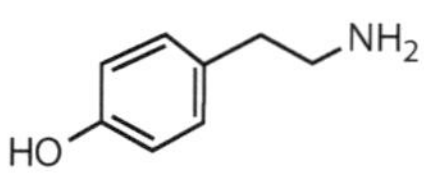	137.3
Spermidine	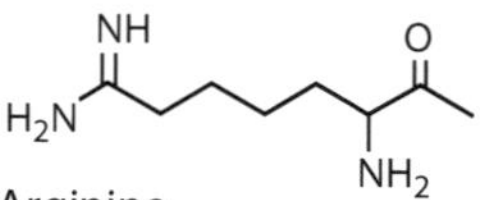 Arginine	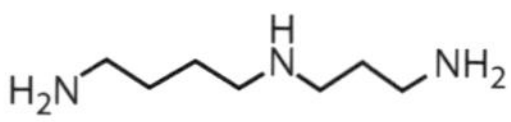	145.3
Spermine	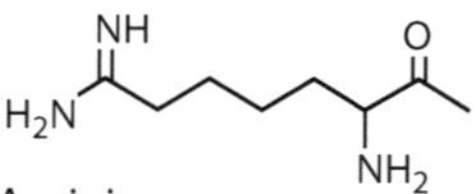 Arginine	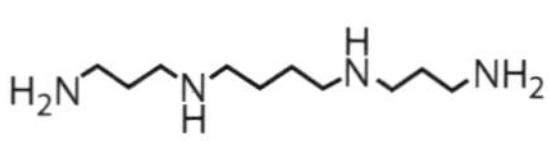	348.1

Source: Shukla, S. et al., *Soybean and Health*, ed. H. El-Shemy, pp. 181–206, InTech, Rijeka, Croatia, 2011.

Table 6.7 Contents of Biogenic Amines in Different Fermented Soybean Products (mg/kg)

FOOD	TYRAMINE	TRYPTAMINE	HISTAMINE	PUTRESCINE	CADAVERINE	PHENYLETHYLAMINE	SPERMINE	SPERMIDINE	REFERENCE
Miso (Japanese soybean paste)	35,680	–	4620	12,340	6340	–	–	–	Shalaby (1996)
Doenjang (Korean – traditional type)	669.5	105.5	596.4	462.6	23.5	244.7	3.8	15.6	Cho et al. (2006)
Doenjang (Korean – modern type)	133.0	22.4	83.6	46.4	3.2	6.5	2.4	7.4	
Miso	48.6	22.6	0.9	19.8	3.0	4.4	2.2	15.7	
Chongkukjang	133.8	69.9	10.1	26.4	9.7	22.0	10.7	52.0	
Chongkukjang powder	68.1	35.0	1.0	10.2	12.1	17.0	15.5	54.6	
Natto (Japanese)	1.2	9.1	35.4	17.1	21.9	ND	8.6	45.0	Tsai et al. (2007a)
Natto (Taiwan)	ND	ND	45.1	1.6	0.5	ND	ND	25.0	Kim et al. (2012)
Natto (Japanese)	47.1	19.2	2.6	8.0	10.3	22.8	45.5	339.7	

Source: Modified from Shukla, S. et al., *Soybean and Health*, ed. H. El-Shemy, pp. 181–206, InTech, Rijeka, Croatia, 2011.
Note: ND, not detected.

(29–1077 mg/kg) and tyramine (13–968 mg/kg) exhibited high values compared with those of other biogenic amines (Lee et al., 2009). According to the results of a comparison between traditional doenjang and commercial doenjang, the levels of biogenic amines in commercial doenjang were significantly reduced to 83.6 mg/kg (histamine), 133.0 mg/kg (tyramine), and 46.4 mg/kg (putrescine) (Cho et al., 2006). The differences in the contents of biogenic amines between the traditional doenjang and the commercial doenjang could be attributed to the variation of target starter and fermentation conditions. Tsai et al. (2007b) tested the biogenic amine levels in seven samples of soybean and 19 samples of black bean douchi. The Chinese ferment black beans used for condiments, which are sold in Taiwan. The test found that four of the soybean douchi samples had high histamine levels (563–808 mg/kg). They are greater than 500 mg/kg, a toxic level to humans. Natto showed particularly high levels of spermidine (339.7 mg/kg) among other biogenic amines (Kim et al., 2012).

The variability and high deviation of biogenic amine levels in the same/similar products are characteristic of all kinds of fermented foods. This means that it is not easy to control fermentation environment such as NaCl concentration, temperature, moisture content, additives, microorganisms including starters, and the environment related to the production of biogenic amines. Nonetheless, there is no doubt that it is important to reduce biogenic amine contents in fermented foods. Several studies suggested that addition of garlic and glycine can suppress amino acid decarboxylase activity in fish sauce. For example, the contents of cadaverine and tyramine were reduced by up to 18.4% and 30.9%, respectively, in the culture treated by garlic extract (Mah et al., 2009). The glycine has the greatest inhibitory activities on biogenic amine production. The contents of putrescine, cadaverine, histamine, tyramine, and spermidine were reduced by 32.6%, 78.4%, 93.2%, 100%, and 100%, respectively, compared with control (Mah and Hwang, 2009).

There are different regulatory limits for biogenic amines by food type. The upper limits of histamine and tyramine in foods were set at 100 and 100–800 mg/kg, respectively. The 30 mg/kg of β-phenylethylamine has been reported to be a toxic dose in food. Total biogenic amine levels of 1000 mg/kg in food are considered harmful to human health (Taylor and Eitenmiller, 1986; Ten Brink et al., 1990).

6.3 Control of Hazards

The AFFs in Korea and Japan are highly commercialized, and industrial quality control systems are well established. On the other hand, many of the African AFFs are still produced at the cottage or household level. Contamination of mycotoxin and/or pathogenic microorganisms in raw materials, especially maize, beans, and seeds are often found in this region. This proves that it is needed to control the harvesting and storage processes by applying good agricultural practices (GAP) and hazard analysis and critical control points (HACCP). In the case of African countries that do not have enough facilities, small-scale industrial systems could be an alternative option.

6.3.1 GAP

Numerous regulations, standards, codes, manuals, and practices have been recently made for the purpose of providing safe foods. The objective of GAP is to ensure safety and quality of produce in the food chain, to capture new market advantages by modifying supply chain governance, to improve natural resources use, workers' health and working conditions, and to create new market opportunities for farmers and exporters in developing countries (FAO GAP principles). These principles can be applied not only to local production, harvest, postharvest processes, and storage but also to soil, water, seed, pesticide, and fertilizer. There are potential benefits when GAP is applied to farming systems because it helps reduce chemical, microbiological, and physical contamination hazards. Applying GAP into the African environment may require a lot of business to control. However, HACCP can be implemented with minimal cost and some changes (e.g., techniques, utensils).

6.3.2 HACCP

The hygienic conditions of cottage level production of fermented foods are generally poor in most of the developing countries, and people in these regions are not well-educated to control their processing procedure in hygienic ways. It is critical to equip clean utensils, working place, clean water, and clean rooms to make safe foods. Most of all, it is important to educate people who are involved in food processing in the developing countries.

The HACCP system has been well-recognized as an approach to eliminate, minimize, or prevent hazards from farm to fork. This control system includes seven principles designed to prevent problems before they occur and to correct deviations as soon as they are found. The critical control points (CCPs) are placed in the flowchart of operational steps, where each practice can be applied to its corresponding steps. Many countries have put effort into developing HACCP models for improving safety and reducing hazards of their fermented foods. The potential hazard lists of fermented products can be summarized as follows: biological hazards (pathogens like *B. cereus* and mycotoxins), chemical hazards (biogenic amines), and physical hazards (foreign substances like grass, seed, metal, string, stick, and wood waste). In the following models, chongkukjang and iru are provided as examples (Figures 6.2 and 6.3).

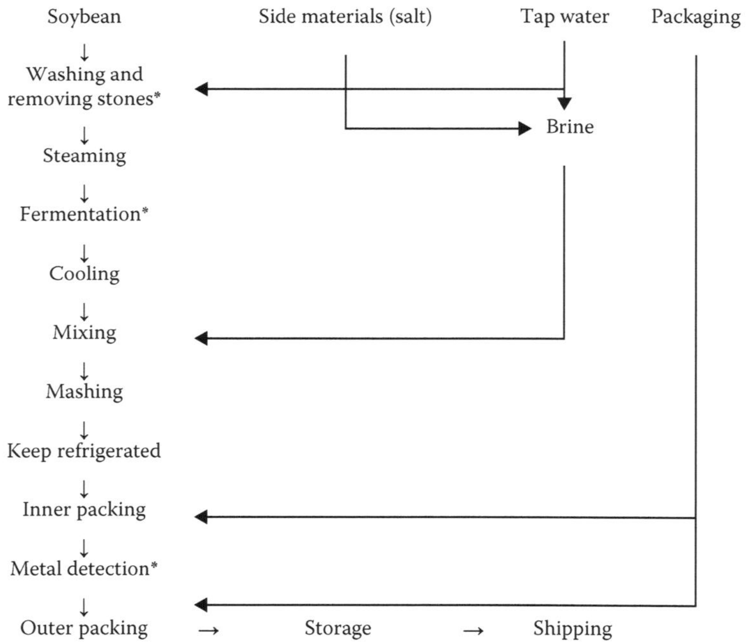

Figure 6.2 Chongkukjang production process marked with CCPs (*). (From Shin, D.H., Research for safety improve and quality improvements of small business and traditional fermented soybean products, *National Institute of Food and Drug Safety Evaluation [Korea NIFDS] research report*, 2011.)

Locust bean seed**
↓
Boiling in water (16–18 h)
(→ *use pressure cooker for 1 h*)
↓
Draining and cooling
(→ *use clean sieve*)
↓
Dehulling by pressing with feet**
(→ *with clean hands or mortar*)
↓
Washing*
↓
Boiling of cotyledons (1–2 h)
(→ *in pressure cooker for 30 min*)
↓
Draining through basket*
(→ *use clean sieve*)
↓
Spreading in calabash trays
↓
Wrapping with jute bags**
↓
Fermentation for up to 36 h
(→ *in screw-capped bottles for 36 h with starter*)
↓
Salting and molding*
↓
Air and/or sun-drying (24 h)*
↓
Iru
(→ *packaged in sealed nylon*)

Figure 6.3 Traditional processing of locust beans into iru showing the CCPs (*). **CCP and food safety risk. The *italic* phrase indicates recommendation for safety process. (From Olayinka, E.M., and Omobayonle, F., *Research Journal of Microbiology*, 1, 503–511, 2006.)

According to the chongkukjang HACCP model suggested by Shin (2011), three CCPs were noticed in their procedures (Figure 6.2). First, he suggests that foreign substances like stones, string, and insects should be removed in the steps of washing and removing stones, with keeping washing speed of 6 rpm. Second, in the fermentation step, proper time (44 ± 1 h) is required to avoid the possibility of contamination with *B. cereus* (<10,000 cfu/g). Lastly, he indicates the second metal detection step just before packaging as another critical control point.

The traditional steps of iru were examined to identify CCPs and then optimize the process (Olayinka and Omobayonle, 2006). The CCPs they identified were the process of including locust beans, dehulling by pressing with feet, and wrapping in the bag (Figure 6.3).

To reduce the chance of contamination, a pressure cooker, dehulling with hands or mortar, and defined culture with known starter were applied. It is evident that these simple changes can reduce microbial contamination and improve nutritional values. As for starters, the mixed culture of *Bacillus* sp. and lactic acid bacteria is recommended. Several studies have found that GAP, GMP (good manufacturing practices), and HACCP are effective as a quality management system for assuring the safety of traditional local foods in Africa (Amoa-Awua et al., 2007).

HACCP systems introduced above are one of the models to verify safety of traditional fermented foods. To ensure both validity and security of the HACCP system, other food safety assurance systems like GAP, GMP, GHP (good hygiene practices), and ISO (International Organization for Standardization) standards can be applied in the food supply chain. These certifications should be encouraged and supported by government authorities. Personal training in terms of sanitation principles and food safety control is definitely required. In addition, universal standards like Codex are required for the global trade of traditional foods because more and more countries are signing FTAs (free trade agreements). Having Codex standards for more traditional foods will facilitate exports since the standards have been applied primarily in international trade and commerce.

6.4 Conclusion

AFFs are very popular in Africa and Asia as a protein source and condiments. Their own indigenous techniques for making fermented foods were defined and succeeded as a food culture. These foods were produced according to traditional methods transferred from their ancestors, and the safety of these foods has been confirmed over the centuries. Recently, fermented foods like natto, doenjang, and chongkukjang, have been produced industrially using their own target starters and exported to other countries.

A number of studies have been performed to identify toxic substances and to find ways to reduce toxin production. The goal for securing safety can be achieved by various actions. First, the whole system from the stage of raw materials (beans and seeds) to storage should be inspected by applying GAP and HACCP. Personal training based on guidelines of GAP, GMP, and HACCP is critical. When looking at the HACCP in detail, each step can be checked whether there are factors to contaminate. Microbial starter cultures should be selected and controlled so they may not produce biogenic amines, mycotoxins, and pathogenic microorganisms during the fermentation process. It needs to be defined which target starter or combination of starters would be better. Biogenic amines can occur during the food production and storage process under unsanitary conditions. Therefore, poor quality of raw materials and inappropriate conditions during food production process result in microbial contamination. Finally, changes to better equipment like mortar, pressure cooker, and appropriate packaging make a big difference to safety.

References

Ahaotu, I., Anyogu, A., Njoku, O.H., Odu, N.N., Sutherland, J.P., and Ouoba, L.I.I. 2013. Molecular identification and safety of *Bacillus* species involved in the fermentation of African oil beans (*Pentaclethra macrophylla* Benth) for production of Ugba. *International Journal of Food Microbiology* 162: 95–104.

Amoa-Awua, W.K., Ngunjiri, P., Anlobe, J., Kpodo, K., Halm, M., Hayford, A.E., and Jakobsen, M. 2007. The effect of applying GMP and HACCP to traditional food processing at a semi-commercial kenkey production plant in Ghana. *Food Control* 18: 1449–1457.

Ann, Y.G. 2011. Changes in components and peptides during fermentation of cheonggookjang. *Korean Journal of Food and Nutrition* 24: 124–131.

Azokpota, P., Hounhouigan, D.J., and Nago, M.C. 2006. Microbiological and chemical changes during the fermentation of African locust bean (*Parkia biglobosa*) to produce afitin, iru and sonru, three traditional condiments produced in Benin. *International Journal of Food Microbiology* 107: 304–309.

Bankole, S.A., Adenusi, A.A., Lawal, O.S., and Adesanya, O.O. 2010. Occurrence of aflatoxin B1 in food products derivable from "egusi" melon seeds consumed in southwestern Nigeria. *Food Control* 21: 974–976.

Bankole, S.A., Ogunsanwo, B.M., and Mabekoje, O.O. 2004. Natural occurrence of moulds and aflatoxin B1 in melon seeds from markets in Nigeria. *Food and Chemical Toxicology* 42: 1309–1314.

Betina, V., Gorst-Allman, C., Vleggaar, R., Cole, R.J., Seitz, L.M., Scott, P.M., and Ueno, Y.A.N. eds. 1984. *Mycotoxins. Production, Isolation, Separation and Purification*. Amsterdam, The Netherlands: Elsevier.

Beutling, D.M. 1996. *Biogene Amine in der Ernährung*. Berlin, Germany: Springer Verlag.

BIOHAZ 2011. (EFSA Panel on Biological Hazards). Scientific opinion on risk based control of biogenic amine formation in fermented foods. *EFSA Journal* 9: 2393.

Blaser, M.J., Hardesty, H.L., Powers, B., and Wang, W.L. 1980. Survival of *Campylobacter fetus* subsp. *jejuni* in biological milieus. *Journal of Clinical Microbiology* 11: 309–313.

Bullerman, L.B. 1986. Mycotoxins and food safety. *Food Technology* 40: 59–66.

Cho, T.Y., Han, G.H., Bahn, K.N., Son, Y.W., Jang, M.R., Lee, C.H., Kim, S.H., Kim, D.B., and Kim, S.B. 2006. Evaluation of biogenic amines in Korean commercial fermented foods. *Korean Journal of Food Science and Technology* 38: 730–737.

Chukeatirote, E., Chainun, C., Siengsubchart, A., Moukamnerd, C., Chantawannaku, L.P., Lumyong, S., and Thakang, P. 2006. Microbiological and biochemical changes in Thua nao fermentation. *Research Journal of Microbiology* 1: 38–44.

Chun, H.S., Ok, H.E., Kim, H.J., Hwang, J.B., and Chung, D.H. 2006. Risk assessment of aflatoxins in food products consumed in South Korea. Proceedings, *IUFoST 13th World Congress of Food Science and Technology*. Nantes, France.

Dakić, I., Morrison, D., Vuković, D., Savić, B., Shittu, A., Jezek, P., Hauschild, T., and Stepanović, S. 2005. Isolation and molecular characterization of *Staphylococcus sciuri* in the hospital environment. *Journal of Clinical Microbiology* 43: 2782–2785.

Egounlety, M., and Aworh, O.C. 2003. Effect of soaking, dehulling, cooking and fermentation with *Rhizopus oligosporus* on the oligosaccharides, trypsin inhibitor, phytic acid and tannins of soybean (*Glycine max* Merr.), cowpea (*Vigna unguiculata* L. Walp) and groundbean (*Macrotyloma geocarpa* Harms). *Journal of Food Engineering* 56: 249–254.

Ehling-Schulz, M., Fricker, M., and Scherer, S. 2004. *Bacillus cereus*, the causative agent of an emetic type of food-borne illness. *Molecular Nutrition & Food Research* 48: 479–487.

Ekundayo, C.A., and Idzi, E. 1990. Mycoflora and nutritional value of shelled melon seeds (*Citrulus vulgaris* Schrad.) in Nigeria. *Plant Foods for Human Nutrition* 40: 215–222.

Evans, E., Musa, A., Abubakar, Y., and Mainuna, B. 2013. Nigerian indigenous fermented foods: Processes and prospects. In *Mycotoxin and Food Safety in Developing Countries*, ed. H.A. Makun, pp. 153–180. Rijeka, Croatia: InTech.

From, C., Pukall, R., Schumann, P., Hormázabal, V., and Granum, P.E. 2005. Toxin-producing ability among *Bacillus* spp. outside the *Bacillus cereus* group. *Applied and Environmental Microbiology* 71: 1178–1183.

Gareis, M., and Wernery, U. 1994. Determination of gliotoxin in samples associated with cases of intoxication in camels. *Mycotoxin Research* 10: 2–8.

Gill, C.O., and Harris, L.M. 1982. Survival and growth of *Campylobacter fetus* subsp. *jejuni* on meat and in cooked foods. *Applied and Environmental Microbiology* 44: 259–263.

Granum, P.E. 2001. *Bacillus cereus*. In *Food Microbiology, Fundamental and Frontiers*, 2nd ed., eds. M.P. Doyle, L.R. Beuchat, and T.J. Montville, pp. 373–381. Washington, DC: ASM Press.

Guo, X.H., Li, D.F., Lu, W.Q., Piao, X.S., and Chen, X.L. 2006. Screening of *Bacillus* strains as potential probiotics and subsequent confirmation of the in vivo effectiveness of *Bacillus subtilis* MA139 in pigs. *Antonie van Leeuwenhoek* 90: 139–146.

Ijabadeniyi, A.O. 2007. Microbiological safety of gari, lafun and ogiri in Akure metropolis, Nigeria. *African Journal of Biotechnology* 6: 2633–2635.

Jay, J.M. 1986. *Modern Food Microbiology*, 3rd ed. New York: Van Nostrand Reinhold.

Khokhar, S., and Chauhan, B.M. 1986. Antinutritional factors in moth bean (*Vigna aconitifolia*): Varietal differences and effects of methods of domestic processing and cooking. *Journal of Food Science* 51: 591–594.

Kim, B., Byun, B.Y., and Mah, J.H. 2012. Biogenic amine formation and bacterial contribution in Natto products. *Food Chemistry* 135: 2005–2011.

Kim, Y.J. 2006. *Study on the safety for traditional foods*: Korea Food and Drug Administration research report.

Ladero, V., Calles-Enríquez, M., Fernández, M., and Alvarez, M.A. 2010. Toxicological effects of dietary biogenic amines. *Current Nutrition and Food Science* 6: 145–156.

Lee, C.H., and Lee, S.S. 2002. Cereal fermentation by fungi. In *Applied Mycology and Biotechnology, vol. 2 Agriculture and Food Production*, eds. G.G. Khachatourians, and D.K. Arora, pp. 151–170. Amsterdam, The Netherlands: Elsevier Science.

Lee, E.J., Seo, J.E., Lee, J.K., Oh, S.W., and Kim, Y.J. 2008. Microbial and chemical properties of ready-to-eat skate in Korean market. *Journal of Food Hygiene and Safety* 23: 137–141.

Lee, G.I., Lee, H.M., and Lee, C.H. 2012. Food safety issues in industrialization of traditional Korean foods. *Food Control* 24: 1–5.

Lee, H.T., Kim, J.H., and Lee, S.S. 2009. Analysis of microbiological contamination and biogenic amines content in traditional and commercial doenjang. *Journal of Food Hygiene and Safety* 24: 102–109 (in Korean).

Lee, S.O. 2010. A study on the hazard analysis and HACCP system for Chongkookjang manufacturing. M.Sc. thesis. Asan, Korea: Hoseo University.

Leejeerajumnean, A. 2000. *Bacillus* fermentation of soybeans: Characterisation of traditional thua nao manufacture. Ph.D. thesis. Reading, UK: University of Reading.

Mah, J.H., and Hwang, H.J. 2009. Effects of food additives on biogenic amine formation in Myeolchi-jeot, a salted and fermented anchovy (*Engraulis japonicas*). *Food Chemistry* 114: 168–173.

Mah, J.H., Kim, Y.J., and Hwang, H.J. 2009. Inhibitory effects of garlic and other spices on biogenic amine production in Myeolchi-jeot, Korean salted and fermented anchovy product. *Food Control* 20: 449–454.

Mensah, P. 1997. Fermentation – the key to food safety assurance in Africa? *Food Control* 8: 271–278.

N'dir, B., Hbid, C., Cornelius, C., Roblain, D., Jacques, P., Vanhentenryck, F., Diop, M., and Thonart, P. 1994. Propriétés antifongiques de la microflore sporulée du nététou. *Cahiers Agricultures* 3: 23–30.

Nout, M.J.R., Bakshi, D., and Sarkar, P.K. 1998. Microbiological safety of kinema, a fermented soya bean food. *Food Control* 9: 357–362.

Odunfa, S.A. 1985. Microbiological and toxicological aspects of fermentation of castor oil seeds for ogiri production. *Journal of Food Science* 50: 1758–1759.

Oguntoyinbo, F.A., and Oni, M.O. 2004. Incidence and characterization of *Bacillus cereus* from traditional fermented meals in Nigeria. *Journal of Food Protection* 67: 2805–2808.

Oh, S.S. 2011. Strategy for the advancement of risk evaluation on food contaminants. *National Institute of Food and Drug Safety Evaluation (Korea NIFDS) research report.*

Olayinka, E.M., and Omobayonle, F. 2006. Evaluation and optimization of critical control points in the production of iru. *Research Journal of Microbiology* 1: 503–511.

Omafuvbe, B.O., Shonukan, O.O., and Abiose, S.H. 2000. Microbiological and biochemical changes in the traditional fermentation of soybean for "soy-daddawa" – Nigerian food condiment. *Food Microbiology* 17: 469–474.

Ouoba, L.I.I., Diawara, B., Jespersen, L., and Jakobsen, M. 2007. Antimicrobial activity of *Bacillus subtilis* and *Bacillus pumilus* during the fermentation of African locust bean (*Parkia biglobosa*) for Soumbala production. *Journal of Applied Microbiology* 102: 963–970.

Ouoba, L.I.I., Parkouda, C., Diawara, B., Scotti, C., and Varnam, A.H. 2008. Identification of *Bacillus* spp. from Bikalga, fermented seeds of *Hibiscus sabdariffa*: Phenotypic and genotypic characterization. *Journal of Applied Microbiology* 104: 122–131.

Owens, J.D., Allagheny, N., Kipping, G., and Ames, J.M. 1997. Formation of volatile compounds during *Bacillus subtilis* fermentation of soya beans. *Journal of the Science of Food and Agriculture* 74: 132–140.

Park, K.Y., Jung, K.O., Rhee, S.H., and Choi, Y.H. 2003. Antimutagenic effects of doenjang (Korean fermented soypaste) and its active compounds. *Mutation Research* 523–524: 43–53.

Parkouda, C., Nielsen, D.S., Azokpota, P., Ouoba, L.I.I., Amoa-Awua, W.K., Thorsen, L., Hounhouigan, J.D., Jensen, J.S., Tano-Debrah, K., Diawara, B., and Jakobsen, M. 2009. The microbiology of alkaline-fermentation of indigenous seeds used as food condiments in Africa and Asia. *Critical Reviews in Microbiology* 35: 139–156.

Parkouda, C., Thorsen, L., Compaore, C.S., Nielsen, D.S., Tano-Debrah, K., Jensen, J.S., Diawara, B., and Jakobsen, M. 2010. Microorganisms associated with Maari, a baobab seed fermented product. *International Journal of Food Microbiology* 142: 292–301.

Rice, S.L., Eitenmiller, R.R., and Koehler, P.E. 1976. Biologically active amines in foods. A review. *Journal of Milk and Food Technology* 39: 353–358.

Roy, A., Moktan, B., and Sarkar, P.K. 2007. Microbiological quality of legume-based traditional fermented foods marketed in West Bengal, India. *Food Control* 18: 1405–1411.

Sanni, A.I., Onilude, A.A., Fadahunsi, I.F., Ogunbanwo, S.T., and Afolabi, R.O. 2002. Selection of starter cultures for the production of ugba, a fermented soup condiment. *European Food Research and Technology* 215: 176–180.

Sarkar, P.K., Cook, P.E., and Owens, J.D. 1993. *Bacillus* fermentation of soybeans. *World Journal of Microbiology and Biotechnology* 9: 295–299.
Sarkar, P.K., Hasenack, B., and Nout, M.J.R. 2002. Diversity and functionality of *Bacillus and* related genera isolated from spontaneously fermented soybeans (Indian kinema) and locust beans (African soumbala). *International Journal of Food Microbiology* 77: 175–186.
Sarkar, P.K., and Tamang, J.P. 1995. Changes in the microbial profile and proximate composition during natural and controlled fermentations of soybeans to produce kinema. *Food Microbiology* 12: 317–325.
Shalaby, A.R. 1996. Significance of biogenic amines to food safety and human health. *Food Research International* 29: 675–690.
Shin, D.H. 2011. Research for safety improve and quality improvements of small business and traditional fermented soybean products: *National Institute of Food and Drug Safety Evaluation (Korea NIFDS) research report.*
Shinagawa, K., Konuma, H., Sekita, H., and Sugii, S. 1995. Emesis of rhesus monkeys induced by intragastric administration with the HEp-2 vacuolation factor (cereulide) produced by *Bacillus cereus*. *FEMS Microbiology Letters* 130: 87–90.
Shinagawa, K., Ueno, Y., Hu, D., Ueda, S., and Sugii, S. 1996. Mouse lethal activity of a HEp-2 vacuolation factor, cereulide, produced by *Bacillus cereus* isolated from vomiting-type food poisoning. *Journal of Veterinary Medical Science* 58: 1027–1029.
Shukla, S., Kim, J.K., and Kim, M. 2011. Occurrence of biogenic amines in soybean food products. In *Soybean and Health*, ed. H. El-Shemy, pp. 181–206. Rijeka, Croatia: InTech.
Steinkraus, K.H. 1997. Classification of fermented foods: Worldwide review of household fermentation techniques. *Food Control* 8: 311–318.
Taylor, S.L., and Eitenmiller, R.R. 1986. Histamine food poisoning: Toxicology and clinical aspects. *Critical Reviews in Toxicolology* 17: 91–128.
Ten Brink, B., Damink, C., Joosten, H.M.L.J., and Huis In 't Veld, J.H.J. 1990. Occurrence and formation of biologically active amines in foods. *International Journal of Food Microbiology* 11: 73–84.
Tsai, Y.H., Chang, S.C., and Kung, H.F. 2007a. Histamine contents and histamine-forming bacteria in natto products in Taiwan. *Food Control* 18: 1026–1030.
Tsai, Y.H., Kung, H.F., Chang, S.C., Lee, T.M., and Wei, C.I. 2007b. Histamine formation by histamine-forming bacteria in douchi, a Chinese traditional fermented soybean product. *Food Chemistry* 103: 1305–1311.
Varga, J., Péteri, Z., Tábori, K., Téren, J., and Vágvölgyi, C. 2005. Degradation of ochratoxin A and other mycotoxins by *Rhizopus* isolates. *International Journal of Food Microbiology* 99: 321–328.

7

Challenges Associated with Technological Aspects for Modernization of Alkaline-Fermented Foods

Among the new generations, the awareness and appreciation of indigenous healthy foods including alkaline-fermented foods (AFFs) needs to be enhanced. AFFs are considered to be old-fashioned. In fact, some of the food processing techniques are outdated with low hygiene and quality standards. Some AFFs have a poor appearance and lack convenience. Are many valuable AFFs threatened with extinction? Upgrading of quality, safety, and convenience is required to keep AFFs on the future menu.

Research and development activities such as upgrading of small-scale food fermentations, however, require capital investment that is often lacking in small and cottage level enterprises. There is an urgent opportunity and challenge for the local and international food industry business, as well as the technical cooperation, to strengthen this culturally important aspect of life.

7.1 Modulation of Product Chains

Anita R. Linnemann

7.1.1 *Key Determinants of Successful Modernization of Traditional Foods*

Delivery of a product that meets the consumers' demands with respect to preferred quality characteristics is the key to the successful commercialization of any food product (Linnemann et al., 2006). This is especially true in markets that offer an abundance of products because it allows consumers to be picky about their purchases. Moreover, the characteristics of the product must be constant over time because consumers expect the same experience from a branded product when they purchase the product repeatedly. The above implies two things, namely that it is necessary to establish the main quality characteristics that make consumers want to buy the product and to assure that the quality is constant over time by controlling the critical points in the production process from raw materials to the product as it is offered to the customers on the shelves of the retailer. To comply with the latter condition, a chain approach is required. Such an approach not only helps to deliver a product of a constant quality but additionally is a way to make the production process efficient and effective, in other words, to obtain the best monetary returns from the efforts to produce the food product.

7.1.2 How to Determine Consumer Preferences

Food products have many quality characteristics. Some of these can be perceived when you hold the product in your hands and are able to smell, touch, and see it, like the color and texture. These, combined with other characteristics that we can perceive with our senses, are called the intrinsic quality characteristics (van Boekel and Linnemann, 2011). They include mouthfeel, viscosity, flavor, size, price, and brand. In addition, there are so-called extrinsic quality characteristics that are related to, for instance, the contents of macronutrients and micronutrients, the presence of genetically modified organisms (GMOs), and the production process (e.g., conventional or organic). These characteristics are difficult to verify for consumers, and therefore require consumers to trust the producer.

When consumers visit a retailer to do their shopping, they do not consider all the quality characteristics that a product may or may not have, but they select on a relatively small number of quality characteristics to buy a certain product. To determine what these five to six quality characteristics of overriding importance are, several research methods exist (van Kleef et al., 2005). When we have an existing product that we can present to consumers, techniques like focus group discussions and consumer panels would be adequate. However, it is important to properly select the respondents for the assessment of consumer preferences from the right target group. Developments like an increasing number of elderly consumers, migration, and household diversification as a result of the increasing number of alternative forms of cohabitation, all have led to market segmentation, i.e., the existence of groups in society with more or less the same needs. If possible, lead users should be invited to take part as respondents for product characterization. Lead users are the ones who are regularly consuming the product that we want to profile, and therefore possess the knowledge and experience to best describe the product under investigation. Be aware to get clear descriptions from the respondents. Consumers tend to say, for example, that "they like the taste." This answer is so generic that a product developer cannot use it. Make sure to get a clear description of the taste that your respondent likes so much in your product, e.g., "a creamy mouthfeel."

It is likely that the consumer research will generate more than the 5 to 6 quality characteristics that we are looking for. Ideally, the participants of focus groups reach consensus on the quality characteristics they prefer most. If not, the top 5 to 6 characteristics need to be assessed by scoring the frequency with which they were mentioned. Next, the answers given by the respondents need to be translated into technological product characteristics. A tool that supports this, is the first matrix of the Quality Function Deployment method, the so-called House of Quality (HoQ) (Hauser and Clausing, 1988; Benner et al., 2003). In its basic form, the HoQ comprises five rooms: (1) the WHATs, which describe the maximally five to six most important quality characteristics of the food product; (2) the HOWs, which are the translations of the consumers' quality preferences into technological product properties; (3) the RELATIONSHIP MATRIX, which indicates the presence and strength of the relationships between a WHAT and one or more HOWs; (4) the CORRELATIONSHIP MATRIX as a roof over the HOWs in which the relations between the technological product characteristics are presented; and (5) the TARGET values in the cellar that describe the minimum values that have to be reached for the entries in the room of the HOWs, so the technological product properties (Figure 7.1).

Figure 7.2 presents an example of a HoQ for meitauza, a Chinese traditionally fermented okara (see Section 2.1.1.1.7 in this book). In

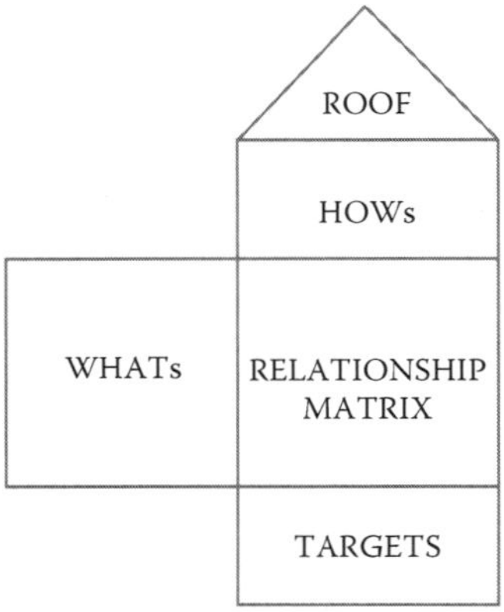

Figure 7.1 The House of Quality, the first matrix of the Quality Function Deployment method, which is used to systematically translate consumer wishes into technological product requirements.

this case, the five most important consumer demands that were to be met in the final product were defined as: smooth mouthfeel, no fishy smell, easy digestion, good shelf life, and healthy. These demands were translated into the following technological product and process requirements: use of *Aspergillus oryzae* starter, use of *Aspergillus elegans* starter, good fermentation conditions, dehydration, and addition of rice bran. The strength of the relations between the consumer demands and their technological translation were indicated in the

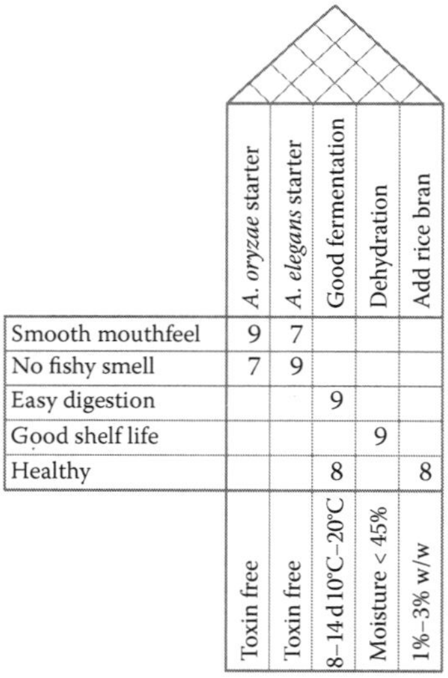

Figure 7.2 A fictitious example of the House of Quality for meitauza, a Chinese traditionally fermented okara.

relationship matrix and target values were set, see the lowest room in Figure 7.2. Finally, all the relationships between each set of HOWs were considered: Is it possible to use *A. oryzae* starter as well as *A. elegans* starter in the same product? Do these two microorganisms tolerate each other's presence? Is it possible to use *A. oryzae* starter in a product to which rice bran is added? In our example no conflicting HOWs were identified, but this is not always the case. Consider, for instance, the development of a refrigerator door. On the one hand, it should be as tight as possible to prevent cold air from inside escaping to the outside. On the other hand, you also want to have a door that opens easily. In such a case of conflicting technological requirements, a compromise is needed.

7.1.3 Applying a Chain Approach to Ensure Constant Product Quality

Once the predominant quality characteristics of a food product as it is to be marketed are established, the production chain needs to be defined. This is necessary because the quality of the end product is influenced by the quality of the inputs as well as the operations and conditions they are subjected to in each step of the chain. In other words, the selected varieties, the growing conditions in the field, the storage conditions of the raw material before processing, the processing steps themselves and the conditions during processing, the packaging, all affect the quality of the end product. An example of a proper mapping of a production chain for a food that is prepared by alkaline fermentation is presented for tayohounta as produced in Benin (Figure 7.3).

At each step of the production chain, handling practices and environmental parameters impact the product, and this can be wanted (i.e., beneficial for the quality of the end product) or unwanted (i.e., favoring spoilage microorganisms). In the case of the production of tayohounta (Figure 7.3), manual decortication will cause physical damage to the seeds, which may favor infestation by spoilage organisms. On the other hand, the leaves that are used to cover the seeds after the cooking and draining of the kernels may be crucial for the quality of the end product by creating the desired conditions for successful fermentation. An example of factors during the fermentation step of tayohounta that may be crucial in achieving the desired product is presented in Figure 7.4; they include the composition of the

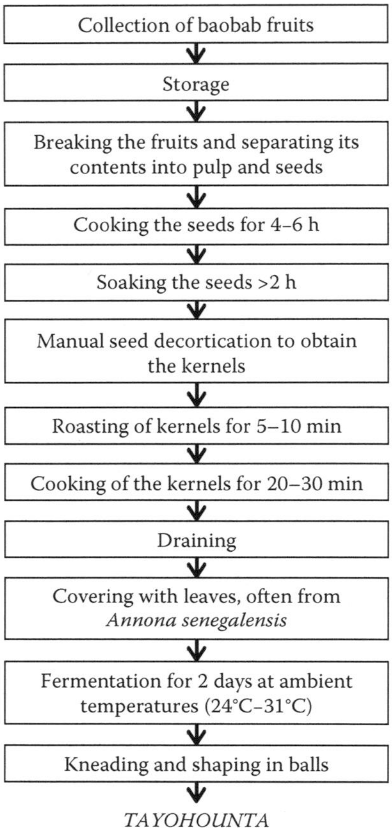

Figure 7.3 The production chain for tayohounta, a fermented food made from baobab kernels. (Adapted from Chadare, F.J. et al., *Ecology of Food and Nutrition*, 49, 279–297, 2010.)

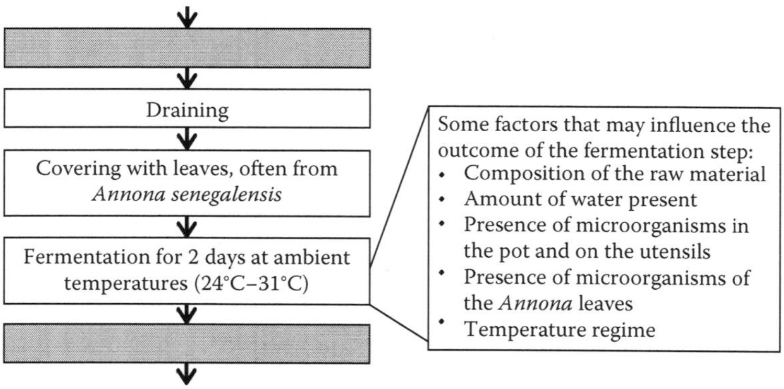

Figure 7.4 Example of some factors of the fermentation step in the production of tayohounta that may be critical to achieve the desired product quality.

raw material, the numbers in which desired and undesired microorganisms are present, and the environmental conditions. Ideally, these quality changes are well understood and can be mathematically described by so-called Quality Change Modeling (QCM) (Verkerk et al., 2007). Such a model relates the conditions along the production chain to changes in quality characteristics of the product, and consequently enables producers to focus on the most crucial factors during each step of the production process. This approach is elaborated for a case study on Nile perch (*Lates niloticus*) from Kenya by Verkerk et al. (2007).

7.1.4 Benchmarking

Once we have made our product according to the main characteristics we aimed for, and know and control the production chain, we may want to test it against competing products. This can again be done with the HoQ, using a consumer panel of lead users or trained panelists. The respondents are asked to score our product and those of our main competitors on the quality characteristics that were defined earlier as the desired consumer WHATs, our unique selling points. Figure 7.5 presents a possible outcome for such a consumer competitive assessment for only two products. The figure shows that our

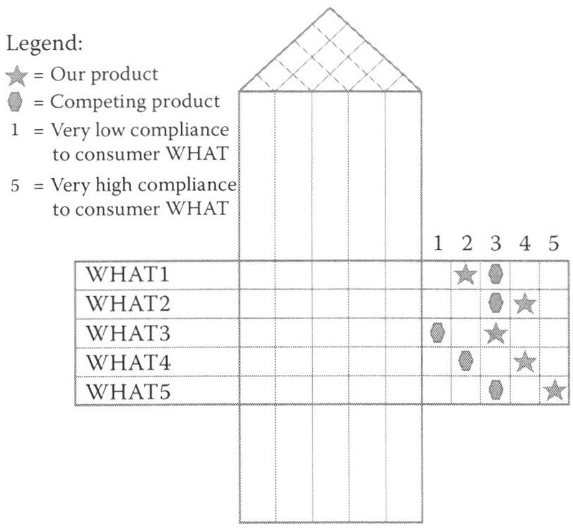

Figure 7.5 The House of Quality as a tool to conduct competitive assessment between products by consumer panels.

product scores higher than the product of the competitor on four of the five consumer requirements. The panel considered our product to be inferior on the first characteristic only, indicating that we need to improve the quality of our product in that respect.

7.1.5 Conclusion

When traditional food products are to be produced on a larger scale and offered as a branded product, the needs and wants of its customers need to be taken into account. Therefore, it is necessary to first assess what customers expect from the product, and next, to make sure that the product never disappoints and always meets the expectations. Tools to assist in achieving these goals are the first matrix of the Quality Function Deployment method, the so-called HoQ, and a systematic analysis of the production chain to determine the factors that are crucial in attaining the targeted end product quality. A tool that may assist in this process is QCM. However, this tool requires in-depth knowledge of the changes in the product at each step of the production chain in relation to handling practices and environmental factors. For many products that are described in this book, this knowledge is still partly lacking.

7.2 Development of Starter Cultures

Egon Bech Hansen

7.2.1 Food Fermentation and Food Cultures

Preservation of perishable food products by fermentation has been a common practice since ancient times. This form of applied biotechnology predates any knowledge about microorganisms. The food products traditionally preserved by fermentation are fruits, vegetables, meat, and milk. Typically, the fermentation converts carbohydrates into alcohols and organic acids, resulting in a drop in pH. AFFs are less common as the increase in pH is not caused directly by the microbial growth but indirectly through ammonia released from enzymatic hydrolysis of proteins. For AFFs, a satisfactory outcome depends on a delicate balance between the enzymatic and the metabolic processes. The widespread use of food fermentations has been extensively covered in the literature, and a few books (Hui et al., 2004; Steinkraus, 2004; Bamforth, 2008) may be considered as excellent entries into most aspects of food fermentations.

7.2.1.1 Microorganisms Used in Food Fermentations A multitude of microbial species has been used for the production of a large variety of fermented foods. Recently, a working group under the International Dairy Federation (IDF) established an inventory of microorganisms with a documented use in food fermentations (Bourdichon et al., 2012). This list is currently the most complete overview of species used in microbial food fermentations. Microorganisms from 264 different species of bacteria, yeasts, and molds have a documented history of use in food. Table 7.1 lists the genera to which 264 species belong. It should be noted that some of the genera listed also contain species not suitable for food fermentations. However, this precaution is also relevant even at the species level as some species with food use also contain strains that are problematic or even pathogenic.

Table 7.1 Genera of Microorganisms with Documented Beneficial Use in Food Fermentations

	BACTERIA		
GRAM NEGATIVE			
Acetobacter	*Gluconacetobacter*	*Gluconobacter*	*Hafnia*
Halomonas	*Zymomonas*		
GRAM POSITIVE			
Arthrobacter	*Bifidobacterium*	*Lactococcus*	*Staphylococcus*
Pediococcus	*Brachybacterium*	*Leuconostoc*	*Streptococcus*
Propionibacterium	*Brevibacterium*	*Macrococcus*	*Streptomyces*
Enterococcus	*Carnobacterium*	*Microbacterium*	*Tetragenococcus*
Oenococcus	*Corynebacterium*	*Micrococcus*	*Weissella*
Bacillus	*Kocuria*	*Lactobacillus*	
	YEASTS AND MOLDS		
Aspergillus	*Geotrichum*	*Mucor*	*Scopulariopsis*
Candida	*Guehomyces*	*Neurospora*	*Starmerella*
Cyberlindnera	*Hanseniaspora*	*Penicillium*	*Torulaspora*
Cystofilobasidium	*Kazachstania*	*Pichia*	*Trigonopsis*
Debaryomyces	*Kluyveromyces*	*Rhizopus*	*Wickerhamomyces*
Dekkera	*Lachancea*	*Saccharomyces*	*Yarrowia*
Fusarium	*Lecanicillium*	*Schizosaccharomyces*	*Zygosacharomyces*
Galactomyces	*Metschnikowia*	*Schwanniomyces*	*Zygotorulaspora*

7.2.1.2 Inoculation for Food Fermentations Microorganisms are ubiquitously present, and food products, if left under conditions allowing microbial growth, spontaneously undergo fermentation. Traditional fermented foods are usually based on such spontaneous fermentations where the physical and chemical conditions are adjusted to favor the desired fermentation process. It is important here to emphasize that "spontaneous fermentation" is a misnomer, as Louis Pasteur showed that microorganisms do not spontaneously arise in sterile media. The term is used here solely to designate that a fermentation process is based on preexisting (indigenous) microbiota in nonsterile raw materials. When relying on such an undefined inoculum, the method to gain some control over the fermentation process is to adjust the conditions to favor the desired types of microorganisms by adjusting access of air, temperature, water activity (a_w), pH, sugar, and salt concentrations. If the fermentation takes the right direction, the metabolites produced often eliminate competing microorganisms and lead to preservation of the food. This is particularly the case for alcoholic fermentations and for lactic fermentations. A large batch-to-batch variation is, nevertheless, inevitable and even complete failures are not uncommon when relying on the indigenous flora to initiate the fermentation.

"Back-slopping" is a first step toward the use of starters. In this procedure, a sample of a successful fermentation serves as an inoculum for subsequent fermentations. Typically, the inoculum consists of complex microbial communities. The microbial composition varies over time, and the quality of the fermented food still shows batch-to-batch variation. However, due to the large biodiversity in the inoculum, the system can provide a remarkable robustness, and complete failures are relatively rare. Some "back-slopping inocula" have been developed to become commercial products and may be regarded as starter cultures.

"Commercial starter cultures" are standardized inocula intended for the production of fermented foods. Starter cultures are produced by specialized manufacturers. Rigorous quality assurance and quality control are conducted to ensure performance, composition, and safety of the culture. The use of commercial starter cultures allows the food manufacturer to maintain control over the production process and to minimize the risk of fermentation failures. Commercial starter cultures range in complexity from complex microbial communities of multiple strains and species, over mixtures of a limited number of microbial strains down to starter cultures of pure single strains (Hansen, 2004).

Inoculation procedures are important for fermentations not relying on "spontaneous fermentation." The scale of operation, the value of the product, the importance of timing, as well as availability and format of starter cultures are the determining factors for the choice of inoculation procedure. There is an inverse correlation between the volume of the inoculum and the time needed for the fermentation. Increasing the volume of the inoculum reduces the time needed for the fermentation. Balancing these two parameters usually leads to an inoculation rate in the range of 1%–10%. The inoculum can be provided in two different ways: bulk starter or direct inoculum.

"Bulk starter systems" are used for the on-site production of inoculum. It does require establishment of a starter room with fermentors for the propagation of the starter culture. It also requires the food manufacturer to be skilled in managing and propagating cultures. The inoculum used to start the bulk starter fermentors is typically supplied from a starter culture company, although some food companies maintain culture systems of their own.

"Direct inocula" are commercial starter cultures prepared in a highly concentrated form. Such cultures are used for inoculation directly into the vessel used for the food fermentation. Direct inoculation gives a good control of timing of the process, and this procedure minimizes batch-to-batch variation as propagation steps in the food factory are minimized. The use of highly concentrated starter cultures allows the desired inoculation rate to be reached using as little as 0.01% relative to the fermentor volume. This is convenient and allows inventories of inoculants for several productions to be maintained. The disadvantage of direct inoculation is a relatively high direct cost of starter cultures.

The physical format of starter cultures can be liquid, frozen, or dry. "Liquid cultures" are easy to handle but suffer from a relatively short shelf life. A liquid culture remains active from a few days up to a couple of weeks. "Deep-frozen cultures" have excellent storage stability and can be relatively easy to handle and dose. Frozen cultures can, if stored under optimal conditions (below -50°C), remain viable for years. Easy handling requires the culture to be frozen as pellets. The disadvantage of frozen cultures is that shipment has to be with dry ice and transport is therefore relatively expensive. "Lyophilized cultures" also have excellent storage stability. They can be kept at moderate freezing temperature and even tolerate periods at ambient temperature. The drawbacks of lyophilized cultures are the need for rehydration and a longer lag phase.

Spores of some bacteria and molds can be prepared in dry format and generally show excellent storage stability. The time needed to germinate spores is even longer than for lyophilized cultures.

7.2.1.3 Requirements for a Starter Culture The advantage of using a starter culture for inoculation is to gain control over the fermentation process including safety and regulatory compliance. A starter culture must therefore meet a set of specifications regarding performance, safety, and price.

The performance of a culture is determined by the combined metabolic activity of the microorganisms in the culture. The metabolic processes add a multitude of properties to the food product in question. Metabolites can contribute to the preservation and provide flavor and texture to the food product. Another important aspect of the metabolism may be the removal of certain compounds. The removal of anti-nutritional factors or toxins is for several fermented foods the main explanation for an increase in the nutritional value compared with the unfermented raw materials. Fermented products represent an enormous potential for innovation in food and feed as the cultures may be composed from microorganisms with quite diverse metabolic pathways. The metabolic activity and the growth rate of the culture determine the time needed for the fermentation and thus be a factor contributing to the process economy and cost of the product. Often, the activity of a starter culture is defined based on a primary fermentation parameter. This activity defines the units of dosing and pricing. The rate of acidification is commonly used to define units for lactic cultures. Obviously, cultures for AFFs need an alternative unit definition. If a unit based on the primary activity of a culture cannot easily be established, the choice will often be to dose and sell the culture based on colony-forming units (cfu).

"Robustness" of a fermentation process is another important performance parameter for a starter culture. With an increase in scale, the value of a lost batch becomes increasingly visible, and a process can therefore only be industrialized if it is reasonably robust. Factors leading to batch-to-batch variation may either be an insufficient standardization of process parameters or variation in the composition of the starter culture. Infection by bacteriophages can be a major cause of culture failure, and it is of extreme importance to eliminate the risk of contaminating the early steps of the process with components from later steps. The problem with bacteriophages can be reduced through the use of cultures selected for phage resistance, and by the rotation with cultures showing different phage susceptibility.

"Safety" of a microbial food culture requires that the constituent microorganisms are safe and that the product is free from contaminating undesirable microorganisms. Contaminants may be avoided through the design of the production process combined with a suitable quality control process. Quality assurance and quality control need to be an ongoing integral part of a starter culture production process, whereas the safety evaluation of the constituent microorganisms is usually conducted once, prior to introduction on the market. The regulatory environment for microbial food cultures is not standardized globally. Regulatory frameworks have been established in some countries, whereas the area is unregulated in others. The countries in the European Union have established a common regulatory framework for food and food ingredients. In the European system, safety assessment of microorganisms is based on the tool "qualified presumption of safety" (QPS). In the United States, the area is also regulated, and here the central principle is "generally recognized as safe" (GRAS). The QPS and GRAS are, however, not synonymous, and the two systems are in fact quite different (Herody et al., 2010). Many countries in the world lean toward either the European or the American system for regulation of microbial food cultures. Irrespective of the framework for the safety assessment, a "history of safe use" must be established. Microbial taxonomy becomes central in safety evaluations as the microorganisms involved need to be defined to document their history of use. Current taxonomy is based on molecular methods such as the nucleotide sequences of ribosomal RNAs and, if needed, supplemented by phenotypic analysis. Taxonomy is usually not enough for the safety assessment as there is only very few microbial species not containing problematic strains. Depending on the type of microorganism and the type of food matrix it might be necessary to establish the absence of virulence genes, absence of antibiotic resistance, the inability to produce toxins, or the absence of other undesirable traits.

Safety of fermented foods should, however, not only be viewed as the potential risk associated with the microorganism as an important aspect of fermented foods is to improve safety and shelf life of the perishable raw materials.

7.2.1.4 Suppliers of Starter Cultures Global suppliers of microbial food cultures are listed in Table 7.2, together with some significant regional starter culture companies. The Danish company Chr. Hansen, which supplies a broad range of microbial food cultures, is the market leader within starter cultures. The established starter culture industry has been focused on products for dairy, meat, and probiotics, whereas the areas of starters for bread, cereals, and vegetables are sparsely covered, and only a few of the starter culture companies have products for these applications. It is to be expected that those areas will be developed significantly during the coming years. Starter cultures for wine are well-established but are covered by only a few of the starter culture companies. The Canadian company Lallemand is the leading company for wine cultures.

Table 7.2 Suppliers of Microbial Food Cultures

COMPANY NAME	COUNTRY OF ORIGIN	PRODUCT FOCUS
Alce	Italy	Dairy cultures
Blessing Biotech	Germany	Meat cultures
Cargill	USA	Wide range of ingredients
Clerici-Sacco	Italy	Dairy cultures
CSL	Italy	Dairy cultures
Chr Hansen	Denmark	All types of cultures for food and feed
CSK Food Enrichment	The Netherlands	Dairy cultures
DSM	The Netherlands	Wide range of ingredients
DuPont	USA	Wide range of ingredients
Kerry	Ireland	Ingredients and flavors
Lallemand	Canada	Wine cultures, baker's yeast, food cultures, probiotics
Naruse	Japan	Starters for natto
Miura	Japan	Starters for natto
Takahashi	Japan	Starters for natto
Valio	Finland	Dairy, probiotics

Baker's yeast is probably the food microorganism produced in the largest volume, however, without much differentiation among the products. Table 7.2 is not complete with respect to yeast-producing companies.

Inocula for the production of alcohol, beer, and vinegar are maintained and propagated by the manufacturers, and such products are therefore insignificant as commercial starter cultures.

7.2.2 Starters for Alkaline-Fermented Foods

The production of various AFFs is common in Asia and Africa. Soybeans have a widespread use, and other seeds rich in protein such as castor and locust beans and seeds from baobab, cotton, melon, and sesame are also used for AFFs. The alkaline-fermented soy product natto is popular in Japan and produced on an industrial scale. Similar products are common in China, Korea, Thailand, and India. The inoculation practice for the traditional indigenous natto product was based on spontaneous fermentation, whereas the industrial production of natto is based on inoculation with carefully selected strains of *Bacillus subtilis*. *Bacillus* spp. are commonly identified as major components in the microbiota of AFFs. In indigenous products, the *Bacillus* spp. are usually accompanied by a multitude of bacteria, yeasts, and molds (Parkouda et al., 2009).

7.2.2.1 Potential Improvements of Inoculation Practice An increase in yield and quality of AFFs is likely to be achieved through the selection of microbial strains with optimal properties for each product. Natto is the only AFF for which optimized starters have been selected and introduced into the production process. In the traditional production process, natto was produced using rice straw to inoculate steamed soybeans. Rice straw is naturally rich in *Bacillus* spores and could therefore serve as an inoculum. However, only a fraction of the *Bacillus* spores naturally present on rice straws are suitable cultures for natto production, and there is a large geographical variation in the natural occurrence of suitable strains (Kubo et al., 2011). Natto is now produced with pure cultures of *B. subtilis* var. *natto* (Nagai, 2012). The *B. subtilis* (*natto*) strains have also been shown to possess probiotic properties in addition to the properties required to produce natto (Cutting, 2011).

Natto could serve as a model for developing a traditional AFF into a safe high-quality product. In the search for a suitable starter culture, it might even be worthwhile initially to test if natto cultures could be used for the product in question. The optimal culture will, however, most likely have to be isolated and composed for each product due to the difference in raw materials and consumer preference. A culture with just one microbial strain, as for natto, might not be sufficient to give the desired product. Thus, it might be necessary to develop a culture consisting of a microbial consortium of bacteria, yeasts, and molds, or it might even be necessary to develop cultures for a multi-step fermentation process where a primary fermentation is followed by a secondary ripening process.

7.2.2.2 Development and Production of Starter Cultures for Alkaline-Fermented Foods As commercial starter cultures are unavailable for AFFs (except for natto), it is necessary to establish a source of inoculum if it is desired to secure a consistent production of the fermented food product. Initially, an improvement of the back-slopping process can be attempted by preserving samples from the fermentations and trying to propagate the consortia preserved from products that subsequently are found to meet expectations. It is possible to develop such inocula into complex starter cultures by applying standardization and quality control.

The development of starter cultures mixed from defined single strains requires a thorough investigation of the microorganisms participating in the fermentation process. If such investigations have already been conducted and published, isolates of the microorganisms might be obtainable from the researcher or from a culture collection. Otherwise, it will be necessary to examine the complexity of the microbiota and isolate the major constituents. Test fermentations with isolated microorganisms are needed to identify suitable strains or combination of strains.

When a satisfactory inoculum has been identified, either complex or defined, the regular production must be organized. It might be desirable to have one of the starter culture companies to take the responsibility for this task. They might, however, due to the relatively small volume of such a specialized product, be reluctant to add extra item numbers to their product range. To a large company, regulatory compliance and approvals may be relatively costly. A small-scale local operation might thus have to be established to produce the starter cultures. Such a unit can be handled from within a microbiology laboratory at a local university or institute. Most starter manufacturers have started from such a modest scale, and it might be inspirational to study the development and recent history of companies such as Wisby, Rudolph Müller, and Gewürtzmüller in Germany; Lactolabo in France; Alce and Centro Sperimentale del Latte (CSL) in Italy; and Australian Starter Culture Research Centre (ASCRC) in Australia. Obviously, a small operation will need to put the same emphasis on safety as a global company and might find it a heavy responsibility to introduce a new microbial food culture into the food chain. A local supplier will, however, be able to take advantage of policies and recommendations from the United States and Europe like GRAS and QPS but need regulatory approval only from the local governments where the products will be marketed.

7.2.3 Conclusion

In the area of AFFs, development of starter cultures is largely a virgin field. The current chapter focuses on advising on what will be the most important issues as this field hopefully will go through a rapid growth during the next few years. The main emphasis is on plant-based products rather than alkaline foods of animal origin such as thousand-year-eggs and fermented fish. The general principles will, however, also apply to those products, even the ones in which the alkaline conditions are not created by the fermentation but through the addition of caustic materials as ash or soda.

The AFF represents a very interesting area likely to become a major contributor to the innovation on an the food and starter culture industries. Mastering AFFs in industrial scale will allow such processes to become applied to a much larger extent in food and feed fermentations.

7.3 Hygienic Design Aspects of Alkaline Fermentation Processes

Huub L. M. Lelieveld

As with all food preparation processes, there is always a risk of harmful contamination of the product as a result of the way the process is executed and the materials that are used in contact with the food. For that reason, food should be handled hygienically in all stages of the manufacturing process, i.e., from the ingredients to the end product, intended for consumption. If the ingredients, intermediate products, and final product are not consistently handled hygienically, the product may become unsafe and cause illness or even death. The hazards can be divided into three main categories, namely, first, contamination with chemicals; second, with "foreign bodies" (physical materials that can be anything from insects and parts of animals that should not be there, to parts of equipment and utensils, such as bolts and nuts). The third category is microbes. In contrast to the chemical and physical contaminations, microbes are likely to multiply in the product and therefore may be present in harmless numbers at the start of the manufacturing process. If conditions allow the multiplication of microbes, which is often the case, their concentrations in the final product may exceed the safety level, either at the moment of production or at the time of consumption. The risk is particularly large if the contaminating microbes grow faster than the ones added as inoculum for the fermentation process.

7.3.1 Chemical Contamination

Chemical contamination can be due to contamination of the ingredients used, including any water that is used. Hence, it is in principle important to know the composition of all ingredients, to be able to decide whether they may present a risk for the consumer. If the supplier cannot provide the required data, chemical analysis is important. One of the largest risks is contamination with mycotoxins, developed in the substrate for fermentations that have been contaminated by molds. Once present, mycotoxins are extremely difficult to remove and are likely to persist until the end of the process, causing severe illness and often painful death after consumption. If the concentration is high, the effect will be immediate. If the concentration is low, the effect may become apparent after months or years. Hence, to have the capacity to detect the concentration of mycotoxins is an essential requirement, unless the supplier can be fully trusted. Another source of chemical contamination can be the presence of residues of the chemicals used for cleaning of the equipment and utensils and for their disinfection. That is why after cleaning and disinfection of a process line with chemical solutions, the equipment must be flushed with clean water to remove residues. This, however, is possible only if the equipment has no areas that cannot be cleaned and cannot be drained from the liquids used. Contamination with cleaning and disinfection agents can also result from the use of containers, such as buckets, for both the chemical solutions and the product. This should never happen and such containers should be dedicated. It is advisable to use different colors, depending on the intended use. It is also essential to label every container very clearly to avoid confusion. Moreover, chemicals should be stored separately from ingredients and during production there should not be any harmful chemicals around. Finally, chemical reaction with the product contact surfaces may cause contamination, in particular with metals of which too high concentrations can be harmful, mainly in the long term. The choice of materials should be such that leaching is avoided or at least limited to the extent that the concentrations in the final product are below the acceptable, safe concentration.

7.3.2 Physical Contamination

There are many potential sources of physical contamination. They may be of biological origin, i.e., it may be stalks, leafs, or seeds from plants, insects and mites or parts thereof, or parts of other animals. It can, however, also be of human origin, such as hairs, nails, or jewels, matches, buttons, and pens. Therefore, the wearing of jewelry and other ornaments is not allowed, unless they are irremovable, and buttons must be avoided or on the inside of the protective garment used by the personnel. In addition, where the product is exposed, hairnets must be worn. The contamination can also be nuts, bolts, and pieces of rubber, wood, or plastic, originating from the utensils or equipment. When equipment has to be taken apart, for whatever reason, care must be taken that no components are left behind when the work is completed. Small components should also never be put into containers (packaging material) intended for the food products, also not if the intention is to take these containers away when the work is finished because they may easily be overlooked.

7.3.3 Microbial Contamination

Microbes are everywhere and some of them can be harmful, for both the quality of the products and for the consumer. Some microbes can cause severe and sometimes deadly diseases. For that reason, the ingredients must have been decontaminated before use, i.e., harmful microbes should be inactivated, usually by heat treatment, but sometimes by other means, such as irradiation. For some ingredients, they may be removed by filtration (particle-free liquid ingredients, water, air, or other gases), or in particular cases, by centrifugation, if the microbes of concern have a higher density than the ingredient. As mentioned before, the additional problem with microbes is that, given suitable conditions, they will multiply, and this multiplication can be rapid. Many of the harmful microbes can double in number in just 10 to 15 min under production conditions. In the case of fermentation, the conditions used often are optimal for the desired microbes, conditions that unfortunately are also likely to be quite optimal for other microbes. If the harmful (pathogenic or toxinogenic) microbes have a shorter doubling time, they may even outnumber the ones that had been used as inoculum (starter culture). This plays a very important role if inoculation is with a small amount of starter culture, such as 0.01% only. When the amount of starter culture is high, e.g., 10%, and the starter culture itself is free from contamination, the undesired microbes may not present a big hazard, but this depends also on the type of microbes. Some can be very harmful even in very low concentrations. If equipment is not cleaned properly, soil residues will protect microbes present in these residues against inactivation. This happens whether the inactivation treatment is by heat or by chemical solutions. The surviving microbes will start growing in the product and may reach harmful concentrations.

7.3.4 Managing Food Safety

To ensure that the final product is safe for consumption, the entire process, from raw materials to final packed product must be managed by staff who understand what may compromise safety of the products produced. This requires sufficient insight in the causes as discussed above, i.e., enough knowledge of chemistry, physics, and microbiology to recognize potential dangers. Training of food safety managers is essential and, in many countries such as the EU, compulsory. Managing food safety, however, is impossible if the tools are inadequate or if the instruments to produce the food cannot be cleaned and/or decontaminated. One of the essential requirements is that the entire process line is of a hygienic design or, if that is impossible under the prevailing circumstances, measures have been devised that compensate the flaws in the design. This usually means that the equipment and utensils have to be taken apart for cleaning, then be decontaminated and reassembled. After reassembling, the inside of the equipment must be decontaminated again. For many food production processes a limited recontamination with microbes is acceptable when the total process time is fairly short (not longer than a few hours) and the product is subjected to an inactivation treatment after packaging. In the case of fermentation, in particular, alkaline fermentation, the process time can be many hours or even days or weeks, giving contaminating microbes the time to multiply to dangerous concentrations, which may be infectious or produce toxin. Therefore, it is also essential that recontamination of the product in the equipment is prevented. This means that it must be impossible that any contaminated liquid or solid can penetrate from the outside to the inside of the equipment.

Managing food safety means taking care of cleanliness of the workspace (floors, walls, and so on), of the equipment and utensils, of packing materials, and of protective garments for personnel. There must also be facilities for cleaning of hands and shoes. It is recommended that shoes that are used outside the food factory are not used inside the factory as well. This can be achieved with a step-over barrier where the outside shoes are left on the "dirty" side and food area shoes are put on the production side of the barrier. The same holds for the garments. Further, management is required to make certain that all staff in the factory have been properly trained for their tasks, which include that they must be taught about the risk of contamination and what their duties are to prevent such contamination.

7.3.5 Hygienic Processing and Plant Design

The design of all parts of the process line must be such that it is possible to, first, remove all contamination from the equipment and, second, to inactivate any microbes that may affect the fermentation process or the safety and quality of the fermented product. For some products, only vegetative microbes need to be absent, for other products microbial spores need to be destroyed as well. Thus, the process line and all its components that will be in contact with ingredients, intermediate of final product must be cleanable, suitable for methods to inactivate undesired microbes and be drainable to allow the complete removal of any residues of chemicals used for cleaning or disinfection. Further, the materials that will be in contact with the product must be such that they will not unacceptably contaminate the product during processing due to, e.g., leaching or wear. The basic requirements, therefore, are the following.

7.3.5.1 Construction Materials Materials of construction must withstand contact with product, individual ingredients, cleaning and disinfection solutions in use, at the temperature and during the duration of contact. It is important to realize that some of the ingredients used may be more corrosive than the final product, such as solutions of acid, caustics, salt, and other chemicals. Depending on the process, the equipment may have to be freed from vegetative microorganisms, implying that the equipment, including any sealing materials used, must withstand a temperature of at least 90°C. When spores are relevant – most likely the case with alkaline fermentations – the equipment must withstand temperatures of 120°C or higher. If chemical disinfectants are used, the corrosivity at the temperatures required must be known and must meet the requirements. In the food industry, generally, stainless steel AISI 304 and 316 are used for, e.g., pipelines, tanks, valves, and pumps. For gaskets (static seals) used between metal surfaces the material of choice often is ethylene propylene diene monomer (EPDM), suitable for temperatures up to 120°C. For higher temperatures, up to 180°C, silicon rubber or fluoroelastomers may be used. In all cases, the supplier should be asked to provide a certificate of approval for food contact applications. Nonresilient materials such as polytetrafluoroethylene (PTFE: Teflon) should not be used because they will not seal permanently. Although they may seal upon installation, after heating and cooling down again, the dimensions will have been changed as the result of the difference in expansion coefficient between PTFE and steel, and the seal will no longer be tight (Lelieveld et al., 2014b).

7.3.5.2 Surface Roughness Too rough surfaces may retain soil residues and microbes, and therefore, product contact surfaces must be smooth enough for adequate and efficient cleaning. Generally, a surface finish of 0.8 μm R_a is recommended. For definition of R_a, see International Standards Organization (1982). Rougher surfaces may be acceptable but will increase the time needed for cleaning.

7.3.5.3 Drainability All equipment and pipelines must be self-drainable. This obviously means that outlets of vessels should be at the lowest point in any vessel or pump and that pipeline should slope slightly to the outlet. Often, however, standard vessels have outlets higher than the lowest point, making complete draining impossible. Also, pumps too are often designed without draining in mind. Centrifugal pumps have the outlet usually pointing upward and the inlet in the center, so that when emptied, the pump housing is still filled with liquid to about its center. Gearwheel and lobe pumps have traditionally both their inlet and outlet in the middle and therefore cannot be drained. If pipelines are not well supported, they will bend, with the lowest point in the middle between two supports and hence the pipes cannot be fully drained either. When in one installation, pipes with different diameters are connected, this usually is done concentrically, again hampering draining.

7.3.5.4 Dead Spaces Areas where product may reside longer than intended are usually designated as "dead spaces." Such spaces provide microbes with a longer residence time during processing, causing higher numbers of microbes in the equipment. They also make cleaning and disinfection, by either hot water, steam, or chemical solutions, ineffective. For effective cleaning, turbulence is required. While turbulence can be fairly easily controlled in the more open space, in dead spaces turbulence will be very low, the result being that the cleaning time used will not clean the dead space. When hot water is circulated through pipelines or being agitated in a tank, the water in a dead space will be at a lower temperature because it will lose heat to the environment but the energy is not replenished by the main bulk of hot water. When using steam, water will condensate in dead spaces and again, the surfaces to be treated will be at too low a temperature, resulting in microbial spores surviving the treatment (Lelieveld et al.,

2014b). Therefore, dead spaces must be avoided, and when they cannot be avoided, measures must be devised to take care of cleaning and disinfection. Often, dead spaces are the result of the installation of sensors to monitor process conditions, such as temperature, pressure, and pH. They often are installed on tees, where the sensor is mounted on one leg. Designs are available that provide the same functionality without creating dead spaces. Another source of large dead areas is the application of safety circuits around pumps that can deliver pressures higher than the equipment can withstand. Such a circuit consists of a loop between the outlet of the pump and its inlet, with a pressure relieve valve in between. The pipes to the inlet and outlet of that valve are then full of product that stands still all the time, allowing microbes to multiply to high concentrations. Only when something goes wrong and the pressure increases to above the maximum value will the valve open and the product – potentially loaded with microbes – circulate through the circuit, preventing mechanical damage to the pump and process line but not the safety of the product. The same can be achieved using valves that have been designed to eliminate dead spaces (Lelieveld et al., 2014b).

7.3.5.5 Crevices and Corners Crevices are narrow dead spaces that will be filled with product that will never be removed again because there will be no turbulence whatsoever in crevices. Microbes in the soil (product residue) in a crevice have a very good chance to survive any decontamination treatment, to start multiplying again after such treatment, infecting the product passing the crevice. Similarly, sharp internal corners have the same effect: there will be very little turbulence in such corners. Crevices often are the result of connecting various pieces of equipment together in the wrong way or using the wrong components, such as pipe couplings with built-in crevices. Many couplings still in use, such as the "dairy coupling," had been designed almost a century ago for easy dismantling, but not for easy cleaning. In such designs, the seal ring is at a significant distance from the main product flow, creating deep crevices. Modern couplings for the food industry are crevice-free (Lelieveld et al., 2014b). Sharp corners are design mistakes. Similarly, lids are often mounted in such a way that a large and deep crevice is created between the lid and the rim of the vessel.

Crevices and sharp corners must be absent from food processing equipment; an internal corner should have a radius of at least 6 mm.

7.3.5.6 Pipe Connections When pipes need to be connected permanently, the best way is to weld them because then problems that are associated with the use of pipe couplings are avoided. The welds must be of high quality, which requires the use of skilled welders and preferably the use of orbital welding equipment that, when properly adjusted, delivers a high-quality crevice-free and corrosion-free weld in a short time.

7.3.5.7 Preventing Ingress of Contamination Before adding starter culture to the material to be fermented, that material needs to be freed from relevant microbes, which for alkaline fermentations will be all microbes, including their spores. Hence, the material and thus also the inside of the equipment will have to be sterilized. Similar to processing of food products that must have a long shelf life at ambient temperatures, reinfection of the product after decontamination must be prevented and therefore the fermenter and all auxiliary must be impervious to microbes from outside. This requires tight seals (see also Section 7.3.5.1). For dynamic seals in valves and pumps, where ingress of microbes must be prevented, double seals are needed, where the space between the seals is continuously flushed with a disinfecting fluid, which can be steam, hot water, or a (safe) antimicrobial liquid. For valves, the seals may be made from elastomers; for pumps, the use of double mechanical seals is recommended because they are more reliable and last longer. Another option is the use of membrane valves and equally membrane pumps, avoiding mechanical seals. Shafts of stirrers in fermenters may also be sealed by double mechanical seals. Another option is the use of hygienically designed magnetic drives, depending on the amount of energy to be transferred. That way seals, and hence the risk of ingress of microbes is completely avoided. For large fermenters or for fermenters and pumps for viscous products, the magnetic momentum (torque) required may be too large. Currently, magnetic drives are available for torques up to 200 Nm.

7.3.5.8 Inoculation and Sampling Usually, preparing the starter culture takes several steps, starting with a pure culture in a culture tube or a conical vessel at laboratory scale. Of course, care must be taken so that during the subsequent transfers of the culture no contamination takes place. The last step, namely the transfer to the fermenter, may be the most critical one. In a clean environment, aseptic transfer to the fermenter may often be successful. Nevertheless, it is recommended to use an aseptic transfer system to minimize the risk of contamination and consequently the loss of product. This involves the use of valves that have a connection to a steam supply. Both the starter culture vessel and the fermenter must be equipped with such a valve. After making the connection between them, the space between the starter culture vessel and the fermenter is flushed with steam. Then, after allowing the system to cool down again, the valves between the vessel and the fermenter can be opened to aseptically transfer the starter culture.

7.3.5.9 Continuous Fermentation Batch-wise fermentation has a number of drawbacks, such as preparation and transfer of starter culture, cleaning and sterilization of the equipment, and size of the equipment in relation to production capacity. Fermentation can also be executed as a continuous process, running for a prolonged time. Production of starter culture and cleaning and sterilization of the fermenter will be required only once during that prolonged period. Moreover, as the microbes in the fermenter can be kept under optimal condition, maintaining a high (often maximum) growth rate makes it possible to produce the same amount with much smaller equipment than needed for the traditional batch process. Such a process is possible using aseptic equipment as discussed above, together with accurate control of the fermentation conditions. The throughput can be controlled by measuring one of the parameters that change as a result of fermentation, such as the pH. Investments and operational costs will be lower that with the batch process. Such processes are applied industrially, e.g., the fermentation of milk (Lelieveld, 1984).

7.3.5.10 Conveyor Belts for Solid-State Fermentations With respect to chemical and physical contamination, solid-state fermentations (SSFs) have similar requirements, although the construction materials can, to a large extent, be different. The process lines may vary considerably depending on the fermentation applied, and therefore, with the principles of hygienic design in mind, every process line will have to be tailor-made. When SSFs are carried out in a continuous way, conveyor belts made of polymer materials, reinforced with fibers (glass or polyamide) may be used. Care must be taken so that the belt material is not damaged at the product contact side because such damages will allow ingress of moisture into the interior of the belt and then microbes will multiply and cannot be inactivated (cf. earlier section about crevices). Product transported on the belt will consequently be contaminated with these microbes, which may include pathogenic ones. Only recently, this type of material in use for the in-line transport of salmon was the cause of a severe *Salmonella* outbreak in the Netherlands, in 2012 (Foppen, 2012; Friesema et al., 2012).

7.3.5.11 Process Control Self-evidently, control of all process conditions is essential to ensure that the required quality of the product is obtained. There may, however, be pitfalls that can easily be avoided. One of them is the response time of control circuits. There is always some lapse between the occurrence of an event between the reactive response of the controller and subsequently of the activator. For instance, if a liquid medium is sterilized by a continuous system, such as a plate or tubular heat-exchanger, temperature and flow control are essential to ensure that all products receive the same treatment to inactivate all relevant microbes. If the temperature sensor detects that the temperature has dropped below the critical value, the flow of product to the fermenter must be diverted back to the balance tank at the beginning of the process line. Switching the diversion valve, however, takes time and therefore it must be ensured that during that period no insufficiently treated product contaminates the product in the fermenter. This means that care must be taken that the distance between the temperature sensor and the diversion valve must be long enough to prevent the insufficiently treated product to enter the pipeline after the diversion valve, leading to the fermenter. A similar principle applies to other controls, such as the pH.

7.3.5.12 Equipment Installation The risk of contamination is always proportional to the concentration of the contaminant in the surroundings of the process equipment. It is therefore important to keep the environment clean. As discussed before, the problem with microbes is that they increase in number with time. This also applies to insects (and other animals). These pests should not be given a chance to do so and therefore the surroundings must be well cleanable and accessible for inspection of cleanliness. This means that there must be no narrow spaces, e.g., between the underside of the equipment and the floor, which should be at least 300 mm, or between control boxes and the wall, which should be at least 30 mm for cleaning and inspection. If this is for other reasons undesirable, the space should be minimized and then sealed to the wall or floor.

7.3.6 Further Reading

In the context of this book, it is not possible to go into details of hygienic design of the fermentation plant. Readers who will be involved in the design and operation of fermentation plants are recommended to consult the following books: *The Handbook of Hygiene Control in the Food Industry* by Lelieveld et al. (2004), *Hygienic Design of Food Factories* by Holah and Lelieveld (2011), *Hygiene in Food Processing – Principles and Practice* by Lelieveld et al. (2014a), and *Food Safety Management – A Practical Guide for the Food Industry* by Motarjemi and Lelieveld (2014), dedicated to hygienic design and operation of food factories.

7.4 Food Packaging

Jenneke K. Heising and Matthijs Dekker

Traditionally, AFFs are often produced on a small scale and the foods are not packed at all or basic packages like plant leaves are used. Modernization of the package will lead to new opportunities, but it also creates technological challenges. Many highly technical developments are going on in the area of food packaging, like the inclusion of sensors in packages for monitoring food quality (Heising et al., 2014), the use of nanomaterials to influence the quality of the packed food actively, and the production of biodegradable packages (Jiménez et al., 2012). However, in this chapter, the focus lies on simple basic and low-cost packages for foods, adapted to the package requirements of manufacturers that produce AFFs on a small scale.

Perhaps, the most important function of a package is to protect a food. A good package can lead to improved food safety and security. Besides, a good package can help to slow down the quality degradation (e.g., appearance, flavor, texture, and also the nutritional value) of foods and prolong their shelf life. This leads to opportunities to increase the distribution possibilities of the foods to other areas and scaling up their production. Packaging can reduce food waste by extending the shelf-life of the foods and prevent damage, leakage, or deterioration that can occur in the supply chain during transportation and distribution. Besides, producers can secure the quality of their products to the consumer by a "use by" or "best before" date and build up a reliable brand. A package will add some extra costs in the production process of the food product, and this should be taken into account in the design of a food packaging. However, all these advantages could lead to increased sales and income for the producers.

This chapter describes the basic technological principles for applying a (low-cost) packaging for AFFs. The general guidelines are illustrated by some specifications and examples for (alkaline-fermented) food products.

7.4.1 Challenges

The implementation of proper packaging can sometimes be difficult due to local constraints and may differ by country. The choice of a suitable package for fermented foods for small-scale producers in developing countries is often restricted by the availability of local packaging materials and suitable packaging equipment. Besides, the equipment and packaging materials can be expensive and can contribute to a large part of the production costs (Rolle and Satin, 2002). Therefore, suboptimal packaging materials are used for fermented products, such as leaves, vegetable fibers, earthen pots, and newspapers, which limit the shelf life and can raise safety issues (Rolle and Satin, 2002).

In more industrialized areas, more modern package materials are available. The food and the package interact with each other, and selection of a good package can have a strong influence on the quality of the packed food. The basic aspects of food packaging, also applicable for AFFs, will be discussed in the next sections.

7.4.2 Basic Functions of Food Packaging

A food package basically has four functions: containment, protection, convenience, and communication (Robertson, 1993). In the sections below, these functions are described in some more details, and examples are given for the application to different foods.

7.4.2.1 Containment When foods are not large discrete items, the foods need to have a primary package that contains the food product to facilitate its distribution. The containment functions by preventing a large amount of product loss, e.g., when powders or liquids are transported. Methods for filling and sealing food containers in small-scale processing are described on the website of practical actions (http://practicalaction.org/media/preview/10700).

The primary package should be strong enough to withstand the conditions in the food supply chain (like shocks and vibrations) and to have minimal interactions with the packed foods, e.g., liquid from a wet food (like soy sauce) should not migrate into the package, but also constituents from the package should not migrate to the food. A secondary package (e.g., a carton box) that contains a number of food package units in their primary package can be applied to improve efficient logistics and to prevent the primary package from shocks and damages. The traditional Nepalese food kinema or other local small-scale fermented products are sometimes packed in leaves (Figure 7.6), which can easily get damaged, leading to leakage of the food from the package.

Figure 7.6 Traditional kinema packages at the market. (Courtesy of Prabir K. Sarkar.)

7.4.2.2 Protection The package should protect the food against influences from outside. These environmental influences can be due to biological factors, e.g., the package has to protect the food from insects, rodents, and other animals, or due to physical influences like bruising (e.g., fruit), from dust and soil, and all other unwanted and possible hazardous constituents. When foods are cooked properly, the products become sterile or at least free of pathogenic bacteria. Without a package, microorganisms from the air or contact surfaces will recontaminate the food and cause rapid spoilage, especially when they are not properly refrigerated. Foods that are processed (heated) inside their hermetically sealed package, e.g., glass jar, can, or plastic pouch, are sterile after production, and safe and stable for a very long period until the package integrity is broken. Also, hygienic processing and packaging that prevent microorganisms from contaminating the foods can help to control food spoilage and food safety. Hygienic processing prevents growth of unknown and possibly hazardous microorganisms from outside, and only the bacteria already present in/on the food can grow.

Other environmental influences are moisture, water vapor, and gases like oxygen. Some of these influences will only affect the sensory attributes of the food, e.g., moisture causing loss of crispiness of cookies, or odor causing off-flavor in the food product (directly from aroma, or from reactions caused by entering gases, e.g., oxygen can cause rancidity in high-fat foods, causing off-flavor and decreasing the nutritional value). These environmental influences also affect the shelf life of foods, e.g., low-a_w products (like dried herbs or cocoa powder) will spoil fast when moisture enters the food and oxygen can increase microbial growth and induce enzymatic reactions (Cutter, 2002). The barrier or permeability properties of packaging materials are usually specified by packaging suppliers with the gas transmission rate (GTR) and the water vapor transmission rate (WVTR). The unit of GTR of films is generally ml/(day $*$ m^2 $*$ bar) and for WVTR g/(day $*$ m^2) and is given for a reference thickness, so it needs to be converted for the actual thickness of the packaging material (Dekker, 2011) (Equation 7.1):

$$\text{GTR}_a = \text{GTR}_{ref} * \frac{l_{ref}}{l_a} \tag{7.1}$$

where *l* is the thickness of the material and the subscript *a* refers to "actual" and *ref* to "reference," and the same equation can be used for calculating the actual WVTR value by replacing the GTR values by WVTR values.

When the GTR and WVTR values of the package material are known, the flux through the package can be calculated from the differences in partial pressure (p) of the gas or relative humidity (RH) inside and outside the package and the surface A through which the migration can occur (Equations 7.2 and 7.3):

$$\varphi_j = \mathrm{GTR}_j * A * (p_{j,1} - p_{j,2}) \quad (7.2)$$

$$\varphi_w = \mathrm{WVTR} * A * (RH_1 - RH_2) \quad (7.3)$$

where φ is the flux through the packaging material in mol/s, A is the surface area in m^2, and subscripts j and w refer to the specific gaseous compound and water, respectively.

To apply a correct film to the package, one has to establish the desired film properties by studying the food degradation and calculating the desired flux of moisture and gases (Siracusa, 2012). The selection of a suitable packaging material can strongly influence the food, e.g., choosing a plastic with good moisture-retaining properties, but enabling permeability for oxygen and carbon dioxide can extensively prolong the shelf life of packed respiring foods, like sliced vegetables.

7.4.2.3 Convenience Packaging can add convenience in distribution and use, but it also gives convenience by the desired portion package. Foods with short shelf life are preferably packed in small-sized portions that are consumed within one or few days after opening of the package. Once the package is opened, the protection of the package from environmental conditions is reduced and the spoilage rate increases and the remaining shelf life decreases if no preservatives are added or other precautionary measures are taken.

7.4.2.4 Communication Text and graphics can be printed on a package, or a label can be applied on it, on which product information is given. The label can provide basic information, like the ingredients used, nutritional information, and a "use by" or "best before" date, but it can also be used for additional information, like preparation instructions and storage information (e.g., recommended storage temperature). An advantage of information on the package is that the specific food item is always presented to the consumer together with its own specific product information. Besides, the package can be used for marketing purposes, e.g., by building up a reliable brand that is recognized by consumers.

7.4.3 Interactions between Food and Package: Flavor Scalping and Migration

One of the main characteristics of fermented foods is their strong and characteristic flavor. Interactions between the package and the food can alter the flavors of the food. The term "flavor scalping" is used in the food packaging industry to describe the loss of quality of a packaged food due to either its volatile flavors being absorbed by the package or from undesirable volatile compounds from the packaging material that are absorbed by the food (Sajilata et al., 2007). The sorption of food aromas by packaging materials can be perceived as a major factor contributing to the quality alteration of many foods during storage. This causes changes in the intensity of the flavors, but also leads to an unbalanced flavor profile of the food. Especially for foods that can be stored for longer periods, flavor scalping can have a high impact on the quality of the food and can decrease consumer acceptance. The sorption of flavors depends on the characteristics of the polymer and the type of volatile compounds, but also on other factors like the storage conditions (e.g., temperature) and on the composition of the packed food (Nielsen and Jägerstad, 1994; Sajilata et al., 2007).

The term "migration" is used to describe the diffusion of (potentially hazardous) compounds from the packaging material into the food. Packaging materials should be approved for food applications by migration testing with suitable food simulants as defined by national and international legislation (Arvanitoyannis and Kotsanopoulos, 2014). The U.S. Food and Drug Administration (FDA) regulates packaging materials under section 409 of the Food, Drug, and Cosmetic Act and all food contact substances (FCS) that may reasonably migrate to food under conditions of intended use are identified and regulated as food additives unless classified as generally recognized as safe (GRAS) substances (Marsh and Bugusu, 2007).

7.4.4 Interactions between Packaging and Environment

Interactions between the food and the package do not only include mass transport between the food and the packaging material (flavor scalping or migration) but also include interactions with the environment (Figure 7.7) (Nielsen and Jägerstad, 1994).

Permeation in food packaging is the penetration of a permeate through the package. The use of polymer materials in food packages is increasing, and a lot of different polymers are present on the market (Siracusa, 2012). These plastic packaging materials have variability in permeability to gases, water vapor, and other low-molecular-weight compounds, ranging from excellent to low-barrier values, and packaging engineers can influence the final properties of the packaging material. It is important for food packaging to prevent contamination from the external environment, but the shelf life of the food can also be controlled by alteration of the gaseous atmosphere of the food product inside the package with controlled atmosphere packaging (CAP), vacuum packaging, and modified atmosphere packaging (MAP).

In CAP, the gas composition of the package is monitored continuously and adapted to obtain a constant (desired) atmosphere, regardless of environmental fluctuations or influences from the food (e.g., respiration) (Cutter, 2002). Vacuum packaging is a method of packaging

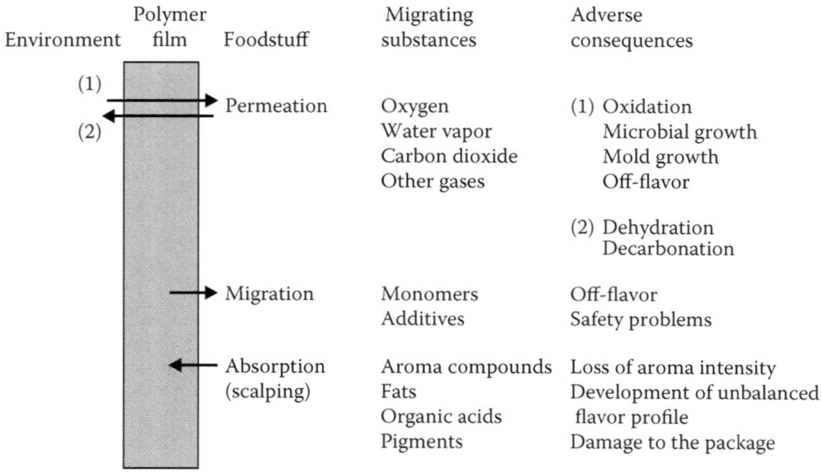

Figure 7.7 Possible interaction among foods, their package materials, and the environment, together with the adverse consequences from these interactions. (From Nielsen, T., and Jägerstad, M., *Trends in Food Science and Technology*, 5, 353–356, 1994.)

that removes air from the package before sealing, but the gaseous atmosphere can change during storage due to respiration of the food or microbial growth. CAP is mainly used for bulk storage of fruits and vegetables, while vacuum packaging is used for primary packages of, for example, meat and cheese. MAP is mainly used in primary packages to prolong the shelf life of fresh or minimally processed foods. The air in the package is replaced initially by a gas mixture, depending on the food type, the packaging material, and the storage conditions (e.g., temperature). Often, oxygen is (partly) removed to slow down oxidation reactions and growth of aerobic microorganisms, and the removed oxygen can be replaced by the often used inert gas nitrogen or other gases. Carbon dioxide has bacteriostatic effects (Cutter, 2002), but this water- and lipid-soluble gas decreases the pH of the food and can induce undesired textural and sensorial effects.

During storage, the gas composition will change due to food reactions and interactions with the environment, but the interaction among the packaging material, the gas atmosphere, and the food can be used in active packaging (e.g., using oxygen or carbon dioxide scavengers) to prolong the quality of foods during storage (Cutter, 2002). Another example of the use of an active packaging to improve the quality of foods or prolong the shelf life is the use of a package material that absorbs undesired volatiles produced by the food product during storage.

7.4.5 Packaging Materials

Traditional or locally available packaging materials such as plant leaves, vegetable fibers, earthen pots, and newsprints for fermented foods and beverages for small-scale producers (Marshall and Mejia, 2011) do not sufficiently protect the food resulting in a restricted shelf-life. Some materials can even induce safety risks. Besides, these materials do not have the desired options for communication of information, marketing, and legislation requirements and are therefore hardly acceptable for distribution into other markets.

The properties of the food and the package materials also need to be taken into account in the design of a package. Besides being an appropriate package at the moment of packaging, a package needs to be versatile to withstand the effects of the composition of the packed food and to withstand (conditions in) the handling process in the whole chain while maintaining physical and chemical integrity (Siracusa, 2012).

Paper has been used as a food packaging material since the 17th century, but it has poor barrier properties and is not heat-sealable; therefore, it often needs to be combined with other packaging materials to protect foods for long periods (Marsh and Bugusu, 2007). For a long time, metal and glass have been popular materials for food packaging. Glass bottles were popular for beverages and prevented a lot of wastage and leakage. However, glass packaging can be expensive, and proper cleaning and sterilization need to be applied before reusing the packages. Nowadays, the rigid metal and glass packages for foods and beverages are partly replaced by plastic packages because plastic packages have the following advantages over metal and glass packages: lower cost, lighter in weight, less susceptible to shatter, transparent, flexible, and convenient to the consumer. The disadvantages of plastic packages are interactions between the food and the package and limited reuse, although new products can be made after melting the plastics (Marsh and Bugusu, 2007).

Not all package materials interact with the packaged food substantially. The transfer of small molecules through package materials such as glass and metal can usually be assumed to be zero because of the extremely low diffusion rate in these materials, while polymer materials have a large variability in interactions with the food and the environment (Marsh and Bugusu, 2007). This can be a disadvantage due to undesired reactions, but in other cases, the interaction is desired (e.g., selecting a packaging material through which ethylene can permeate that is formed from packed fruits and vegetables during storage).

Flavor scalping is mainly a problem with polymeric packaging materials, and therefore, not all polymer films are good choices for AFFs. Among the plastic materials, polyethylenes (PE) and polypropylenes (PP) are often used for food packaging, also for fermented foods, e.g., pickled limes are packed in small sealed ethylene bags (Battcock and Azam-Ali, 1998). PE and PP are frequently used materials because of their good chemical resistance and inertness to most foods, good barrier properties to water, and thermosealability; however, their oxygen barrier is limited and nonpolar compounds (such as many aroma compounds) can be absorbed in these materials (Sajilata et al., 2007).

Marsh and Bugusu (2007) have given an overview of properties of many food packaging materials. In Table 7.3, some basic characteristics of commonly used packaging polymers are summarized. An option to improve the properties of a package material is to combine layers of materials on top of each other. These multilayered packages or laminates can benefit from the desired properties of the individual materials, e.g., when a package material with good moisture barrier properties is combined with a material with good gas barrier properties. To estimate the GTR or WVTR of the new package material, the values of the individual layers need to be taken into account (Equation 7.4):

$$\frac{1}{\text{GTR}_{laminate}} = \frac{1}{\text{GTR}_1} + \frac{1}{\text{GTR}_2} + \ldots + \frac{1}{\text{GTR}_n} = \sum_{i=1}^{n} \frac{1}{\text{GTR}_i} \quad (7.4)$$

where i refers to the number of the layers in the total of n layers.

Furthermore, environmental factors, like temperature and RH, can influence the chemical–physical behavior of the package materials (Siracusa, 2012). Therefore, the storage conditions need to be known.

Although food packaging prevents food wastage, the package materials and production of the packages do have an environmental impact. To minimize environmental impact, possibilities for reusing or recycling, the packaging needs to be studied or biodegradable

Table 7.3 Characteristics of Commonly Used Plastics

POLYMER	WATER BARRIER	OXYGEN BARRIER	APPEARANCE	APPLICATION EXAMPLE
Low-density polyethylene (LDPE)	High	Low	Soft, translucent, flexible	Bag, bottle, film
High-density polyethylene (HDPE)	High	Low/medium	White, flexible	Bag, bottle, jar, box
Polypropylene (PP)	High	Low	Clear, brittle	Bag, tub
Polystyrene (PS)	Medium	Medium	Transparent	Tray, cup
Polyvinyl chloride (PC)	High	High	Clear, glass like	Bottle
Polyethylene terephthalate (PET)	High	High	Transparent	Bottle, tray, film

polymers or edible films can be considered as possible packaging materials (Cutter, 2002; Marsh and Bugusu, 2007).

7.4.6 *Fermentation in Package*

Traditional fermentation processes are typically uncontrolled and depend on microorganisms from the environment or the fermentation substrate (Marshall and Mejia, 2011). A packaging material should protect the food from external contamination with undesired microorganisms because the growth of the starter culture only is desirable. An example of fermentation in the package is fermentation of wine in bottles, but it is also used in traditional methods in which ages ago foods were already fermented in pottery jars (Marshall and Mejia, 2011). Important for fermentation in a package is the strength of the packaging material, since gas produced during fermentation might lead to a high pressure inside the package. Nowadays, at-home production of beer is quite popular in Western countries. The final stage of fermentation is performed in sealed glass bottles to mimic the beer carbonation process performed at industrial plant. This practice led to the explosion of some bottles with evident potential risks.

Fermentation is used as a preservation technique, and failure of the fermentation process causes spoilage and/or a safety risk from the growth of pathogenic microorganisms (Holzapfel, 2002). Packaging has a large influence on the direct environmental conditions of the food. For example, if the fermentation needs anaerobic conditions (Marshall and Mejia, 2011), a porous package is not suitable and is a potential hazard. Or if the products needs to be protected from sunlight during or after fermentation (e.g., banana beer; Marshall and Mejia, 2011), a nontransparent package can be chosen to maintain a high-quality product. Packaging can be part of the HACCP and GMP principles to secure a high-quality safe product.

7.4.7 *Case: Application of a Package on the Product Kinema*

7.4.7.1 Product Description Kinema is a low-cost, non-salted, flavorful AFF with a sticky texture, similar to the Japanese fermented product natto (Sarkar and Tamang, 1995). Kinema is produced from whole soybeans fermented predominantly by *Bacillus subtilis* and is mainly consumed in a dish with boiled rice in Nepal, India, and Bhutan (Shrestha et al., 2010). Although fermentation is a method to preserve foods, kinema is still highly prone to spoilage due to its high moisture content (~62%) and low salt content and high pH of ~7.9 (Sarkar et al., 1994). Kinema is highly nutritious (Shrestha et al., 2010) and rich in proteins (48% dry matter basis [DM]); it contains 17% (DM) fat and 29% (DM) carbohydrates (Sarkar et al., 1994), and its physical texture can be described as highly viscous and sticky.

7.4.7.2 Traditional Package Kinema is still produced in a traditional way and sold at local markets. It is packed in plant leaves and then tied loosely with straw (Figure 7.6). Such a package does not protect the food product from drying out and against gases, and the interaction between the food and the package could lead to safety hazards due to pathogenic bacteria from the plant-based package material.

7.4.7.3 Package Requirements Since the package and the food are one system influencing and interacting with each other, the first step in the design of a package is to determine the most important food quality determining reactions of a food product. Some important features of kinema are that it is a high-moisture low-cost food product, prone to spoilage. To prevent moisture migration to the package and prevent drying out of the product, a package with good moisture barrier properties is necessary. So, the WVTR should be relatively low and the package properties should not be highly influenced by moisture. For example, some polymer materials are prone to swelling from water from the food, and this will change the structure of the materials, leading to an increase in the permeability for gases or vapor (Siracusa, 2012). A plastic package, tin, or glass would be a good option. The disadvantage of glass or metal are its heavy weight and high cost. A plastic foil can be applied as a low-cost packaging and can easily be adapted to a desired portion-size for one person. The unique flavor of kinema should be retained in the product, so the polymer material should not be highly prone to flavor scalping. However, since the shelf life of kinema when stored at chilled conditions is still limited, flavor scalping will be limited due to the time and spoilage. It is probably the main issue to tackle by the appropriate package.

The introduction of MAP could help to prolong the shelf life of the fresh kinema, since unpacked kinema has a shelf life of only 2–3 days during summer (Sarkar et al., 1994). The gas composition inside the package will influence the microbial growth, e.g., the addition of carbon dioxide can reduce or inhibit the rate of microbial growth in foods (Sivertsvik et al., 2002). However, this technique requires appropriate processing and packaging control with necessary cost and investment.

7.4.8 Conclusion

The quality of foods is influenced by the package and the implementation of a package on a food brings technological challenges, but it also offers opportunities. A package, the packed food, and the environment interact with each other, which can cause (undesired) changes in the food and in the properties of the packaging materials. The selection of the right packaging can help to maintain the quality and safety of the packed food, can help to reduce food losses and waste, and can improve distribution of the food, offering opportunities for increasing sales and building up a reliable brand. When selecting an appropriate material, not only the properties of the food and packaging material need to be studied, but also the regulations need to be taken into account. The moisture and gas barrier properties of packaging materials are important characteristics in the selection of the optimal packaging material for a food; for wet-fermented foods, a low-cost polymer with low WVTR might be an appropriate packaging material.

References

Arvanitoyannis, I.S., and Kotsanopoulos, K.V. 2014. Migration phenomenon in food packaging. Food-package interactions, mechanisms, types of migrants, testing and relative legislation—A review. *Food and Bioprocess Technology* 7: 21–36.

Bamforth, C.W. 2008. *Food Fermentation and Micro-organisms*. New York: John Wiley & Sons.

Battcock, M., and Azam-Ali, S. 1998. Fermented fruits and vegetables: A global perspective. Agricultural Services Bulletin No. 134, http://www.fao.org/docrep/x0560e/x0560e00.HTM. Rome: FAO.

Benner, M., Linnemann, A.R., Jongen, W.M.F., and Folstar, P. 2003. Quality Function Deployment (QFD)—Can it be used to develop food products? *Food Quality and Preference* 14: 327–339.

Bourdichon, F., Casaregola, S., Farrokh, C., Frisvad, J.C., Gerds, M.L., Hammes, W.P., Harnett, J., Huys, G., Laulund, S., Ouwehand, A., Powell, I.B., Prajapati, J.B., Seto, Y., Ter Schure, E., van Boven, A., Vankerckhoven, V., Zgoda, A., Tuijtelaars, S., and Hansen, E.B. 2012. Food fermentations: Microorganisms with technological beneficial use. *International Journal of Food Microbiology* 154: 87–97.

Chadare, F.J., Gayet, D.P., Azokpota, P., Nout, M.J.R., Linnemann, A.R., Hounhouigan, J.D., and van Boekel, M.A.J.S. 2010. Three traditional fermented baobab foods from Benin, *Mutchayan*, *Dikouanyouri* and *Tayohounta*: Preparation, properties and consumption. *Ecology of Food and Nutrition* 49: 279–297.

Cutter, C.N. 2002. Microbial control by packaging: A review. *Critical Reviews in Food Science and Nutrition* 42: 151–161.

Cutting, S.M. 2011. *Bacillus* probiotics. *Food Microbiology* 28: 214–220.

Dekker, M. 2011. Food packaging design. In *Food Product Design*, 2nd ed., eds. M.A.J.S. van Boekel, A.R. Linnemann, and C.P.G.H. Schroën, pp. 197–206. Wageningen, The Netherlands: Wageningen Academic Publishers.

Foppen, J. 2012. Het kan één mug zijn geweest. In *NRC, 2012-11-17*, ed. R. Postma, pp. 12–13.

Friesema, I.H., de Jong, A.E., Fitz James, I.A., Heck, M.E., van den Kerkhof, J.H., Notermans, D.W., van Pelt, W., and Hofhuis, A. 2012. Outbreak of *Salmonella* Thompson in The Netherlands since July 2012. *Eurosurveillance* 17: 2–5. Available online: http://www.eurosurveillance.org/ViewArticle.aspx?ArticleId=20303.

Hansen, E.B. 2004. Microorganisms. In *Handbook of Food and Beverage Fermentation Technology*, eds. Y.H. Hui, L. Meunier-Goddik, A.S. Hansen, J. Josephsen, W.-K. Nip, P.S. Stanfield, and F. Toldrá, pp. 9–21. New York: Marcel Dekker.

Hauser, J.R., and Clausing, D. 1988. The House of Quality. *The Harvard Business Review* 66: 63–73.

Heising, J.K., Dekker, M., Bartels, P.V., and van Boekel, M.A.J.S. 2014. Monitoring the quality of perishable foods: Opportunities for intelligent packaging. *Critical Reviews in Food Science and Nutrition* 54: 645–654.

Herody, C., Soyeux, Y., Hansen, E.B., and Gillies, K. 2010. The legal status of microbial food cultures in the European Union: An overview. *European Food and Feed Law Review* 5: 258–269.

Holah, J., and Lelieveld, H.L.M. eds. 2011. *Hygienic Design of Food Factories*. Cambridge: Woodhead Publishing.

Holzapfel, W.H. 2002. Appropriate starter culture technologies for small-scale fermentation in developing countries. *International Journal of Food Microbiology* 75: 197–212.

Hui, Y.H., Meunier-Goddik, L., Hansen, A.S., Josephsen, J., Nip, W.-K., Stanfield, P.S., and Toldrá, F. eds. 2004. *Handbook of Food and Beverage Fermentation Technology*. New York: Marcel Dekker.

ISO 1982. *Surface Roughness—Parameters, their Values and General Rules for Specifying Requirements. ISO standard no. 468*. Geneva, Switzerland: International Organization for Standardization.

Jiménez, A., Fabra, M.J., Talens, P., and Chiralt, A. 2012. Edible and biodegradable starch films: A review. *Food and Bioprocess Technology* 5: 2058–2076.

Kubo, Y., Rooney, A.P., Tsukakoshi, Y., Nakagawa, R., Hasegawa, H., and Kimura, K. 2011. Phylogenetic analysis of *Bacillus subtilis* strains applicable to natto (fermented soybean) production. *Applied and Environmental Microbiology* 77: 6463–6469.

Lelieveld, H.L.M. 1984. Mixed-strain continuous milk fermentation. *Process Biochemistry* 19: 112–113.

Lelieveld, H.L.M., Holah, J., and Napper, D. eds. 2014a. *Hygiene in Food Processing—Principles and Practice*, 2nd ed. Cambridge: Woodhead Publishing.

Lelieveld, H.L.M., Mostert, M.A., and Curiel, G.J. 2014b. Hygienic equipment design for food processing. In *Hygiene in Food Processing—Principles and Practice*, 2nd ed., eds. H.L.M. Lelieveld, J. Holah, and D. Napper. Cambridge: Woodhead Publishing.

Lelieveld, H.L.M., Mostert, M.A., and Holah, J. eds. 2004. *The Handbook of Hygiene Control in the Food Industry*. Cambridge: Woodhead Publishing.

Linnemann, A.R., Benner, M., Verkerk, R., and van Boekel, M.A.J.S. 2006. Consumer-driven food product development. *Trends in Food Science and Technology* 17: 184–190.

Marsh, K., and Bugusu, B. 2007. Food packaging—Roles, materials, and environmental issues. *Journal of Food Science and Nutrition* 72: 39–55.

Marshall, E., and Mejia, D. 2011. Traditional fermented food and beverages for improved livelihoods. Diversification booklet number 21. http://www.fao.org/docrep/015/i2477e/i2477e00.pdf. Rome: FAO.

Motarjemi, Y., and Lelieveld, H.L.M., eds. 2014. *Food Safety Management—A Practical Guide for the Food Industry*. Amsterdam, The Netherlands: Elsevier Science.

Nagai, T. 2012. Bacteriophages of *Bacillus subtilis (natto)* and their contamination in natto factories. In *Bacteriophages*, ed. I. Kurtböke, pp. 95–110. Rijeka, Croatia: InTech.

Nielsen, T., and Jägerstad, M. 1994. Flavour scalping by food packaging. *Trends in Food Science and Technology* 5: 353–356.

Parkouda, C., Nielsen, D.S., Azokpota, P., Ouoba, L.I.I., Amoa-Awua, W.K., Thorsen, L., Hounhouigan, J.D., Jensen, J.S., Tano-Debrah, K., Diawara, B., and Jakobsen, M. 2009. The microbiology of alkaline-fermentation of indigenous seeds used as food condiments in Africa and Asia. *Critical Reviews in Microbiology* 35: 139–156.

Robertson, G.L. 1993. *Food Packaging: Principles and Practice*. New York: Marcel Dekker.

Rolle, R., and Satin, M. 2002. Basic requirements for the transfer of fermentation technologies to developing countries. *International Journal of Food Microbiology* 75: 181–187.

Sajilata, M.G., Savitha, K., Singhal, R.S., and Kanetkar, V.R. 2007. Scalping of flavors in packaged foods. *Comprehensive Reviews in Food Science and Food Safety* 6: 17–35.

Sarkar, P.K., and Tamang, J.P. 1995. Changes in the microbial profile and proximate composition during natural and controlled fermentations of soybeans to produce kinema. *Food Microbiology* 12: 317–325.

Sarkar, P.K., Tamang, J.P., Cook, P.E., and Owens, J.D. 1994. Kinema—A traditional soybean fermented food: Proximate composition and microflora. *Food Microbiology* 11: 47–55.

Shrestha, A.K., Dahal, N.R., and Ndungutse, V. 2010. *Bacillus* fermentation of soybean: A review. *Journal of Food Science and Technology Nepal* 6: 1–9.

Siracusa, V. 2012. Food packaging permeability behaviour: A report. *International Journal of Polymer Science* 2012: 302029.

Sivertsvik, M., Jeksrud, W.K., and Rosnes, J.T. 2002. A review of modified atmosphere packaging of fish and fishery products—Significance of microbial growth, activities and safety. *International Journal of Food Science and Technology* 37: 107–127.

Steinkraus, K.H. ed. 2004. *Industrialization of Indigenous Fermented Foods*, 2nd ed. New York: Marcel Dekker.

van Boekel, M.A.J.S., and Linnemann, A.R. 2011. The need for food product design. In *Food Product Design—An Integrated Approach*, eds. A.R. Linnemann, K. Schroën, and M.A.J.S. van Boekel, pp. 15–25. Wageningen, The Netherlands: Wageningen Academic Publishers.

van Kleef, E., van Trijp, H.C.M., and Luning, P. 2005. Consumer research in the early stages of new product development: A critical review of methods and techniques. *Food Quality and Preference* 16: 181–201.

Verkerk, R., Linnemann, A.R., and van Boekel, M.A.J.S. 2007. Quality analysis critical control points in consumer-oriented agro-food chains. In *Tropical Food Chains—Governance Regimes for Quality Management*, eds. R. Ruben, M.A.J.S. van Boekel, A. van Tilburg, and J. Trienekens, pp. 241–255. Wageningen, The Netherlands: Wageningen Academic Publishers.

8

Value-Added Products from Alkaline-Fermented Foods or from Microorganisms Predominating Therein

The alkaline-fermented foods and the microorganisms predominating therein are an excellent source of value-added items, thereby promoting the principle of sustainable development. These have immense industrial potential.

8.1 Alkaline Proteases

C. Ganesh Kumar

Enzymes have been used indirectly since antiquity by man. The beer and wine production involving the fermentation of sugar to ethanol, vinegar production achieved by the oxidation of ethanol, and curdling of milk by lactose fermentation are thousands of years old traditional processes where catalytic activities of enzymes are responsible for chemical transformations. The protease rennin derived from calf stomach was the first used cell-free enzyme for the coagulation of milk protein in the cheese-making process. Currently, there are more than 3000 different enzymes identified from a wide variety of organisms finding a wide range of commercial application in detergents, food as well as biotechnological, chemical, and pharmaceutical industries. The use and function of the majority of these enzymes have been limited by several factors such as enzyme instability at high temperatures, extreme pH, ionic strength, presence of organic solvents, cofactors requirement, and narrow range of substrate specificity. To overcome these limitations, it is necessary to search for new microbial sources by tapping the microbial diversity. Extremozymes from extremophiles have been isolated from a diverse array of exotic and harsh environments that are inhospitable to support life and their specific properties can be explored for the development of novel process applications (Schiraldi and De Rosa, 2002).

Proteases are ubiquitous in nature and important from both physiological and commercial perspective and are present in diverse sources including plants, animals, and microorganisms. The International Union of Biochemistry and Molecular Biology has broadly classified proteases into seven distinct classes: serine, cysteine, aspartate, glutamate, threonine, metalloproteases, and the newly identified asparagine peptide lyases (Rawlings et al., 2011), based on the critical amino acid or ion that catalyzes the peptide bond cleavage and the mechanistic aspects of substrate cleavage (Deu et al., 2012). Although proteases carry out a wide array of functions, the basic theme of peptide bond cleavage exhibited by all proteases is identical, which involves the polarization of the scissile amide C=O bond and the activation of a nucleophilic group to attack the carbonyl carbon resulting in the hydrolysis. The estimated value of the world enzyme market in 2005 was approximately US$2.0 billion, and it is expected to reach US$3.0 billion by the year 2015 (Reddy et al., 2008). The group of alkaline proteases (or subtilisins, E.C. 3.4.21.14) occupies 30% of the enzyme market and caters for their major use as laundry detergent additives.

8.1.1 Alkalophilic Microorganisms and Their Habitats

Alkalophiles exhibit growth at pH above 10, with an optimum at around pH 9, and cannot grow at pH 7 or lower. The extreme alkalophiles are further subdivided as facultative and obligate alkalophiles. Facultative alkalophiles exhibit optimal growth at pH 10 or above as well as at neutrality, while obligate alkalophiles do not grow at neutral pH values. The basis for adaptation for the survival and multiplication of alkalophiles under such adverse conditions is the alteration of the bioenergetic membrane properties and transport mechanisms (Krulwich et al., 1990).

Alkalophilic microorganisms are ubiquitous with wide distribution in both natural and anthropogenic (man-made) alkaline environments. Soda soils and lakes are the very well studied natural habitats from both ecological and chemical perspective (Grant et al., 1990). Other environments are forest soils, dilute alkaline springs, and desert soils (Horikoshi and Akiba, 1982). Anthropogenic environments such as effluents generated from tanneries, detergents, potato processing units, paper industries, $CaCO_3$ kilns, textiles and food industries have pH values around and above 10 that also harbor alkalophiles (Kumar and Takagi, 1999). Haloalkalophiles, which have gained renewed interest due to their tolerance to two extreme environmental conditions, salinity and high pH, were isolated from soda lakes like Wadi Natrum Lake in Egypt, Magadi Lake in Kenya (Grant et al., 1990), and Lonar and Sambhar Lakes in India (Karan et al., 2011). Alkalophilic psychrotolerant bacteria have also been isolated from the alpine permafrost environment, Qinghai-Tibet Plateau with an average altitude of 4500 m (Zhang et al., 2007) and the Gangotri glacier in Western Himalaya, India (Baghel et al., 2005). Table 8.1 lists different alkaline protease-producing microorganisms isolated from different habitats.

Table 8.1 Some Examples of Alkaline Protease Producers Isolated from Different Unique Habitats

MICROBIAL STRAIN	ISOLATION SOURCE	REFERENCE
ALKALINE AND ARID SOIL		
Bacillus lentus MK5-6	Soil, Madhuban, Karnal, India	Kumar et al. (1999b)
Streptomyces malaysiensis AMT-3	Brazilian cerrado soil	Nascimento et al. (2005)
Aspergillus clavatus CCT2759	Atlantic forest soil, Brazil	Tremacoldi and Carmona (2005)
Bacillus patagoniensis PAT 05^T	Rhizosphere soil of *Atriplex lampa* – a perennial shrub, Patagonian Monte, Argentina	Olivera et al. (2006)
Streptomyces clavuligerus Mit-1	Saline soil, Mithapur, Gujarat, India	Thumar and Singh (2007)
Chryseobacterium taeanense TKU001	Taiwan soil	Wang et al. (2008)
Halobacillus karajensis MA-2	Saline soil (5% w/v total salt), Iran	Karbalaei-Heidari et al. (2009)
Bacillus pseudofirmus Mn6	Olive field soil sample, El-Menia governorate, Egypt	Abdel-Fattah et al. (2009)
Salimicrobium halophilum LY20	Saline soil, Yuncheng, China	Li and Yu (2012)

(*continued*)

Table 8.1 (Continued) Some Examples of Alkaline Protease Producers Isolated from Different Unique Habitats

MICROBIAL STRAIN	ISOLATION SOURCE	REFERENCE
HOT SPRING		
Alkalophilic *Bacillus subtilis* CHZ1	Zimbabwean hot spring	Zvidzai and Zvauya (2001)
Thermoplasma volcanicum GSS1	Solfataras at Vulcano Island, Italy	Kocabiyik and Özdemir (2006)
Geobacillus sp. YMTC 1049	Hotspring in Rehai, Tengchong, Yunnan Province, China	Zhu et al. (2007)
Saccharomonospora viridis SJ-21	Hot water spring	Jani et al. (2012)
MARINE ENVIRONMENT		
Teredinobacter turnirae	Marine shipworm, *Psiloteredo healdi* (symbiotic relationship)	Elibol and Moreira (2005)
Bacillus horikoshii and *Bacillus* sp. I-103	Body fluid of Korean polychaete, *Periserrula leucophryna*, Tidal mud flats of Yellow sea, Kwangwha island, Inchon, South Korea	Joo et al. (2002, 2004)
Bacillus clausii strains I-52 and I-58	Tidal mud flat, Yellow sea, Inchon, South Korea	Joo et al. (2003); Kumar et al. (2004)
γ-proteobacterium DGII	Marine sediment, Lothian Island, The Sundarbans, India	Sana et al. (2006)
Engyodontium album BTMFS10	Marine sediment, Cochin, India	Chellappan et al. (2006)
Vibrio fluvialis VM10 and *Bacillus circulans* BM15	Mangrove sediment, Cochin estuary, India	Venugopal and Saramma (2006, 2007)
Roseobacter sp. MMD040	Marine sponge *Fasciospongia cavernosa*, Vizhinjam, Indian peninsular coast	Shanmughapriya et al. (2008)
Streptomyces fungicidicus MML1614	Sea water sample, Bay of Bengal, India	Ramesh et al. (2009)
Bacillus mojavensis A21	Water sample, Sfax city, Tunisia	Haddar et al. (2010)
SODA LAKE		
Nesterenkonia sp. AL-20	Feather sample, shore of an alkaline soda lake, Rift Valley of Ethiopia	Bakhtiar et al. (2002)

(*continued*)

Table 8.1 (Continued) Some Examples of Alkaline Protease Producers Isolated from Different Unique Habitats

MICROBIAL STRAIN	ISOLATION SOURCE	REFERENCE
Arthrobacter ramosus MCM B-351	Alkaline soda lake sediment, Lonar, India	Nilegaonkar et al. (2002)
Vibrio metschnikovii DL 33	Water sample of Dali lake (soda lake, pH 9.6, alkalinity of 42.7–47.2 mmol/l), Neimenggu Province of North China	Mei and Jiang (2005)
Halogeometricum borinquense strain TSS-101	Solar salterns of Tuticorin, Tamilnadu, India	Vidyasagar et al. (2006)
Salinivibrio proteolyticus strain AF-2004	Bakhtegan lake (hypersaline lake, south of Iran)	Amoozegar et al. (2008)
Bacillus licheniformis KBDL-4	Water sample, Lonar (alkaline soda lake), India	Pathak and Deshmukh (2012)
FERMENTED FOOD		
Bacillus sp. CK11-4	Chungkook-jang, a traditional Korean fermented soybean sauce	Kim et al. (1996)
Bacillus amyloliquefaciens	Soybean fermentation starter	Cho et al. (2003)
Bacillus subtilis JM-3	Anchovy sauce	Kim and Kim (2005)
Amylomyces rouxii Calmette	Starter of chiu-niang, a traditional Chinese fermented rice product	Yu and Chou (2005)
Filobacillus sp. RF2-5	Nam-pla, Thai fish sauce	Hiraga et al. (2005)
Halobacillus sp. SR5-3	Fermented fish sauce	Namwong et al. (2006)
Virgibacillus sp. SK37	Fermented fish sauce	Sinsuwan et al. (2007)
Halobacterium sp. SP1(1)	Fermented fish sauce	Akolkar et al. (2009)
Tetragenococcus halophilus	Fermented fish sauce	Udomsil et al. (2010)
ANTHROPOGENIC (MAN-MADE) ENVIRONMENT		
Bacillus pumilus JB05	Cement industry effluents	Johnvesly and Naik (2001a)
Geobacillus caldoproteolyticus	Sewage sludge	
Bacillus sp. L21	Tannery waste	Tari et al. (2006)
Aspergillus clavatus ES1	Wastewater	Hajji et al. (2008)
Cellulosimicrobium cellulans	Fermentation industrial waste	Luciana and Sato (2009)
Pseudomonas thermaerum GW1	Poultry waste site, Ghaziabad, Delhi, India	Gaur et al. (2010)
Bacillus licheniformis RP1	Polluted water, Sfax, Tunisia	Haddar et al. (2011)
Streptomyces ambofaciens DP2	Milk processing plant	Bajaj and Sharma (2011)

8.1.2 Isolation and Screening of Alkalophiles

Alkalophilic microorganisms have gained significant attention from a bioindustrial perspective since they produce various alkaline enzymes like proteases, amylases, lipases, cellulases, xylanases, pectinases, and mannanases, which find wide application as laundry and automatic dishwashing detergent additives, biopolishing agents, textile desizing, enzymatic bleaching of soft pulps in paper and pulp industries, starch saccharification process, and food, pharmaceutical, and cosmetic industries (Kumar and Takagi, 1999; Gupta et al., 2002). From an application viewpoint, the isolation and screening of alkalophiles from various natural and anthropogenic environments have gained importance. Alkalophiles are fairly widespread in both neutral and acidic soils. Sodium carbonate contributes to alkalinity in natural environments and its supplementation in the isolation medium favors the growth of alkalophiles (Grant et al., 1990). However, the isolation of thermophilic alkalophiles is difficult, since the sodium carbonate present in the isolation medium causes browning and cracking of the medium, and the agar in the medium usually loses its gel strength and exhibits water of syneresis at temperatures of >70°C (Kitada et al., 1987). Gelrite™ (Gellan gum), due to its good thermal stability, has been identified as suitable agar substitute for the isolation of alkalophilic thermophiles (Lin and Casida, 1984). Horikoshi and Akiba (1982) first demonstrated the isolation of alkalophiles on an alkaline medium. Later several alkaline media used for isolation of alkalophiles including alkaline casein agar, wheat meal agar, glucose-yeast extract-asparagine agar, glucose-yeast extract-casein agar, sucrose-peptone agar, casein-Norberg agar, and casein-Tindal's agar were reported.

Different microorganisms have been reported to produce alkaline protease, and among them the members of the genera *Bacillus* and *Aspergillus* have been identified as the major industrial workhorses. Various *Bacillus* species, including *B. subtilis*, *B. licheniformis*, *B. amyloliquefaciens*, *B. alcalophilus*, *B. circulans*, *B. coagulans*, and *B. mojavensis* isolated from different exotic environments, have been explored for alkaline protease production (Kumar and Takagi, 1999; Gupta et al., 2002). Different subtilisin variants produced by *Bacillus* spp. are subtilisin Carlsberg from *B. licheniformis*, subtilisin amylosacchariticus from *B. subtilis* var. *amylosacchariticus*, subtilisin BPN' from

B. amyloliquefaciens, subtilisin E from *B. subtilis*, nattokinase from *B. subtilis* (*natto*), and alzwiprase from *B. subtilis* DM-04. Gram-negative bacterial strains reported to produce alkaline proteases included different strains of *Pseudomonas aeruginosa*, *Flavobacterium arborescens*, *Vibrio metschnikovii*, *Vibrio fluvialis*, *Stenotrophomonas maltophilia*, *Brevundimonas diminuta*, and *Chryseobacterium taeanense*. Among the fungi, *Aspergillus*, *Penicillium*, *Tritirachium*, and *Rhizopus* were widely explored for alkaline protease production (Rao et al., 1998; Kumar and Takagi, 1999). Literature on alkalophilic actinomycetes is very scanty and alkaline protease-producing alkalophilic actinomycetes identified were different strains of *Streptomyces* sp., *Oerskovia xanthineolytica*, *Nocardiopsis dassonvillei*, *Nesterenkonia* sp., and *Cellulosimicrobium cellulans*. Some of the yeasts reported to produce alkaline proteases included *Candida lipolytica*, *Saccharomyces carlsbergensis*, *Saccharomyces lipolytica* CX161-IB, *Yarrowia lipolytica*, and *Aureobasidium pullulans*. In addition, some microbes surviving in some extreme environments, like high salinity (halophiles), low temperatures (psychrophiles), or high temperatures (thermophiles and hyperthermophiles) were also known to produce alkaline proteases (Kumar and Takagi, 1999).

8.1.3 Production of Alkaline Proteases

Alkalophilic microorganisms are an important bioresource for alkaline protease production in view of their short generation time, and easiness in bulk handling, genetic, and environmental manipulations. Many microbial strains produce alkaline proteases; however, those producing sufficient quantities of the enzyme are interesting. In general, the protease production is either constitutive or inducible in nature depending on the culture conditions. It is, therefore, necessary that optimal growth conditions are provided to the microbial strains to increase the enzyme production (Moon and Parulekar, 1991). Numerous recent efforts in this direction made to develop a cost-effective feasible technology were based on two approaches: (i) improvement in alkaline protease yields and (ii) optimization of fermentation conditions.

8.1.3.1 Improvement in Alkaline Protease Yields In general, the wild strains produce low enzyme yields that find commercial application. Therefore, microbial strain improvement is important from a commercial perspective to enhance the protease-producing ability of the strain. Mutants can be generated by either conventional mutagenesis (exposure to ultraviolet (UV) or chemical mutagen) or recombinant DNA (rDNA) technology. In conventional practice, simple selection methods are used like spread plating the culture on specific medium supplemented with skim milk and then picking up the colonies that produce substantial yields. A cysteine auxotroph of *B. licheniformis* was developed by UV treatment for improving the protease production (Shah et al., 1986). A mutant *V. metschnikovii* DL 33-51, developed by a combination of UV and diethyl sulfate treatment, exhibited 4.74-fold increase in alkaline protease yield (Mei and Jiang, 2005). Research efforts were also made to develop asporogenous mutant strains of *Bacillus* spp. for commercial use by enzyme industries, since they lack the ability to sporulate and produce extracellular alkaline proteases for a longer duration and in increased quantities (Zamost et al., 1990). The splicing by overlap extension PCR method (Gene SOEing) was used to create a specific *spoIIAC* gene deletion to produce an asporogenic strain of *B. licheniformis* expressing increased levels of keratinase (Wang et al., 2005).

rDNA technology has opened new avenues in the recent past for construction of recombinant strains to improve enzyme yields. The isolation and cloning of the alkaline protease-encoding gene is a primary and essential step for the genetic engineering of efficient alkalophiles. *Escherichia coli* and *B. subtilis* are ideal hosts for cloning and overexpression of different alkaline protease-encoding genes resulting in a number of high-yielding variants. Further, the protein engineering technique has been widely explored for the improvement of subtilisins and/or alkaline proteases beyond its current limitations (Kumar and Takagi, 1999). Protein engineering is an art of introducing deliberate changes in the primary structure of the protein to alter its properties and function that are required for a specific application. This approach has been successfully employed for generating unique variants that have enhanced stability toward harsh environments like chemical oxidants, surfactants, etc. and/or substrate specificity at high pH and temperatures required in detergents, tanning, and chemical industries. From a basic research perspective, it has also been applied in studies to establish the structure–function relationship of alkaline proteases.

Currently, two strategies commonly employed for producing protein engineered variants are random and site-directed mutagenesis. Both of these strategies have been used optimally in combination to generate the desired mutants (Bryan, 2000). Many alkaline proteases have been protein engineered to improve their robustness and versatility with respect to their utilization in detergents and other industrial applications, and many studies have been reported in literature. Random mutagenesis was adopted to improve the alkaline and thermal stabilities of subtilisins (Cunningham and Wells, 1987) and different site-directed mutagenesis strategies have been used, like change in the electrostatic potential (Mitchinson and Wells, 1989), binding affinity improvement, changes in substrate specificity, and productivity (Takagi et al., 1997), protein stability by stabilizing loops (Kumar et al., 2000), introduction of disulfide bonds to improve thermal stability, and substitution of oxidation labile residues to improve the resistance to chemical oxidants, a special requirement in oxygen bleach detergents (Yang et al., 2000). Cold-active alkaline proteases from psychrophiles exhibit high catalytic efficiency at low temperatures and high thermal sensitivity as compared with their mesophilic counterparts (Cavicchioli et al., 2002). Rational mutagenesis strategy was adopted to improve the thermostability of cold active-alkaline protease from psychrotrophic *Shewanella* sp. Ac10 (Kulakova et al., 2003).

In recent years, directed evolution (DE) has been identified as a powerful protein engineering tool for expanding the number of existing enzymes as well as for tailoring them for novel and improved properties as well as to elucidate the protein's structure–function relationship (Arnold, 1996). DE employs the basic principle of selective Darwinian pressure to generate enzyme variants of increased fitness with respect to some desirable properties, such as activity, stability, substrate specificity, or stereoselectivity. DE follows three approaches, namely DNA shuffling, random recombination, and the staggered extension process (StEP). The DE approach begins with the selection of the target enzyme having unsatisfactory reactivity, specificity, or selectivity. The gene of this target enzyme is subjected to in vitro random mutagenesis to generate a library of mutants. This DNA library is then cloned and expressed in a bacterial host, thereby creating a library of enzymes followed by high-throughput screening and identification of enzyme variants with desired properties. The promising mutants are subjected to a further round of mutagenesis, which can generate enzymes of increasing quality. The process can be further iterated until a suitable catalyst of desired property is evolved (Petrounia and Arnold, 2000).

In the case of screening, the protease variants generated through DE are subjected to the microplate-based detection systems with screening capacities for more complex cases. This system is associated with low mutational loads (changes observed are 1–3 amino acids per mutated gene) so that a medium throughput is acceptable to find improved variants (Wong et al., 2006). Very recently, a high-throughput screening approach of protease variants was developed involving the DE of the surface membrane protease OmpT to alter its substrate specificities by employing an *E. coli* OmpT-deficient strain (UT5600). Multiple substrates, each containing a different cleavage sequence and a different colored fluorophore, can be screened using a multicolor fluorescence-activated cell sorting (FACS) system for highly specific and improved variants that cleave only the desired substrate sequences. The fluorescence intensity is proportional to the catalytic turnover, which is measured by FACS, a means to quantify and pick up the efficient OmpT protease variants that specifically prefer substrates with novel P_1 and/or P_1' residues in a high throughput mode from the diverse mutant libraries (Varadarajan et al., 2008b). Further, different OmpT variants were engineered for substrate specificity against post-translationally modified tyrosine residues at P_1, namely phosphotyrosine (Knight et al., 2007), sulfotyrosine (Varadarajan et al., 2008a), and nitrotyrosine (Varadarajan et al., 2009). In either of the cases, the engineered protease variants preferred the modified substrates over unmodified tyrosine. Some examples on the application of directed evolution approach on alkaline proteases are shown in Table 8.2.

8.1.3.2 Regulation and Medium Optimization for Protease Biosynthesis Different strains of *Bacillus* and *Aspergillus* identified as industrial workhorses produce extracellular alkaline proteases during the late-logarithmic or stationary phases of microbial growth. The protease synthesis is either constitutive or partially inducible, which is controlled by numerous complex mechanisms operational during the

Table 8.2 Some Examples on the Application of Directed Evolution Approach on Alkaline Proteases

PROTEASE TYPE (MICROBIAL SOURCE)	STRATEGIES USED	CHARACTERISTICS OF ENZYME VARIANT	REFERENCE
Subtilisin E (*Bacillus subtilis*)	Error-prone PCR, screening	100-fold increase of activity in dimethylformamide	Chen and Arnold (1993)
Subtilisin E (*B. subtilis*)	Error-prone PCR, screening	500-fold increase of activity in dimethylformamide	You and Arnold (1996)
Subtilisin E (*B. subtilis*)	DNA shuffling, screening	50-fold increase in $t_{1/2}$ at 60°C	Zhao et al. (1998)
Subtilisin E (*B. subtilis*)	Random priming, screening	8-fold increase in $t_{1/2}$ at 65°C	Shao et al. (1998)
Cold-adapted protease (*B. subtilis*)	Sequential random mutagenesis, screening	Decrease in K_m value rather than k_{cat} value	Taguchi et al. (1998)
Subtilisin E (*B. subtilis*)	Error-prone PCR, DNA shuffling, screening	Increased activity at different temperatures and 17°C increase in T_m	Zhao and Arnold (1999)
Subtilisin S41 (*B. subtilis*)	Error-prone PCR, cassette mutagenesis, screening	Improved thermostability up to 100-fold	Miyazaki and Arnold (1999)
Subtilisins (*B. subtilis*)	DNA family shuffling, screening	Evolution of multiple properties of enzyme	Ness et al. (1999)
Subtilisin SSII (*Bacillus sphaericus*)	Random mutagenesis, screening	Increased catalytic activity (9.6 times) and increased turnover number (k_{cat} at 10°C increased 6.6-fold)	Wintrode et al. (2000)
Subtilisin BPN' (*Bacillus amyloliquefaciens*)	Random mutagenesis, mutagenesis, screening	15,000-fold increase in $t_{1/2}$ at 75°C of calcium-free subtilisin	Strausberg et al. (2005)
WF146 protease (*Bacillus* sp. WF146)	Error-prone PCR, DNA shuffling, screening	Increased activity up to 6-fold; decrease in optimum temperature to 15°C and thermostability	Zhong et al. (2009)
Nattokinase (*Bacillus subtilis* (*natto*))	DNA family shuffling, screening	Increased catalytic efficiency by 2.3 times	Cai et al. (2011)
Alkaline protease (*Bacillus gibsonii*)	SeSaM random mutagenesis, screening	100-fold increase in $t_{1/2}$ at 60°C	Martinez et al. (2013)

transition state between exponential and the stationary growth phases (Strauch and Hoch, 1993). Carbon and nitrogen stress also induce protease synthesis during the stationary phase. In spore-formers, the onset of stationary phase is manifested by the changes from vegetative to sporulation stages. While evaluating various isolated or mutant strains for their production ability, the fermentation medium is an important and critical factor that influences the strain performance. The strain screening process for medium design is an iterative process that involves the testing of different isolates in different media to identify the suitable medium. However, there is no best defined medium that is deemed fit for alkaline protease synthesis by all the strains. Each microbial strain has its own idiosyncratic physiochemical and nutritional requirement for optimal growth and enzyme production. Nearly 30%–40% of enzyme production cost is accounted by the growth medium composition, and thus its optimization is vital (Joo et al., 2003; Kumar et al., 2004). Under commercial practice, submerged fermentation is preferred over solid-state fermentation for alkaline protease production. Medium optimization in different microorganisms is often dictated by various environmental (pH, temperature, aeration, agitation, dissolved oxygen, inoculum density, and incubation time), and nutritional factors (carbon and nitrogen sources, and trace elements) that effect alkaline protease secretion.

8.1.3.2.1 Nitrogen Source Organic N-sources have proved favorable for alkaline protease production as compared with inorganic N-sources. Peptones of plant and animal origin are water soluble and non-heat-coagulable protein hydrolysates comprising free amino acids and peptides. They are marketed in different forms, such as bactopeptone, biopeptone, neopeptone, protease peptone, tryptone, soy peptone, meat peptone, and fish peptone. However, peptones and yeast extract were considered as expensive medium components for large-scale industrial fermentations, and hence identifying cheaper substitutes will address the economic issues (Reddy et al., 2008). Soybean meal, a by-product derived after oil extraction, comprising approximately 40% protein and rich in other organic and inorganic components, is suggested as a cost-effective N-source for alkaline protease production (Kumar et al., 2004; Joo and Chang, 2005). Soybean meal in combination with bonito extract also improved alkaline protease yields. Corn steep liquor (CSL) contains several micronutrients, vitamins, and growth-promoting factors, and is another low-cost N-source (Kumar, 2002; Reddy et al., 2008). However, the major limitation with CSL is its seasonal and interbatch variability, which results in variation in protease productivity. Other promising low-cost N-sources identified for increasing the alkaline protease productivity were nug meal (Gessesse, 1997), amaranth seed meal (Pastor et al., 2001), pigeon pea waste (Johnvesly et al., 2002), green gram husk (Prakasham et al., 2006), defatted cotton seed meal (Jaswal and Kocher, 2007), sardinella peptone (Haddar et al., 2010), and mustard cake (Bajaj and Sharma, 2011). Addition of inorganic N-sources, such as ammonium sulfate or sodium nitrate, also favored high levels of protease production (Johnvesly and Naik, 2001b).

8.1.3.2.2 Carbon Source Simple sugars such as glucose, lactose, maltose, sucrose, and fructose favored increased alkaline protease yields. However, addition of these sugars at high concentrations causes a decrease in pH values due to the rapid growth rates of microorganisms resulting in catabolite repression in enzyme synthesis (Zamost et al., 1990). Complex carbohydrates, such as tapioca (*Manihot esculenta* Crantz) starch (Chandrasekaran and Dhar, 1983), arrowroot (*Maranta* arundinacea L.) starch (Kumar and Parrack, 2003b), corn starch (Kumar et al., 2004), *Mirabilis jalapa* L. tuber powder (Hajji et al., 2008), and hulled wheat grains (Haddar et al., 2010) were reported as principal C-sources for increased alkaline protease production. Different organic acids (acetic acid, methyl acetate, and citric acid or sodium citrate) also favored increased levels of alkaline proteases (Kumar and Takagi, 1999). These organic acids were not only economical, but also controlled the pH variations during the fermentation process. Different industrial effluents or wastes explored for alkaline protease production included cheese whey (Kumar et al., 1999a), wheat bran (Chellappan et al., 2006), cogon grass (*Imperata cylindrica* L.) Raeusch., potato peel (Mukherjee et al., 2008), shrimp wastes (Haddar et al., 2011), olive oil cake (Moftah et al., 2012), and cow dung (Vijayaraghavan and Vincent, 2012).

8.1.3.2.3 Trace Salt Requirement Supplementation of different trace metal salts of calcium, magnesium, manganese, iron, cobalt, copper, and molybdenum in the fermentation medium favors optimal production of alkaline protease. However, different microorganisms have a requirement for specific metal ions. Potassium phosphate acted as a buffering agent, and its supplementation was optimal for protease production. However, its addition in higher amounts caused precipitation of the medium after autoclaving, which inhibited cell growth and lowered protease production (Moon and Parulekar, 1993).

8.1.3.2.4 pH and Temperature Alkalophiles have a specific requirement of alkaline pH of the medium for increased alkaline protease productivity, which could be achieved with the addition of sodium carbonate (Horikoshi and Akiba, 1982). The culture pH mainly influences different enzymatic processes and transport mechanisms across the cell membrane. Further, a drop in pH during the fermentation process may be the result of formation of acidic products (Moon and Parulekar, 1991). The medium turns acidic when ammonium ions were used, while utilization of organic N-sources, like amino acids or peptides turned the medium alkaline. A close correlation between the protease synthesis, the nitrogenous nutrient utilization, and the pH variations during fermentation presents a kinetic profile on the start and end of the protease production (Kumar et al., 1999a). Different studies have shown that the temperature varied between 30°C and 40°C for alkaline protease production (Kumar and Takagi, 1999).

8.1.3.2.5 Statistical Design-Based Optimization Process optimization is a central theme in many industrial production processes, wherein the fermentation medium design is critical since the medium composition significantly affects product concentration, yield, and volume productivity. In addition, it indirectly affects the cost and ease of downstream processing for product separation. The classical optimization approach is simple, easy, and most commonly practiced which involves the alteration of "one variable at a time," while keeping all others at a predetermined level. However, in this approach a series of experiments needs to be carried out when a large number of variables needs to be tested to determine the optimum level, which is time-consuming, labor-intensive, open-ended, expensive, and the interaction between the variables is ignored (Kennedy and Krouse, 1999). On the other hand, the production medium improvement using an experimental design approach requires both a design and an optimization method. The design indicates the medium variables to test within the experiment, including the number of replicates and the arrangement of tests into homogenous "blocks." Based on the experimental data generated, the optimization method is used to predict the improved medium composition employing a mathematical model. The statistical approaches can thus offer cost-effective solutions in process optimization by providing well-planned individual experiments through which the interaction between the variables can be determined (Stowe and Mayer, 1966).

Medium optimization using statistical designs generally requires screening a large number of variables when a researcher is unsure as to which settings may result in optimal or nearly optimal response(s). Therefore, the identification of the key response(s) and all possible process factors are important key steps to be followed while employing Design of Experiments (DOE). In addition, the published reports and prior experience can be additional valuable resources for choosing the correct factor settings (Tari et al., 2006). Various statistical approaches, such as Plackett-Burman design, Box-Behnken design, central composite design (CCD), full or fractional factorial design, response surface methodology (RSM), Evolutionary Operation (EVOP) (Kvist and Thyregod, 2005), Taguchi methodology (Rao et al., 2008), and artificial neural networks (Subba Rao et al., 2007) based on statistical software programs were used to optimize the fermentation conditions for increasing the alkaline protease production yields. Some statistical-based studies used for alkaline protease production are shown in Table 8.3.

8.1.4 Purification and Characterization of Alkaline Proteases

The primary step in the purification of alkaline proteases is the cell separation from the fermented broth by filtration or centrifugation and/or flocculating agents such as hydrocolloids and organic polyelectrolytes. Further, the cell-free supernatant can be concentrated by means of ultrafiltration, temperature-sensitive hydrogel ultrafiltration, salting out using ammonium sulfate, or solvents like acetone or ethanol (Kumar and Takagi, 1999). Concentration of alkaline proteases could be achieved using PEG-35,000 (Larcher et al., 1996), activated charcoal (Kumar and Parrack, 2003a), lyophilization, and heat

Table 8.3 Statistical Designs Employed for Improving Alkaline Protease Production from Different Microbial Sources

MICROBIAL STRAIN	STATISTICAL DESIGN[a]	SOFTWARE USED[b]	MEDIUM DESIGNED/OPTIMIZED PARAMETERS	PROTEASE YIELD IMPROVEMENT	REFERENCE
Bacillus sp. RGR-14	FCCCD	Design-Expert (Statease, Minneapolis, MN, USA)	Starch, 1.52%; casamino acids, 0.8%; soybean meal, 1.25%	12.8-fold	Chauhan and Gupta (2004)
Bacillus mojavensis	FCCCD	Design-Expert	Casamino acids, 12 g/l; inoculum age, $A_{550} = 0.25$; agitation, >250 rpm	4.2-fold	Beg et al. (2003)
Bacillus sp. L21	Box-Behnken design; CCD	Minitab (Minitab, State College, PA, USA)	Soybean meal, 2.0 g/l; maltose, 20 g/l; Tween 80, 0.35 g/l; initial pH, 8.0	10-fold	Tari et al. (2006)
Bacillus circulans	Taguchi orthogonal array (L-27) DOE	Qualitek-4 (Nutek, MI, USA)	pH, 10; temperature, 36°C; particle size, 0.3–1.0 mm; moisture, 140%; maltose, 1.5%; yeast extract, 2.0%; inoculum size, 3.0%; incubation time, 60 h	51%	Prakasham et al. (2005)
Bacillus circulans	Feed-forward neural network and genetic algorithm	MATLAB® (Mathworks, Natick, MA, USA)	Glucose, 12 g/l; soybean meal, 12.1 g/l; temperature, 34.6°C; inoculum level, 5.1 ml	>2.5-fold	Subba Rao et al. (2007)
Bacillus sp. RKY3 (KCTC 10412BP)	Plackett-Burman design; CCD	Design-Expert	Corn starch, 10 g/l; corn steep liquor, 30 g/l; inoculum size, 2.99%	2.3-fold	Reddy et al. (2008)
Bacillus pseudofirmus Mn6	Plackett-Burman factorial design; Box-Behnken design	Microsoft Excel-*Solver*	Peptone, 2%; yeast extract, 1%; glucose, 0.2%; starch, 0.2%; KH_2PO_4, 0.1%; $MgSO_4$, 0.01% (at pH 9.5, 30°C, 60 h)	6.4-fold	Abdel-Fattah et al. (2009)

Microbacterium sp.	RSM	Experimental Design Module of the Statistica 5.0 (Statsoft, Inc., Tulsa, OK, USA)	Temperature, 25°C; initial pH, 7.0; 12.5 g/l of feather meal	3.6-fold	Thys et al. (2006)
Shewanella oneidensis MR-1	Box-Behnken design; RSM	n.s.	Incubation period, 36 h; pH, 9.0; temperature, 30°C; glucose, 12.5 g/l; tryptone, 2.5 g/l	60%	Anbu et al. (2009)
Rhizopus oryzae RO IIT RN-13 (NRRL21498)	EVOP-factorial design	n.s.	Biotin, 2.5 mg/l; $CaCl_2$, 20 mg/l; 1-napthanelic acid, 12.5 mg/l	1.9-fold	Tunga et al. (1999)
Aspergillus clavatus ES1	Plackett-Burman design; RSM	Design-Expert	*Mirabilis jalapa* tuber powder, 10 g/l; pH 8.0; temperature, 30°C	14.0-fold	Hajji et al. (2008)
Bacillus subtilis	CCD; RSM	Design-Expert	Glucose, 1%; peptone, 5.5%; $MgSO_4$, 0.2%; $CaCl_2$, 0.5%	2-fold	Deepak et al. (2008)
Candida utilis	CCD; RSM	MATLAB	Moisture, 70%; inoculum size, 1 ml; fermentation time, 5 d	1.9-fold	Moftah et al. (2012)

[a] FCCCD, face-centered central composite design; CCD, central composite design; RSM, response surface methodology; EVOP, evolutionary operation.

[b] n.s., not specified.

treatment at 98°C for 24 h with 1% SDS (Blumentals et al., 1990). Further, a combination of one or more chromatographic techniques, such as ion exchange, hydrophobic interaction chromatography, affinity, and gel filtration were used to purify alkaline proteases. In general, alkaline proteases are positively charged and do not bind to anion exchangers, and as a result, cation exchangers are the preferred choice (Kumar and Takagi, 1999). The different affinity resins used are *N*-benzoyloxycarbonyl phenylalanine agarose (Larcher et al., 1996), aprotinin-agarose (Petinate et al., 1999), hydroxyapatite (Chakrabarti et al., 2000), and α-casein-agarose (Kocabiyik and Özdemir, 2006; Sareen and Mishra, 2008). Other purification methods employed on a smaller scale were reversed micellar extraction (Rahaman et al., 1988), affinity precipitation (Pecs et al., 1991), foam fractionation (Banerjee et al., 1993), dye ligand chromatography (Cowan and Daniel, 1996), and aqueous two-phase systems (Zhu et al., 2007). Extensive studies were carried out on the physicochemical properties of several microbial alkaline proteases to understand their suitability for different industrial applications. Alkaline proteases exhibit optimal activity in the alkaline pH range (8–12) and thermostability (50°C–70°C), to best meet the conditions prevalent in laundry detergents (Kumar et al., 1998). Specific metal ions, like Ca^{2+}, Mg^{2+}, and Mn^{2+} are required for enhanced activity and thermal stability, and these metal ions protect the enzyme from denaturation and maintain the active enzyme conformation at high temperatures. They also exhibit broad substrate specificity toward many synthetic substrates and natural proteins like casein, azocasein, bovine serum albumin, hemoglobin, or keratin (Kumar et al., 1999b; Kumar, 2002).

8.1.5 Applications of Alkaline Proteases

Alkaline proteases are robust enzymes that contribute to complete or partial enzymatic hydrolysis of proteins resulting in different value-added products. They are applied in a wide array of different industrial applications such as the following:

(a) *Laundry detergent additives*: Alkaline proteases cater to a major share of the world enzyme market for application in laundry detergents that contributes to the release of protein-based stains from the fabrics on washing (Kumar et al., 1998).

For achieving optimal results, the wash performance of the enzyme detergent is dictated by many factors including the test fabrics, detergent composition and dosage, activity and stability over a broad pH and temperature range, washing time and temperature, soiling agents, water hardness, broad substrate specificity, extended enzyme shelf life with compatibility to various detergent components, oxidizing and sequestering agents, and the effective low use concentrations of 0.4%–0.8% (Kalisz, 1988). Several commercial alkaline proteases produced by different alkalophilic *Bacillus* strains exhibit a good stability at high pH and temperature values that find use as laundry detergent additives. They are Alcalase, Savinase, Esperase, Durazyme, and Kannase (Novozymes, Bagsvaerd, Denmark), Opticlean, Optimase, and Maxapem (Solvay Enzymes GmbH, Hannover, Germany), Corolase 7089 and Biotouch® ROC 250L (AB Enzymes GmbH, Darmstadt, Germany), Maxacal and Maxatase (DSM, The Netherlands), Purafect (Genencor International, Rochester, NY, USA), Bioprase (Nagase Biochemicals, Japan), Godo-Bap (Godo Shusei, Japan), Alkaline protease "Wuxi" (Wuxi Synder Bioproducts, China), and Protosol (Advanced Biochemicals, Thane, India). These commercial enzyme preparations find application in cleaning of clothes, dishwashing, surgical instruments, and contact lenses (Kumar et al., 1998; Gupta et al., 2002). A novel subtilisin from *Bacillus gibsonii* showed a better activity profile as compared with other known subtilisins like *Bacillus lentus* alkaline protease, Properase and Savinase, based on the performance tests and was identified as a potential additive in fabric washing and dishwashing cleaners (Siegert et al., 2009). Further, the cold-active alkaline proteases have gained renewed interest in the recent years due to the change in the preferences toward low-temperature (15°C–25°C) washing to reduce the energy consumption (Cavicchioli et al., 2002; Saeki et al., 2007). However, the cold-adapted enzymes have not been commercially used due to several constraints with regard to their stability performance in detergent formulation and large-scale production (Saeki et al., 2007).

(b) *Ultrafiltration membrane cleaning*: Biofouling is a major problem due to improper cleaning of food-contact surfaces, which may favor the formation of biofilms as a result of accumulation of particulate food materials (Kumar and Anand, 1998). All dairy and food products have their own specific fouling problem depending on the soiling agents, like proteins, fat, minerals, and bacteria. During ultrafiltration, a gradual decline in permeate flux rate is attributed to the buildup of rejected protein on the membrane surface and/or within the membrane pore resulting in membrane fouling, which limits process efficiency, loss in the product quality, and increase in energy and production costs (Grandison and Lewis, 1996). Fouling studies of ultrafiltration (UF) and/or reverse osmosis (RO) membrane systems suggest that protein-based deposits are formed. Whey proteins, like α-lactalbumin (strongest gel-forming tendency and existing as granules) and β-lactoglobulin (major and long-term foulant and formed either as strands or sheets) were responsible for 95% of the filtration interruptions during UF of whole milk. In addition, impermeable calcium combines with colloidal phosphate, which tends to crystallize and precipitate in the membrane pores resulting in flux decline (Goosen et al., 2004). The use of shorter cleaning cycles and extending the life time of membranes has been the primary goal in membrane processing research. The advantages of using enzyme-based cleaners are due to their biodegradable nature, ability to function under mild conditions, improve the cleaning efficiency with lower amounts of detergent, and savings in energy costs by working under mild conditions of pH 8.5–9.5 and temperature range of 50°C–55°C (Smith and Bradley, 1987). Thermophilic alkaline proteases from *Thermus* sp. Rt41A (Coolbear et al., 1992) and alkalophilic *B. lentus* MK5-6 (Kumar and Tiwari, 1999) were successfully demonstrated to clean UF membranes fouled with milk and whey proteins. Some of the commercial enzyme-based membrane cleaners available for use are Alkazym (Novodan A/S, Copenhagen, Denmark), Bioklenz (Economics Laboratories NZ, Hamilton, New Zealand), Maxatase® XL (Genencor International), Terg-a-zyme (Alconox, New York, NY, USA), and Ultrasil 53 and P3-paradigm (Henkel-Ecolab GmbH, Düsseldorf, Germany).

(c) *Silver recovery*: Waste X-ray or photographic films contain 1.5%–2.0% silver by weight in their gelatin layer. Conventional method of silver recovery is by incineration of these films and the polyester base film cannot be recovered, which causes undesirable environmental pollution. However, the protease treatment enables the breakdown of gelatin layer with the recovery of silver and recycling of polyester film base. Different *Bacillus* alkaline proteases were reported for silver recovery from used X-ray films (Kumar and Takagi, 1999).

(d) *Cheese making*: Rennet is a classic milk-clotting protease derived from the stomach (abomasum) of young suckling calves, which has been used in the cheese-making process since ancient times. Rennet contains chymosin (EC 3.4.23.4), a pepsin-like aspartic protease. Chymosin is a bilobular enzyme having Asp 32 and Asp 215, which act as catalytic residue, and exists as chymosin A and chymosin B with acidic pH optima of 4.2 and 3.8, respectively, which selectively cleaves the Phe_{105}-Met_{106} bond in κ-casein into para-κ-casein and a glycomacropeptide, which stabilizes the casein micelle in milk, eventually leading to the gelation of casein micelles and the formation of the cheese curd (Dalgleish, 1992). During cheese processing, rennet not only possesses milk-clotting activity, but it also contributes to a balanced flavor and texture development during cheese ripening. However, in view of the worldwide shortage of calf chymosin and preferences for vegetarian cheese, attempts led to the search for substitutes for this enzyme (Neelakantan et al., 1999). Fungal proteases of *Endothia parasitica*, *Mucor miehei*, and *Mucor pusillus* have received wide acceptance from an industrial perspective. However, the rennet produced from these fungal sources contained low specific proteases that resulted in bitter flavor and low cheese yields (Neelakantan et al., 1999). The recombinant chymosin produced using genetically modified strains like *E. coli* K12 (Emtage et al., 1983), *Saccharomyces cerevisiae* (Mellor et al., 1983), *Kluyveromyces lactis* (Van den Berg et al., 1990), and *Aspergillus oryzae* (Tsuchiya et al., 1993) containing the bovine (*Bos taurus*) chymosin gene gained consumer acceptance in about 20 different countries. Despite the fact that recombinant bovine chymosin is the widely used chymosin in the cheese-making industry, new sources of recombinant chymosin are now available from alternate animal sources like goat (Vallejo et al., 2012), camel (Kappeler et al., 2006), and buffalo (Vallejo et al., 2008) to meet the growing commercial market demands.

In addition, alkaline proteases from different microbial sources like *Penicillium oxalicum* (Hashem, 2000), *Myxococcus xanthus* (Poza et al., 2004), *Amylomyces rouxii* (Yu and Chou, 2005), *Nocardiopsis* sp. (Cavalcanti et al., 2005), *B. licheniformis* strain USC13 (Ageitos et al., 2007), *B. subtilis* (*natto*) Takahashi (Shieh et al., 2009), and *Aspergillus niger* (Fazouane-Naimi et al., 2010) showing good milk-coagulating abilities were identified as promising calf rennet substitutes. Alkaline proteases also coagulated the milk from soybean to obtain curds with smooth texture and high flavor-potentiating activity (umami in Japanese) without bitterness (Yusada et al., 1999).

(e) *Food fermentations*: Certain indigenous fermented foods of Asia, like Japanese natto, Thai thua nao, Indian kinema, and African dawadawa, kpaye, soumbara, iru, or ogiri are prepared from various beans, seeds, and leaves involving an alkaline fermentation by different species of *Bacillus*, principally *B. subtilis, B. licheniformis*, and *B. pumilus* (Steinkraus, 1996). The majority of the proteases produced by these microbes have been purified and characterized and these enzymes have been identified as serine alkaline proteases. These alkaline proteases extensively hydrolyze the food proteins to amino acids and low-molecular-weight peptides with liberation of free ammonia. These fermented foods find use for direct consumption or as flavoring condiments in soups or sauces (Aderibigbe et al., 1990). Soy sauce, the traditional fermented food of Japan (shoyu) and China, is prepared from soybean and wheat flour and fermented with koji molds, *Aspergillus sojae* and/or *A. oryzae*, and the fungal fermented product (koji) is placed in brine of 23% NaCl (brine fermentation to yield moromi). During fermentation, the microbiological changes observed are mainly due to the alkaline and neutral proteases produced by these koji molds, which are the important hydrolyzing enzymes responsible for increasing the rate of protein solubilization, and the brine fermentation is unique as it uses halophilic lactic acid bacteria and salt-tolerant yeasts, which metabolize the formed amino acids (O'Toole, 1997). The different volatile compounds comprising carbonyls, acids, nitrogen- and sulfur-containing compounds generated during the fermentation process by the microbial action are responsible for the flavor and/or aroma development (Lee et al., 2006). Some efforts have been made to accelerate the fish sauce fermentation process by decreasing the salt content, lowering the pH, increasing the temperature, and exogenous addition of commercial protease preparations and/or efficient starter cultures (Akolkar et al., 2009; Udomsil et al., 2010). During ripening of dry fermented sausages, several changes are observed, and proteolysis is one such important change that affects both texture and flavor formation. The fermented sausage manufacture involves high costs of storage until a suitable maturation state is reached in the final product and a shortened ripening time would be convenient and beneficial to the food industry. The combined use of pronase E and fungal extracts of *Mucor racemosa* forma *sphaerosporus* (Bruna et al., 1999) and *Penicillium aurantiogriseum* (Bruna et al., 2000) were demonstrated for accelerated ripening with desirable flavor and textural attributes in the fermented sausages.

In the blood coagulation process, fibrin is the most important protein component normally formed from fibrinogen by thrombin. During thrombosis, the fibrin gets accumulated in the blood vessels resulting in myocardial infarction. Urokinase, streptokinase, and recombinant tissue-type plasminogen activator are the currently used versions of fibrinolytic agents that convert plasminogen to plasmin, which then degrades the fibrin. These fibrinolytic agents are of human origin; however, the enzyme therapy (intravenous administration) using these agents is very expensive and patients may suffer from side effects such as acute coronary re-occlusion and bleeding complications (Blann et al., 2002). Many traditional fermented foods are gaining renewed interest from a medical perspective as promising and adjunct microbial fibrinolytic enzyme sources. Several alkaline proteases have been identified as potential fibrinolytic regimens from a wide variety of fermented foods such as datsuwokinase from skipjack, shiokara (Sumi et al., 1995), *Bacillus* sp. KA38 from Korean salty fermented fish, jeot-gal (Kim et al., 1997a), *Bacillus* sp. from kimichi (Noh et al., 1999), *Bacillus* sp. strain DJ-4 from Korean fermented soybean, doenjang (Kim and Choi, 2000), *B. subtilis* IMR-NK1 (Chang et al., 2000), and *B. subtilis* (Urano et al., 2001) from Japanese fermented soybean, natto, *B. subtilis* BK-17 (Jeong et al., 2001), *B. subtilis* strain KCK-7 from Korean fermented soybean sauce, chongkukjang (Paik et al., 2004), fermented shrimp paste (Wong and Mine, 2004), *B. amyloliquefaciens* DC-4 (Peng et al., 2003) and *B. subtilis* DC33 (Wang et al., 2006) from Chinese fermented soybean, douchi, *B. subtilis* TP-6 from an Indonesian fermented soybean, tempeh (Kim et al., 2006), and nattokinase from *B. subtilis* natto B-12 (Wang et al., 2009). Fibrinolytic enzymes have also been identified from other sources like *Streptomyces megasporus* SD5 (Chitte and Dey, 2002), *Rhizopus chinensis* 12 from brewing rice wine (Liu et al., 2005), *Armillaria mellea* (Lee et al., 2005), *Fusarium* sp. BLB (Ueda et al., 2007), and *Cordyceps militaris* (Cui et al., 2008). The enzymes from these foods and/or microbial sources have been purified, characterized for their physiochemical properties, and they find significant application in food fortification and as nutraceuticals for effective prevention of cardiovascular diseases (Mine et al., 2005).

(f) *Protein processing*: Commercial alkaline proteases, like Alcalase® (Novo), Optimase™ (Novo), and Proleather® (Amano) with the ability to achieve maximum degree of hydrolysis have been demonstrated for the production of protein hydrolysates with well-defined peptide profiles and specific functional properties (Clemente, 2000). As the proteins are hydrolyzed, they form small peptides and free amino acids. However, the enzymatic hydrolysis reaction must be carefully controlled to avoid undesirable proteolysis resulting in bitter-tasting peptides or losses of desirable functionality. When food proteins are enzymatically hydrolyzed, the changes in flavor attributes are due to the development of bitterness, which is attributed to the hydrophobic nature of oligopeptides liberated rather than free amino acids. The bitter peptides account for only 5%–10% of the hydrolysate weight with molecular masses ranging from 1000 to 6000 Da and contain C-terminal positioned hydrophobic amino acid residues such as leucine, valine, phenylalanine, and tyrosine (FitzGerald and O'Cuinn, 2006). Ultrafiltration membranes can be used for the fractionation or concentration of protein hydrolysates to improve the bioactive peptidic fractions (Chabeaud et al., 2009). Debittered protein hydrolysates were produced from proteins of different sources, like sunflower, soybean, wheat gluten, chickpea, and brewer's spent grain have been found to possess antioxidant activity and they find use in nutritional fortification of fruit juices or soft drinks, flavor enhancers, and dietetic supplements (Kotlar et al., 2011). Many antihypertensive peptides derived from milk protein, dried bonito (katsuo-bushi), α-zein, wheat germ, soy sauce, and soy protein have the ability to inhibit angiotensin I-converting enzyme (ACE) resulting in the lowering of blood pressure and these ACE-inhibitory peptides have the ability to reduce the risk of the cardiovascular diseases and are marketed as nutraceuticals (Zhu et al., 2008). ACE-inhibitory peptides have also been derived from different marine organisms that are potential therapeutic candidates in the pharmaceutical industry (Wijesekara and Kim, 2010). In addition, many of these peptides are also derived from enzymatic hydrolysates of animal muscle protein (Ryan et al., 2011). Cryotin (North, Höfn, Iceland) a commercial protease mixture derived from a cold-adapted Atlantic cod (*Gadus morhua*), was explored for the production of fish protein hydrolysates exhibiting high ACE-inhibitory activity (Raghavan et al., 2008). Deseasin MCP-01, a cold-active alkaline protease with collagenolytic activity, secreted by a cold-adapted bacterium *Pseudoalteromonas* sp. SM9913 exhibited broad substrate specificity to various types of collagens. Therefore, MCP-01 could find use to digest fish skin of fish processing by-products to produce marine collagen peptides with ACE-inhibitory activity (Zhao et al., 2008). Protein hydrolysates with good physiological and organoleptic attributes, nutritional properties like balanced amino acid composition and high digestibility were produced by controlled enzymatic hydrolysis of lean meat, poultry wastes, and bone residues after mechanical deboning using commercial proteases, such as Alcalase and Flavourzyme, for use in food industry as flavor enhancers, beverage stabilizers, milk replacers, surimi production, and as special nutritional supplements for poultry and calves (Nolsoe and Undeland, 2009). Fish protein hydrolysates prepared using Alcalase and Flavourzyme from fish sauce and fish silage contain bioactive peptides with immunostimulating properties (Gildberg, 2004). Treatment of maize or rice starch with commercial proteases like Promod 25P or protease N enabled the protein removal and modification of starch by solubilization of α-glucans and thus contributed in improving the rheological properties in foods (Li et al., 2009).

(g) *Meat tenderization*: Tenderness is an important quality, palatability, and rheological attribute of meat that is dependent on the molecular structure of collagen and elastin in the connective tissues. Plant proteases, like papain from papaya (*Carica papaya* L.), ficin from fig (*Ficus* spp.), and bromelain from pineapple (*Ananas comosus* [L.] Merr. and *Ananas bracteatus* [Lindl.] Schult. & Schult.f.), were used traditionally for meat tenderization with advantages of established safety, availability, and economy. However, the major limitation is their nonspecific proteolysis and capacity to degrade both connective and muscle proteins somewhat indiscriminately, resulting in unfavorable taste due to overtenderization. Lately, some fungal proteases from *A. oryzae* and *Aspergillus flavus* have gained clearance for meat tenderization. In addition, alkaline elastase (Takagi et al., 1992), thermophilic alkaline protease (Wilson et al., 1992), and elastase from *Bacillus* sp. EL31410 (Chen et al., 2006) also proved as promising meat tenderizers due to their greater specificity and hydrolyzing activity toward both collagen and elastin in connective tissues and a relatively lower hydrolyzing activity toward myofibrillar proteins, thus reducing the possibility of overtenderization. Commercial enzymes, like Savorase and Flavourzyme (Novozymes), are mainly used to hydrolyze meat before cooking that improve the meat taste by releasing the taste amino acids, nucleotides, and short peptides. These taste enzymes function at an optimum temperature of above 50°C for protein hydrolysis (Imm and Lee, 1999). A significant increase in tenderness was also observed until 2 weeks of storage (Ashie et al., 2002). A cold-adapted protease derived from *Pseudoalteromonas* sp. SM9913 was used for tenderization and taste improvement of refrigerated meats kept in cold storage (He et al., 2004).

(h) *Baking*: The use of enzymes in the milling and baking processes stems from the deficiency of some naturally occurring enzymes in wheat and flour. Amylases and proteases are used in the baking industry to improve dough handling properties and enhance dough quality. The proteases are generally added during the fermentation stage to permit breakdown of wheat flour proteins (gluten). The extent of gluten degradation is an important determinant of the rheological properties of the dough. Limited enzymatic hydrolysis induces breakage of intermolecular bonds of the gluten network to make the dough more pliable, while a prolonged hydrolysis can destroy the same (Pedersen et al., 2005). Proteases hydrolyze the gluten and thus improve the rheological properties of the dough by reducing mixing times without affecting loaf volume or crumb quality of the finished product (Kong et al., 2007). The protease treatment enhanced the nutritional and functional values of the resulting gluten-reduced or gluten-free products, which are amenable for the diet of patients with celiac disease (gluten intolerance) (Clement, 2011). Further, the incorporation of dairy ingredients, like enzymatic hydrolysates of acid casein and sodium caseinate in the dough improved the dough rheological properties and bread quality during the bread-making process. These functional benefits are mainly the result of the effects of milk fat and protein (Kenny et al., 2000).

(i) *Brewing*: Brewers have identified two applications for proteolytic enzymes: chill-proofing and low/no-malt brewing. Beer on long-term storage develops colloidal hazes that are generally complexes of polyphenols, carbohydrates, and proteins, which result from the interaction of polyphenols with peptides derived originally from the malt during mashing. The haze found at 0°C is known as chill-haze and that formed at room temperature is permanent haze. The proteins are responsible for haze formation; however, their presence is also essential for foam formation and retention. It is thus important that sufficient proteins remain in the chill-proofed beer to maintain the desirable head properties (Siebert, 2009). Chill-proofing is a proteolytic treatment, which prevents the haze formation on extended storage at cold temperatures during or after the maturation of beer. Limited proteolysis degrades the protein component of haze and prevents these complexes from precipitation during chill-proofing. The treatment is generally carried out at 4°C for 3–4 days under controlled conditions, since excessive proteolysis can affect the foaming and other qualities of the beer. The addition of proteases during the fermentation of beer also exhibits a high affinity and breakdown of the chill-haze precursors resulting in greater haze stability than control beers (Piddocke et al., 2011).

(j) *Other applications*: Marine crustacean wastes including shrimp and crab shells are very rich in chitin, a homopolymer of *N*-acetyl-D-glucosamine, protein, and calcium carbonate. Chitin and its derivatives are of high economic value due to their versatile biological activities and agrochemical applications (Hirano, 1996). Chemical processing has been the conventional method for chitin extraction from crustacean wastes, which involves demineralization and deproteinization steps using strong acids and bases to remove calcium carbonate and proteins, respectively. However, these chemical treatment steps generate a partially deacetylated product having variable and inconsistent physiological properties and also generate large quantities of waste water required for further neutralization and detoxification. Several alkaline proteases from *P. aeruginosa* K-187 (Oh et al., 2000), *A. niger* (Teng et al., 2001), and *B. cereus* SV1 (Manni et al., 2010) were explored for the deproteinization of crustacean wastes for the preparation of chitin. Alkaline proteases were also used for the economic extraction and purification of a carotenoprotein pigment, astaxanthin, from shrimp wastes, which finds application as a functional food additive (Lee et al., 1999).

Biodegradable dextran hydrogels were prepared by enzymatic transesterification by acylating dextran with Divinyladiapte in dimethylsulfoxide using an alkaline protease, Proleather FG-F (Amano). These hydrogels find special use in tissue engineering and colonic drug-delivery applications (Ferreira et al., 2005; Van Tomme and Hennink, 2007). A thermophilic alkaline protease Rt41A from *Thermus* sp. Rt41A, marketed as PreTaq™ (Pacific Enzymes, Hamilton, New Zealand), has been applied for PCR amplification of mRNA directly from the crude cell lysate of blood or tissue samples (Fung and Fung, 1991). Alkaline protease Q from *B. pumilus* has been used for inactivating RNase during purification of RNA from crude cell lysates (Han and Damodaran, 1998). A thermophilic proteinase EA1, marketed as *ForensicGEM*™ (ZyGEM Corporation, Hamilton, New Zealand), was found suitable for DNA extraction from forensic samples like blood, saliva, semen, and tissues (Moss et al., 2003; Lounsbury et al., 2012). Versazyme™, a keratinase produced by *B. licheniformis*, has been demonstrated to degrade prions (Langeveld et al., 2003) and finds use for inactivating prions in meat and bone meal (Coll et al., 2007). Nattokinase purified from *B. subtilis* (*natto*) was found to degrade amyloid fibrils and could find use in the therapy of amyloid-related diseases like Alzheimer's disease, prion-mediated disease, and systematic amyloidosis (Hsu et al., 2009). Alkaline proteases, like TrypLE™ Express, a recombinant fungal trypsin-like protease (Nestler et al., 2004), and *Vibrio* sp. V26 (Manjusha et al., 2013), have been used as a trypsin substitute for use in animal cell cultures.

8.2 Poly-γ-Glutamic Acid

Toshirou Nagai

Poly-γ-glutamic acid (PGA) is a polymer composed of glutamic acids, linked through a carboxylic group on γ-carbon and an amino group of a neighboring glutamic acid (Figure 8.1a), unlike peptide bonds of proteins, in which a carboxylic group on α-carbon linked to an amino group of a neighboring amino acid (Figure 8.1b). *Bacillus subtilis* (including *B. subtilis* (*natto*) and *B. subtilis* var. *polyglutiminicum*), *Bacillus licheniformis*, *Bacillus megaterium*, and *Bacillus anthracis* produce PGA, the molecular weights of which range from 2000 Da to more than 10^6 Da, depending on species or strains (Murao, 1967). PGA produced by *B. anthracis* is composed of exclusively D-glutamic acid, but PGA of other *Bacillus* species is composed of both D- and L-glutamic acid. The ratio of D- to L-glutamic acid varies corresponding to culture conditions (especially, the concentration of Mn^{2+} in the culture) and incubation time (Fujii, 1963a, 1963c).

The quality and quantity of PGA are very important factors for preparing natto especially in Japan, where natto is eaten directly with cooked rice. Many *B. subtilis* strains, isolated from the soybean foods in Southeast Asia, could produce PGA (Tamang and Nikkuni, 1996; Mitsuboshi et al., 2007). The sticky material, PGA, produced during the preparation of kinema (non-salted fermented soybeans produced

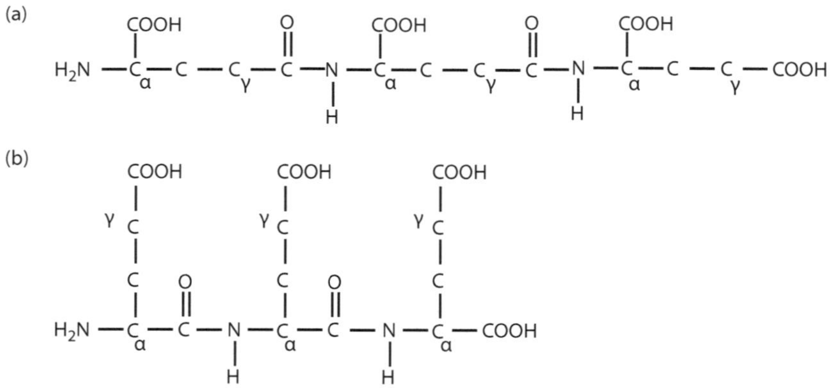

Figure 8.1 Chemical formulas of polyglutamates: (a) poly-γ-glutamic acid (trimer) and (b) poly-α-glutamic acid (trimer). Carbons on a backbone of glutamic acid are marked alphabetically with Greek characters.

in the eastern Himalayas) gave a good quality to the final products, so PGA production may be an important factor at least in the course of preparing soybean foods fermented by *Bacillus* strains (Tamang and Nikkuni, 1996; Tamang, 1999). In pe pok (fermented soybeans in Myanmar; also called peegapi) fermentation, production of PGA is also important for its quality (Mitsuboshi et al., 2007).

8.2.1 PGA Produced on Natto

Although some trials to elucidate a chemical entity of strings of natto (viscous material, also see Figure 2.2c in Section 2.1.1.1.1) were carried out in Japan, the trials failed because of insufficient purification. Viscous material of natto had already been thought to contain both sugars and peptides in the early days of investigation on natto (Sawamura, 1905). Matsumoto and Iwahara (1961) speculated that the viscous material of natto was PGA based on their chemical analyses on the purified material and earlier studies on PGA produced by *B. subtilis* strains in foreign countries (it is unclear whether the strains were *B. subtilis* (*natto*)-type strains or not; Bovarnick, 1942). Fujii (1963a, 1963b) showed that the viscous material of natto was composed of PGA and levan (a polymer of fructose), both of which were found not to link together by experiments of ultracentrifugal sedimentation. High viscosity of the material produced on natto was ascribed to PGA, and levan, which also had very low viscosity in itself, contributed to stabilization of the viscosity of PGA. The ratio of PGA and levan produced by *B. subtilis* (*natto*) varied depending on the conditions of experiments, and levan was not produced in the semi-synthetic media containing glucose instead of sucrose, indicating that PGA free of levan can be prepared simply by changing compositions of media.

8.2.2 Role of Natto Plasmid and γ-Glutamyl Transferase in PGA Production

In the 1980s, the theory of the role of natto plasmid on PGA production was proposed and accepted by the scientific community; however, the theory was refuted in 1990s. The story is summarized in this section, with an introduction to some studies on γ-glutamyl transferase (γ-GTP) of *B. subtilis* (*natto*).

Aumayr et al. (1981) showed experimentally that productivity of PGA was transformable to *B. subtilis* Marburg (not a *B. subtilis* (*natto*)-type strain) at a high frequency of 30% in comparison with that of transformation of auxotroph genes on the chromosomal DNA. The transformants of *B. subtilis* Marburg, which produces a low level of γ-GTP by nature, also could produce a high level of γ-GTP, implicating a relationship between PGA production and γ-GTP. Acridine orange, a fluorescent dye used for the curing of plasmids, deadened the productivity of PGA in *B. subtilis* (*natto*) Asahikawa (Hara et al., 1981). The *B. subtilis* (*natto*) strains treated with the dye also decreased the activity of γ-GTP and, interestingly, lost their plasmids in cells at the same time. The plasmid (named pUH1, 5.7 kb in length), a so-called natto plasmid, was transformed into *B. subtilis* Marburg at a high frequency and the resulting transformants acquired the productivity of PGA (Hara et al., 1982a). The experiments above with their additional studies on PGA production of *B. subtilis* (*natto*) strongly indicated that PGA was produced by γ-GTP and that the gene of γ-GTP was coded on the natto plasmid (Hara et al., 1982b, 1985; Hara and Ueda, 1982).

Although PGA-producing *B. subtilis* (*natto*) is susceptible to bacteriophages, the cured strains that lost the productivity of both PGA and γ-GTP became resistant to phage infection (Hara et al., 1984). *Bacillus subtilis* Marburg, producing no PGA and a low activity of γ-GTP, was resistant to phage infection, but the transformants that acquired productivity of both PGA and γ-GTP became susceptible to phages. From the results of these experiments, PGA was presumed to be an essential factor for phage absorption. However, the opposite results were obtained that PGA protected *B. subtilis* (*natto*) cells from phage absorption and therefore from phage infection (Kimura and Itoh, 2003). In the other studies, cured *B. subtilis* (*natto*) strains could be infected with phages, indicating that the natto plasmid had no association with phage infection (Nagai and Itoh, 1997). This discrepancy should be investigated carefully.

The natto plasmid pUH1 from *B. subtilis* (*natto*) Asahikawa was sequenced and its replication origin and *rep* gene were determined (Hara et al., 1982b, 1991). Two long open reading frames (ORFs), which were supposed to correspond to genes for light and heavy chains of γ-GTP, were revealed on the pUH1, but the homology between the DNA sequence and that of *Escherichia coli* or mammalian γ-GTPs were quite low (Hara et al., 1982a, 1992). After cloning and sequencing a gene of γ-GTP in *B. subtilis* (*natto*) as described below, the gene for PGA production on pUH1 was thought to be a factor stimulating PGA production (γ-PGA–stimulating factor, PSF) (Hara et al., 1994).

Contrary to the results of these experiments, it was observed that there was no correlation between enzymatic activities of γ-GTP and production of PGA (Fujii and Shiraishi, 1987). γ-GTP was purified from the culture broth of *B. subtilis* (*natto*) NR-1, isolated from commercial natto (Ogawa et al., 1991). Although the γ-GTP polymerized glutamine to some extent in vitro, the enzyme failed to produce viscous PGA (Ogawa et al., 1991). On the contrary, the purified γ-GTP hydrolyzed PGA to glutamic acid: γ-GTP might function in incorporation of glutamic acid from PGA in starvation (Ogawa et al., 1997). The gene, *ggt*, of the γ-GTP was also cloned from the chromosomal DNA of NR-1 stain and sequenced (Ogawa et al., 1997). The sequence of *ggt* did not have a homology with those of ORFs on pUH1, but the deduced amino acid sequence from *ggt* had similarities of 40% with those of bacterial γ-GTPs (Ogawa et al., 1997).

Nagai et al. (1997) cured *B. subtilis* (*natto*) of two plasmids, pUH1 and a larger plasmid, and showed that the elimination of plasmids did not affect the productivity (quantity, molecular size, and D- and L-glutamate ratio) of PGA. However, before their elimination experiment, Koehler and Thorne (1987) had already pointed out no correlation between the presence of natto plasmids and PGA production. Genes for PGA production are described in the next section.

γ-GTP is not directly concerned with the PGA synthesis, but weak positive correlations between stringiness and γ-GTP activity (r = 0.638) and between quality score and γ-GTP activity (r = 0.487) were observed by sensory evaluation on natto products from a market (Akimoto et al., 1990). Nagai et al. (1994) observed that *B. subtilis* (*natto*) strains produced more protease and PGA also produced more γ-GTP. Although no direct evidence is yet available, a high activity of γ-GTP in natto might simply reflect a high productivity of proteins or enzymes, one of which is merely γ-GTP.

8.2.3 Genes Related to the Production of PGA

Although PGA was found to be produced by a membranous enzyme complex using *B. licheniformis* ATCC 9945a producing D-poly-γ-glutamic acid (Troy, 1973a, 1973b), the detailed mechanism of PGA production was unclear. γ-GTP had been thought for a long time to be a PGA synthetase of *B. subtilis* (*natto*). However, the theory was denied as described above.

Genes responsible for PGA production, *pgsBCA*, of *B. subtilis* IFO 3336 (originally identified as *B. natto* = NBRC 3336) were introduced and expressed in *E. coli* cells by using a low-copy number vector (Ashiuchi et al., 1999). The genes were homologous to *capBCA* for capsular PGA production *of B. anthracis*, and to *ywsC*, *ywtA*, and *ywtB* of *B. subtilis* 168, which produces no PGA (Figure 8.2). Interestingly, DL-ratios of glutamic acid in PGA produced in *E. coli* were not influenced by Mn^{2+} ion, although early studies of PGA production by *B. subtilis* (*natto*) showed that Mn^{2+} in culture broth influenced the ratios in PGA. D-Glutamic acid residues in PGA are supplied from L-glutamic acid by glutamate recemase (Ashiuchi et al., 1999). Only a complete gene product assembly, PgsBCA, produced PGA in *E. coli*. On the other hand, a complex of two gene products translated from *ywsC* (= *pgsB*) produced PGA (Urushibata et al., 2002). This discrepancy remains to be elucidated. The membranous enzyme complex PgsBCA was presumed to polymerize glutamic acid to PGA by binding a γ-carboxyl group on elongating PGA and an amino group of a free glutamic acid via a hyperphosphorylated PGA intermediate (Figure 8.3) (Ashiuchi et al., 1999). PgsB and Pgs A were supposed to be an amide ligase and a transporter moving PGA out of the cells, respectively (Misono and Ashiuchi, 2005).

PGA production in *B. subtilis* (*natto*) is controlled by comP, which is a sensor protein kinase in the ComP-ComA two-component signal transduction system (Nagai et al., 2000; Tran et al., 2000). During growth and preservation, *B. subtilis* (*natto*) loses PGA productivity at a high rate. This instability of PGA production is caused by insertional mutation by IS*4Bsu*1 to the *comP* region. This IS (insertion sequence) was the first IS found in *B. subtilis*.

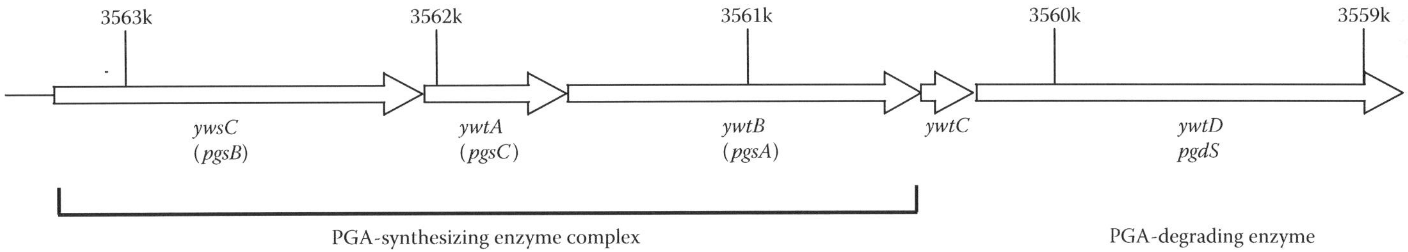

Figure 8.2 Genes for PGA production. The genes and the coordinate are originated from a database on the genome of *Bacillus subtilis* (*natto*) BEST195 (Nishito et al., BMC Genomics, 11, 243 2010) (Natto Genome Project, http://natto-genome.org/). Names of genes designated by Ashiuchi et al. (1999) are described in parentheses.

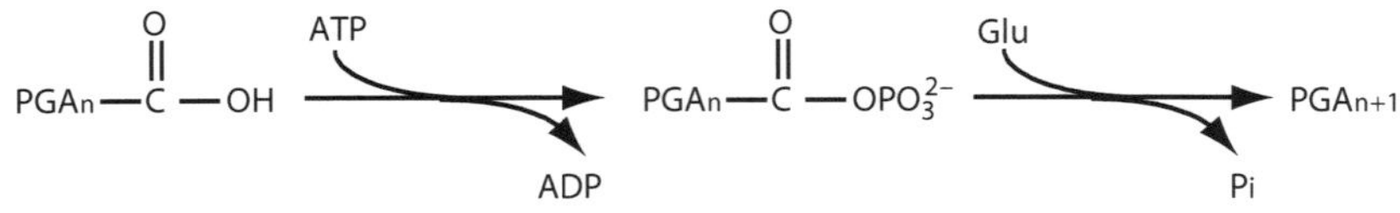

Figure 8.3 Deduced reaction scheme of PGA synthesis.

8.2.4 PGA Resin

PGA was linked together by γ-ray radiation to make transparent water-absorbing resin (Kunioka, 1993). The PGA resin absorbs 3500 times as much water as its own weight and is a little stronger than 2% agar gel. Swelling of the resin was stable at pH more than 5, but it degraded at pH 1.

Soybean seeds can be embedded in the PGA resin pellets, and the pellets are spread on desert areas with water. After cropping soybeans grown from the pellets, natto was produced again from the soybeans, and PGA was purified from the natto, and PGA resin was synthesized from it. The cycle was named "a green recycling system" for desert greening (Hara, 1997). Although the PGA resin is biodegradable by nature, researchers are trying to isolate microorganisms that can depolymerize the PGA resin to facilitate the green recycling system after the resin served its purpose.

PGA is becoming an important industrial material as biodegradable resin. Production of PGA resin costs too much at present, but the problem may be solved in the near future.

8.2.5 PGA-Decomposing Enzymes

Although PGA-decomposing enzymes have been found from bacteria to plants, the enzymes related to natto or *B. subtilis* (*natto*) are reviewed in this section.

8.2.5.1 Bacillus subtilis (natto) *Bacteriophages and Their PGA-Depolymerizing Enzymes* Before modernization of natto factories, natto without PGA were often produced. From such an abnormal natto, a bacteriophage was detected, and it was proven that the loss of PGA on natto was caused by the phage from the experiments of fermentation of cooked soybeans with a mixture of the phage and *B. subtilis* (*natto*) (Fujii et al., 1967). Many natto factories in Japan were contaminated with *B. subtilis* (*natto*) phages and the isolated phages from the factories were classified into two groups based on antiserum reactions (Yoshimoto et al., 1970). The two groups were different from each other in morphology, digestion pattern of genomic DNA with restriction enzymes, and DNA homology by Southern blot analysis (Nagai and Yamasaki, 2009).

A lysogenic *B. subtilis* (*natto*) strain harboring a *B. subtilis* (*natto*) bacteriophage could not accumulate PGA in culture broth, whereas a non-lysogenic strain could do so (Hongo and Yoshimoto, 1968). The lysogenic strain appeared to produce a PGA depolymerase in culture broth with phage induction. The PGA depolymerase was also produced extracellularly after infection with *B. subtilis* (*natto*) phage NP-1 cl (Hongo and Yoshimoto, 1970a; Yoshimoto and Hongo, 1970). The PGA depolymerase digested PGA by endo-type reaction and therefore decreased the viscosity of PGA solution rapidly. The products of the digestion were di-γ-glutamate and tri-γ-glutamate (Hongo and Yoshimoto, 1970b). The optimal pH and temperature of the PGA depolymerase were 6–8 and 40°C–50°C, respectively, and stable at pH 4–9 or below 70°C. The enzyme was not linked to phage particles. The molecular size of the enzyme was not reported. The research group found that *B. subtilis* phages PBS1 and SPOI, which could infect Marburg strain, and *B. subtilis* (*natto*) phage NP-38 also induced PGA depolymerase after infection (Yoshimoto et al., 1973). However, the PGA depolymerases produced by these phages were not studied in details, except that of *B. subtilis* (*natto*) phage NP-38, which had a molecular weight of 20 kDa.

Among the PGA depolymerases produced by *B. subtilis* (*natto*) phage, the enzyme from the culture supernatant of *B. subtilis* (*natto*) infected with a phage ΦNIT1 was investigated in detail. The culture broth of infected *B. subtilis* (*natto*) by ΦNIT1 decreased the viscosity of PGA solution rapidly (Kimura and Itoh, 2003). The enzyme, named PghP for "γ-PGA hydrolase of phage," was a 25-kDa monomeric enzyme and digested PGA to tri-γ-glutamate, tetra-γ-glutamate, and penta-γ-glutamate. The hydrolyzed products are different from that produced by PGA depolymerase above: di-γ-glutamate and tri-γ-glutamate (Hongo and Yoshimoto, 1970b). PghP was inhibited with monoiodoacetate and EDTA and the inhibition was restored by Zn^{2+} or Mn^{2+}, indicating that a cysteine residue(s) of the enzyme and Zn^{2+}

or Mn^{2+} participated in the PGA hydrolysis. The gene of PghP on the genomic DNA of *B. subtilis* (*natto*) phage ΦNIT1 was cloned and sequenced (Kimura and Itoh, 2003). The molecular weight of PghP was estimated to be 22.9 kDa, with 203 amino acid residues and the first methionine of the PghP was eliminated post-translationally. The BLAST search program showed that PghP was a unique protein because any similar proteins to PghP were not retrieved from the database. The homologous genes to *PghP* were distributed in *B. subtilis* (*natto*) phages and *B. subtilis* phages (Kimura and Itoh, 2003). PghP was subjected to crystallographic analysis and found to belong to space group $P3_121$ or P_3221 (Fujimoto et al., 2009). PghP may be useful in the food industry. PGA is an obstacle to usage of natto as an ingredient of foods, such as fried rice or snack foods, because of its high viscosity. PghP can remove the obstacle by decomposing PGA effectively (Kimura, 2008).

PGA, produced by *B. subtilis* (*natto*), protected its own cells from being infected by *B. subtilis* (*natto*) phages (Kimura and Itoh, 2003). The phage ΦNIT1 could amplify in *B. subtilis* (*natto*) whether it produced PGA or not because ΦNIT1 produced PghP. PghP decomposes PGA on *B. subtilis* (*natto*) cells and facilitate infection of phages to the cells. *Bacillus subtilis* phage BS5 (Ackermann et al., 1995), which produced no PGA depolymerase, could amplify only in *B. subtilis* (*natto*), producing no PGA. Addition of PghP made it possible for BS5 to amplify in PGA-producing *B. subtilis* (*natto*).

8.2.5.2 YwtD, γ-DL-Glutamyl Hydrolase of Bacillus subtilis (natto) An enzyme YwtD (44 kDa), of which gene (*ywtD*) is located downstream of the genes related to PGA production in *B. subtilis* (*natto*) (Figure 8.2), hydrolyzes PGA to 490- and 2-kDa peptides (Suzuki and Tahara, 2003; Chunhachart et al., 2006). A large peptide consists of only L-Glu, and small pepetide both D- and L-Glu at a ratio of 80:20. Detailed investigation to 2-kDa peptide showed that YwtD cut a γ-glutamyl bond between D- and D-Glu in PGA. On the other hand, the 490-kDa peptide had L-glutamic acid residues on both termini and the mechanism of the production was not clear (Chunhachart et al., 2006).

Bacillus licheniformis ATCC 9945a, closely related to *B. subtilis*, also produced an endo-type PGA depolymerase at the late stationary phase (Troy, 1973a). Whether the enzyme was translated from the gene corresponding to *ywtD* or not is obscure. Moreover, *Bacillus* sp. No 153 also produced PGA-decomposing enzyme of 28 kDa (Mori et al., 2005). These results show that *B. subtilis* or *Bacillus* sp. produces a variety of PGA-decomposing enzyme.

8.2.6 PGA-Decomposing Enzymes from Other Sources

An unknown factor, which reduced viscosity of PGA, and therefore, undermined physical quality of natto, was detected in radish (*Raphanus raphanistrum* subsp. *sativus* (L.) Domin, which was used as a spice for natto (Ikeda et al., 2000). The factor was thought to be an enzyme decomposing PGA by an endo-type reaction because of rapid reduction of the viscosity, and its optimum pH and temperature were 5.5°C and 37°C, respectively. The activities reducing viscosity of PGA were distributed in many vegetables and fruits (spinach, cabbage, asparagus, tomato, melon, pineapple, etc.). Products after digestion of PGA by the factors are not identified and details of the PGA-decomposing enzyme remain to be elucidated.

In northern Japan, natto was further fermented by koji mold (*Aspergillus oryzae*) in the presence of salt (named "goto natto" in Japanese). *Aspergillus oryzae* N-2 produced a PGA-decomposing enzyme acting on γ-L-glutamic acid bonds in PGA by an exo-type reaction (Ueda et al., 1988). The enzyme did not digest PGA purified from natto so well, only giving a faint spot of L-glutamic acid on a paper chromatogram. Ueda et al. (1988) also observed that almost all the strains of *A. oryzae*, used as starters for soy sauce and miso fermentation, produced PGA-decomposing enzymes in varying degrees of activities. The PGA-decomposing enzymes, however, might not so greatly contribute to enhancement of umami taste because of its low activity on PGA of natto.

These findings, including studies on fungal PGA hydrolases (Tanaka et al., 1993a, 1993b), indicate that PGA-decomposing enzymes are widely distributed in nature, acting on PGA in varied manners of reaction.

8.2.7 Conclusion

PGA is a highly viscous polymer of glutamic acid liked together via γ-carboxyl groups and amino groups. It is one of the determinants of the quality of fermented soybeans, especially, natto. In the production of fermented soybeans using *Bacillus*, the generation of such a viscous material seems to be regarded as a sign of successful fermentation, even though the viscous material is not important in the final products. The gene(s) for production of PGA was thought to be on a plasmid, a natto plasmid, which is not harbored in the type strain of *B. subtilis*, and to code the PGA synthetase, γ-glutamyl transpeptidase. Today, this postulation is refuted: the genes for the PGA production are cloned from the chromosomal DNA of *B. subtilis* (*natto*), and code a PGA synthetase complex, which could be expressed in *E. coli*. The important property of PGA is its powerful water-absorbing activity, and greening of deserts is tried using the PGA resin as nursery beds for soybeans. Natto is produced from the soybeans growing on the bed, and PGA is purified from the natto produced from the cropped soybeans. These processes form a recycling system. There are many microorganisms decomposing PGA, including *B. subtilis* in our environment, indicating that PGA is biodegradable, and therefore, it is a "nature-friendly" material.

8.3 Lipopeptides

Philippe Jacques and Aly Savadogo

8.3.1 Microbial Lipopeptides

Microbial lipopeptides are molecules consisting of a fatty acid chain connected to a peptide moiety via ester or amide bonds or both. They have surfactant properties and different biological activities in function of their structure. Lipopeptides have been proposed for possible roles such as participation in the acquisition of hydrophobic water-insoluble nutrients and influencing the attachment or detachment of bacteria to and from surfaces (Rosenberg and Ron, 1999). They could also inhibit growth of some competitors. These compounds are synthesized by a complex ribosome-independent mechanism involving nonribosomal peptide synthetases (NRPS) (Sieber and Marahiel, 2005). These huge modular proteins are able to incorporate more than 500 different monomers in their peptide products including fatty acid chains (Caboche et al., 2010). In some cases, a hybrid protein complex combining both NRPS and polyketide synthases (PKS), another family of modular proteins involved in the biosynthesis of fatty acid-derived compounds, is required for the biosynthesis of these lipopeptides. NORINE is the main database gathering most of the peptides and lipopeptides synthesized by NRPS (Caboche et al., 2008). A request for "*Bacillus*," "*Paenibacillus*," and "lipopeptides" in NORINE shows a result (Table 8.4) of 24 molecules gathered into five main families: surfactins, iturins, fengycins, lichenysins, and polymyxins. Two other families less studied need also to be mentioned: fusaricidins (Kajimura and Kaneda, 1996) and kurstakins (Béchet et al., 2012). The *Bacillus* and *Paenibacillus* strains producing lipopeptides were mainly isolated from soils, composts, or fermented food products.

Table 8.4 List of Lipopeptides Produced by *Bacillus* and *Paenibacillus* Strains

FAMILY OF LIPOPEPTIDES	VARIANTS IN THE PEPTIDE MOIETY
Fengycin/plipastatin	Fengycin A (Plipastatin A)
	Fengycin B (Plipastatin B)
Iturin	Bacillomycin D
	Bacillomycin F
	Bacillomycin L
	Iturin A
	Iturin C
	Mojavensin A[a]
	Mycosubtilin
	[Gln3]mycosubtilin[b]
Lichenysin	Lichenysin
Polymyxin	Polymyxin B
	Ile-polymyxin B
	Polymyxin E
	Ile-polymyxin E
	Val-polymyxin E
	Nva-polymyxin E
	Polymyxin M
Surfactin	Surfactin
	[Ala4]surfactin
	[Ile2.4.7]surfactin
	[Ile4.7]surfactin
	[Ile4]surfactin
	[Ile7]surfactin
	[leu4]surfactin
	[Val7]surfactin

(*continued*)

Table 8.4 (Continued) List of Lipopeptides Produced by *Bacillus* and *Paenibacillus* Strains

FAMILY OF LIPOPEPTIDES	VARIANTS IN THE PEPTIDE MOIETY
Fusaricidin	Fusaricidin A[c]
	Fusaricidin B[d]
	Fusaricidin C[d]
	Fusaricidin D[d]
Kurstakin[e]	Kurstakin (cyclic form)
	Kurstakin (linear form)

Source: Data mainly extracted from NORINE database (http://bioinfo.lifl.fr/norine/) except [a] Ma, Z. et al., *The Journal of Antibiotics*, 65, 317–322, 2012; [b] Béchet, M. et al., *Bioresource Technology*, 145, 264–270, 2013; [c] Kajimura, Y., and Kaneda, M., *The Journal of Antibiotics*, 49, 129–135, 1996; [d] Kajimura, Y., and Kaneda, M., *The Journal of Antibiotics*, 50, 220–228, 1997; and [e] Béchet, M. et al., *Applied Microbiology and Biotechnology*, 95, 593–600, 2012. For each variant, a set of homologous compounds differing in length and isomer of the fatty acid chain is produced.

8.3.2 Lipopeptides and Fermented Food Products

The first example of a lipopeptide-producing *Bacillus* strain isolated from fermented foods was given by Kameda et al. (1974) who identified the presence of surfactin in the culture medium of *Bacillus subtilis natto* KMD 2311 isolated from natto, a Japanese soybean-fermented product. Other examples of lipopeptide-producing strains isolated from food products were then given in literature: *Bacillus* strains isolated from nététu, a condiment obtained in Senegal by the fermentation of seeds of African locust bean (*Parkia biglobosa* (Jacq.) G.Don), produce compounds showing similar behavior on thin layer chromatography to surfactin and iturin A (N'Dir et al., 1994). *Bacillus subtilis* JKK238 isolated from jeotkal, a Korean traditional fermented fish, is a co-producer of surfactin, iturin, and fengycin (Yoon et al., 2005). *Bacillus subtilis* 20B isolated from Indian idli (rice-black gram mixed) batter produces an antifungal biosurfactant; its structure was not precisely determined but could correspond to surfactin and iturin (Joshi et al., 2008). *Bacillus subtilis* HY1 isolated from Korean soy sauce (kanjang) produces iturin (Cho et al., 2009). In Korea, fermented soybean foods such as doenjang (soybean paste), kanjang (soybean sauce), and chongkukjang (soybean cook) are served as a major source of protein.

Bacillus strains isolated from bikalga and soumbala, condiments produced in Burkina Faso by traditional uncontrolled alkaline fermentation of roselle (*Hibiscus sabdariffa* L.) and locust bean, respectively, produce lichenysin or surfactin (Savadogo et al., 2011). *Bacillus subtilis* CSY191 isolated from doenjang produces surfactin (Lee et al., 2012). *Bacillus subtilis* subsp. *subtilis* H4 isolated from bikalga co-produces surfactin and a bacteriocin (Compaoré et al., 2013).

Two other studies are indicators of the potential presence of lipopeptide compounds into fermented food products. On one side, Ohno et al. (1995) showed that *B. subtilis* RB14 produces iturin A and surfactin when cultivated in solid-state fermentation of okara and tofu. On the other side, Ouoba et al. (2007) reported antimicrobial activity of *B. subtilis* and *Bacillus pumilus* during the fermentation of locust bean for soumbala production.

To our knowledge, only three studies have been dedicated to the detection of the lipopeptide compounds in the fermented food product itself: N'Dir et al. (1994) have extracted from nététu compounds showing similar behavior on TLC to surfactin and iturin A. Lee et al. (2012) showed an increase of the presence of surfactin from 0.3 to 48.2 mg/kg over 48 h of chongkukjang production by the surfactin-producing strain *B. subtilis* CSY191. Savadogo et al. (2011) have analyzed by MALDI-TOF MS (matrix-assisted laser desorption/ionization time-of-flight mass spectrometry) crude extract of soumbala and detected the presence of surfactin.

8.3.3 Structures of Lipopeptides Produced by Strains Isolated from Fermented Products

As previously mentioned, surfactin, lichenysin, iturin A, and fengycin are the four main lipopeptides produced by *Bacillus* strains isolated from fermented products. Surfactin is a heptapeptide with a chiral sequence LLDLLDL interlinked with a β-hydroxy fatty acid (Figure 8.4). It contains two positively charged amino acid residues (glutamic acid and aspartic acid in positions 1 and 5, respectively) and five amino acid residues belonging to the aliphatic group including alanine, valine, leucine, and isoleucine in the other positions (Bonmatin et al., 1995; Peypoux et al., 1999; Jacques, 2011). The β-hydroxy fatty acid chain can contain 12–16 C atoms and shows *n*, *iso*, and *anteiso*

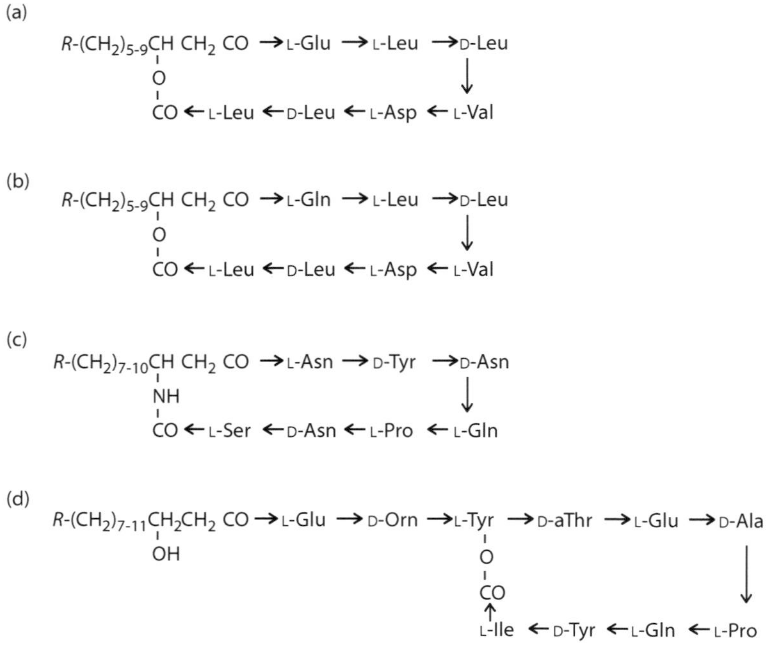

Figure 8.4 Structure of the main lipopeptides produced by *Bacillus* strains isolated from fermented food products: (a) surfactin, (b) lichenysin, (c) iturin a, (d) fengycin A/plipastatin A. R = $CH_3CH_2CH_2CH_2$ (*n*), $CH_3CH(CH_3)CH_2$ (*iso*), $CH_3CH_2CH(CH_3)$ (*anteiso*).

configurations. The main fatty acid chains are usually C14 and C15. Lichenysin only differs from surfactin by the presence of a glutamine residue instead of glutamic acid in the first position of the peptide moiety (Horowitz et al., 1990).

Iturin A is the main studied lipopeptide of the iturin family. This heptapeptide is interlinked with a β-amino acid fatty acid with a length from C14 to C17 (Peypoux et al., 1978). The sequence of the peptide moiety is L-Asn-D-Tyr-D-Asn-L-Gln-L-Pro-D-Asn-L-Ser. The main lengths of iturin A fatty acyl chain are 14 and 15 C.

The family of fengycin lipopeptides includes fengycins A and B, which are also called plipastatins. These lipodecapeptides differ by their amino acid residue in position 6 that can be alanine (form A) or valine (form B). They display an internal lactone ring in the peptide moiety between the C-terminal amino acid (isoleucine) and the hydroxyl group in the side chain of the tyrosine residue in position 3. Different β-hydroxy fatty acid chains (C14 to C18) are linked with an amide bond to the N-terminal amino acid residue (glutamic acid) (Nishikiori et al., 1986; Vanittanakom et al., 1986). The main representative fatty acid chains are C15, C16, and C17.

8.3.4 Biological Activities of Lipopeptides

Surfactin is considered as one of the most powerful biosurfactants with exceptional emulsifying and foaming properties (Razafindralambo et al., 1996). It reduces surface tension of water to 27 mN/m and shows a low critical micellar concentration (CMC) of about 10 mg/l. Due to its amphiphilic nature, surfactin can also interfere with biological membrane integrity in a dose-dependent manner. Different authors have demonstrated biological activities of surfactin that could be related to this property: Itokawa et al. (1994) have isolated novel [Ile7] surfactin, which showed anti-human immunodeficiency virus activity and has been isolated from *B. subtilis (natto)*. Vollenbroich et al. (1997) have reported antimycoplasma properties of surfactin. Kracht et al. (1999) showed that surfactin *iso*-forms have differential antiviral and hemolytic activities. Ongena et al. (2007) have reported that surfactin stimulates immune-related responses in plant tissues, an important phenomenon for biological control of plant diseases. More recent results from Henry et al. (2011) indicate that surfactin perception relies on a lipid-driven process at the plasma membrane level.

Many other biological activities were shown for surfactin such as anti-inflammatory (Kim et al., 1998), anticancer activity (Kameda et al., 1974; Cao et al., 2010), and immunomodulatory effects (Park and Kim, 2009). Surfactin causes potassium leakage trigger multicellularity (pellicle at the air–water interface and/or biofilm) in *B. subtilis* (Hofemeister et al., 2004; Chollet-Imbert et al., 2009; Lopez et al., 2009). It is also involved in the swarming of the strain (Julkowska et al., 2005; Debois et al., 2008). Surfactin is able to modify not only the surface properties of the producing strain (Ahimou et al., 2000) but also the hydrophobicity of stainless steel or Teflon substrata and the *Bacillus cereus* spore attachment to these surfaces (Shakerifard et al., 2009).

Iturin A is a less potent biosurfactant than surfactin (Deleu et al., 1999). It reduces the surface tension of water to 43 mN/m and shows a CMC of about 20 mg/l. At concentrations slightly higher than CMC, iturin probably forms multilamellar vesicles with an average size of 150 nm (Grau et al., 2001). Iturin A displays a strong in vitro and in vivo antifungal action against a large variety of yeasts and fungi (Bonmatin et al., 2003; Mizumoto and Shoda, 2007; Romero et al., 2007). Iturin A is also hemolytic. These activities against eukaryotic cells almost certainly rely on osmotic perturbation due to the formation of ion-conducting pores into membrane (Aranda et al., 2005).

Fengycins/plipastatins are characterized by a strong fungitoxic activity, more specifically against filamentous fungi, accompanied by a less hemolytic activity than iturins and surfactins. Their mode of action is less known but they also alter cell membrane structure and permeability in a dose-dependent way. They also showed anticancer (Yin et al., 2013) and immunosuppressive activity (Umezawa et al., 1988) and are able to induce systemic resistance against pathogens in plants (Ongena and Jacques, 2008).

8.3.5 Biosynthesis

Lipopeptides from *Bacillus* spp. are synthesized by NRPS. These large multienzymatic proteins are organized in modules (Sieber and Marahiel, 2005). Each module is responsible for the incorporation of one amino acid residue into the growing polypeptide chain. It can be subdivided into domains involved in a specific enzyme activity. Four main domains are present in most of the synthetases: adenylation domain (A), peptidyl-carrier protein (PCP), condensation domain (C), and thioesterase domain (Te). The A domain selects more and less specifically an amino acid residue and activates it as aminoacyl adenylate. This activated amino acid is then transferred to the PCP as it is the transport unit of the activated intermediate. In this PCP domain, the activated amino acid residue is covalently tethered to its 4′-phosphopantetheinic (4′-PP) cofactor as thioester. This cofactor is post-translationally transferred to a serine of the PCP with the help of a phosphopantetheinyl transferase (encoded in *B. subtilis* by the *sfp* gene). The 4′-PP cofactor acts as a flexible arm to allow the bound aminoacyl and peptidyl substrate to travel between different catalytic centers. The C domain catalyzes the formation of the peptide bond between aminoacyl substrate bound to PCPs of adjacent modules. The termination of synthesis is catalyzed by the terminal enzyme of the last module. In most cases and for lipopeptide synthesis, this reaction is performed by a Te domain. This allows the release of the peptide and is also frequently involved in the formation of a macrocyclic product (lactones and lactams) or the oligomerization of peptide units (Kopp and Marahiel, 2007).

Additional enzymes can be involved in the biosynthesis of the peptide to modify the structure of the monomer involved in the primary structure or to add some external compounds to the peptide. Among these tailoring domains epimerization (E) and addition of fatty acid chain are involved in lipopeptide biosynthesis in *Bacillus* spp. The E-domain catalyzes the epimerization of the PCP-bound L-amino acid of the growing polypeptide chain. The addition of the fatty acid chain to the first amino acid of the peptide moiety is catalyzed by a first specific condensation domain, also called starter condensation domain. Like in iturin biosynthesis, the added fatty acid chain can itself be partially synthesized by another main group of modular enzymes, the PKS (Tsuge et al., 2001). In this last case, a hybrid PKS/NRPS is required for the synthesis of the biomolecules (Du et al., 2001) (Figure 8.5).

8.3.6 Regulation and Influence of Environmental Factors on Lipopeptide Production

Surfactin and mycosubtilin (a member of the iturin family) production is quorum-sensing-dependent. Its regulation involves several peptide pheromones, which are first secreted in the culture medium and then processed and re-internalized through specific membrane receptors. These modified compounds are involved in a complex cascade governing multiple differentiation pathways such as sporulation and competence (Hamoen et al., 2003; Hayashi et al., 2006; Duitman et al., 2007).

The first culture medium proposed for lipopeptide production was nutrient broth or the Landy medium (Landy et al., 1948) with glucose as C-source and glutamic acid as a N-source. In subsequent studies, a minimal mineral salts medium, containing NH_4NO_3 as the inorganic N-source and glucose (4%) as the C-source, was defined by Cooper et al. (1981). The replacement of Cooper's N-source and the introduction of O_2 limitation, which redirects the energy flux into product synthesis, have led to an increased productivity of surfactin (Kim et al.,

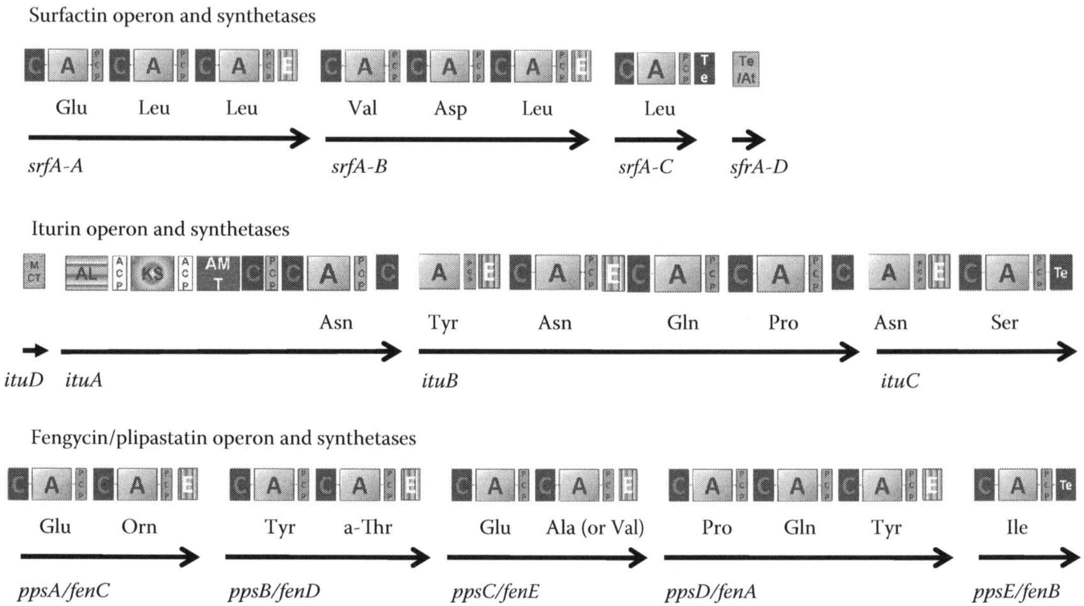

Figure 8.5 Schematic representation of operons and multienzymatic proteins (PKS or NRPS) responsible for the biosynthesis of surfactin (Menkhaus et al., *Journal of Biological Chemistry*, 268, 7678–7684, 1993), iturin A (Tsuge et al., *Journal of Bacteriology*, 183, 6265–6273, 2001), and fengycin (Hofemeister et al., *Molecular Genetics and Genomics*, 272, 363–378, 2004; Roongsawang et al., *International Journal of Molecular Sciences*, 12, 141–172, 2011). Each box corresponds to one domain and one enzyme activity. PKS domains: AL, acyl-CoA ligase; ACP, acyl carrier protein; AMT, amino transferase; KS, α-keto acyl synthetase; MCT, malonyl-CoA transacylase. NRPS domains: A, adenylation; C condensation; E, epimerase; PCP, peptidyl carrier protein; Te, thioesterase; Te/At, thioesterase/acyltransferase.

1997b). Such a process appears not to be adapted to *B. subtilis* S499, which produced higher lipopeptide yield in better aeration conditions (Jacques et al., 1999).

Other studies have been performed to optimize lipopeptide production. Glucose, sucrose, and fructose have been mentioned as efficient C-sources for surfactin production, while the presence of glycerol and hexadecane greatly decreased the biosynthesis of this lipopeptide (Peypoux et al., 1999; Akpa et al., 2001). Ammonium sulfate, peptones, and yeast extract can be used as N-source. The two last sources are more favorable to the lipopeptide production by *B. subtilis* S499 (Jacques et al., 1999). The presence of salts is of importance too. Studies of the mineral requirement clearly established the need and the stimulatory effect of iron and manganese for surfactin biosynthesis (Wei et al., 2004). Another important parameter is temperature. The most commonly used temperatures are 30°C and 37°C. Higher temperatures (37°C) favored surfactin synthesis of *B. subtilis* RB14 isolated from compost (Ohno et al., 1995) and ATCC 6633, but not of *B. subtilis* S499 (Jacques et al., 1999). Lower temperatures are more adapted to iturin A production.

Environmental conditions can drastically modify the ratio between surfactin and fengycin produced by a derivative strain of *B. subtilis* ATCC 21332. Cell immobilization on polypropylene carriers (Gancel et al., 2009) or on polypropylene membrane (Coutte et al., 2010) increases the biosynthesis of fengycin.

8.3.7 Potential Role of Lipopeptide in Fermented Products

Surfactin could play different roles in the food products. Indeed, its remarkable emulsifying and biosurfactant properties could, for example, impact the organoleptic properties of the products by modifying, even at low concentrations, the solubility of different hydrophobic compounds. In addition, its different biological activities in relation to health could give an added value to the fermented food products.

Iturin and fengycin can play an important role for maintaining the harmlessness of the food product by reducing fungal growth. Iturin, for example, has been reported by different studies to inhibit aflatoxin production (Ona and Kimura, 1991; Klich et al., 1993; Cho et al., 2009).

In addition, these different lipopeptides can act in a synergistic manner as suggested by several studies on surfactin with iturin (Maget-Dana et al., 1992; Razafindralambo et al., 1997), surfactin with fengycin (Ongena et al., 2007), and iturin with fengycin (Romero et al., 2007) or amplified the biocontrol activities of other antagonistic compounds. *Bacillus subtilis* is indeed also able to produce several bacteriocins such as subtilin and subtilosin and, as already mentioned, the co-production of surfactin and a bacteriocin by a strain of *B. subtilis* isolated from traditional bikalga was recently demonstrated (Compaoré et al., 2013).

8.4 Short-Chain Volatile Organic Acids and Various Other Organic Compounds

Maureen-Theodore Chinwe Ojinnaka

Alkaline, proteolytic fermented foods play an important role in the diets of many people in developing and a few developed countries. Alkaline-fermented foods (AFFs), generally regarded as safe, constitute a group of food products popularly consumed in several Asian and African countries (Steinkraus, 2002).

Organic acids are formed in food as a result of metabolism of large molecular mass compounds (Shukla et al., 2010) and play an important role in the taste and aroma of fermented foods. These have a long history of being utilized as food additives and preservatives for preventing food deterioration and extending shelf life of perishable food ingredients. Specific organic acids have been used to control microbial contamination and dissemination of food-borne pathogens in pre- and post-harvest food production and processing. Organic acids affect stability, color, and flavor of the final product, but their greatest role comes from their ability to stop or at least retard the growth of many potentially harmful microorganisms that cause spoilage in food products (Zotou et al., 2004). Biopreservation has gained increasing attention as a means of naturally controlling the shelf life and safety of food products (Olaoye and Onilude, 2009).

Organic acids are some of the most widely researched antimicrobials. These, like lactic, acetic, and citric acids, can enhance or contribute to the flavor of acidified or fermented foods such as sausages, cheeses, and pickles. The use of any antimicrobial depends on several factors such as desired effect, legal limits of use, and effect on food (Jamilah et al., 2008).

Short-chain organic acids (C_1–C_2) contribute significantly to the aroma and taste of consumer products (Clark and Bunch, 1997). Acetic and propionic acids are used by food manufacturers as antimicrobial preservatives or acidulants in a variety of food products (Davidson and Juneja, 1990; Sholberg, 1998). Short-chain organic acids lead to poor recoveries due to excessive sample volatility (Clark and Bunch, 1997). Formic acid, containing only one carbon atom, is the smallest short-chain organic acid, but due to its toxicity, it is of limited use in foods, although it is still used in many industrial processes. Very little pure acetic acid is used in foods, but as the principal component of vinegar more than 100,000 tons are produced annually in the United States for food (Gardner, 1981). It is commonly known as vinegar, which has antimicrobial capabilities due to its ability to lower the pH and cause instability of bacterial cell membranes (Jay, 1992). The use of propionic acid and propionates has been directed primarily against molds, although some yeasts and bacteria, particularly Gram-negative strains, may also be inhibited (Davidson and Juneja, 1990). Formic, acetic, and propionic acids have melting points of 8, 17, and -24°C, boiling points of 100°C, 118°C, and 141°C, and vapor pressures of 35.0, 11.8, and 2.4 mm Hg at 20°C, respectively, making them relatively volatile liquids (Sholberg, 1998). Acetic acid is volatile as well as odorous, detectable as the smell of vinegar (Zotou et al., 2004). High concentration of acetic acid can result in an undesirable vinegar flavor in fermented dairy foods (Chick et al., 2001). Higher concentration of acetic acid can result in a pungent off-flavor (Bückenhuskes, 1993). Harper and Collins (1992) reported an increased production of acids such as lactic acid, and especially acetic acid, accompanied with a decrease in pH during the production of Sudanese furundu a fermented product of roselle (*Hibiscus sabdariffa* L.) seeds.

The quantitative data for volatile and nonvolatile organic acid contents can be determined by comparing the abundances of each peak by gas chromatography (GC) and high-performance liquid chromatography (HPLC). The mean values of volatile organic acids (acetic, propionic, and butyric acids) determined using HPLC in fermented castor oil bean samples were 0.52, 0.78, and 1.31 μg/ml, respectively. The mean values of non-volatile organic acids (lactic acid, citric acid, succinic acid, and oxalic acids) were 0.87, 0.13, 0.73, and 0.05 μg/ml, respectively (Ojinnaka et al., 2013). It has been observed that 3% lime promoted the production of organic acids in Nigerian ogiri, the fermented castor (*Ricinus communis* L.) seed product, suggesting a superior preservative effect (Ojinnaka et al., 2013). The metabolism of carbohydrates in the raw seeds would have resulted in the accumulation of organic acids (Enujiugha, 2003). In almost all fermentations involving *Bacillus* spp., there is production of primary organic acids, especially butyric acid (Moat, 1979). Organic acids occur in fermented products as a result of hydrolysis, biochemical metabolism, and microbial activity (Andersson and Hedlund, 1983). The advantages of using HPLC in the analysis of organic acids are speed, accuracy, and precision (Fernandez-Garcia and McGregor, 1994).

Lactic acid, which is a weak acid, may occur in both dissociated and undissociated forms depending on the pH. It is the undissociated form of lactic acid that inhibits bacterial growth. The inhibitory capacity of this acid lies in its reduction of pH to levels below which bacteria cannot initiate growth. Lactic acid is an excellent inhibitor of spore-forming bacteria at pH 5.0 (Jamilah et al., 2008).

Citric acid is a white powder extracted from the juice of citrus and other acidic fruits such as lemons, limes, pineapples, and gooseberries. It is also produced by the fermentation of glucose. Citric acid is highly soluble in water and primarily insoluble in fat.

The production of organic acids is undoubtedly the determining factor on which the shelf life and the safety of the final product depend, while the inhibition of pathogenic and spoilage flora is also dependent on a rapid and adequate formation of these organic acids (Lindgren and Dobrogosz, 1990; Olaoye et al., 2008). The mode of action of organic acids is that the undissociated acids penetrate the cell of microorganisms by means of diffusion and then dissociate and acidify the cell interior, thus interfering with cellular metabolism or decreasing the biological activity as a result of pH changes of the cell's environment (Cherrington et al., 1991).

Some other volatile constituents have been studied in AFFs. Leejeerajumnean et al. (2001) reported that microbial fermentation of soybeans led to an enormous increase in the concentration of volatile compounds, and presumed that it could be as a result of the metabolic activities of the bacterium *Bacillus subtilis*, which produces diverse extracellular hydrolases, including amylases, several proteases, lipase, and sucrase. Pyrazines and sulfur-containing compounds have been reported to be the main contributors to the characteristic odor of Japanese natto (Kanno et al., 1984). Volatile compounds of Thai thua nao have been identified to include esters, pyrazines, carbonyl compounds, and various other compounds that contribute to the aroma of the product (Leejeerajumnean, 2003).

Aroma volatiles analyzed using Likens-Nikerson method coupled with GC-MS (mass spectrometry) have been reported by Ouoba et al. (2005) on soumbala (fermented African locust bean; *Parkia biglobosa* (Jacq.) G.Don). The volatiles identified include pyrazines, aldehydes, alkenes, amines, pyridines, benzenes, phenols, sulfur, furans, and other compounds. Leejeerajumnean et al. (2001) reported that the fermentation led to a large increase in the concentration of total volatile compounds in *Bacillus*-fermented soybeans; thua nao, while sun drying, resulted in the loss of 65% of the total volatiles. The volatile compounds identified at different stages of thua nao production, commercial thua nao, and commercial natto include alcohols, aldehydes, keetones, acids, esters, pyrazines, aromatic compounds, furans, and sulfur-containing compounds. Azokpota et al. (2008) in their study of volatile compounds of afitin, iru, and sonru (three fermented locust bean condiments from Benin) reported a total of 13 chemical groups of volatile compounds that included pyrazines and benzene derivatives. They reported pyrazines as the dominant constituent group in the three condiments. Dajanta et al. (2011b) also reported pyrazines as being the predominant volatile compound in their study of thua nao. Onyenekwe et al. (2012) reported aldehydes as the dominant volatile compound in three AFF condiments (melon [*Citrullus lanatus* (Thunb.) Matsum. & Nakai] seed ogiri, soybean dawadawa, and locust bean dawadawa). Other volatile compounds identified were alcohols, ketones, esters, pyrazines, alkanoic acids, alkanes, alkenes, and others. Onyenekwe et al. (2012) also reported that alcohols constitute 15% of the total volatiles in soybean dawadawa with undecanol as the major one, constituting about 9% of the total volatiles. The alcohols present in the three condiments help to prevent them from spoilage since alcohols are known to act as an antimicrobial and prevent food spoilage. Ojinnaka and Ojimelukwe (2013) reported a total of 17 volatile compounds in *Bacillus*-fermented castor oil seed samples. The volatile compounds, namely esters, acids, alcohol, furan, and ketone were identified in ogiri (castor oil bean condiment). They, however, reported acids as the dominant volatile constituent group in the ogiri samples.

The importance of ester contributions toward food aroma is undisputed with the fact that esters with low carbon atoms are highly volatile at ambient temperature and the perception thresholds are 10 times lower than their alcohol precursors (Izco and Torre, 2000; Nogueira et al., 2005). The esters have attractive aromas that are presumably the consequence of chemical reactions between microbial acidic and alcoholic metabolites, although the reaction might as well be catalyzed by microbial esterases. In addition to imparting a fruity floral character, esters can diminish or mask the sharpness of unpleasant free fatty acid (FFA)-derived notes. These esters are formed by esterification between the short-chain FFAs and the alcohols (Qin and Ding, 2007b). Esters, mainly formed by esterification of carboxylic acids and alcohols, were reported to determine the characteristic pleasant aromatic notes (Rodriguez-Bernaldo De Quiros et al., 2000; Klesk and Qian, 2003). Furans have been reported as aroma constituents in some Asian sauces.

Pyrazines, especially alkylpyrazines, are known to have nutty aroma (Ho and Carlin, 1989) and could be generated naturally during the aging process by the condensation of amino ketones formed through the Maillard reaction and Strecker degradation. The formation of pyrazine is associated with heating and metabolic activities of microorganisms (Owens et al., 1997), and pyrazines are the major volatile compounds in various fermented foods (Sugawara et al., 1985; Azokpota et al., 2010).

8.5 Flavor Additives

Maureen-Theodore Chinwe Ojinnaka

In most of the countries of Africa and Asia, alkaline condiments are common and form an important part of the diet (Dakwa et al., 2005). Because of their sensory characteristics and high nutritional value, these products are largely used to flavor soups and stews and constitute a significant low-cost source of protein foods in some parts of the world (Odunfa, 1988; Azokpota et al., 2006). Among these are Nigerian dawadawa, Ivory Coast soumbara, or Beninese iru, afitin, and sonru made by fermentation of soaked, cooked African locust bean (*Parkia biglobosa* (Jacq.) G.Don) using bacteria belonging to the genus *Bacillus*, specifically *B. subtilis*, Nigerian ogiri made by fermentation of melon [*Citrullus lanatus* (Thunb.) Matsumi & Nakai] seeds, Nigerian ugba made by fermentation of African oil bean (*Pentaclethra macrophylla* Benth.), Sierra Leone ogiri-saro made by fermentation of sesame (*Sesamum indicum* L.) seeds, Nigerian ogiri-igbo made by fermentation of castor (*Ricinus communis* L.) seeds, and Nigerian ogirnwan made by fermentation of fluted pumpkin (*Telfairia occidentalis* Hook.f.) seeds. Soybeans can be substituted for locust beans. This group also includes Japanese natto, Thai thua nao, and Indian kinema all based upon soybeans. In AFFs, the protein of the raw materials is broken down into peptides and amino acids. Ammonia is released during the fermentation, increasing the pH of the final product and giving the food a strong ammoniacal smell.

The preservative and flavor characteristics of alkaline fermenters are derived in part from the liberation of ammonia and increased pH (Beaumont, 2002). The fermentation of legumes changes the texture and organoleptic properties of the original legume seeds. The unpalatable beany flavor of the unfermented legume is eliminated. The fermented product has a different flavor and requires less cooking time. Most of these products are prepared by solid-state fermentation in which the substrate is allowed to ferment "spontaneously" or by adding *Bacillus* inoculum (Leejeerajumnean, 2003). The fermented products have a distinct odor. Some products such as natto or dawadawa became commercialized in the market. Dawadawa is sold in cubed packages, as are Nestlé Maggi soup cubes in the African markets. In Asia, thua nao and natto are very similar in terms of proteolytic *Bacillus* fermentation but these products have different aroma. Natto is a Japanese fermented whole soybean product that is eaten uncooked as a relish with rice, whereas Thai thua nao is generally sold as dried disks of ground fermented mass and is used as a flavoring agent in soups and curries (Leejeerajumnean, 2003).

Amino acids are known to play a major role in the taste and flavor development of foods (Kpikpi et al., 2009). Free amino acids such as glutamic and aspartic acids may contribute to pleasant umami taste or savory enhancement in foods (Yasuda et al., 1994; Katsura, 1996). The results of amino acid analysis indicated a high concentration of all amino acids especially at 96 h of fermentation of castor oil bean into ogiri (Ojinnaka and Ojimelukwe, 2013). Thus, the flavor and aroma of ogiri may be due to the production of fatty acids and amino acids, especially glutamic acid during the fermentation process. According to Anosike and Egwuatu (1980) glutamine, which is the amide of glutamic acid, is one of the compounds that contribute to the characteristic flavor of ogiri.

Proteolysis has been reported as the main metabolic activity during the fermentation of locust bean, which also contributes to the development of texture and flavor of fermented products (Ouoba et al., 2003a, 2003b; Olajuyigbe and Ajele, 2008). During soybean fermentation, protein is hydrolyzed to low-molecular-weight components such as peptides and amino acids due to the action of enzymes produced by bacteria (Kiers et al., 2000). This outcome is favorable to the development of flavor, absorbability, digestibility, and functionality of traditional Chinese fermented soybean product used as seasoning. Glutamic acid is the most important flavor-enhancing amino acid (Davids et al., 2004), and glycine and alanine give the sweet flavor (Norziah and Ching, 2000).

Volatile compounds in alkaline-fermented foods have been found to contribute to the flavor and aroma of this category of fermented foods. Various volatile constituents have been identified in castor seed ogiri using *B. subtilis* as starter culture; melon seed ogiri, soybean and locust bean dawadawa, Burkinabé soumbala, Beninese afitin, iru and sonru, Japanese natto, Korean chongkukjang, and Thai thua nao (Tanaka et al., 1998; Leejeerajumnean et al., 2000; Dakwa et al., 2005; Ouoba et al., 2005; Azokpota et al., 2008, 2010; Parkouda et al., 2011; Onyenekwe et al., 2012; Ojinnaka and Ojimelukwe, 2013). Some of the volatile compounds identified in these AFFs include esters, acids, alcohols, pyrazines, ketones, aldehydes, phenols, hydrocarbons, pyridines, pyrimidines, and sulfur-containing compounds. Since these compounds contribute to the flavor of most AFFs, they can as well be used as flavor additive. Some of the volatile compounds that contribute to the flavor of AFFs are discussed below.

8.5.1 Esters

Esters have been reported to constitute a major volatile compound in some AFFs. These form one of the major classes of compounds detected in three commercial douchies. Douchi is a traditional fermented soybean product originated in China and consumed as seasoning for food since ancient times (Wang et al., 2010). Some of the ester compounds found in the three douchi samples were phenylacetic acid methyl ester, phenylacetic acid ethyl ester, 2-phenylethylacetate, hexadecanoic acid methyl ester, hexadecanoic acid ethyl ester, 9,12,15-octadecatrienoic acid methyl ester, and ethyl 9-octadecenoate (Wang et al., 2010).

Esters are mostly formed by esterification of alcohols with fatty acids during the fermentation process (Van der Sluis et al., 2002); many of which have attractive aroma. The esters are presumably the consequence of chemical reactions between microbial acidic and alcoholic metabolites (Leejeerajumnean et al., 2000, 2001), although it is also possible that the reactions may be catalyzed by microbial esterases (Eskin, 1990).

Dajanta et al. (2011a) reported the presence of esters in thua nao. Some of the esters identified in thua nao were methyl acetate, ethy acetate, methyl 2-methylpropionate, ethyl 2-methylpropionate, methyl 2-methylbutanoate, methyl 3-methylbutanaote, ethyl 2-methylbutanoate, and ethyl 3-methylbutanoate. Ojinnaka and Ojimelukwe (2013) identified cyclohexane-carboxylic acid, 4-octyl esters at retention time of 12.29 min in *Bacillus*-fermented castor seed (ogiri).

The importance of ester contributions toward food aroma is undisputed with the fact that esters with low-carbon atoms are highly volatile at ambient temperatures, and the perception thresholds are 10 times lower than their alcohol precursors (Izco and Torre, 2000; Nogueira et al., 2005). In addition to imparting a fruity floral character, esters can diminish or mask the sharpness of unpleasant FFA-derived notes. These esters are formed by esterification between the short-chain FFAs and the alcohols (Qin and Ding, 2007a).

8.5.2 Acids

Ojinnaka and Ojimelukwe (2013) reported short-chain organic acids as the predominant volatile compound in fermented castor oil bean (ogiri) using *B. subtilis* as starter culture. They reported major carboxylic acids in one of their samples as butanoic acid, 2-methyl-oleic acid, *n*-hexadecanoic acid, and ricinoleic acid. Butanoic acid and

3-methylbutanoic acid, compounds with a sweaty odor, have been determined as major aroma compounds in Korean soy sauces and barley bran sauces (Choi et al., 2007; Steinhaus and Schieberle, 2007).

Acids identified in douchi extracts, 3-methylbutanoic acid, tetradecanoic acid, and *n*-hexadecanoic acid, were found in all the three brands of douchi samples (Wang et al., 2010). Some acids found in douchi have been reported in other soy products. Acetic acid, propanoic acid, 2-methylpropanoic acid, and 2-methlybutanoic acid have been reported in thua nao (Dajanta et al., 2011a). Wittanalai et al. (2011) also reported the presence of acids in kapi, a traditional fermented shrimp paste used as food condiment in Thailand. Some of the acid compounds reported were acetic acid, propanoic acid, 2-methylpropanoic acid, butanoic acid, 2-methylbutanoic acid, pentanoic acid, and 2-methylpentanoic acid. 2-Methylbutanoic acid is the major flavor component of natto (Leejeerajumnean et al., 2001).

8.5.3 Pyrazines

Azokpota et al. (2008) identified pyrazines as the dominant constituent group in afitin, iru, and sonru. Dajanta et al. (2011a) also reported pyrazines as the dominant volatile compound in thua nao. Wittanalai et al. (2011) reported the presence of pyrazines in kapi. Pyrazines were found to be among the volatile compounds that were the major compounds in the N-containing group and have been reported to be good precursors of volatile compounds (Allagheny et al., 1996).

The formation of pyrazines is associated with heating and metabolic activities of microorganisms (Owens et al., 1997). Pyrazines have been reported to be the major volatile compounds in various fermented foods (Sugawara et al., 1985; Azokpota et al., 2010). Pyrazines produced during the fermentation of soybeans into natto are the main contributors to the characteristic odor of natto (Sugawara et al., 1985). Tetramethylpyrazine, trimethylpyrazine, and 2,5-dimethylpyrazine are the major volatile compounds present in highest amounts in natto. The preparation of soybean paste involves several processes that can cause the formation of pyrazines mainly from amino acids by heating (Mori et al., 1983). The pyrazines, which are important contributors to the characteristic odor of douchi, could be generated naturally during the aging process by the condensation of amino-ketones formed through the Maillard reaction and Strecker degradation (Chung, 1999). Pyrazines are typical aroma components of heated food to which they give a characteristic roasty or nutty flavor. Metabolic activities of microorganisms can generate various precursors such as amino acids, monosaccharides, and ammonia that are converted to pyrazines (Lasekan et al., 2001). Formation of pyrazines is generally associated with non-enzymatic reaction as heating (Owens et al., 1997).

8.5.4 Aldehydes

Onyenekwe et al. (2012) reported aldehydes as the dominant volatile compounds in three fermented food flavor enhancers: melon seed ogiri, soybean dawadawa, and locust bean dawadawa. Seven aldehydes were identified in fermented soybean paste samples too (Lee and Ahn, 2009), although all the compounds have been previously detected in various fermented soybean foods (Ku et al., 2000; Park et al., 2003). Aldehydes are considered to enhance flavor quality, which is mainly associated with sweat fruity, nutty, and caramel-like odor (Fors, 1983).

8.5.5 Furan

Furans have been reported as aroma constituents in some Asian sauces. *Bacillus*-fermented castor bean (ogiri) samples showed the presence of dihydro-3-hydroxy-4,4-dimethyl-2(3*H*)-furanone. The compounds 4-HDMF (4-hydroxy-2,5-dimethyl-3(2*H*)-furanone) and 4-HEMF (4-hydroxy-2-ethyl-5-methyl-3(2*H*)-furanone) have been suggested as the two key aroma constituents in Korean soy sauces, Japanese soy sauces, and Japanese barley miso (Baek and Kim, 2004; Steinhaus and Schieberle, 2007).

8.5.6 Ketones

Ketones could be generated by fungal enzymatic actions on lipids and/or amino acids or by the Maillard reaction (Su, 1986). Nine ketones were identified in fermented soybean pastes. Although the flavor notes of ketones are generally desirable, their aroma contributions might have been minimal since low levels of these compounds were quantified in fermented soybean paste (Lee and Ahn, 2009).

Some ketones found in douchi have also been reported in other soy products. Thirteen ketones were detected in douchi extracts and only 1-(2-furanyl)ethanone was detected in all three brands. The volatile ketones were most likely the product of lipid and/or amino acid degradation, and several were reported among the volatile compounds in a fermenting room of natto (Tanaka and Shoji, 1993).

8.5.7 Sulfur-Containing Compounds

Sulfur-containing compounds have a great influence on product odor although they exist at very low concentrations (Dajanta et al., 2011a). Sugawara et al. (1985) listed five sulfur-containing compounds as contributing to cooked soybean odor in beans cooked for 5–8 h. Tanaka et al. (1998) revealed that homemade natto had higher amounts of sulfur-containing compounds than the commercial natto, owing to its milder cooking condition and the presence of different microbial strains. However, most volatiles increase after fermentation. Sugawara et al. (1985) attributed the odor of natto to nine pyrazine derivatives and four sulfur-containing compounds. The sulfur compounds, 4-ethyl-2-methyl-thiazole, 3,5-dimethyl-1,2,4-trithiolane, (*Z*)-3,5-dimethyl-1,2,4-trithiolane, and thialdine identified by Sugawara et al. (1985) as contributing to natto odor, were the products of cooking but were not detected in the fermented soybeans.

References

Abdel-Fattah, Y.R., El-Enshasy, H.A., Soliman, N.A., and El-Gendi, H. 2009. Bioprocess development for production of alkaline protease by *Bacillus pseudofirmus* Mn6 through statistical experimental designs. *Journal of Microbiology and Biotechnology* 19: 378–386.

Ackermann, H., Azizbekyan, R., Bernier, R., de Barje, H., Saindouk, S., Valéro, J., and Yu, M. 1995. Phage typing of *Bacillus subtilis* and *B. thuringensis*. *Research in Microbiology* 146: 643–657.

Aderibigbe, E.Y., Schink, B., and Odunfa, S.A. 1990. Extracellular proteinases of *Bacillus* sp. isolated from fermented African locust bean "iru". *Food Microbiology* 7: 281–294.

Ageitos, J.M., Vallejo, J.A., Sestelo, A.B.F., Poza, M., and Villa, T.G. 2007. Purification and characterization of a milk-clotting protease from *Bacillus licheniformis* strain USC13. *Journal of Applied Microbiology* 103: 2205–2213.

Ahimou, F., Jacques, P., and Deleu, M. 2000. Surfactin and iturin A effects on *Bacillus subtilis* surface hydrophobicity. *Enzyme and Microbial Technology* 27: 749–754.

Akimoto, T., Yamada, S., and Matsumoto, I. 1990. The relation between protease, and γ-glutamyltranspeptidase activities and qualities of natto. *Nippon Shokuhin Kogyo Gakkaishi* 37: 872–877 (in Japanese).

Akolkar, A.V., Durai, D., and Desai, A.J. 2009. *Halobacterium* sp. SP1(1) as a starter culture for accelerating fish sauce fermentation. *Journal of Applied Microbiology* 109: 44–53.

Akpa, E., Jacques, P., Wathelet, B., Paquot, M., Fuchs, R., Budzikiewicz, H., and Thonart, P. 2001. Influence of culture conditions on lipopeptide production by *Bacillus subtilis*. *Applied Biochemistry and Biotechnology* 91–93: 551–561.

Allagheny, N., Obanu, Z.A., Campbell-Platt, G., and Owens, J.D. 1996. Control of ammonia formation during *Bacillus subtilis* fermentation of legumes. *International Journal of Food Microbiology* 29: 321–333.

Amoozegar, M.A., Schumann, P., Hajighasemi, M., and Fatemi, A.Z. 2008. *Salinivibrio proteolyticus* sp. nov., a moderately halophilic and proteolytic species from a hypersaline lake in Iran. *International Journal of Systematic and Evolutionary Microbiology* 58: 1159–1163.

Anbu, P., Annadurai, G., Lee, J.F., and Hur, B.K. 2009. Optimization of alkaline protease production from *Shewanella oneidensis* MR-1 by response surface methodology. *Journal of Chemical Technology and Biotechnology* 84: 54–62.

Andersson, R., and Hedlund, B. 1983. HPLC analysis of organic acids in lactic acid fermented vegetables. *Zeitschrift für Lebensmittel-Untersuchung und-Forschung* 176: 440–443.

Anosike, E.O., and Egwuatu, C.K. 1980. Biochemical changes during the fermentation of castor oil (*Ricinus communis*) seeds for use as a seasoning agent. *Plant Foods for Human Nutrition* 30: 181–185.

Aranda, F.J., Teruel, J.A., and Ortiz, A. 2005. Further aspects on the hemolytic activity of the antibiotic lipopeptide iturin A. *Biochimica et Biophysica Acta* 1713: 51–56.

Arnold, F.H. 1996. Directed evolution: Creating biocatalysts for the future. *Chemical Engineering Sciences* 51: 5091–5102.

Ashie, I.N.A., Sorensen, T.L., and Nielsen, P.M. 2002. Effect of papain and a microbial enzyme on meat proteins and beef tenderness. *Journal of Food Science* 67: 2138–2142.

Ashiuchi, M., Soda, K., and Misono, H. 1999. A poly-γ-glutamate synthetic system of *Bacillus subtilis* IFO 3336: Gene cloning and biochemical analysis of poly-γ-glutamate produced by *Escherichia coli* clone cells. *Biochemical and Biophysical Research Communications* 263: 6–12.

Aumayr, A., Hara, T., and Ueda, S. 1981. Transformation of *Bacillus subtilis* in polyglutamate production by deoxyribonucleic acid from *B. natto*. *Journal of General and Applied Microbiology* 27: 115–123.

Azokpota, P., Hounhouigan, D.J., and Nago, M.C. 2006. Microbiological and chemical changes during the fermentation of African locust bean (*Parkia biglobosa*) to produce afitin, iru and sonru, three traditional condiments produced in Benin. *International Journal of Food Microbiology* 107: 304–309.

Azokpota, P., Hounhouigan, J.D., Annan, N.T., Nago, M.C., and Jakobsen, M. 2008. Diversity of volatile compounds of afitin, iru and sonru, three fermented food condiments from Benin. *World Journal of Microbiology and Biotechnology* 24: 879–885.

Azokpota, P., Hounhouigan, J.D., Annan, N.T., Odjo, T., Nago, M.C., and Jakobsen, M. 2010. Volatile compounds profile and sensory evaluation of Beninese condiments produced by inocula of *Bacillus subtilis*. *Journal of the Science of Food and Agriculture* 90: 438–444.

Baek, H.H., and Kim, H.J. 2004. Solid phase microextraction–Gas chromatography–olfactometry of soy sauce based on sample dilution analysis. *Food Science and Biotechnology* 13: 90–95.

Baghel, V.S., Tripathi, R.D., Ramteke, P.W., Gopal, K., Dwivedi, S., Jain, R.K., Rai, U.N., and Singh, S.N. 2005. Psychrotrophic proteolytic bacteria from cold environment of Gangotri glacier, Western Himalaya, India. *Enzyme and Microbial Technology* 36: 654–659.

Bajaj, B.K., and Sharma, P. 2011. An alkali-thermotolerant extracellular protease from a newly isolated *Streptomyces* sp. DP2. *New Biotechnology* 28: 725–732.

Bakhtiar, S., Andersson, M., Gessesse, A., Mattiasson, B., and Kaul, R. 2002. Stability characteristics of a calcium-independent alkaline protease from *Nesterenkonia* sp. *Enzyme and Microbial Technology* 32: 525–531.

Banerjee, R., Agnihotri, R., and Bhattacharyya, B.C. 1993. Purification of alkaline protease of *Rhizopus oryzae* by foam fractionation. *Bioprocess Engineering* 9: 245–248.

Beaumont, M. 2002. Flavouring composition prepared by fermentation with *Bacillus* spp. *International Journal of Food Microbiology* 75: 189–196.

Béchet, M., Caradec, T., Hussein., W., Abderrahmani, A., Chollet, M., Leclère, V., Dubois, T., Lereclus, D., Pupin, M., and Jacques, P. 2012. Structure, biosynthesis, and properties of kurstakins, nonribosomal lipopeptides from *Bacillus* spp. *Applied Microbiology and Biotechnology* 95: 593–600.

Béchet, M., Guy-Castera, J., Guez, J.S., Chihib, N., Coucheney, F., Coutte, F., Fickers, P., Leclère, V., Wathelet, B., and Jacques, P. 2013. Production of a novel mixture of mycosubtilins by mutants of *Bacillus subtilis*. *Bioresource Technology* 145: 264–270.

Beg, Q.K., Sahai, V., and Gupta, R. 2003. Statistical media optimization and alkaline protease production from *Bacillus mojavensis* in a bioreactor. *Process Biochemistry* 39: 203–209.

Blann, A.D., Landray, M.J., and Lip, G.Y. 2002. An overview of antithrombotic therapy. *British Medical Journal* 325: 762–765.

Blumentals, I.I., Robinson, A.S., and Kelly, R.M. 1990. Characterization of sodium dodecyl sulfate-resistant proteolytic activity in the hyperthermophilic archaebacterium *Pyrococcus furiosus*. *Applied and Environmental Microbiology* 56: 1255–1262.

Bonmatin, J.M., Labbe, H., Grangemard, I., Peypoux, F., Maget-Dana, R., Ptak, M., and Michel, G. 1995. Production, isolation and characterization of [Leu(4)]surfactins and [Ile(4)]surfactins from *Bacillus subtilis*. *Letters in Peptide Science* 2: 41–47.

Bonmatin, J.M., Laprévote, O., and Peypoux, F. 2003. Diversity among microbial cyclic lipopeptides: Iturins and surfactins. Activity-structure relationships to design new bioactive agents. *Combinatorial Chemistry and High Throughput Screening* 6: 541–556.

Bovarnick, M. 1942. The formation of extracellular d(−)-glutamic acid polypeptiede by *Bacillus subtilis*. *Journal of Biological Chemistry* 145: 415–424.

Bruna, J.A., Fernandez, M., Hierro, E.M., De La Hoz, L., and Ordonez, J.A. 1999. Effect of the combined use of Pronase E and a fungal extract (*Mucor racemosus* forma *sphaerosporus*) on the ripening of dry fermented sausages. *Food Science and Technology International* 5: 327–337.

Bruna, J.A., Fernandez, M., Hierro, E.M., Ordonez, J.A., and De La Hoz, L. 2000. Combined use of Pronase E and a fungal extract (*Penicillium aurantiogriseum*) to potentiate the sensory characteristics of dry fermented sausages. *Meat Science* 54: 135–145.

Bryan, P.N. 2000. Protein engineering of subtilisin. *Biochimica et Biophysica Acta* 1543: 203–222.

Bückenhuskes, H.J. 1993. Selection criteria for lactic acid bacteria to be used as starter cultures for various food commodities. *FEMS Microbiology Reviews* 12: 253–272.

Caboche, S., Leclère, V., Pupin, M., Kucherov, G., and Jacques, P. 2010. Diversity of monomers in nonribosomal peptides: Towards the prediction of origin and biological activity. *Journal of Bacteriology* 192: 5143–5150.

Caboche, S., Pupin, M., Leclère, V., Fontaine, A., Jacques, P., and Kucherov, G. 2008. NORINE: A database of nonribosomal peptides. *Nucleic Acids Research* 36: D326–D331.

Cai, Y., Bao, W., Jiang, S., Weng, M., Jia, Y., Yin, Y., Zheng, Z., and Zou, G. 2011. Directed evolution improves the fibrinolytic activity of nattokinase from *Bacillus natto*. *FEMS Microbiology Letters* 325: 155–161.

Cao, X.H., Wang, A.H., Wang, C.L., Mao, D.Z., Lu, M.F., Cui, Y.Q., and Jiao, R.Z. 2010. Surfactin induces apoptosis in human breast cancer MCF-7 cells through a ROS/JNK-mediated mitochondrial/caspase pathway. *Chemico-Biological Interactions* 183: 357–362.

Cavalcanti, M.T.H., Martinez, C.R., Furtaso, V.C., Neto, B.B., Teixeira, M.F.S., Lima Filho, J.L., and Porto, A.L.F. 2005. Milk-clotting protease production by *Nocardiopsis* sp. in an inexpensive medium. *World Journal of Microbiology and Biotechnology* 21: 151–154.

Cavicchioli, R., Siddiqui, K.S., Andrews, D., and Sowers, K.R. 2002. Low-temperature extremophiles and their applications. *Current Opinion in Biotechnology* 13: 253–261.

Chabeaud, A., Vandanjon, L., Bourseau, P., Jaouen, P., and Chaplain-Derouiniot, M. 2009. Performances of ultrafiltration membranes for fractionating a fish protein hydrolysate: Application to the refining of bioactive peptidic fractions. *Separation and Purification Technology* 66: 463–471.

Chakrabarti, S.K., Matsumura, N., and Ranu, R.S. 2000. Purification and characterization of an extracellular alkaline serine protease from *Aspergillus terreus* (IJIRA 6.2). *Current Microbiology* 40: 239–244.

Chandrasekaran, S., and Dhar, S.C. 1983. A low cost method for the production of extracellular alkaline proteinase using tapioca starch. *Journal of Fermentation Technology* 61: 511–514.

Chang, C.T., Fan, M.H., Kuo, F.C., and Sung, H.Y. 2000. Potent fibrinolytic enzyme from a mutant of *Bacillus subtilis* IMR-NK1. *Journal of Agricultural and Food Chemistry* 48: 3210–3216.

Chauhan, B., and Gupta, R. 2004. Application of statistical experimental design for optimization of alkaline protease production from *Bacillus* sp. RGR-14. *Process Biochemistry* 39: 2115–2122.

Chellappan, S., Jasmin, C., Basheer, S.M., Elyas, K.K., Bhat, S.G., and Chandrasekaran, M. 2006. Production, purification and partial characterization of a novel protease from marine *Engyodontium album* BTMFS10 under solid state fermentation. *Process Biochemistry* 41: 956–961.

Chen, K., and Arnold, F.H. 1993. Tuning the activity of an enzyme for unusual environments: Sequential random mutagenesis of subtilisin E for catalysis in dimethyl formamide. *Proceedings of the National Academy of Sciences of the United States of America* 90: 5618–5622.

Chen, Q.H., He, G.Q., Jiao, Y.C., and Ni, H. 2006. Effects of elastase from a *Bacillus* strain on the tenderization of beef meat. *Food Chemistry* 98: 624–629.

Cherrington, C.A., Hinton, M., Person, G.R., and Chopra, I.J. 1991. Organic acids as microorganisms decontaminators. *Journal of Applied Bacteriology* 70: 156–172.

Chick, H., Shin, H.S., and Ustunol, Z. 2001. Growth and acid production by lactic acid bacteria and bifidobacteria growth in skim milk containing honey. *Journal of Food Science* 66: 478–481.

Chitte, R.R., and Dey, S. 2002. Production of a fibrinolytic enzyme by thermophilic *Streptomyces* species. *World Journal of Microbiology and Biotechnology* 18: 289–294.

Cho, K.M., Math, R.K., Hong, S.Y., Islam, S.M.A., Mandanna, D.K., Cho, J.J., Yun, M.G., Kim, J.M., and Yun, H.D. 2009. Iturin produced by *Bacillus pumilus* HY1 from Korean soybean sauce (kanjang) inhibits growth of aflatoxin producing fungi. *Food Control* 20: 402–406.

Cho, S.J., Oh, S.H., Pridmore, R.D., Juillerat, M.A., and Lee, C.H. 2003. Purification and characterization of proteases from *Bacillus amyloliquefaciens* isolated from traditional soybean fermentation starter. *Journal of Agricultural and Food Chemistry* 51: 7664–7670.

Choi, U.K., Kim, M.H., Kwon, O.J., Lee, T.J., and Lee, N.H. 2007. Characterization of aroma components in barley bran sauce using statistical analysis. *Food Science and Biotechnology* 16: 23–28.

Chollet-Imbert, M., Gancel, F., Slomianny, C., and Jacques, P. 2009. Differentiated pellicle organization and lipopeptide production in standing culture of *Bacillus subtilis* strains. *Archives of Microbiology* 191: 63–71.

Chung, H.-Y. 1999. Volatile components in fermented soybean (*Glycine max*) curds. *Journal of Agricultural and Food Chemistry* 47: 2690–2696.

Chunhachart, O., Hanayama, T., Hidesaki, M., Tanimoto, H., and Tahara, Y. 2006. Structure of the hydrolyzed product (F-2) released from γ-polyglutamic acid by γ-glutamyl hydrolase YwtD of *Bacillus subtilis*. *Bioscience, Biotechnology, and Biochemistry* 70: 2289–2291.

Clark, T.J., and Bunch, J.E. 1997. Determination of volatile acids in tobacco, tea, and coffee using derivatization–purge and trap gas chromatography–selected ion monitoring mass spectrometry. *Journal of Chromatographic Science* 35: 206–208.

Clement, J. 2011. *Composition for preparing improved gluten-free or gluten-reduced bakery products.* U.S. Patent 0117246A1. Wayzata, MN: Cargill.

Clemente, A. 2000. Enzymatic protein hydrolysates in human nutrition. *Trends in Food Science and Technology* 11: 254–262.

Coll, B.A., Garcia, R.A., and Marmer, W.N. 2007. Diffusion of protease into meat and bone meal for solubility improvement and potential inactivation of the BSE prion. *PLoS ONE* 2: e245.

Compaoré, C.S., Nielsen, D.S., Ouoba, L.I.I., Berner, T.S., Nielsen, K.F., Sawadogo-Lingani, H., Diawara, B., Ouédraogo, G.A., Jakobsen, M., and Thorsen, L. 2013. Co-production of surfactin and a novel bacteriocin by *Bacillus subtilis* subsp. *subtilis* H4 isolated from Bikalga, an African alkaline *Hibiscus sabdariffa* seed fermented condiment. *International Journal of Food Microbiology* 162: 297–307.

Coolbear, T., Monk, C., Peek, K., Morgan, H.W., and Daniel, R.M. 1992. Laboratory-scale investigations into the use of extremely thermophilic proteinases for cleaning ultrafiltration membranes fouled during whey processing. *Journal of Membrane Science* 67: 93–101.

Cooper, D.G., Macdonald, C.R., Duff, S.J., and Kosaric, N. 1981. Enhanced production of surfactin from *Bacillus subtilis* by continuous product removal and metal cation additions. *Applied and Environmental Microbiology* 42: 408–412.

Coutte, F., Lecouturier, D., Ait Yahia, S., Leclère, V., Béchet, M., Jacques, P., and Dhulster, P. 2010. Production of surfactin and fengycin by *Bacillus subtilis* in a bubbleless membrane bioreactor. *Applied Microbiology and Biotechnology* 87: 499–507.

Cowan, D.A., and Daniel, R.M. 1996. Rapid purification of two thermophilic proteinases using dye-ligand chromatography. *Journal of Biochemical and Biophysical Methods* 31: 31–37.

Cui, L., Dong, M.S., Chen, X.H., Jiang, M., Lv, X., and Yan, G. 2008. A novel fibrinolytic enzyme from *Cordyceps militaris*, a Chinese traditional medicinal mushroom. *World Journal of Microbiology and Biotechnology* 24: 483–489.

Cunningham, B.C., and Wells, A.W. 1987. Improvement in the alkaline stability of subtilisin using an efficient random mutagenesis and screening procedure. *Protein Bioengineering* 1: 319–325.

Dajanta, K., Apichartsrangkoon, A., Chukeatirote, E., and Frazier, R.A. 2011a. Free-amino acid profiles of thua nao, a Thai fermented soybean. *Food Chemistry* 125: 342–347.

Dajanta, K., Chukeatirote, E., and Apichartsrangkoon, A. 2011b. Analysis and characterisation of amino acid contents of thua nao, a traditionally fermented soybean food of Northern Thailand. *International Food Research Journal* 18: 588–592.

Dakwa, S., Sakyi Dawson, E., Diako, C., Annan, N.T., and Amoa Awua, W.K. 2005. Effect of boiling and roasting on the fermentation of soybeans into dawadawa (soy-dawadawa). *International Journal of Food Microbiology* 104: 69–82.

Dalgleish, D.G. 1992. The enzymatic coagulation of milk. In *Advanced Dairy Chemistry – 1: Proteins*, ed. P.F. Fox, pp. 579–619. London: Elsevier Applied Science.

Davids, S.J., Yaylayan, V.A., and Turcotte, G. 2004. Use of unusual storage temperatures to improve the amino acid profile of potatoes for novel flavoring applications. *LWT – Food Science and Technology* 37: 619–626.

Davidson, P.M., and Juneja, V.K. 1990. Antimicrobial agents. In *Food Additives*, eds. A.L. Branen, P.M. Davidson, and S. Salminen, pp. 83–137. New York: Marcel Dekker.

Debois, D., Kassem, H., Guérineau, V., Le Caër, J.P., Holland, I.B., Lopes, P., Ouazzani, J., Séror, S.J., Brunelle, A., and Laprévote, O. 2008. In situ localisation and quantification of surfactins in a *Bacillus subtilis* swarming community by imaging mass spectrometry. *Proteomics* 18: 3682–3691.

Deepak, V., Kalishwaralal, K., Ramkumarpandian, S., Babu, S.V., Senthilkumar, S.R., and Sangiliyandi, G. 2008. Optimization of media composition for nattokinase production by *Bacillus subtilis* using response surface methodology. *Bioresource Technology* 99: 8170–8174.

Deleu, M., Razafindralambo, H., Popineau, Y., Jacques, P., Thonart, P., and Paquot, M. 1999. Interfacial and emulsifying properties of lipopeptides from *Bacillus subtilis*. *Colloids and Surfaces A* 152: 3–10.

Deu, E., Verdoes, M., and Bogyo, M. 2012. New approaches for dissecting protease functions to improve probe development and drug discovery. *Nature Structural and Molecular Biology* 19: 9–16.

Du, L., Sanchez, C., and Shen, B. 2001. Hybrid peptide-polyketide natural products: Biosynthesis and prospects toward engineering novel molecules. *Metabolic Engineering* 3: 78–95.

Duitman, E.H., Wyczawski, D., Boven, L.G., Venema, G., Kuipers, O.P., and Hamoen, L.W. 2007. Novel methods for genetic transformation of natural *Bacillus subtilis* isolates used to study the regulation of the mycosubtilin and surfactin synthetases. *Applied and Environmental Microbiology* 73: 3490–3496.

Elibol, M., and Moreira, A.R. 2005. Optimizing some factors affecting alkaline protease production by a marine bacterium *Teredinobacter turnirae* under solid substrate fermentation. *Process Biochemistry* 40: 1951–1956.

Emtage, J.S., Angal, S., Doel, M.T., Harris, T.J.R., Jenkins, B., Lilley, G., and Lowe, P.A. 1983. Synthesis of calf prochymosin (prorennin) in *Escherichia coli*. *Proceedings of the National Academy of Sciences of the United States of America* 80: 3671–3675.

Enujiugha, V.N. 2003. Nutrient changes during the fermentation of African oil bean (*Pentaclethra macrophylla* Benth) seeds. *Pakistan Journal of Nutrition* 2: 320–323.

Eskin, N.A.M. 1990. *Biochemistry of Foods*, 2nd ed. San Diego, CA: Academic Press.

Fazouane-Naimi, F., Mechakra, A., Abdellaoui, R., Nouani, A., Daga, S.M., Alzouma, A.M., Gais, S., and Penninckx, M.J. 2010. Characterization and cheese-making properties of rennet-like enzyme produced by a local Algerian isolate of *Aspergillus niger*. *Food Biotechnology* 24: 258–269.

Fernandez-Garcia, E., and McGregor, J.U. 1994. Determination of organic acids during the fermentation and cold storage of yoghurt. *Journal of Dairy Science* 11: 2934–2939.

Ferreira, L., Gil, M.H., Cabrita, A.M.S., and Dordick, J.S. 2005. Biocatalytic synthesis of highly ordered degradable dextran-based hydrogels. *Biomaterials* 26: 4707–4716.

FitzGerald, R.J., and O'Cuinn, G. 2006. Enzymatic debittering of food protein hydrolysates. *Biotechnology Advances* 24: 234–237.

Fors, S. 1983. Sensory properties of volatile Maillard reaction products and related compounds. In *The Maillard Reaction in Foods and Nutrition*, eds. G.R. Waller, and M.S. Feather, pp. 185–286. Washington, DC: American Chemical Society.

Fujii, H. 1963a. On the formation of mucilage by *Bacillus natto*. Part III. Chemical constituents of mucilage in natto. (1). *Journal of the Agricultural Chemical Society of Japan* 37: 407–411 (in Japanese).

Fujii, H. 1963b. On the formation of mucilage by *Bacillus natto*. Part IV. Chemical constituents of mucilage in natto. (2). *Journal of the Agricultural Chemical Society of Japan* 37: 474–477 (in Japanese).

Fujii, H. 1963c. On the formation of mucilage by *Bacillus natto*. Part V. Effects of some cultural conditions on the chemical constituents of mucilage. *Journal of the Agricultural Chemical Society of Japan* 37: 615–618 (in Japanese).

Fujii, H., Oki, M., Makihara, M., Keshino, J., and Takeya, R. 1967. On the formation of mucilage by *Bacillus natto*; Part VII. Isolation and characterization of a bacteriophage active against "natto"-producing bacteria. *Journal of the Agricultural Chemical Society of Japan* 41: 39–43 (in Japanese).

Fujii, H., and Shiraishi, A. 1987. On γ-glutamyltranspeptidase of *Bacillus subtilis*. *Bulletin of the Faculty of Home Life Science, Fukuoka Women's University* 8: 1–5 (in Japanese).

Fujimoto, Z., Shiga, I., Itoh, Y., and Kimura, K. 2009. Crystallization and preliminary crystallographic analysis of poly-γ-glutamate hydrolase from bacteriophage ΦNIT1. *Acta Crystallographica Section F: Structural Biology and Crystallization Communications* 65: 913–916.

Fung, M.C., and Fung, K.Y.M. 1991. PCR amplification of mRNA directly from a crude cell lysate prepared by thermophilic protease digestion. *Nucleic Acids Research* 19: 4300.

Gancel, F., Montastruc, L., Liu, T., Zhao, L., and Nikov, I. 2009. Lipopeptide overproduction by cell immobilization on iron-enriched light polymer particles. *Process Biochemistry* 44: 975–978.

Gardner, W.H. 1981. Acidulants in food processing. In *CRC Handbook of Food Additives*, ed. T.E. Furia, pp. 225–270. Boca Raton, FL: CRC Press.

Gaur, S., Agrahari, S., and Wadhwa, N. 2010. Purification of protease from *Pseudomonas thermaerum* GW1 isolated from poultry waste site. *The Open Microbiology Journal* 4: 67–74.

Gessesse, A. 1997. The use of nug meal as a low-cost substrate for the production of alkaline protease by the alkaliphilic *Bacillus* sp. AR-009 and some properties of the enzyme. *Bioresource Technology* 62: 59–61.

Gildberg, A. 2004. Enzymes and bioactive peptides from fish waste related to fish silage, fish feed and fish sauce production. *Journal of Aquatic Food Product Technology* 13: 3–11.

Goosen, M.F.A., Sablani, S.S., Al-Hinai, H., Al-Obeidani, S., Al-Belushi, R., and Jackson, D. 2004. Fouling of reverse osmosis and ultrafiltration membranes: A critical review. *Separation Science and Technology* 39: 2261–2298.

Grandison, A.S., and Lewis, M.J. 1996. *Separation Processes in the Food and Biotechnology Industries: Principles and Applications*. Cambridge: Woodhead Publishing.

Grant, W.D., Mwatha, W.E., and Jones, B.E. 1990. Alkaliphiles: Ecology, diversity and applications. *FEMS Microbiology Reviews* 75: 255–270.

Grau, A., Gómez-Fernández, J.C., Peypoux, F., and Ortiz, A. 2001. Aggregational behavior of aqueous dispersions of the antifungal lipopeptide iturin A. *Peptides* 22: 1–5.

Gupta, R., Beg, Q.K., and Lorenz, P. 2002. Bacterial alkaline proteases: Molecular approaches and industrial applications. *Applied Microbiology and Biotechnology* 59: 15–32.

Haddar, A., Fakhfakh-Zouari, N., Hmidet, N., Frikha, F., Nasri, M., and Kamoun, A.S. 2010. Low-cost fermentation medium for alkaline protease production by *Bacillus mojavensis* A21 using hulled grain of wheat and sardinella peptone. *Journal of Bioscience and Bioengineering* 110: 288–294.

Haddar, A., Hmidet, N., Ghorbel-Bellaaj, O., Fakhfakh-Zouari, N., Sellami-Kamoun, A., and Nasri, M. 2011. Alkaline proteases produced by *Bacillus licheniformis* RP1 grown on shrimp wastes: Application in chitin extraction, chicken feather-degradation and as a dehairing agent. *Biotechnology and Bioprocess Engineering* 16: 669–678.

Hajji, M., Rebai, A., Gharsallah, N., and Nasri, M. 2008. Optimization of alkaline protease production by *Aspergillus clavatus* ES1 in *Mirabilis jalapa* tuber powder using statistical experimental design. *Applied Microbiology and Biotechnology* 79: 915–923.

Hamoen, L.W., Venema, G., and Kuipers, O.P. 2003. Controlling competence in *Bacillus subtilis*: Shared use of regulators. *Microbiology* 149: 9–17.

Han, X.Q., and Damodaran, S. 1998. Purification and characterization of protease Q: A detergent- and urea-stable serine endopeptidase from *Bacillus pumilus*. *Journal of Agricultural and Food Chemistry* 46: 3596–3603.

Hara, T. 1997. "Desert greening" with natto. *Green Age* 11: 8–9.
Hara, T., Aumayr, A., Fujio, Y., and Ueda, S. 1982a. Elimination of plasmid-linked polyglutamate production by *Bacillus subtilis* (*natto*) with acridine orange. *Applied and Environmental Microbiology* 44: 1456–1458.
Hara, T., Aumayr, A., and Ueda, S. 1981. Characterization of plasmid deoxyribonucleic acid in *Bacillus natto*: Evidence for plasmid-linked PGA production. *Journal of General and Applied Microbiology* 27: 299–305.
Hara, T., Fujio, Y., and Ueda, S. 1982b. Polyglutamate production by *Bacillus subtilis* (*natto*). *Journal of Applied Biochemistry* 4: 112–120.
Hara, T., Fujio, Y., and Ueda, S. 1985. *Bacillus subtilis* (*natto*) plasmid responsible for polyglutamate production encoding γ-glutamyltranspeptidase. *Journal of the Faculty of Agriculture Kyushu University* 30: 95–105.
Hara, T., Nagatomo, S., Iwamoto, N., Tadokoro, Y., Kaneko, S., Orima, M., and Ogata, S. 1994. Cloning and nucleotide sequence of γ-polyglutamate production stimulating factor on *Bacillus subtilis* (*natto*) plasmid, pUH1. *Journal of the Faculty of Agriculture Kyushu University* 39: 43–51.
Hara, T., Nagatomo, S., Ogata, S., and Ueda, S. 1991. Molecular structure of the replication origin of a *Bacillus subtilis* (*natto*) plasmid, pUH1. *Applied and Environmental Microbiology* 57: 1838–1841.
Hara, T., Shiraishi, A., Fujii, H., and Ueda, S. 1984. Specific host range of *Bacillus subtilis* (*natto*) phages associated with polyglutamate production. *Agricultural and Biological Chemistry* 48: 2373–2374.
Hara, T., and Ueda, S. 1982. Regulation of polyglutamate production in *Bacillus subtilis* (*natto*): Transformation of high PGA productivity. *Agricultural and Biological Chemistry* 46: 2275–2281.
Hara, T., Yasuda, S., Ogata, S., Soeda, E., and Ueda, S. 1992. Complete nucleotide sequence of *Bacillus subtilis* (*natto*) plasmid responsible for γ-polyglutamate synthesis. *Journal of the Faculty of Agriculture Kyushu University* 36: 209–218.
Harper, D.B., and Collins, M.A. 1992. Leaf and seed fermentations of Western Sudan. In *Applications of Biotechnology in Traditional Fermented Foods*, eds. E.L. Gaden, M. Bokanga, S. Harlander, C.W. Hesseltine, and K.H. Steinkraus, pp. 105–113. Washington, DC: National Academic Press.
Hashem, A.M. 2000. Purification and properties of a milk clotting enzyme produced by *Penicillum oxalicum*. *Bioresource Technology* 75: 219–222.
Hayashi, K., Kensuke, T., Kobayashi, K., Ogasawara, N., and Ogura, M. 2006. *Bacillus subtilis RghR* (*YvaN*) represses *rapG* and *rapH*, which encode inhibitors of expression of the *srfA* operon. *Molecular Microbiology* 59: 1714–1729.
He, H., Chen, X., Li, J., Zhang, Y., and Gao, P. 2004. Taste improvement of refrigerated meat treated with cold-adapted protease. *Food Chemistry* 84: 307–311.
Henry, G., Deleu, M., Jourdan, E., Thonart, P., and Ongena, M. 2011. The bacterial lipopeptide surfactin targets the lipid fraction of the plant plasma membrane to trigger immune-related defence responses. *Cellular Microbiology* 13: 1824–1837.

Hiraga, K., Nishikata, Y., Namwong, S., Tanasupawat, S., Takada, K., and Oda, K. 2005. Purification and characterization of serine proteinase from a halophilic bacterium, *Filobacillus* sp. RF2–5. *Bioscience, Biotechnology, and Biochemistry* 69: 38–44.

Hirano, S. 1996. Chitin biotechnology application. *Biotechnology Annual Reviews* 2: 237–258.

Ho, C.T., and Carlin, J.T. 1989. Formation and aroma characteristics of heterocyclic compounds in foods. In *Flavor Chemistry, Trends and Developments*, eds. R. Teranishi, R.G. Buttery, and F. Shahidi, pp. 92–104. Washington, DC: American Chemical Society.

Hofemeister, J., Conrad, B., Adler, B., Hofemeister, B., Feesche, J., Kucheryava, N., Steinborn, G., Franke, P., Grammel, N., Zwintscher, A., Leenders, F., Hitzeroth, G., and Vater, J. 2004. Genetic analysis of the biosynthesis of non-ribosomal peptide- and polyketide-like antibiotics, iron uptake and biofilm formation by *Bacillus subtilis* A1/3. *Molecular Genetics and Genomics* 272: 363–378.

Hongo, M., and Yoshimoto, A. 1968. Formation of phage-induced γ-polyglutamic acid depolymerase in lysogenic strain of *Bacillus natto*. *Agricultural and Biological Chemistry* 32: 525–527.

Hongo, M., and Yoshimoto, A. 1970a. Bacteriophages of *Bacillus natto*. Part II. Induction of γ-polyglutamic acid depolymerase following phage infection. *Agricultural and Biological Chemistry* 34: 1047–1054.

Hongo, M., and Yoshimoto, A. 1970b. Bacteriophages of *Bacillus natto*. Part III. Action of phage-induced γ-polyglutamic acid depolymerase on γ-polyglutamic acid and the enzymatic hydrolyzates. *Agricultural and Biological Chemistry* 34: 1055–1063.

Horikoshi, K., and Akiba, T. 1982. *Alkalophilic Microorganisms: A New Microbial World*. Tokyo: Japan Scientific Societies Press.

Horowitz, S., Gilbert, J.N., and Griffin, W.M. 1990. Isolation and characterization of a surfactant produced by *Bacillus licheniformis* 86. *Journal of Industrial Microbiology and Biotechnology* 6: 243–248.

Hsu, R.L., Lee, K.T., Wang, J.H., Lee, L.Y.L., and Chen, R.P.Y. 2009. Amyloid-degrading ability of nattokinase from *Bacillus subtilis natto*. *Journal of Agricultural and Food Chemistry* 57: 503–508.

Ikeda, R., Hosoi, Y., and Hara, T. 2000. Study on the stringiness-decreasing factor in radish (*Raphanus sativus* L.) on natto. *Journal for the Integrated Study of Dietary Habits* 11: 50–56 (in Japanese).

Imm, J.Y., and Lee, C.M. 1999. Production of seafood taste from red hake (*Urophycis chuss*) by enzymatic hydrolysis. *Journal of Agricultural and Food Chemistry* 47: 2360–2366.

Itokawa, H., Miyashita, T., Morita, H., Takeya, K., Homma, T., and Oka, K. 1994. Structural and conformational studies of [Ile7] and [Leu7] surfactins from *Bacillus subtilis natto*. *Chemical and Pharmaceutical Bulletin* 42: 604–607.

Izco, J.M., and Torre, P. 2000. Characterization of volatile flavour compounds in Roncal cheese extracted by the "purge and trapi" method and analysed by GC-MS. *Food Chemistry* 70: 409–417.

Jacques, P. 2011. Surfactin and other lipopeptides from *Bacillus* spp. In *Biosurfactants*, Microbiology Monographs 20, ed. G. Soberon-Chavez, pp. 57–91. Heidelberg: Springer-Verlag.

Jacques, P., Hbid, C., Destain, J., Razafindralambo, H., Paquot, M., DePauw, E., and Thonart, P. 1999. Optimization of biosurfactant lipopeptide production from *Bacillus subtilis* S499 by Plackett-Burman design. *Applied Biochemistry and Biotechnology* 77: 223–233.

Jamilah, M.B., Abbas, K.A., and Abdul Rahman, R. 2008. A review on some organic acids additives as shelf life extenders of fresh beef cuts. *American Journal of Agricultural and Biological Sciences* 3: 566–574.

Jani, S.A., Chudasama, C.J., Patel, D.B., Bhatt, P.S., and Patel, H.N. 2012. Optimization of extracellular protease production from alkali thermotolerant actinomycetes: *Saccharomonospora viridis* SJ-21. *Bulletin of Environment, Pharmacology and Life Sciences* 1: 84–92.

Jaswal, R.K., and Kocher, G.S. 2007. Partial characterization of a crude alkaline protease from *Bacillus circulans* and its detergent compatability. *The Internet Journal of Microbiology* 4: 1–9.

Jay, J.M. 1992. Intrinsic and extrinsic parameters of foods that affect microbial growth. In *Modern Food Microbiology*, ed. J.M. Jay, pp. 38–62. New York: Chapman & Hall.

Jeong, Y.K., Park, J.U., Baek, H., Park, S.H., Kong, I.S., Kim, D.W., and Joo, W.H. 2001. Purification and biochemical characterization of a fibrinolytic enzyme from *Bacillus subtilis* BK-17. *World Journal of Microbiology and Biotechnology* 17: 89–92.

Johnvesly, B., Manjunath, B.R., and Naik, G.R. 2002. Pigeon pea waste as a novel, inexpensive, substrate for production of a thermostable alkaline protease from thermoalkalophilic *Bacillus* sp. JB-99. *Bioresource Technology* 82: 61–64.

Johnvesly, B., and Naik, G.R. 2001a. Production of bleach-stable and halotolerant alkaline protease by an alkaliphilic *Bacillus pumilus* JB05 isolated from cement industry effluents. *Journal of Microbiology and Biotechnology* 11: 558–563.

Johnvesly, B., and Naik, G.R. 2001b. Studies on production of thermostable alkaline protease from thermophilic and alkaliphilic *Bacillus* sp. JB-99 in a chemically defined medium. *Process Biochemistry* 37: 139–144.

Joo, H.S., and Chang, C.S. 2005. Production of protease from a new alkalophilic *Bacillus* sp. I-312 grown on soybean meal: Optimization and some properties. *Process Biochemistry* 40: 1263–1270.

Joo, H.S., Kumar, C.G., Park, G.C., Kim, K.T., Paik, S.R., and Chang, C.S. 2002. Optimization of the production of an extracellular alkaline protease from *Bacillus horikoshii*. *Process Biochemistry* 38: 155–159.

Joo, H.S., Kumar, C.G., Park, G.C., Kim, K.T., Paik, S.R., and Chang, C.S. 2004. Bleach-resistant alkaline protease produced by a *Bacillus* sp. isolated from the Korean polychaete, *Periserrula leucophryna*. *Process Biochemistry* 39: 1441–1447.

Joo, H.S., Kumar, C.G., Park, G.C., Paik, S.R., and Chang, C.S. 2003. Oxidant and SDS-stable alkaline protease from *Bacillus clausii* I-52: Production and some properties. *Journal of Applied Microbiology* 95: 267–272.

Joshi, S., Bharucha, C., and Desai, A.J. 2008. Production of biosurfactant and antifungal compound by fermented food isolate *Bacillus subtilis* 20B. *Bioresource Technology* 99: 4603–4608.

Julkowska, D., Obuchowski, M., Holland, I.B., and Seror, S.J. 2005. Comparative analysis of the development of swarming communities of *Bacillus subtilis* 168 and a natural wild type: Critical effects of surfactin and the composition of the medium. *Journal of Bacteriology* 187: 65–76.

Kajimura, Y., and Kaneda, M. 1996. Fusaricidin A, a new depsipeptide antibiotic produced by *Bacillus polymyxa* KT-8. Taxonomy, fermentation, isolation, structure elucidation and biological activity. *The Journal of Antibiotics* 49: 129–135.

Kajimura, Y., and Kaneda, M. 1997. Fusaricidins B, C, and D, new depsipeptide antibiotics produced by *Bacillus polymyxa* KT-8: Isolation, structure elucidation and biological activity. *The Journal of Antibiotics* 50: 220–228.

Kalisz, H.M. 1988. Microbial proteinases. *Advances in Biochemical Engineering and Biotechnology* 36: 1–65.

Kameda, Y., Ouhira, S., Matsui, K., Kanatomo, S., Hase, T., and Atsusaka, T. 1974. Antitumor activity of *Bacillus natto*. 5. Isolation and characterization of surfactin in culture medium of *Bacillus natto* KMD 2311. *Chemical and Pharmaceutical Bulletin* 22: 938–944.

Kanno, A., Takamatsu, H., and Takano, N. 1984. Determination of several volatile components produced by *Bacillus natto* in commercial natto. *Nippon Shokuhin Kogyo Gakkaishi* 31: 587–595.

Kappeler, S.R., van den Brink, H.M., Rahbek-Nielsen, H., Farah, Z., Puhan, Z., Hansen, E.B., and Johansen, E. 2006. Characterization of recombinant camel chymosin reveals superior properties for the coagulation of bovine and camel milk. *Biochemical and Biophysical Research Communications* 342: 647–654.

Karan, R., Singh, R.K.M., Kapoor, S., and Khare, S.K. 2011. Gene identification and molecular characterization of solvent-stable protease from a moderately haloalkaliphilic bacterium, *Geomicrobium* sp. EMB2. *Journal of Microbiology and Biotechnology* 21: 129–135.

Karbalaei-Heidari, H.R., Amoozegar, M.A., Hajighasemi, M., Ziaee, A.A., and Ventosa, A. 2009. Production, optimization and purification of a novel extracellular protease from the moderately halophilic bacterium *Halobacillus karajensis*. *Journal of Industrial Microbiology and Biotechnology* 36: 21–27.

Katsura, M. 1996. Tofuyo. *Journal of Cookery Science of Japan* 29: 314–322 (in Japanese).

Kennedy, M., and Krouse, D. 1999. Strategies for improving fermentation medium performance: A review. *Journal of Industrial Microbiology and Biotechnology* 23: 456–475.

Kenny, S., Wehrle, K., Stanton, C., and Arendt, E.K. 2000. Incorporation of dairy ingredients into wheat bread: Effects on dough rheology and bread quality. *European Food Research Technology* 210: 391–396.

Kiers, J.L., van Laeken, A.E.A., Rombouts, F.M., and Nout, M.J.R. 2000. In vitro digestibility of *Bacillus* fermented soya bean. *International Journal of Food Microbiology* 60: 163–169.
Kim, H.K., Kim, G.T., Kim, D.K., Choi, W.A., Park, S.H., Jeong, Y.K., and Kong, I.S. 1997a. Purification and characterization of a novel fibrinolytic enzyme from *Bacillus* sp. KA38 originated from fermented fish. *Journal of Fermentation and Bioengineering* 84: 307–312.
Kim, H.S., Lee, C.H., and Suh, H.H. 1997b. A lipopeptide biosurfactant produced by *Bacillus subtilis* C9 selected through the Oil Wlm-collapsing assay. *Journal of Microbiology and Biotechnology* 7: 180–188.
Kim, S.B., Lee, D.W., Cheigh, C.I., Choe, E.A., Lee, S.J., Hong, Y.H., Choi, H.J., and Pyun, Y.R. 2006. Purification and characterization of a fibrinolytic subtilisin-like protease of *Bacillus subtilis* TP-6 from an Indonesian fermented soybean, Tempeh. *Journal of Industrial Microbiology and Biotechnology* 33: 436–444.
Kim, S.H., and Choi, N.S. 2000. Purification and characterization of subtilisin DJ-4 secreted by *Bacillus* sp. strain DJ-4 screened from Doen-Jang. *Bioscience, Biotechnology, and Biochemistry* 64: 1722–1725.
Kim, W., Choi, K., Kim, Y., Park, H., Choi, J., Lee, Y., Oh, H., Kwon, I., and Lee, S. 1996. Purification and characterization of a fibrinolytic enzyme produced from *Bacillus* sp. strain CK11–4 screened from Chungkook-Jang. *Applied and Environmental Microbiology* 62: 2482–2488.
Kim, W.J., and Kim, S.M. 2005. Purification and characterization of *Bacillus subtilis* JM-3 protease from anchovy sauce. *Journal of Food Biochemistry* 29: 591–610.
Kim, Y.O., Lee, J.K., Kim, H.K., Yu, J.H., and Oh, T.K. 1998. Cloning of the thermostable phytase gene (*phy*) from *Bacillus* sp. DS11 and its overexpression in *Escherichia coli*. *FEMS Microbiology Letters* 162: 185–191.
Kimura, K. 2008. γ-PGA hydrolase of phage. In *Advanced Science on Natto: Japanese Soybean Fermented Foods*, eds. K. Kiuchi, T. Nagai, and K. Kimura, pp. 268–270. Tokyo: Kenpakusha (in Japanese).
Kimura, K., and Itoh, Y. 2003. Characterization of poly-γ-glutamate hydrolase encoded by a bacteriophage genome: Possible role in phage infection of *Bacillus subtilis* encapsulated with poly-γ-glutamate. *Applied and Environmental Microbiology* 69: 2491–2497.
Kitada, M., Wijayanti, L., and Horikoshi, K. 1987. Biochemical properties of a thermophilic alkalophile. *Agricultural and Biological Chemistry* 51: 2429–2435.
Klesk, K., and Qian, M. 2003. Preliminary aroma comparison of marion (*Rubus* spp. Hyb) and evergreen (*R. laciniatus* L.) blackberries by dynamic headspace/OSME technique (RH Marion and Evergreen Volatiles). *Journal of Food Science* 68: 697–700.
Klich, M.A., Lax, A.R., Bland, J.M., and Scharfenstein, L.L. 1993. Influence of iturin-A on mycelial weight and aflatoxin production by *Aspergillus flavus* and *Aspergillus parasiticus* in shake culture. *Mycopathologia* 123: 35–38.

Knight, Z., Garrison, J., Chan, K., King, D., and Shokat, K. 2007. A remodeled protease that cleaves phosphotyrosine substrates. *Journal of the American Chemical Society* 129: 11672–11673.

Kocabiyik, S., and Özdemir, I. 2006. Purification and characterization of an intracellular chymotrypsin-like serine protease from *Thermoplasma volcanium*. *Bioscience, Biotechnology, and Biochemistry* 70: 126–134.

Koehler, T.M., and Thorne, C.B. 1987. *Bacillus subtilis* (*natto*) plasmid pLS20 mediates interspecies plasmid transfer. *Journal of Bacteriology* 169: 5271–5278.

Kong, X., Zhou, H., and Qian, H. 2007. Enzymatic hydrolysis of wheat gluten by proteases and properties of the resulting hydrolysates. *Food Chemistry* 102: 759–763.

Kopp, F., and Marahiel, M.A. 2007. Macrocyclization strategies in polyketide and nonribosomal peptide biosynthesis. *Natural Product Reports* 24: 735–749.

Kotlar, C.E., Belagardi, M., and Roura, S.I. 2011. Brewer's spent grain: Characterization and standardization procedure for the enzymatic hydrolysis by *Bacillus cereus* strain. *Biotechnology and Applied Biochemistry* 58: 464–475.

Kpikpi, E.N., Dzogbefia, V.P., and Glover, R.K. 2009. Enzymatic and some biochemical changes associated with the production of "kantong," a traditional fermented condiment in northern Ghana. *Journal of Food Biochemistry* 33: 61–73.

Kracht, M., Rokos, H., Ozel, M., Kowall, M., Pauli, G., and Vater, J. 1999. Antiviral and hemolytic activities of surfactin isoforms and their methyl ester derivatives. *The Journal of Antibiotics* 52: 613–619.

Krulwich, T.A., Guffanti, A.A., and Seto-Young, D. 1990. pH homeostasis and bioenergetic work in alkalophiles. *FEMS Microbiology Reviews* 75: 271–278.

Ku, K.L., Chen, T.P., and Chiou, R.Y.Y. 2000. Apparatus used for small-scale volatile extraction from ethanol-supplemented low-salt miso and GC-MS characterization of the extracted flavors. *Journal of Agricultural and Food Chemistry* 48: 3507–3511.

Kulakova, L., Galkin, A., Nakayama, T., Nishino, T., and Esaki, N. 2003. Improvement of thermostability of cold-active serine alkaline protease from the psychrotrophic bacterium *Shewanella* sp. strain Ac10 by rational mutagenesis. *Journal of Molecular Catalysis B: Enzymatic* 22: 113–117.

Kumar, C.G. 2002. Purification and characterization of a thermostable alkaline protease from alkalophilic *Bacillus pumilus*. *Letters in Applied Microbiology* 34: 13–17.

Kumar, C.G., and Anand, S.K. 1998. Significance of microbial biofilms in food industry: A review. *International Journal of Food Microbiology* 42: 9–27.

Kumar, C.G., Joo, H.S., Koo, Y.M., Paik, S.R., and Chang, C.S. 2004. Thermostable alkaline protease from a novel marine haloalkalophilic *Bacillus clausii* isolate. *World Journal of Microbiology and Biotechnology* 20: 351–357.

Kumar, C.G., Malik, R.K., and Tiwari, M.P. 1998. Novel enzyme-based detergents: An Indian perspective. *Current Science* 75: 1312–1318.
Kumar, C.G., Malik, R.K., Tiwari, M.P., and Jany, K.D. 1999a. Optimal production of *Bacillus* alkaline protease using a cheese whey medium. *Microbiologie-Aliments-Nutrition* 17: 39–48.
Kumar, C.G., and Parrack, P. 2003a. Activated charcoal: A versatile decolorization agent for the recovery and purification of alkaline protease. *World Journal of Microbiology and Biotechnology* 19: 243–246.
Kumar, C.G., and Parrack, P. 2003b. Arrowroot (*Marantha arundinacea*) starch as a new low-cost substrate for alkaline protease production. *World Journal of Microbiology and Biotechnology* 19: 757–762.
Kumar, C.G., and Takagi, H. 1999. Microbial alkaline proteases: From a bioindustrial viewpoint. *Biotechnology Advances* 17: 561–594.
Kumar, C.G., and Tiwari, M.P. 1999. Use of alkaline proteases for ultrafiltration membrane cleaning. *Biotechnology Techniques* 13: 235–238.
Kumar, C.G., Tiwari, M.P., and Jany, K.D. 1999b. Novel alkaline serine proteases from an alkalophilic *Bacillus* spp.: Purification and some properties. *Process Biochemistry* 34: 441–449.
Kumar, S., Tsai, C.J., and Nussinov, R. 2000. Factors enhancing protein thermostability. *Protein Engineering* 13: 179–191.
Kunioka, M. 1993. Properties of hydrogels prepared by γ-irradiation in microbial poly(γ-glutamic acid) aqueous solutions. *Kobunshi Ronbunshu* 50: 755–760 (in Japanese).
Kvist, T., and Thyregod, P. 2005. Using evolutionary operation to improve yield in biotechnological processes. *Quality and Reliability Engineering International* 21: 457–463.
Landy, M., Warren, G.H., Roseman, S.B., and Colio, L.G. 1948. Bacillomycin: An antibiotic from *Bacillus subtilis* active against pathogenic fungi. *Proceedings of the Society of Experimental Biology in Medicine* 67: 539–541.
Langeveld, J.P.M., Wang, J.J., van de Wiel, D.F.M., Shih, G.C., Garssen, G.J., Bossers, A., and Shih, J.C.H. 2003. Enzymatic degradation of prion protein in brain stem from infected cattle and sheep. *Journal of Infectious Diseases* 188: 1782–1789.
Larcher, G., Cimon, B., Symoens, F., Tronchin, G., Chabasse, D., and Bouchara, J.P. 1996. A 33 kDa serine proteinase from *Scedosporium apiospermum*. *Biochemical Journal* 315: 119–126.
Lasekan, O.O., Feijiao Teixeira, J.P., and Salva, T.J.G. 2001. Volatile flavour compounds of cooked acha (*Digitaria exilis* Stapf). *Food Chemistry* 75: 333–337.
Lee, J.H., Nam, S.H., Seo, W.T., Yun, H.D., Hong, S.Y., Kim, M.K., and Cho, K.M. 2012. The production of surfactin during the fermentation of cheonggukjang by potential probiotic *Bacillus subtilis* CSY191 and the resultant growth suppression of MCF-7 human breast cancer cells. *Food Chemistry* 131: 1347–1354.
Lee, S.H., Roh, S.K., Park, K.H., and Yoon, K.R. 1999. Effective extraction of astaxanthin pigment from shrimp using proteolytic enzymes. *Biotechnology and Bioprocess Engineering* 4: 199–204.

Lee, S.J., and Ahn, B. 2009. Comparison of volatile components in fermented soybean pastes using simultaneous distillation and extraction (SDE) with sensory characterisation. *Food Chemistry* 114: 600–609.

Lee, S.M., Seo, B.C., and Kim, Y.S. 2006. Volatile compounds in fermented and acid hydrolyzed soy sauces. *Journal of Food Science* 71: C146–C156.

Lee, S.Y., Kim, J.S., Kim, J.E., Sapkota, K., Shen, M.H., Kim, S., Chun, H.S., Yoo, J.C., Choi, H.S., Kim, M.K., and Kim, S.J. 2005. Purification and characterization of fibrinolytic enzyme from cultured mycelia of *Armillaria mellea*. *Protein Expression and Purification* 43: 10–17.

Leejeerajumnean, A. 2003. Thua nao: Alkali fermented soybeans. *International Journal of Silpakorn University* 3: 277–292.

Leejeerajumnean, A., Ames, J.M., and Owens, J.D. 2000. Effect of ammonia on the growth of *Bacillus* species and some other bacteria. *Letters in Applied Microbiology* 5: 385–389.

Leejeerajumnean, A., Duckham, S.C., Owens, J.D., and Ames, J.M. 2001. Volatile compounds in *Bacillus* fermented soybeans. *Journal of the Science of Food and Agriculture* 81: 525–529.

Li, X., and Yu, H.Y. 2012. Purification and characterization of novel organic-solvent tolerant β-amylase and serine protease from a newly isolated *Salimicrobium halophilum* strain LY20. *FEMS Microbiology Letters* 329: 204–211.

Li, Y., Shoemaker, C.F., Ma, J., Luo, C., and Zhong, F. 2009. Effects of Alcalase/Protease N treatments on rice starch isolation and their effects on its properties. *Food Chemistry* 114: 821–828.

Lin, C.C., and Casida Jr., L.E. 1984. Gelrite as a gelling agent in media for growth of thermophilic microorganisms. *Applied and Environmental Microbiology* 47: 427–429.

Lindgren, S.E., and Dobrogosz, W.J. 1990. Antagonistic activities of lactic acid bacteria in food and feed fermentations. *FEMS Microbiology Reviews* 87: 149–164.

Liu, X.L., Du, L.X., Lu, F.P., Zheng, X.Q., and Xiao, J. 2005. Purification and characterization of a novel fibrinolytic enzyme from *Rhizopus chinensis* 12. *Applied Microbiology and Biotechnology* 67: 209–214.

Lopez, D., Fischbach, M.A., Chu, F., Losick, R., and Kolter, R. 2009. Structurally diverse natural products that cause potassium leakage trigger multicellularity in *Bacillus subtilis*. *Proceedings of the National Academy of Sciences of the United States of America* 106: 280–285.

Lounsbury, J.A., Coult, N., Miranian, D.C., Cronk, S.M., Haverstick, D.M., Kinnon, P., Saul, D.J., and Landers, J.P. 2012. An enzyme-based DNA preparation method for application to forensic biological samples and degraded stains. *Forensic Science International: Genetics* 6: 607–615.

Luciana, F.S., and Sato, H.H. 2009. Production of alkaline protease from *Cellulosimicrobium cellulans*. *Brazilian Journal of Microbiology* 40: 54–60.

Ma, Z., Wang, N., Hu, J., and Wang, S. 2012. Isolation and characterization of a new iturinic lipopeptide, mojavensin A produced by a marine-derived bacterium *Bacillus mojavensis* B0621A. *The Journal of Antibiotics* 65: 317–322.

Maget-Dana, R., Thimon, L., Peypoux, F., and Ptak, M. 1992. Surfactin/iturinA interactions may explain the synergistic effect of surfactin on the biological properties of iturin A. *Biochimie* 74: 1047–1051.

Manjusha, K., Jayesh, P., Jose, D., Sreelakshmi, B., Priyaja, P., Gopinath, P., Saramma, A.V., and Bright Singh, I.S. 2013. Alkaline protease from a non-toxigenic mangrove isolate of *Vibrio* sp. V26 with potential application in animal cell culture. *Cytotechnology* 65: 199–212.

Manni, L., Jellouli, K., Ghorbel-Bellaaj, O., Agrebi, R., Haddar, A., Sellami-Kamoun, A., and Nasri, M. 2010. An oxidant- and solvent-stable protease produced by *Bacillus cereus* SV1: Application in the deproteinization of shrimp wastes and as a laundry detergent additive. *Applied Biochemistry and Biotechnology* 160: 2308–2321.

Martinez, R., Jakob, F., Tu, R., Siegert, P., Maurer, K.H., and Schwaneberg, U. 2013. Increasing activity and thermal resistance of *Bacillus gibsonii* alkaline protease (BgAP) by directed evolution. *Biotechnology and Bioengineering* 110: 711–720.

Matsumoto, M., and Iwahara, S. 1961. On the production of slime substance by natto bacteria (I) Chemical composition of the slime substance. *Bulletin of the Shimane Agricultural College* 9: 172–174 (in Japanese).

Mei, C., and Jiang, X. 2005. A novel surfactant- and oxidation-stable alkaline protease from *Vibrio metschnikovii* DL 33–51. *Process Biochemistry* 40: 2167–2172.

Mellor, J., Dobson, M.J., Roberts, N.A., Tuite, M.F., Emtage, J.S., White, S., Lowe, P.A., Patel, T., Kingsman, A.J., and Kingsman, S.M. 1983. Efficient synthesis of enzymatically active calf chymosin in *Saccharomyces cerevisiae*. *Gene* 24: 1–14.

Menkhaus, M., Ullrich, C., Kluge, B., Vater, J., Vollenbroich, D., and Kamp, R.M. 1993. Structural and functional organization of the surfactin synthetase multienzyme system. *Journal of Biological Chemistry* 268: 7678–7684.

Mine, Y., Wong, A.H.K., and Jiang, B. 2005. Fibrinolytic enzymes in Asian traditional fermented foods. *Food Research International* 38: 243–250.

Misono, H., and Ashiuchi, M. 2005. Function and biosynthetic mechanism of poly-γ-glutamate, a main component of natto mucilage. *Vitamins (Japan)* 79: 71–78 (in Japanese).

Mitchinson, C., and Wells, J.A. 1989. Protein engineering of disulfide bonds in subtilisin BPN'. *Biochemistry* 28: 4807–4815.

Mitsuboshi, S., Tanaka, T., Murahashi, A., Muramatsu, K., and Kiuchi, K. 2007. Characteristics of Burmese fermented soy food, pe pok, and development of itohiki-natto employing its bacterial isolates. *Nippon Shokuhin Kagaku Kogaku Kaishi* 54: 528–538 (in Japanese).

Miyazaki, K., and Arnold, F.H. 1999. Exploring non-natural evolutionary pathways by saturation mutagenesis: Rapid improvement of protein function. *Journal of Molecular Evolution* 49: 716–720.

Mizumoto, S., and Shoda, M. 2007. Medium optimization of antifungal lipopeptide, iturinA, production by *Bacillus subtilis* in solid-state fermentation by response surface methodology. *Applied Microbiology and Biotechnology* 76: 101–108.

Moat, A.G. 1979. Carbohydrate metabolism and energy production. In *Microbial Physiology*, ed. A.G. Moat, pp. 123–189. New York: John Wiley and Sons.

Moftah, O.A.S., Grbavcic, S., Zuza, M., Luković, N., Bezbradica, D., and Knežević-Jugović, Z. 2012. Adding value to the oil cake as a waste from oil processing industry: Production of lipase and protease by *Candida utilis* in solid state fermentation. *Applied Biochemistry and Biotechnology* 166: 348–364.

Moon, S.H., and Parulekar, S.J. 1991. A parametric study of protease production in batch and fed-batch cultures of *Bacillus firmus*. *Biotechnology and Bioengineering* 37: 467–483.

Moon, S.H., and Parulekar, S.J. 1993. Some observations on protease production in continuous suspension cultures of *Bacillus firmus*. *Biotechnology and Bioengineering* 41: 43–54.

Mori, H., Sugawa-Katayama, Y., and Shiraishi, A. 2005. Partial purification of a poly-γ-glutamate lytic enzyme of a strain of *Bacillus* species from soil and its some enzymatic characterization. *Journal of Home Economics of Japan* 56: 737–742.

Mori, Y., Kiuchi, K., and Tabei, H. 1983. Flavor components of miso: Basic fraction. *Agricultural and Biological Chemistry* 47: 1487–1492.

Moss, D., Harbison, S.A., and Saul, D.J. 2003. An easily-automated, closed-tube forensic DNA extraction procedure using a thermostable proteinase. *International Journal of Legal Medicine* 117: 340–349.

Mukherjee, A.K., Adhikari, H., and Rai, S.K. 2008. Production of alkaline protease by a thermophilic *Bacillus subtilis* under solid-state fermentation (SSF) condition using *Imperata cylindrica* grass and potato peel as low-cost medium: Characterization and application of enzyme in detergent formulation. *Biochemical Engineering Journal* 39: 353–361.

Murao, S. 1967. On the polyglutamic acid fermentation. *Kobunshi* 16: 1204–1212 (in Japanese).

N'Dir, B., Hbid, C., Cornelius, C., Roblain, D., Jacques, P., Vanhentenryck, F., Diop, M., and Thonart, P. 1994. Propriétés antifongiques de la microflore sporulée du nététou. *Cahiers Agricultures* 3: 23–30.

Nagai, T., and Itoh, Y. 1997. Characterization of a generalized transducing phage of poly-γ-glutamic acid-producing *Bacillus subtilis* (*natto*) *and* its application for analysis of Tn*917*-LTV1 insertional mutants defective in poly-γ-glutamic acid production. *Applied and Environmental Microbiology* 63: 4087–4089.

Nagai, T., Koguchi, K., and Itoh, Y. 1997. Chemical analysis of poly-γ-glutamic acid produced by plasmid-free *Bacillus subtilis* (*natto*): Evidence that plasmids are not involved in poly-γ-glutamic acid production. *Journal of General and Applied Microbiology* 43: 139–143.

Nagai, T., Nishimura, K., Suzuki, H., Banba, Y., Sasaki, H., and Kiuchi, K. 1994. Isolation and characterization of a *Bacillus subtilis* strain producing natto with strong umami-taste and high viscosity. *Nippon Shokuhin Kogyo Gakkaishi* 41: 123–128 (in Japanese).

Nagai, T., Phan Tran, L.-S., Inatsu, Y., and Itoh, Y. 2000. A new IS*4* family insertion sequence, IS*4Bsu*1, responsible for genetic instability of poly-γ-glutamic acid production in *Bacillus subtilis*. *Journal of Bacteriology* 182: 2387–2392.

Nagai, T., and Yamasaki, F. 2009. *Bacillus subtilis* (*natto*) bacteriophages isolated in Japan. *Food Science and Technology Research* 15: 293–298.

Namwong, S., Hiraga, K., Takada, K., Tsunemi, M., Tanasupawat, S., and Oda, K. 2006. A halophilic serine proteinase from *Halobacillus* sp. SR5–3 isolated from fish sauce: Purification and characterization. *Bioscience, Biotechnology, and Biochemistry* 70: 1395–1401.

Nascimento, R.P., d'Avila-Levy, C.M., Souza, R.F., Branquinha, M.H., Bon, E.P.S., Pereira Jr., N., and Coelho, R.R.R. 2005. Production and partial characterization of extracellular proteinases from *Streptomyces malaysiensis*, isolated from a Brazilian cerrado soil. *Archives of Microbiology* 184: 194–198.

Neelakantan, S., Mohanty, A.K., and Kaushik, J.K. 1999. Production and use of microbial enzymes for dairy processing. *Current Science* 77: 143–148.

Ness, J.E., Welch, M., Giver, L., Bueno, M., Cherry, J.R., Borchert, T.V., Stemmer, W.P., and Minshull, J. 1999. DNA shuffling of subgenomic sequences of subtilisin. *Nature Biotechnology* 17: 893–896.

Nestler, L., Evege, E., McLaughlin, J., Munroe, D., Tan, T., Wagner, K., and Stiles, B. 2004. TrypLETM Express: A temperature stable replacement for animal trypsin in cell dissociation applications. *Quest* 1: 42–47.

Nilegaonkar, S., Kanekar, P., Sarnaik, S., and Kelkar, A. 2002. Production, isolation and characterization of extracellular protease of an alkaliphilic strain of *Arthrobacter ramosus*, MCM B-351 isolated from the alkaline lake of Lonar, India. *World Journal of Microbology and Biotechnology* 18: 785–789.

Nishikiori, T., Naganawa, H., Muraoka, Y., Aoyagi, T., and Umezawa, H. 1986. Plipastatins: New inhibitors of phospholipase A2, produced by *Bacillus cereus* BMG302-fF67. III. Structural elucidation of plipastatins. *The Journal of Antibiotics* 39: 755–761.

Nishito, Y., Osana, Y., Hachiya, T., Popendorf, K., Toyoda, A., Fujiyama, A., Itaya, M., and Sakakibara, Y. 2010. Whole genome assembly of a natto production strain *Bacillus subtilis natto*. *BMC Genomics* 11: 243.

Nogueira, M.C.L., Lubachevsky, G., and Rankin, S.A. 2005. A study of the volatile composition of Minas cheese. *LWT – Food Science and Technology* 38: 555–563.

Noh, K.A., Kim, D.H., Choi, N.S., and Kim, S.H. 1999. Isolation of fibrinolytic enzyme producing strains from Kimichi. *Korean Journal of Food Science and Technology* 31: 219–223.

Nolsoe, H., and Undeland, I. 2009. The acid and alkaline solubilization process for the isolation of muscle proteins. *Food and Bioprocess Technology* 2: 1–27.

Norziah, M.H., and Ching, C.Y. 2000. Nutritional composition of edible seaweed *Gracilaria changgi*. *Food Chemistry* 68: 69–76.

O'Toole, D.K. 1997. The role of microorganisms in soy sauce production. *Advances in Applied Microbiology* 45: 87–152.

Odunfa, S.A. 1988. Review: African fermented foods, from art to science. *Mircen Journal* 4: 259–273.

Ogawa, Y., Hosoyama, H., Hamano, M., and Motai, H. 1991. Purification and properties of γ-glutamyltranspeptidase from *Bacillus subtilis* (*natto*). *Agricultural and Biological Chemistry* 55: 2971–2977.

Ogawa, Y., Sugiura, D., Motai, H., Yuasa, K., and Tahara, Y. 1997. DNA sequence of *Bacillus subtilis* (*natto*) NR-1 γ-glutamyltranspeptidase gene, *ggt*. *Bioscience, Biotechnology, and Biochemistry* 61: 1596–1600.

Oh, Y.S., Shih, I.L., Tzeng, Y.M., and Wang, S.L. 2000. Protease produced by *Pseudomonas aeruginosa* K-187 and its application in the deproteinization of shrimp and crab shell wastes. *Enzyme and Microbial Technology* 27: 3–10.

Ohno, A., Ano, T., and Shoda, M. 1995. Effect of temperature on production of lipopeptide antibiotics, iturin A and surfactin by a dual producer, *Bacillus subtilis* RB14, in solid-state fermentation. *Journal of Fermentation Bioengineering* 80: 517–519.

Ojinnaka, M.C., and Ojimelukwe, P.C. 2013. Study of the volatile compounds and amino acid profile in *Bacillus* fermented castor oil bean condiment. *Journal of Food Research* 2: 191–203.

Ojinnaka, M.C., Ojimelukwe, P.C., and Ezeama, C.F. 2013. Effect of fermentation period on the organic acid and amino acid contents of ogiri from castor oil bean seeds. *Malaysian Journal of Microbiology* 9: 201–212.

Olajuyigbe, F.M., and Ajele, J.O. 2008. Some properties of extracellular protease from *Bacillus licheniformis* LbbI-11 isolated from "iru", a traditionally fermented African locust bean condiment. *Global Journal of Biotechnology and Biochemistry* 3: 42–46.

Olaoye, O.A., and Onilude, A.A. 2009. A study on isolation of presumptive technologically important microorganisms from Nigerian beef. *American-Eurasian Journal of Sustainable Agriculture* 3: 75–83.

Olaoye, O.A., Onilude, A.A., and Dodd, C.E.R. 2008. Identification of *Pediococcus* spp. from beef and evaluation of their lactic acid production in varying concentrations of different carbon sources. *Advances in Natural and Applied Sciences* 2: 197–207.

Olivera, N., Sequeiros, C., Siñeriz, F., and Breccia, J.D. 2006. Characterization of alkaline proteases from a novel alkali-tolerant bacterium *Bacillus patagoniensis*. *World Journal of Microbiology and Biotechnology* 22: 737–743.

Ona, M., and Kimura, N. 1991. Antifungal peptides produced by *Bacillus subtilis* for the biological control of aflatoxin contamination. *Proceedings of the Japanese Association of Mycotoxicology* 34: 23–28.

Ongena, M., and Jacques, P. 2008. *Bacillus* lipopeptides: Versatile weapons for plant disease control. *Trends in Microbiology* 16: 115–125.

Ongena, M., Jourdan, E., Adam, A., Paquot, M., Brans, A., Joris, B., Arpigny, J.L., and Thonart, P. 2007. Surfactin and fengycin lipopeptides of *Bacillus subtilis* as elicitors of induced systemic resistance in plants. *Environmental Microbiology* 9: 1084–1090.

Onyenekwe, P.C., Odeh, C., and Nweze, C.C. 2012. Volatile constituents of ogiri, soybean daddawa and locust bean daddawa three fermented Nigerian food flavour enhancers. *Electronic Journal of Environmental, Agricultural and Food Chemistry* 11: 15–22.

Ouoba, L.I.I., Cantor, M.D., Diawara, B., Traore, A.S., and Jakobsen, M. 2003a. Degradation of African locust bean oil by *Bacillus subtilis* and *Bacillus pumilus* isolated from soumbala, a fermented African locust bean condiment. *Journal of Applied Microbiology* 95: 868–873.

Ouoba, L.I.I., Diawara, B., Annan, N.T., Poll, L., and Jakobsen, M. 2005. Volatile compounds of Soumbala, a fermented African locust bean (*Parkia biglobosa*) food condiment. *Journal of Applied Microbiology* 99: 1413–1421.

Ouoba, L.I.I., Diawara, B., Jespersen, L., and Jakobsen, M. 2007. Antimicrobial activity of *Bacillus subtilis* and *Bacillus pumilus* during the fermentation of African locust bean (*Parkia biglobosa*) for soumbala production. *Journal of Applied Microbiology* 102: 963–970.

Ouoba, L.I.I., Rechinger, K.B., Barkholt, V., Diawara, B., Traore, A.S., and Jakobsen, M. 2003b. Degradation of proteins during the fermentation of African locust bean *Parkia biglobosa*, by strains of *Bacillus subtilis* and *Bacillus pumilus* for production of soumbala. *Journal of Applied Microbiology* 94: 396–402.

Owens, J.D., Allagheny, N., Kipping, G., and Ames, J.M. 1997. Formation of volatile compounds during *Bacillus subtilis* fermentation of soya beans. *Journal of the Science of Food and Agriculture* 74: 132–140.

Paik, H.D., Lee, S.K., Heo, S., Kim, S.Y., Lee, H., and Kwon, T.J. 2004. Purification and characterization of the fibrinolytic enzyme produced by *Bacillus subtilis* KCK-7 from Chungkookjang. *Journal of Microbiology and Biotechnology* 14: 829–835.

Park, H.K., Gil, B., and Park, J.K. 2003. Characteristic flavor compounds of commercial soybean paste. *Food Science and Biotechnology* 12: 377–607.

Park, S.Y., and Kim, Y.H. 2009. Surfactin inhibits immunostimulatory function of macrophages through blocking NK-jB, MAPK and Akt pathway. *International Immunopharmacology* 9: 886–893.

Parkouda, C., Diawara, B., Lowor, S., Diako, C., Kwesi Saalia, F., Annan, N.T., Jensen, J.S., Tano-Debrah, K., and Jakobsen, M. 2011. Volatile compounds of maari, a fermented product from baobab (*Adansonia digitata* L.) seeds. *African Journal of Biotechnology* 10: 4197–4206.

Pastor, M.D., Lorda, G.S., and Balatti, A. 2001. Protease obtention using *Bacillus subtilis* 3411 and amaranth seed meal medium at different aeration rates. *Brazilian Journal of Microbiology* 32: 6–9.

Pathak, A.P., and Deshmukh, K.B. 2012. Alkaline protease production, extraction and characterization from alkaliphilic *Bacillus licheniformis* KBDL4: A Lonar soda lake isolate. *Indian Journal of Experimental Biology* 50: 569–576.

Pecs, M., Eggert, M., and Schügerl, K. 1991. Affinity precipitation of extracellular microbial enzymes. *Journal of Biotechnology* 21: 137–142.

Pedersen, L., Kaack, K., Bergsoe, M.N., and Adler-Nissen, J. 2005. Effects of chemical and enzymatic modification on dough rheology and biscuit characteristics. *Journal of Food Science* 70: 152–158.

Peng, Y., Huang, Q., Zhang, R.H., and Zhang, Y.Z. 2003. Purification and characterization of a fibrinolytic enzyme produced by *Bacillus amyloliquefaciens* DC-4 screened from douchi, a traditional Chinese soybean food. *Comparative Biochemistry and Physiology* 134B: 45–52.

Petinate, S.D.G., Branquinha, M.H., Coelho, R.R.R., Vermelho, A.B., and Giovanni-De-Simone, S. 1999. Purification and partial characterization of an extracellular serine-proteinase of *Streptomyces cyaneus* isolated from Brazilian cerrado soil. *Journal of Applied Microbiology* 87: 557–563.

Petrounia, I.P., and Arnold, F.H. 2000. Designed evolution of enzymatic properties. *Current Opinion in Biotechnology* 11: 325–330.

Peypoux, F., Bonmatin, J.M., and Wallach, J. 1999. Recent trends in the biochemistry of surfactin. *Applied Microbiology and Biotechnology* 51: 553–563.

Peypoux, F., Guinand, M., Michel, G., Delcambe, L., Das, B.C., and Lederer, E. 1978. Structure of iturin A, a peptidolipid antibiotic from *Bacillus subtilis*. *Biochemistry* 17: 3992–3996.

Piddocke, M.P., Fazio, A., Vongsangnak, W., Wong, M.L., Heldt-Hansen, H.P., Workman, C., Nielsen, J., and Olsson, L. 2011. Revealing the beneficial effect of protease supplementation to high gravity beer fermentations using "-omics" techniques. *Microbial Cell Factories* 10: 27. Available at http://www.microbialcellfactories.com/content/10/21/27.

Poza, M., Prieto-Alcedo, M., Sieiro, C., and Villa, T.G. 2004. Cloning and expression of *clt* genes encoding milk-clotting proteases from *Myxococcus xanthus* 422. *Applied and Environmental Microbiology* 70: 6337–6341.

Prakasham, R.S., Subba Rao, C., and Sarma, P.N. 2006. Green gram husk – an inexpensive substrate for alkaline protease production by *Bacillus* sp. in solid-state fermentation. *Bioresource Technology* 97: 1449–1454.

Prakasham, R.S., Subba Rao, C., Sreenivasa Rao, R., and Sarma, P.N. 2005. Alkaline protease production by an isolated *Bacillus circulans* under solid-state fermentation using agroindustrial waste: Process parameters optimization. *Biotechnology Progress* 21: 1380–1388.

Qin, L., and Ding, X. 2007a. Evolution of proteolytic tasty components during preparation of douchiba, a traditional Chinese soy-fermented appetizer. *Food Technology and Biotechnology* 45: 85–90.

Qin, L.K., and Ding, X.L. 2007b. Formation of taste and odor compounds during preparation of Douchiba, a Chinese traditional soy-fermented appetizer. *Journal of Food Biochemistry* 31: 230–251.

Raghavan, S., Kristinsson, H.G., and Leeuwenburgh, C. 2008. Radical scavenging and reducing ability of tilapia (*Oreochromis niloticus*) protein hydrolysates. *Journal of Agricultural and Food Chemistry* 56: 10359–10367.

Rahaman, R.S., Chee, J.Y., Cabral, J.M.S., and Hatton, T.A. 1988. Recovery of an extracellular alkaline protease from whole fermentation broth using reversed micelles. *Biotechnology Progress* 4: 218–224.

Ramesh, S., Rajesh, M., and Mathivanan, N. 2009. Characterization of a thermostable alkaline protease produced by marine *Streptomyces fungicidicus* MML1614. *Bioprocess and Biosystems Engineering* 32: 791–800.
Rao, M.B., Tanksale, A.M., Ghatge, M.S., and Deshpande, V.V. 1998. Molecular and biotechnological aspects of microbial proteases. *Microbiology and Molecular Biology Reviews* 62: 597–635.
Rao, R.S., Kumar, C.G., Prakasham, R.S., and Hobbs, P. 2008. The Taguchi methodology as a statistical tool for biotechnological applications: A critical appraisal. *Biotechnology Journal* 3: 510–523.
Rawlings, N.D., Barrett, A.J., and Bateman, A. 2011. Asparagine peptide lyases: A seventh catalytic type of proteolytic enzymes. *Journal of Biological Chemistry* 286: 38321–38328.
Razafindralambo, H., Paquot, M., Baniel, A., Popineau, Y., Hbid, C., Jacques, P., and Thonart, P. 1996. Foaming properties of surfactin, a lipopeptide biosurfactant from *Bacillus subtilis*. *Journal of the American Oil Chemists' Society* 73: 149–151.
Razafindralambo, H., Popineau, Y., Deleu, M., Hbid, C., Jacques, P., Thonart, P., and Paquot, M. 1997. Surface-active properties of surfactin/iturin A mixtures produced by *Bacillus subtilis*. *Langmuir* 13: 6026–6031.
Reddy, L.V.A., Wee, Y.-J., Yun, J.-S., and Ryu, H.-W. 2008. Optimization of alkaline protease production by batch culture of *Bacillus* sp. RKY3 through Plackett-Burman and response surface methodological approaches. *Bioresource Technology* 99: 2242–2249.
Rodriguez-Bernaldo De Quiros, A.I., Lopez-Hernandez, J., Gonzalez-Castro, M.J., De la Cruz-Garcia, C., and Simal-Lozano, J. 2000. Comparison of volatile components in raw and cooked green beans by GC-MS using dynamic headspace sampling and microwave desorption. *European Food Research and Technology* 210: 226–230.
Romero, D., de Vicente, A., Rakotoaly, R.H., Dufour, S.E., Veening, J.V.V., Arrebola, E., Cazoria, F.M., Kuipers, O.P., Paquot, M., and Pérez-García, A. 2007. The iturin and fengycin families of lipopeptides are key factors in antagonism of *Bacillus subtilis* toward *Podosphaera fusca*. *Molecular Plant-Microbe Interactions* 20: 430–440.
Roongsawang, N., Washio, K., and Morikawa, M. 2011. Diversity of nonribosomal peptide synthetases involved in the biosynthesis of lipopeptide biosurfactants. *International Journal of Molecular Sciences* 12: 141–172.
Rosenberg, E., and Ron, E.Z. 1999. High- and low-molecular-mass microbial surfactants. *Applied Microbiology and Biotechnology* 52: 154–162.
Ryan, J.T., Ross, R.P., Bolton, D., Fitzgerald, G.F., and Stanton, C. 2011. Bioactive peptides from muscle sources: Meat and fish. *Nutrients* 3: 765–791.
Saeki, K., Ozaki, K., Kobayashi, T., and Ito, S. 2007. Detergent alkaline proteases: Enzymatic properties, genes, and crystal structures. *Journal of Bioscience and Bioengineering* 103: 501–508.
Sana, B., Ghosh, D., Saha, M., and Mukherjee, J. 2006. Purification and characterization of a salt, solvent, detergent and bleach tolerant protease from

a new gamma-Proteobacterium isolated from the marine environment of the *Sunderbans*. *Process Biochemistry* 41: 208–215.

Sareen, R., and Mishra, P. 2008. Purification and characterization of organic solvent stable protease from *Bacillus licheniformis* RSP-09–37. *Applied Microbiology and Biotechnology* 79: 399–405.

Savadogo, A., Tapi, A., Chollet, M., Wathelet, B., Traoré, A.S., and Jacques, P. 2011. Identification of surfactin producing strains in soumbala and bikalga fermented condiments using polymerase chain reaction and matrix assisted laser desorption/ionization-mass spectrometry methods. *International Journal of Food Microbiology* 151: 299–306.

Sawamura, S. 1905. On the bacteria of natto. *Nogakukai Kaiho* 67: 1–6 (in Japanese).

Schiraldi, C., and De Rosa, M. 2002. The production of biocatalysts and biomolecules from extremophiles. *Trends in Biotechnology* 20: 515–521.

Shah, D.N., Shah, V.D., Nehete, P.N., and Kothari, R.M. 1986. Isolation of *Bacillus licheniformis* mutants for stable production profiles of alkaline protease. *Biotechnology Letters* 8: 103–106.

Shakerifard, P., Gancel, F., Jacques, P., and Faille, C. 2009. Effect of different *Bacillus subtilis* lipopeptides on surface hydrophobicity and adhesion of *Bacillus cereus* 98/4 spores to stainless steel and Teflon. *Biofouling* 25: 533–541.

Shanmughapriya, S., Krishnaveni, J., Selvin, J., Gandhimathi, R., Arunkumar, M., Thangavelu, T., Kiran, G.S., and Natarajaseenivasan, K. 2008. Optimization of extracellular thermotolerant alkaline protease produced by marine *Roseobacter* sp. (MMD040). *Bioprocess and Biosystems Engineering* 31: 427–433.

Shao, Z., Zhao, H., Giver, L., and Arnold, F.H. 1998. Random-priming in vitro recombination: An effective tool for directed evolution. *Nucleic Acids Research* 26: 681–683.

Shieh, C.J., Thi, L.A.P., and Shih, I.L. 2009. Milk-clotting enzymes produced by culture of *Bacillus subtilis natto*. *Biochemical Engineering Journal* 43: 85–91.

Sholberg, P.I. 1998. Fumigation of fruit with short-chain organic acids to reduce the potential of postharvest decay. *Plant Disease* 82: 689–693.

Shukla, S., Choi, T.B., Park, H.K., Kim, M., Lee, I.K., and Kim, J.K. 2010. Determination of non-volatile and volatile organic acids in Korean traditional fermented soybean paste (Doenjang). *Food and Chemical Toxicology* 48: 2005–2010.

Sieber, S.A., and Marahiel, M.A. 2005. Molecular mechanisms underlying nonribosomal peptide synthesis: Approaches to new antibiotics. *Chemical Reviews* 105: 715–738.

Siebert, K.J. 2009. Haze in beverages. *Advances in Food Nutrition and Research* 57: 53–86.

Siegert, P., Wieland, S., Engelskirchen, J., Merkel, M., Maurer, K.H., and Bessler, C. 2009. *Novel alkaline protease from Bacillus gibsonii and washing and cleaning agents containing said novel alkaline protease.* Patent U.S. 20090275493 (assigned to Henkel AG & Co. KGaA).

Sinsuwan, S., Rodtong, S., and Yongsawatdigul, J. 2007. NaCl-activated extracellular proteinase from *Virgibacillus* sp. SK37 isolated from fish sauce fermentation. *Journal of Food Science* 72: C264–C269.

Smith, K.E., and Bradley Jr., R.L. 1987. Activity of four enzyme-based cleaners for ultrafiltration systems against proteins in skim milk and whey. *Journal of Dairy Science* 70: 243–251.

Steinhaus, P., and Schieberle, P. 2007. Characterization of the key aroma compounds in soy sauce using approaches of molecular sensory science. *Journal of Agricultural and Food Chemistry* 55: 6262–6269.

Steinkraus, K.H. 1996. Indigenous fermented foods involving an alkaline fermentation. In *Handbook of Indigenous Fermented Foods*, 2nd edn., ed. K.H. Steinkraus, pp. 349–362. New York: Marcel Dekker.

Steinkraus, K.H. 2002. Fermentation in world food processing. *Comprehensive Reviews in Food Science and Food Safety* 1: 23–32.

Stowe, R.A., and Mayer, R.P. 1966. Efficient screening of process variables. *Industrial Engineering and Chemistry* 58: 36–40.

Strauch, M.A., and Hoch, J.A. 1993. Transition-state regulators: Sentinels of *Bacillus subtilis* post-exponential phase gene expression. *Molecular Microbiology* 7: 337–342.

Strausberg, S.L., Ruan, B., Fisher, K.E., Alexander, P.A., and Bryan, P.N. 2005. Directed coevolution of stability and catalytic activity in calcium-free subtilisin. *Biochemistry* 44: 3272–3279.

Su, Y.-C. 1986. Sufu. In *Legume-Based Fermented Foods*, eds. N.R. Reddy, M.D. Pierson, and D.K. Salunkhe, pp. 69–83. Boca Raton, FL: CRC Press.

Subba Rao, C., Sathish, T., Mahalaxmi, M., Laxmi, G.S., Rao, R.S., and Prakasham, R.S. 2007. Modelling and optimization of fermentation factors for enhancement of alkaline protease production by isolated *Bacillus circulans* using feed-forward neural network and genetic algorithm. *Journal of Applied Microbiology Symposium Supplement* 104: 889–898.

Sugawara, E., Ito, T., Odagiri, S., Kubota, K., and Kobayashi, A. 1985. Comparison of compositions of odour components of natto and cooked soybeans. *Agricultural and Biological Chemistry* 49: 311–317.

Sumi, H., Nakajima, N., and Yatagai, C. 1995. A unique strong fibrinolytic enzyme (datsuwokinase) in skipjack "Shiokara", a Japanese traditional fermented food. *Comparative Biochemistry and Physiology* 112B: 543–547.

Suzuki, T., and Tahara, Y. 2003. Characterization of the *Bacillus subtilis ywtD* gene, whose product is involved in γ-polyglutamic acid degradation. *Journal of Bacteriology* 185: 2379–2382.

Taguchi, S., Ozaki, A., and Momose, H. 1998. Engineering of a cold-adapted protease by sequential random mutagenesis and a screening system. *Applied and Environmental Microbiology* 64: 492–495.

Takagi, H., Kondou, M., Hisatsuka, T., Nakamori, S., Tsai, Y.C., and Yamasaki, M. 1992. Effects of an alkaline elastase from an alkalophilic *Bacillus* strain on the tenderization of beef meat. *Journal of Agricultural and Food Chemistry* 40: 2364–2368.

Takagi, H., Ohtsu, I., and Nakamori, S. 1997. Construction of novel subtilisin E with high specificity, activity and productivity through multiple amino acid substitutions. *Protein Engineering* 10: 985–989.

Tamang, J.P. 1999. Development of pulverised starter for kinema production. *Journal of Food Science and Technology* 36: 475–478.

Tamang, J.P., and Nikkuni, S. 1996. Selection of starter cultures for the production of kinema, a fermented soybean food of the Himalaya. *World Journal of Microbiology and Biotechnology* 12: 629–635.

Tanaka, T., Hiruta, O., Futamura, T., Uotani, K., Satoh, A., Taniguchi, M., and Oi, S. 1993a. Purification and characterization of poly(γ-glutamic acid) hydrolase from a filamentous fungus, *Myrothecium* sp. TM-4222. *Bioscience, Biotechnology, and Biochemistry* 57: 2148–2153.

Tanaka, T., Muramatsu, K., Kim, H.R., Watanabe, T., Takeyasu, M., Kanai, Y., and Kiuchi, K. 1998. Comparison of volatile compounds from chungkuk-jang and itohiki-natto. *Bioscience, Biotechnology, and Biochemistry* 62: 1440–1444.

Tanaka, T., and Shoji, Z. 1993. Analysis of volatile compounds in the natto-fermenting room by gas chromatography–mass spectrometry. *Nippon Shokuhin Kogyo Gakkaishi* 40: 656–660.

Tanaka, T., Yaguchi, T., Hiruta, O., Futamura, T., Uotani, K., Satoh, A., Taniguchi, M., and Oi, S. 1993b. Screening for microorganisms having poly(γ-glutamic acid) endohydrolase activity and the enzyme production by *Myrothecium* sp. TM-4222. *Bioscience, Biotechnology, and Biochemistry* 57: 1809–1810.

Tari, C., Genckal, H., and Tokatli, F. 2006. Optimization of a growth medium using a statistical approach for the production of an alkaline protease from a newly isolated *Bacillus* sp. L21. *Process Biochemistry* 41: 659–665.

Teng, W.L., Khor, E., Tan, T.K., Lim, L.Y., and Tan, S.C. 2001. Concurrent production of chitin from shrimp shells and fungi. *Carbohydrate Research* 332: 305–316.

Thumar, J.T., and Singh, S.P. 2007. Secretion of an alkaline protease from salt-tolerant and alkaliphilic, *Streptomyces clavuligerus* strain Mit-1. *Brazilian Journal of Microbiology* 38: 1–9.

Thys, R.C.S., Guzzon, S.O., Claera-Olivera, F., and Brandelli, A. 2006. Optimization of protease production by *Microbacterium* sp. in feather meal using response surface methodology. *Process Biochemistry* 41: 67–73.

Tran, L.-S., Nagai, T., and Itoh, Y. 2000. Divergent structure of the ComQXPA quorum-sensing components: Molecular basis of strain-specific communication mechanism in *Bacillus subtilis*. *Molecular Microbiology* 37: 1159–1171.

Tremacoldi, C.R., and Carmona, E.C. 2005. Production of extracellular alkaline proteases by *Aspergillus clavatus*. *World Journal of Microbiology and Biotechnology* 21: 169–172.

Troy, F.A. 1973a. Chemistry and biosynthesis of the poly(γ-D-glutamyl) capsule in *Bacillus licheniformis*: I. Properties of the membrane-mediated biosynthetic reaction. *Journal of Biological Chemistry* 248: 305–315.

Troy, F.A. 1973b. Chemistry and biosynthesis of the poly(γ-D-glutamyl) capsule in *Bacillus licheniformis*: II. Characterization and structural properties of the enzymatically synthesized polymer. *Journal of Biological Chemistry* 248: 316–324.

Tsuchiya, K., Gomi, K., Kitamoto, K., Kumagai, C., and Tamura, G. 1993. Secretion of calf chymosin from the filamentous fungus *Aspergillus oryzae*. *Applied Microbiology and Biotechnology* 40: 327–332.

Tsuge, K., Akiyama, T., and Shoda, M. 2001. Cloning, sequencing, and characterization of the iturin A operon. *Journal of Bacteriology* 183: 6265–6273.

Tunga, R., Banerjee, R., and Bhattacharyya, B.C. 1999. Optimization of *n* variable biological experiments by evolutionary operation-fractional design technique. *Journal of Bioscience and Bioengineering* 87: 224–230.

Udomsil, N., Rodtong, S., Tanasupawat, S., and Yongsawatdigul, J. 2010. Proteinase-producing halophilic lactic acid bacteria isolated from fish sauce fermentation and their ability to produce volatile compounds. *International Journal of Food Microbiology* 141: 186–194.

Ueda, M., Kubo, T., Miyatake, K., and Nakamura, T. 2007. Purification and characterization of fibrinolytic alkaline protease from *Fusarium* sp. BLB. *Applied Microbiology and Biotechnology* 74: 331–338.

Ueda, S., Ohba, R., and Ijiri, S. 1988. Screening of koji mold with hydrolytic activity on γ-poly glutamic acid. *Nippon Shokuhin Kogyo Gakkaishi* 35: 302–308 (in Japanese).

Umezawa, H., Aoyagi, T., Takeuchi, T., Hamada, M., Naganawa, H., Muraoka, Y., and Nishikiori, T. 1988. Plispastin and method for preparation thereof. Patent US4742155.

Urano, T., Ihara, H., Umemura, K., Suzuki, Y., Oike, M., Akita, S., Tsukamoto, Y., Suzuki, I., and Takada, A. 2001. The profibrinolytic enzyme subtilisin NAT purified from *Bacillus subtilis* cleaves and inactivates plasminogen activator inhibitor type 1. *Journal of Biological Chemistry* 276: 24690–24696.

Urushibata, Y., Tokuyama, S., and Tahara, Y. 2002. Characterization of the *Bacillus subtilis ywsC* gene, involved in γ-polyglutamic acid production. *Journal of Bacteriology* 184: 337–343.

Vallejo, J.A., Ageitos, J.M., Poza, M., and Villa, T.G. 2008. Cloning and expression of buffalo active chymosin in *Pichia pastoris*. *Journal of Agricultural and Food Chemistry* 56: 10606–10610.

Vallejo, J.A., Ageitos, J.M., Poza, M., and Villa, T.G. 2012. *Short communication*: A comparative analysis of recombinant chymosins. *Journal of Dairy Science* 95: 609–613.

Van den Berg, J.A., van der Laken, K.J., van Ooyen, A.J., Renniers, T.C.H.M., Rietveld, K., Schaap, A., Brake, A.J., Bishop, R.J., and Schultz, K. 1990. *Kluyveromyces* as a host for heterologous gene expression: Expression and secretion of prochymosin. *Biotechnology* 8: 135–139.

Van der Sluis, C., Tramper, J., and Wijffels, R.H. 2002. Enhancing and accelerating flavour formation by salt tolerant yeasts in Japanese soy-sauce processes. *Trends in Food Science and Technology* 12: 322–327.

Van Tomme, S.R., and Hennink, W.E. 2007. Biodegradable dextran hydrogels for protein delivery applications. *Expert Review of Medical Devices* 4: 147–164.

Vanittanakom, N., Loeffler, W., Koch, U., and Jung, G. 1986. Fengycin a novel antifungal lipopeptide antibiotic produced by *Bacillus subtilis* F-29-3. *The Journal of Antibiotics* 39: 888–901.

Varadarajan, N., Georgiou, G., and Iverson, B.L. 2008a. An engineered protease that cleaves specifically after sulfated tyrosine. *Angewandte Chemie International Edition in English* 47: 7861–7863.

Varadarajan, N., Pogson, M., Georgiou, G., and Iverson, B.L. 2009. Proteases that can distinguish among different post-translational forms of tyrosine engineered using multicolor flow cytometry. *Journal of the American Chemical Society* 131: 18186–18190.

Varadarajan, N., Rodriguez, S., Hwang, B.Y., Georgiou, G., and Iverson, B.L. 2008b. Highly active and selective endopeptidases with programmed substrate specificities. *Nature Chemical Biology* 4: 290–294.

Venugopal, M., and Saramma, A.V. 2006. Characterization of alkaline protease from *Vibrio fluvialis* strain VM10 isolated from a mangrove sediment sample and its application as a laundry detergent additive. *Process Biochemistry* 41: 1239–1243.

Venugopal, M., and Saramma, A.V. 2007. An alkaline protease from *Bacillus circulans* BM15, newly isolated from a mangrove station: Characterization and application in laundry detergent formulations. *Indian Journal of Microbiology* 47: 298–303.

Vidyasagar, M., Prakash, S., Litchfield, C., and Sreeramulu, K. 2006. Purification and characterization of a thermostable, haloalkaliphilic extracellular serine protease from the extreme halophilic archaeon *Halogeometricum borinquense* strain TSS101. *Archaea* 2: 51–57.

Vijayaraghavan, P., and Vincent, S.G.P. 2012. Cow dung as a novel, inexpensive substrate for the production of a halo-tolerant alkaline protease by *Halomonas* sp. PV1 for eco-friendly applications. *Biochemical Engineering Journal* 69: 57–60.

Vollenbroich, D., Pauli, G., Ozel, M., and Vater, J. 1997. Antimycoplasma properties and application in cell culture of surfactin, a lipopeptide antibiotic from *Bacillus subtilis*. *Applied and Environmental Microbiology* 63: 44–49.

Wang, C., Du, M., Zheng, D., Kong, F., Zu, G., and Feng, Y. 2009. Purification and characterization of nattokinase from *Bacillus subtilis* natto B-12. *Journal of Agricultural and Food Chemistry* 57: 9722–9729.

Wang, C.T., Ji, B.P., Li, B., Nout, M.J.R., Li, P.L., Ji, H., and Chen, L.F. 2006. Purification and characterization of a fibrinolytic enzyme of *Bacillus subtilis* DC33, isolated from Chinese traditional Douchi. *Journal of Industrial Microbiology and Biotechnology* 33: 750–758.

Wang, J.J., Greenhut, W.B., and Shih, J.C.H. 2005. Development of an asporogenic *Bacillus licheniformis* for the production of keratinase. *Journal of Applied Microbiology* 98: 761–767.

Wang, L.J., Mu, H.L., Liu, H.J., Bhandari, B., Saito, M., and Li, L.T. 2010. Volatile components in three commercial douchies, a Chinese traditional salt-fermented soybean food. *International Journal of Food Properties* 13: 1117–1133.

Wang, S.L., Yang, C.H., Liang, T.W., and Yen, Y.H. 2008. Optimization of conditions for protease production by *Chryseobacterium taeanense* TKU001. *Bioresource Technology* 99: 3700–3707.

Wei, Y.H., Wang, L.F., and Chang, J.S. 2004. Optimizing iron supplement strategies for enhanced surfactin production with *Bacillus subtilis*. *Biotechnology Progress* 20: 979–983.

Wijesekara, I., and Kim, S.K. 2010. Angiotensin-I-converting enzyme (ACE) inhibitors from marine resources: Prospects in the pharmaceutical industry. *Marine Drugs* 8: 1080–1093.

Wilson, S.A., Young, D.A., Coolbear, T., and Daniel, R.M. 1992. The use of proteases from extreme thermophiles for meat tenderization. *Meat Science* 32: 93–103.

Wintrode, P.L., Miyazaki, K., and Arnold, F.H. 2000. Cold adaptation of a mesophilic subtilisin-like protease by laboratory evolution. *Journal of Biological Chemistry* 275: 31635–31640.

Wittanalai, S., Rakariyatham, N., and Deming, R.L. 2011. Volatile compounds of vegetarian soybean kapi, a fermented Thai food condiment. *African Journal of Biotechnology* 10: 821–830.

Wong, A.H.K., and Mine, Y. 2004. Novel fibrinolytic enzyme in fermented shrimp paste, a traditional Asian fermented seasoning. *Journal of Agricultural and Food Chemistry* 52: 980–986.

Wong, T.S., Zhurina, D., and Schwaneberg, U. 2006. The diversity challenge in directed protein evolution. *Combinatorial Chemistry and High Throughput Screening* 9: 271–288.

Yang, Y., Jiang, L., Zhu, L., Wu, Y., and Yang, S. 2000. Thermal stable and oxidation resistant variant of subtilisin E. *Journal of Biotechnology* 81: 113–118.

Yasuda, M., Matsumoto, T., Sakaguchi, M., and Kinjyo, S. 1994. Changes in protein and nitrogen compounds of tofuyo prepared by *Aspergillus oryzae* during fermentation. *Nippon Shokuhin Kogyo Gakkaishi* 41: 184–190 (in Japanese).

Yin, H., Guo, C., Wang, Y., Liu, D., Lv, Y., Lv, F., and Lu, Z. 2013. Fengycin inhibits the growth of the human lung cancer cell line 95D through reactive oxygen species production and mitochondria-dependent apoptosis. *Anti-Cancer Drugs* 24: 587–598.

Yoon, S.H., Kim, J.B., Lim, Y.H., Hong, S.R., Song, J.K., Kim, S.S., Kwon, S.W., Park, I.C., Kim, S.J., Yeo, Y.S., and Koo, B.S. 2005. Isolation and characterization of three kinds of lipopeptides produced by *Bacillus subtilis* JKK238 from Jeot-Kal of Korean traditional fermented fishes. *Korean Journal of Microbiology and Biotechnology* 33: 295–301.

Yoshimoto, A., and Hongo, M. 1970. Bacteriophages of *Bacillus natto*. 1. Some characteristics of phage NP-1. *Journal of the Faculty of Agriculture, Kyushu University* 16: 141–158.

Yoshimoto, A., Nomura, S., and Hongo, M. 1970. Bacteriophages of *Bacillus natto* (IV) Natto plant pollution by bacteriophages. *Journal of Fermentation Technology* 48: 660–668 (in Japanese).

Yoshimoto, A., Nomura, S., and Hongo, M. 1973. γ-Polyglutamic acid depolymerase induced by infections of *natto* and *subtilis* phages and its further properties. *Agricultural and Biological Chemistry* 37: 83–90.

You, L., and Arnold, F.H. 1996. Directed evolution of subtilisin E in *Bacillus subtilis* to enhance total activity in aqueous dimethylformamide. *Protein Engineering* 9: 77–83.

Yu, P.J., and Chou, C.C. 2005. Factors affecting the growth and production of milk-clotting enzymes by *Amylomyces rouxii* in rice liquid medium. *Food Technology and Biotechnology* 43: 283–288.

Yusada, M., Aoyama, M., Sakaguchi, M., Nakachi, K., and Kolamoto, N. 1999. Purification and characterization of soybean-milk coagulating enzyme from *Bacillus pumilus* TYO-67. *Applied Microbiology and Biotechnology* 51: 474–479.

Zamost, B.L., Brantley, Q.I., Elm, D.D., and Beck, C.M. 1990. Production and characterization of a thermostable protease produced by an asporogenous mutant of *Bacillus stearothermophilus*. *Journal of Industrial Microbiology* 5: 303–312.

Zhang, G., Ma, X., Niu, F., Dong, M., Feng, H., An, L., and Cheng, G. 2007. Diversity and distribution of alkaliphilic psychrotolerant bacteria in the Qinghai-Tibet Plateau permafrost region. *Extremophiles* 11: 415–424.

Zhao, G.Y., Chen, X.L., Zhao, H.L., Xie, B.B., Zhou, B.C., and Zhang, Y.Z. 2008. Hydrolysis of insoluble collagen by deseasin MCP-01 from deep-sea *Pseudoalteromonas* sp. SM9913: Collagenolytic characters, collagen-binding ability of C-terminal polycystic kidney disease domain, and implication for its novel role in deep-sea sedimentary particulate organic nitrogen degradation. *Journal of Biological Chemistry* 283: 36100–36107.

Zhao, H., and Arnold, F.H. 1999. Directed evolution converts subtilisin E into a functional equivalent of thermitase. *Protein Engineering* 12: 47–53.

Zhao, H., Giver, L., Shao, Z., Affholter, J.A., and Arnold, F.H. 1998. Molecular evolution by StEP in vitro recombination. *Nature Biotechnology* 16: 258–261.

Zhong, C.Q., Song, S., Fang, N., Liang, X., Zhu, H., Tang, X.F., and Tang, B. 2009. Improvement of low-temperature caseinolytic activity of a thermophilic subtilase by directed evolution and site-directed mutagenesis. *Biotechnology and Bioengineering* 104: 862–870.

Zhu, W., Cha, D., Cheng, G., Peng, Q., and Shen, P. 2007. Purification and characterization of a thermostable protease from a newly isolated *Geobacillus* sp. YMTC 1049. *Enzyme and Microbial Technology* 40: 1592–1597.

Zhu, X.L., Watanabe, K., Shiraishi, K., Ueki, T., Noda, Y., Matsui, T., and Matsumoto, K. 2008. Identification of ACE-inhibitory peptides in salt-free soy sauce that are transportable across caco-2 cell monolayers. *Peptides* 29: 338–344.

Zotou, Z., Loukou, Z., and Karava, O. 2004. Method development for the determination of seven organic acids in wines by reversed-phase high performance liquid chromatography. *Chromatographia* 60: 39–44.

Zvidzai, C.J., and Zvauya, R. 2001. Purification of a protease from an alkalophllic *Bacillus subtilis* CHZ1 isolated from a Zimbabwean hot spring. *Journal of Food Biochemistry* 25: 1–13.

9

Future of Alkaline-Fermented Foods for Traditional Markets

M. J. Robert Nout and Prabir K. Sarkar

The previous chapters in this book illustrated the wide range of different types of alkaline-fermented foods (AFFs), chiefly produced in many Asian and African countries. We also noticed that whereas some products are produced at industrial scale with a significant level of sophistication, many others are still produced at artisanal scale in small workshops or at home.

What does the future hold in store for these products? Will consumers stay with their food habits and choices or will the demand for AFFs decline? The future of AFFs depends very much on how these products fulfill the expectations of the consumer. Even when we speak of "traditional" markets, these are not static; thus, it is important to be aware of the dynamics of the market.

Consumers' lifestyle is changing due to migration (urbanization), economic, and political developments, communication infrastructures (transport of goods, Internet, and cell phones), and others. Other factors may play a role, such as food sovereignty (Kloppenburg, 2008), i.e., the right of the people to produce, process, and consume the food they prefer.

These aspects point toward the need for a sound knowledge and understanding of the structure and dynamics of national, regional, and local markets. It is important to know and understand the consumers' perception of quality (Hansen, 2005), i.e., what does the consumer expect from a food product in terms of appearance, flavor, taste, nutrition, safety, convenience, etc.? In addition, with what kind of other products must this compete?

Obviously, market research will be essential to establish requirements for future quality traits of AFFs. Easy availability and conveniently and hygienically packed products at an affordable price will be the keywords.

It was mentioned that there is a diversity of sophistication of AFFs. The existence of commercially successful sophisticated products could inspire product developers to upgrade less developed processes or products into future bestsellers.

Taking Japanese natto as an example (Kiuchi and Watanabe, 2004), we see a product that is produced at an industrial scale, fermented with selected starter cultures under hygienic conditions, and packaged to protect the product and to facilitate supermarket retailing at an affordable price level. Combined with its perceived healthiness, it is hard to imagine that natto would disappear from the Japanese market. In contrast, the level of quality of this product enables it to travel around and get customers in other continents. Like natto, several other AFFs of high standard can be mentioned. On the other hand, a range of AFFs is still manufactured at household- or small industrial-scale with a minimum of packaging, quality, and safety assurance. It will be a business opportunity to upgrade such products and prepare them for a next-generation market.

Upgrading of traditional fermented products has been advocated (Nout, 1992; Achi, 2005). This can have various purposes such as achieving a constant and superior quality, a more convenient (time-saving) product, safe and produced under hygienic conditions, having a prolonged shelf life.

A critical evaluation of the product's "unique selling points" should be part of such upgrading. What is the essence: a particular and complex flavor, perceived healthiness, or a special texture? Product versions for the future should deliver authenticity (Casey et al., 2008) in a modern package comprising safety, shelf life, and convenience at an affordable price. They should not be associated with negative aspects such as poverty or amateurishness.

Food legislation, which is one of the most severe trade barriers, is a great burden for the AFF industry, and the governments in each of the producer countries will have to assist the producers to guarantee the quality and safety of such foods by every possible means. It is also necessary to protect the indigenous technologies in producing these foods by applying legal tools like intellectual property right (IPR) and sharing of benefits with the people producing the indigenous foods in respective countries (Pardey, 2001; Van der Meulen and Van der Velde, 2004; Zilberman and Kim, 2011).

References

Achi, O.K. 2005. The upgrading of traditional fermented foods through biotechnology. *African Journal of Biotechnology* 4: 375–380.

Casey, M.G., Isolini, D., Amrein, R., Wechsler, D., and Berthoud, H. 2008. Naturally occurring genetic markers in lactobacilli and their use to verify the authenticity of Swiss Emmental PDO cheese. *Dairy Science and Technology* 88: 457–466.

Hansen, T. 2005. Understanding consumer perception of food quality: The cases of shrimps and cheese. *British Food Journal* 107: 500–525.

Kiuchi, K., and Watanabe, S. 2004. Industrialization of Japanese natto. In *Industrialization of Indigenous Fermented Foods*, 2nd ed., ed. K.H. Steinkraus, pp. 193–246. New York: Marcel Dekker.

Kloppenburg, J. 2008. Seeds, Sovereignty, and the Vía Campesina: Plants, Property, and the Promise of Open Source Biology. Proceedings, *Workshop on Food Sovereignty: Theory, Praxis and Power*. St. Andrews College, University of Saskatchewan, Saskatoon, Saskatchewan, Canada: November 17–18, 2008.

Nout, M.J.R. 1992. Upgrading traditional biotechnological processes. In *Applications of Biotechnology to Traditional Fermented Foods. Report of an ad-hoc panel of the Board on Science and Technology for International Development*, eds. E.L. Gaden, M. Bokanga, S. Harlander, C.W. Hesseltine, and K.H. Steinkraus, pp. 11–19. Washington, DC: National Research Council.

Pardey, P.G. 2001. *The Future of Food*. Washington, DC: International Food Policy Research Institute.

Van der Meulen, B., and Van der Velde, M. 2004. *Food Safety Law in the European Union*. Wageningen, The Netherlands: Wageningen Academic Publishers.

Zilberman, D., and Kim, E. 2011. The lessons of fermentation for the new bioeconomy. *AgBioForum* 14: 97–103.

10
Outlook

M. J. Robert Nout and Prabir K. Sarkar

To achieve the requirements for future markets that were mentioned in the previous chapter, research and development will be essential at various levels of application. A wealth of data describing microbial and chemical aspects of composition and changes taking place during fermentations is available.

The creation of a comprehensive authentic database is needed to understand the trends in the production of alkaline-fermented foods (AFFs) and to fix the reasons for no growth or decline in consumption of AFFs in some countries (Borresen et al., 2012; Ojiako et al., 2012). This is because the actual amounts of foods consumed usually are not known and some of the data are questionable.

Authenticity comprises a definition of a process or product that can be used for marketing and for protecting it from copying. Such definition includes the basics of ingredients, their provenance, the locality or region where the product is prepared, and the detailed process. Modern analytical techniques allow profiling of microbiota (Casey et al., 2008) and chemical composition, and multivariate analysis software allows comparison of profiles (Le et al., 2011). For future markets, such authenticity should be nurtured, and it should be "packaged," which comprises product safety, shelf life, and convenience at an affordable price. The development of convenient, durable, and safe authentic products is a challenge that can only be successfully overcome by an interdisciplinary approach.

It is remarkable how little specific mention is made about the taste aspects of AFFs. Although the smells are often described as strong, pungent, sweet, sometimes even as offensive, the smell does not necessarily contribute to the flavor of a cooked dish. For example, some of the African products such as iru or soumbala and bikalga have a characteristic but not very pleasant smell. However, during cooking, much of this smell disappears, and then the question remains: what does the odor contribute to the taste of the cooked dish? A short survey among the contributing authors revealed that "umami," meaty, cheesy, sometimes fishy tastes are perceived as characteristic for the African products. On the other hand, the fermented fish products are primarily described as salty, "umami," enhancing the taste of prepared dishes. It is obvious that considerable work remains to identify the chemical components that determine the taste principles of specific fermented products. This knowledge would be extremely useful in the description of product quality traits for protection of their authenticity and to direct quality improvements when using specific microbial starters and/or enzymes.

An important role should be reserved for consumer research to obtain a quantitative understanding of consumers' wishes and perceptions relating to convenience, shelf life, packaging, retailing, and willingness to pay for the product. Preferably such aspects should be combined in mathematical models that facilitate choosing the targets for optimization, such as was done for lafun, a traditional fermented cassava flour in Nigeria (Ojiako et al., 2012).

Knowing the consumers' response is a first step to define modified or novel products. Next, the traditional process needs to be critically reviewed, remodeled where needed, taking into account aspects such as optimum use of water and energy consumption (Carvajal-Larenas et al., 2013). When the production needs up-scaling, an economical study and business plan require specific expertise.

From the food science and technology point of view, more expansion of existing knowledge is needed to understand mechanisms underlying the evolution of product quality. How can robust starter cultures be selected and used to achieve predictable and controllable fermentation outcomes? Can the proteolytic enzymes in these fermentations be purified and characterized so that the process conditions can be optimized? Can added purified enzymes be used to enhance the fermentation outcome? How can final products be portioned, packaged, and distributed for retail?

An integral part of developing processing lines is related to quality and hygiene. Good manufacturing practice (GMP) (Amoa-Awua et al., 2007) comprises specifications of ingredient grades, operational care, etc., to achieve high-quality products. For each product, GMP needs to be defined and documented for future evaluation.

Good hygienic practice (GHP) includes all operational care ensuring that the product is protected from chemical and microbiological contamination. Although general principles of GHP are available, these must be "translated" to specific products and processes to enable their implementation. Again, these practices must be documented to enable future reviewing and evaluation.

Hazard analysis and critical control points (HACCP) (Lee, 2010) aims specifically at microbiological safety of the final product to be obtained, through a proactive control of critical steps in the process. HACCP has been described as a management tool, but the establishment of a HACCP protocol for a specific product or process requires knowledge from various disciplines including epidemiology, microbiology, process engineering, and food science. A protocol describes the process and all its ingredients, operations, and its organization. Steps are identified that aim at reducing microbial load, and simple in-line measures are defined that need to be performed and documented at prescribed intervals. Such documentation enables trend analyses of the process performance and allows assurance of product safety so that unpleasant surprises at the final product stage can be minimized.

Another important aspect is the functional aspect of AFFs: some health aspects have been tested well, but others are still in early phases of in vitro or animal experimentation. To make claims, more double-blind, placebo-controlled human experiments should be undertaken for the effects and mechanisms of action need to be elucidated.

References

Amoa-Awua, W.K., Ngunjiri, P., Anlobe, J., Kpodo, K., Halm, M., Hayford, A.E., and Jakobsen, M. 2007. The effect of applying GMP and HACCP to traditional food processing at a semi-commercial kenkey production plant in Ghana. *Food Control* 18: 1449–1457.

Borresen, E.C., Henderson, A.J., Kumar, A., Weir, T.L., and Ryan, E.P. 2012. Fermented foods: Patented approaches and formulations for nutritional supplementation and health promotion. *Recent Patents on Food, Nutrition and Agriculture* 4: 134–140.

Carvajal-Larenas, F.E., Nout, M.J.R., van Boekel, M.A.J.S., Koziol, M.J., and Linnemann, A.R. 2013. Modelling of the aqueous debittering process of *Lupinus mutabilis* Sweet. *LWT – Food Science and Technology* 53: 507–516.

Casey, M.G., Isolini, D., Amrein, R., Wechsler, D., and Berthoud, H. 2008. Naturally occurring genetic markers in lactobacilli and their use to verify the authenticity of Swiss Emmental PDO cheese. *Dairy Science and Technology* 88: 457–466.

Le, V.-D., Zheng, X.-W., Ma, K., Chen, J.-Y., Han, B.-Z., and Nout, M.J.R. 2011. Characterization of *Fen-Daqu* through multivariate statistical analysis of ^{1}H NMR spectroscopic data. *Journal of the Institute of Brewing* 117: 516–522.

Lee, S.O. 2010. A study on the hazard analysis and HACCP system for Chongkookjang manufacturing. M.Sc. thesis. Asan, Korea: Hoseo University.

Ojiako, I.A., Ezedinma, C., Asumugha, G.N., and Nkang, N.M. 2012. Spatial integration and price transmission in selected cassava products' markets in Nigeria: A case of lafun. *World Applied Sciences Journal* 18: 1209–1219.

Index

Page numbers followed by f and t indicate figures and tables, respectively.

B

D

E

F

M

O

Q